Acta Neurochirurgica Supplement 124

Series Editor
H.-J. Steiger

For further volumes:
http://www.springer.com/series/4

Massimiliano Visocchi
H. Maximilian Mehdorn
Yoichi Katayama • Klaus R.H. von Wild
Editors

Trends in Reconstructive Neurosurgery

Neurorehabilitation, Restoration and Reconstruction

Springer

Editors
Massimiliano Visocchi
Institute of Neurosurgery
Medical School
Catholic University of Rome
Rome, Italy

Yoichi Katayama
Nihon University
Department of Neurological Surgery
Tokyo, Japan

H. Maximilian Mehdorn
Dept. of Neurosurgery
Christian-Albrechts-University of Kiel
Kiel, Germany

Klaus R.H. von Wild
Medical Faculty University of Münster
Münster, Germany

ISSN 0065-1419
Acta Neurochirurgica Supplement
ISBN 978-3-319-39545-6 ISBN 978-3-319-39546-3 (eBook)
DOI 10.1007/978-3-319-39546-3

Library of Congress Control Number: 2016959468

Printed on acid-free paper

This Springer imprint is published by Springer Nature
The registered company is Springer International Publishing AG
The registered company address is: Gewerbestrasse 11, 6330 Cham, Switzerland

Contents

Reconstructive Neurosurgery: A Challenge

Massimiliano Visocchi

Abstract The International Society of Reconstructive Neurosurgery (ISRN) is an "open" multidisciplinary Society in evolution. Many different members of the Society inspire new trends in many different neurosurgical fields, all dealing with neurosurgical reconstruction.

Spine and peripheral nerve reconstructive surgery, central nervous system revascularization (via surgery and interventional radiology), neuromodulation, bioengineering, and transplantation are recent tools used to promote reconstruction, restoration, and rehabilitation.

These are the three key words of our creed and all fulfill the aim of the ISRN, dealing with mechanical, morphological, and functional restoration.

Spinal, functional, vascular, radiological, and oncologic neurosurgeons are those to whom our proposals are addressed, along with biologists, bioengineers, anatomists, physiologists, and physiotherapists, who are precious and irreplaceable inspirers.

Keywords Neurosurgical reconstruction • Neurosurgical restoration • Neurosurgical rehabilitation

The World Federation of Neurosurgical Societies (WFNS) aspires to promote global improvement in neurosurgical care, training, and research to benefit patients. Founded in 1955, the WFNS is a professional, scientific, nongovernmental organization comprising 5 continental associations, 115 national neurosurgical societies, and 7 affiliate societies, representing some 30,000 neurosurgeons worldwide. The WFNS is governed by an executive committee (EC), consisting of two delegates from each member society, and an administrative council, composed of the officers of the Federation, who are elected every 4 years. The EC meets every 2 years and is guided by the administrative council, which meets at least annually. The goals of the WFNS are deliberated and pursued through scientific, standing, and *ad hoc* committees, and during the International Congress of Neurological Surgery, which takes place every 4 years. *The Neurorehabilitation and Reconstructive Neurosurgery Committee* is a special section of the WFNS that promotes all those WFNS activities aimed at implementing and promoting all the restorative, reconstructive, and augmentative neurosurgical procedures that were grossly identified in the past as the functional neurosurgery subspecialty. But now this subspecialty is updating, evolving, and merging with neurooncology, spine surgery, and neuroradiological and neurophysiological intraoperative assistance, and, in a broad sense, new technologies. *The Neuromodulation Committee is another different section of the WFNS; it is administered by a completely different Board working in an independent way and pursuing different objectives.* Two years after the institution of the WFNS the members of the Board *Neurorehabilitation and Reconstructive Neurosurgery Committee* felt the need to found a new Society that would be more free to deal with other medical and/or surgical societies and expert researchers from different branches of biology, physiology, and physiotherapy. The IV International Congress of the International Society of Reconstructive Neurosurgery (ISRN), along with the VII Neurorehabilitation and Reconstructive Neurosurgery Symposium (WFNS), was held in Cerveteri (Rome) on September 12 to 14, 2015. The President of the Congress was Professor Franco Tomasello, former Rector of the University of Messina, Italy, and Vice President of the WFNS; the present author, Massimiliano Visocchi, Past President of the Italian Society of Neurosonology and Cerebral Hemodynamics, former Secretary of the Spine Section of the Italian Society of Neurosurgery, Associate Professor in Neurosurgery at the Catholic University of Rome, and Visiting Professor at the Shanghai Jiao Tong University School of Medicine, was

M. Visocchi, MD
Institute of Neurosurgery, Medical School,
Catholic University of Rome, Rome, Italy
e-mail: mvisocchi@hotmail.com

M. Visocchi et al. (eds.), *Trends in Reconstructive Neurosurgery*, Acta Neurochirurgica Supplement 124,
DOI 10.1007/978-3-319-39546-3_1, © Springer International Publishing Switzerland 2017

appointed President Elect of the ISRN and soon after I started to ask to myself more questions dealing with reconstructive neurosurgery and also to try to better investigate the true roots of our society and the supposed mission of the ISRN. In other words: "*Where are we coming from? Who are we and where are we going?*"

My first impulse was to have a look at the current literature on the topic. If we search for "Reconstructive Neurosurgery" on PubMed. gov., the United States National Library of Medicine-National Institutes of Health, we can find 2810 papers updated as at September 30, 2015. The papers span from the latest, printed in September 2015 and harboring a very intriguing title: *Identification of Circulating miRNAs in a Mouse Model of Nerve Allograft Transplantation under FK506 Immunosuppression by Illumina Small RNA Deep Sequencing* [1], to the earliest, the very first recognized paper on the subject, printed in November 1947 and entitled: *A report of the early results in tantalum cranioplasty* [2]. In between these papers there are many others, dealing with craniofacial endoscopy [3], spinal instrumentation and fusion procedures [4], peripheral nerve reconstructive procedures [5], vascular reconstructive and cerebral blood flow (CBF) restorative surgical procedures [5], neuronavigation and video-assisted neurosurgical procedures [6], complex craniofacial surgical procedures [7], and intraoperative neuromonitoring [8]. Surprisingly, many papers dealing with genetics, and biomolecular and cytochemical studies dealing with the central and peripheral nervous systems, are available as well [9]. Reconstruction is intended in a *purely* mechanical way, and no concepts dealing with restoration or rehabilitation arise from the majority of the manuscripts, but just simple expositions of surgical techniques and procedures aiming at just *repairing* something. Even "dysfunctional" syndromes, such as Raynaud's syndrome, are treated with mechanical procedures instead of neuroaugmentative ones [10].

Neurosurgeons know very well that the central nervous system (CNS), along with the spine and the skull bone components, have an intrinsic pattern of complex physiological nature, both from the neurochemical and the biomechanical points of view. Starting from the experience of the Neurorehabilitation Committee of the WFNS, we very soon turned our gaze toward restoration and rehabilitation and, since the ISRN is an "open" multidisciplinary society in evolution, we first intended the idea of neurosurgical reconstruction *to be interpreted, broadly speaking,* in a new way, so that: *reconstruction also means rehabilitation and restoration.*

Nevertheless, in the common literature, restoration is strongly linked to the concept of reconstruction, as demonstrated by the title of the following paper included in the PubMed list classified as "Reconstruction": *Restoration of the orbital aesthetic subunit with the thoracodorsal artery system of flaps in patients undergoing radiation therapy* [11].

The concept of neurorehabilitation has different meanings and significances in the culture of neurosurgeons; although it still deals with reconstruction, e.g.,: *Functional restoration of diaphragmatic paralysis: an evaluation of phrenic nerve reconstruction* [12] and also with replantation, as described in the paper: *Six years of follow-up after bilateral hand replantation* [13]; functional neurorehabilitation merges with the concept of neuromodulation, and neuromodulation is the core of functional neurosurgery. So, more simply, spine and peripheral nerve reconstructive surgery, CNS revascularization (surgical and interventional radiology), neuromodulation, bioengineering, and transplantation are recent tools used to promote reconstruction in the special sense intended by our society.

Classically, neuromodulation deals with the physiological process by which a given neuron uses one or more neurotransmitters to regulate diverse populations of neurons. This is in contrast to classical synaptic transmission, in which one presynaptic neuron directly influences a single postsynaptic partner. Neuromodulators secreted by a small group of neurons diffuse through large areas of the nervous system, affecting multiple neurons. However, in surgical praxis, the meaning of neuromodulation has shifted toward the armamentarium of surgical tools used with all the procedures involved in CNS electrical and chemical stimulation, as performed with spinal cord stimulation (SCS), deep brain stimulation (DBS), cortical brain stimulation (CBS), and drug delivery system (DDS) implantation, aimed at treating pain; movement disorders; spasticity; bowel and bladder dysfunction; and peripheral, heart, and cerebral vasculopathies. Interestingly, an effect of SCS on CBF was first reported by Hosobuchi [14] in 1985; he reported the intriguing effect of SCS on CBF in human beings, along with a demonstration that SCS could improve peripheral blood flow. Following these initial clinical and experimental observations, he first described the use of cervical SCS for the treatment of cerebral ischemia in humans in 1991.

It has been shown that SCS improves the clinical symptoms of patients in persistent vegetative states, improves CBF in stroke patients, suppresses the hemodynamic mechanism underlying headache attacks in migraneous patients, and increases locoregional blood flow in high-grade brain tumors in humans. In animals, SCS has been shown to prevent the progression of cerebral infarction, reduce infarct volume, reduce ischemic brain edema, and improve vasospasm [15]. Studies from our group have produced variable results: SCS can increase CBF, reduce CBF, or have no effect. In patients studied with both single-photon emission cerebral tomography (SPECT) flowmetry and transcranial Doppler sonography (TCD), the size of the induced variations, when present in both groups, was the same. Cervical stimulation more frequently produced an increase in CBF (61 % of cervical stimulations) [16–18]. Experimental studies by our group have confirmed that SCS: (1) interacts with CO_2 via the competitive regulation of CBF, producing a reversible functional sympathectomy; (2) produces similar flowmetric changes in the brain and in the eyes; (3) can improve both clinical and hemodynamic outcomes in ischemic stroke in humans; (4) prevents hemodynamic deterioration in experimental combined ischemic and traumatic brain injury; and (5) prevents experimental early vasospasm [19–21]. On the other hand, trigeminal ganglion stimulation can have the opposite effects [22].

But when we speak about reconstruction we cannot forget spinal cord and spine surgery, both being surgical challenges, from a biofunctional point of view in the former and from a biomechanical point of view in the latter. Instrumentation and fusion procedures from the upper to the lower levels of the spine have been widely published [23–28]. Nevertheless, the choice of alternative minimally invasive video-assisted surgical routes for spine reconstruction have also generated a great deal of interest and opened new perspectives in reconstruction, rehabilitation, and restoration; these being the three key words of our creed, all fulfilling the aim of the ISRN, dealing with mechanical, morphological, and functional restoration [29–35].

Spinal, functional, vascular, radiological, and oncologic neurosurgeons are those to whom our proposals are addressed, along with biologists, bioengineers, anatomists, physiologists, and physiotherapists, who are precious and irreplaceable inspirers.

Conflict of Interest Statement We declare that we have no conflict of interest.

References

1. Wu SC, Rau CS, Yang JC, Lu TH, Wu YC, Chen YC, Tzeng SL, Wu CJ, Lin CW, Hsieh CH (2015) Identification of circulating miRNAs in a mouse model of nerve allograft transplantation under FK506 immunosuppression by illumina small RNA deep sequencing. Dis Markers 2015:863192. doi:10.1155/2015/863192, Epub 2015 Sep 8
2. Lane S, Webster JE (1947) A report of the early results in tantalum cranioplasty. J Neurosurg 4(6):526–529. No abstract available
3. Sreenath SB, Recinos PF, McClurg SW, Thorp BD, McKinney KA, Klatt-Cromwell C, Zanation AM (2015) The endoscopic endonasal approach to the hypoglossal canal: the role of the Eustachian tube as a landmark for dissection. JAMA Otolaryngol Head Neck Surg 17:1–8. doi:10.1001/jamaoto.2015.1749 [Epub ahead of print]
4. Lim SH, Jo DJ, Kim SM, Lim YJ (2015) *Reconstructive* surgery using dual U-shaped rod instrumentation after posterior en bloc sacral hemisection for metastatic tumor: case report. J Neurosurg Spine 31:1–5 [Epub ahead of print]
5. Hendry JM, Alvarez-Veronesi MC, Snyder-Warwick A, Gordon T, Borschel GH (2015) Side-to-side nerve bridges support donor axon regeneration into chronically denervated nerves and are associated with characteristic changes in schwann cell phenotype. Neurosurgery 77(5):803–813
6. Takanari K, Araki Y, Okamoto S, Sato H, Yagi S, Toriyama K, Yokoyama K, Murotani K, Matsui S, Wakabayashi T, Kamei Y (2015) Operative wound-related complications after cranial revascularization surgeries. J Neurosurg 123(5):1145–1150
7. Andrews BT, Thurston TE, Tanna N, Broer PN, Levine JP, Kumar A, Bradley JP (2015) A multicenter experience with image-guided surgical navigation: broadening clinical indications in complex craniomaxillofacial surgery. J Craniofac Surg 26(4):1136–1139
8. Biscevic M, Biscevic S, Ljuca F, Smrke BU, Ozturk C, Tiric-Campara M (2014) Motor evoked potentials in 43 high risk spine deformities. Med Arch 68(5):345–349. doi:10.5455/medarh.2014.68.345-349. Epub 2014 Oct 15
9. Lopez-Bertoni H, Lal B, Li A, Caplan M, Guerrero-Cázares H, Eberhart CG, Quiñones-Hinojosa A, Glas M, Scheffler B, Laterra J, Li Y (2015) DNMT-dependent suppression of microRNA regulates the induction of GBM tumor-propagating phenotype by Oct4 and Sox2. Oncogene 34(30):3994–4004. doi:10.1038/onc.2014.334. Epub 2014 Oct 20
10. Merritt WH (2015) Role and rationale for extended periarterial sympathectomy in the management of severe Raynaud syndrome: techniques and results. Hand Clin 31(1):101–120. doi:10.1016/j.hcl.2014.09.011. Epub 2014 Nov 25. Review
11. Chanowski EJ, Casper KA, Eisbruch A, Heth JA, Marentette LJ, Prince ME, Moyer JS, Chepeha DB (2013) Restoration of the orbital aesthetic subunit with the thoracodorsal artery system of flaps in patients undergoing radiation therapy. J Neurol Surg B Skull Base 74(5):279–285. doi:10.1055/s-0033-1347366. Epub 2013 Jun 20
12. Kaufman MR, Elkwood AI, Colicchio AR, CeCe J, Jarrahy R, Willekes LJ, Rose MI, Brown D (2014) Functional restoration of diaphragmatic paralysis: an evaluation of phrenic nerve reconstruction. Ann Thorac Surg 97(1):260–266
13. Marques M, Correia-Sá I, Festas MJ, Silva S, Silva AI, Silva A, Amarante J (2013) Six years of follow-up after bilateral hand replantation. Chir Main 32(4):226–234. doi:10.1016/j.main.2013.06.005. Epub 2013 Jul 26
14. Hosobuchi Y (1985) Electrical stimulation of the cervical spinal cord increases cerebral blood flow in humans. Appl Neurophysiol 48:372–376
15. Visocchi M (2008) Neuromodulation of cerebral bloof flow by spinal cord elctrical stimulation: the role of the Italian school and state of art. J Neurosurg Sci 52:41–47
16. Visocchi M (2006) Spinal cord stimulation and cerebral haemodynamics. Acta Neurochir Suppl 99:111–116
17. Meglio M, Cioni B, Visocchi M, Nobili F, Rodriguez G, Rosadini G, Chiappini F, Sandric S (1991) Spinal cord stimulation and cerebral haemodynamics. Acta Neurochir (Wien) 111:43–48
18. Visocchi M, Argiolas L, Meglio M, Cioni B, Dal Basso P, Rollo M, Cabezas Cuevas D (2001) Spinal cord stimulation and early experimental cerebral spasm: the functional monitoring and the preventing effect. Acta Neurochir (Wien) 143:177–185
19. Visocchi M, Tartaglione T, Romani R, Meglio M (2001) Spinal cord stimulation prevents the effects of combined ischemic and traumatic brain injury. An MR study. Stereotact Funct Neurosurg 76:276–281
20. Visocchi M, Cioni B, Pentimalli L, Meglio M (1994) Increase of cerebral blood flow and improvement of brain motor control following spinal cord stimulation in ischemic spastic hemiparesis. Stereotact Funct Neurosurg 62:103–107
21. Visocchi M (2008) Sympathetic activity does influence cerebral blood flow. J Appl Physiol 105(4):1369
22. Visocchi M, Chiappini F, Cioni B, Meglio M (1996) Cerebral autoregulation and trigeminal stimulation. A TCD study. Ster Funct Neurosurg 66:184–192
23. Visocchi M, Di Rocco F, Meglio M (2003) Craniocervical junction instability: instrumentation and fusion with titanium rods and sublaminar wires. Effectiveness and failures in personal experience. Acta Neurochir 145:265–272

24. Visocchi M, Pietrini D, Tufo T, Fernandez E, Di Rocco C (2009) Pre-operative irreducible C1-C2 dislocations: intra-operative reduction and posterior fixation. The "always posterior strategy". Acta Neurochir (Wien) 151(5):551–560
25. Visocchi M, Fernandez EM, Ciampini A, Di Rocco C (2009) Reducible and irreducible os odontoideum treated with posterior wiring, instrumentation and fusion. Past or present? Acta Neurochir (Wien) 151(10):1265–1274
26. Visocchi M, Response to Wang et al., re Letter re Visocchi M, Pietrini D, Tufo T, Fernandez E, Di Rocco C (2009) Pre-operative irreducible C1-C2 dislocations: intra-operative reduction and posterior fixation. The "always posterior strategy". Acta Neurochir (Wien) 151(5):551–559; discussion 560
27. Visocchi M, Conforti G, Roselli R, La Rocca G, Spallone A (2015) From less to maximally invasiveness in cervical spine surgery: a "nightmare" case who deserve consideration. Int J Surg Case Rep 9:85–88
28. Visocchi M, Conforti G, La Rocca G, Roselli R, Spallone A, Barbagallo G (2014) Invasive and mini-invasive lumbar fusions. Does exist a state of art? An enigma in search of solutions. J Neurosurg Sci 58(2 Suppl 1):113–117
29. Visocchi M, Masferrer R, Sonntag VHK (1998) Dickman: thoracoscopic approaches to the thoracic spine. Acta Neurochir (Wien) 140:737–744
30. Visocchi M, La Rocca G, Della Pepa GM, Stigliano E, Costantini A, Di Nardo F, Maira G (2014) Anterior video-assisted approach to the craniovertebral junction: transnasal or transoral? A cadaver study. Acta Neurochir (Wien) 156(2):285–292
31. Visocchi M, Doglietto F, Della Pepa GM, Esposito G, La Rocca G, Di Rocco C, Maira G, Fernandez E (2011) Endoscope-assisted microsurgical transoral approach to the anterior craniovertebral junction compressive pathologies. Eur Spine J 20(9):1518–1525
32. Visocchi M et al (2011) Video-assisted microsurgical transoral approach to the craniovertebral junction: personal experience in childhood. Childs Nerv Syst 27(5):825–831
33. Visocchi M, Trevisi G, Iacopino DG, Tamburrini G, Caldarelli M, Barbagallo GM (2015) Odontoid process and clival regeneration with Chiari malformation worsening after transoral decompression: an unexpected and previously unreported cause of "accordion phenomenon". Eur Spine J 24(Suppl 4):S564–S568
34. Visocchi M, Di Martino A, Maugeri R, González Valcárcel I, Grasso V, Paludetti G (2015) Videoassisted anterior surgical approaches to the craniocervical junction: rationale and clinical results. Eur Spine J 24(12):2713–2723
35. Visocchi M (2015) Transnasal and transoral approach to the clivus and the craniovertebral junction. J Neurosurg Sci, 2015 March PMD 25737364 [Epub ahead of print]

Music and Mind: In Memoriam Professor Carlo Alberto Pagni, MD, PhD: February 13, 1931 –March 1, 2009

Klaus R.H. von Wild

Abstract Carlo Alberto Pagni, born in La Spezia, Italy, on February 13, 1931, was an eminent and respected professor of neurosurgery and chairman of the neurosurgical clinic of the University of Turin from 1980 to 2003. He died on March 1, 2009. As a professor of neurology and neurological surgery he was renowned as an expert on vascular, tumor, and functional neurosurgery. Beyond the Italian Neurosurgical Society, he was the doyen of functional neurosurgery, specializing in motor cortex stimulation for the treatment of focal dystonia, Parkinson's disease, and postictal spasticity and pain. His home was his castle, and his family was fundamental to his life. He shared with his wife, Sandra, his passion for piano playing and for their remarkable library, and together with friends, he and his wife enjoyed dinners with fine food and Barolo wines. Listening to this Grand Seigneur talking about and explaining the music of, above all, Ludwig van Beethoven, and Richard Wagner, one felt he was emotionally just "music and mind". You can imagine this from his books on music, chess, and neuroscience. Indeed, he adored playing correspondence chess worldwide. A sportsman too, he loved hiking, mountaineering, skiing, swimming, and fishing. Nature was his source for slowing down, for regenerating, and for collecting his strength for new projects and new challenges. Friends will remember Dr. Pagni as a Grand Seigneur.

Keywords Carlo Alberto Pagni • Doyen of Italian functional neurosurgery • Ludwig van Beethoven and Richard Wagner • EMN, WFNS Committee, and World-AMN • Correspondence chess

K.R.H. von Wild
Medical Faculty, Westphalian Wilhelms-University, Frauenburgstrasse 32, Munster D- 48155, Germany
e-mail: kvw@neurosci.de

Introduction

In 1993 the annual congress of the German Society of Neurotraumatology and Clinical Neuropsychology (DGNKN) was organized in the assembly hall of the beautiful baroque castle of our Westphalia University in Münster, Germany, when I was the congress president. The main topic was "The Spectrum of Neurorehabilitation", focusing on early neurological-neurosurgical rehabilitation and, secondly, on the neurorehabilitation of children and young adults after traumatic brain injuries (TBIs) [1]. One hot spot issue was the "Early pharmacological treatment of posttraumatic epileptic seizures and the prevention of posttraumatic epilepsy in children and adults", as well as the question of long-term medication and/or neurosurgical treatment. To this day these subjects are still controversial and therefore the neurosurgeon's approach as a genuine neurotraumatologist remains controversial. Our main concern is about the best time window for treatment; the type of medication, the daily dosage, and the duration over years; and the late onset of seizures. Clinical evidence for the treatment of posttraumatic epilepsy is limited, regarding the long-term results and pitfalls, when patients do not respond sufficiently to medication. Little is known about the pros and cons of subsequent functional surgical treatment for epilepsy, e.g., by epidural electrical stimulation, microsurgical resection of scar tissue, or partial lobectomy. In a recent Cochrane review, the authors found only low-quality evidence that early treatment with an antiepileptic drug (AED), compared with placebo or standard care, reduced the risk of early posttraumatic seizures. There was no evidence to support a reduction in the risk of late seizures or mortality. There still is insufficient evidence to reach any conclusions regarding the effectiveness or safety of other neuroprotective agents compared with placebo, or for the comparison of phenytoin, a traditional AED, with another AED [2]. This is in accordance with what my friend, the renowned neurosurgeon Johannes Schramm, at the University of Bonn [3], explained to me in the early 1990s;

M. Visocchi et al. (eds.), *Trends in Reconstructive Neurosurgery*, Acta Neurochirurgica Supplement 124,
DOI 10.1007/978-3-319-39546-3_2,

Fig. 1 March 18, 1993. C.A. Pagni (*left*) and his wife Sandra with Klaus von Wild, congress president, at the president's welcome reception at his home in Münster, Germany

to his best knowledge, he could not give us any evidence-based recommendation for functional surgical indications in pharmacology-resistant posttraumatic epilepsy (personal communication). When reviewing the neurosurgical literature in preparation for our scientific congress program, I came across some earlier publications of Carlo Alberto Pagni, who, at that time, was the Professor and Director of the Neurosurgical Clinic at the University of Turin, Italy [4]. I invited him to present at our congress, as he was a well-known professional with personal expertise in lecturing. Happily, he accepted, and presented a fabulous talk on posttraumatic epilepsy [5]. His visit was the beginning of our friendship.

At our international congress, Dr. Pagni was, as he loved to be, accompanied by his wife Mrs Sandra Pagni, who had studied piano at the conservatorium in Milan together with Claudio Abbado, who later became a world-renowned conductor. When I met these two elegant citizens of the world for the first time "there was music in the air". Music and Mind! In fact during our presidential welcome dinner at our home (Fig. 1) our friend, Dr. Ivar H. Pawlowitzki, Professor of Human Genetics at Münster University Institute, told me that he had met with Dr. Pagni and his wife before, at a concert in Italy some years ago, when Dr. Pagni had told him to play the piano enthusiastically at his home, in the way that Pawlowitzki himself was already doing, quite professionally! And Dr. Pawlowitzki also told me that Dr. Pagni was very much in favor of the music of Ludwig van Beethoven, and had even written books on him. No wonder, when we became friends, Dr. Pagni spoke many times about his beloved favorite composer and showed me many books about Beethoven in his library at home. I guess his last book on Beethoven's late sonatas remained unfinished. But I am not sure.

Years of Academic Interest and Friendship

In the following years I met Dr. Pagni several times at his department in Turin, and I carefully watched and enjoyed seeing some of his major neurosurgical procedures, sharing his passion for aneurysm surgery and functional neurosurgery. By taking part in his clinical rounds and by following his discussions with his staff members and colleagues from other disciplines I saw the acceptance of Dr. Pagni's authority both as a neurosurgeon and as a real gentleman. It was at the time when he had started his extradural motor cortex stimulation (EMCS) for Parkinson's disease. Once he told me about his unexpected observation in the operating room, that one of his patients, who had undergone cortical electrical stimulation for pain control [6], showed a dramatic improvement of his contralateral Parkinson's symptomatology during stimulation. This improvement lasted for months with the implanted stimulator. This was the start of Dr.

Fig. 2 C.A. Pagni (*right*) and his wife Sandra with Klaus von Wild and Professor Rita Levi-Montalcini (22.04.1909 – 30.12.2012), Italian Nobel Laureate 1986 (Nobel Prize for Physiology or Medicine; awarded for the discovery of nerve growth factor ([NGF])) and lifetime Italian Senator, at Brescia Town Hall—Salone Vanvitelliano, Palazzo della Loggia. Presidential Reception of Professor G. A. Brunelli, 1st AMN Congress, Brescia, Italy. See Preface pp V–VII by von Wild K.RH (ed) Reengineering of the damaged brain and spinal cord. Evidence based neurorehabilitation. Acta neurochir. Suppl. 93 Springer Wien New York 2006

Pagni's final scientific project as a skilled neurosurgeon and as a pioneer of epidural electrical stimulation in Parkinson's disease. He was very proud, as he told me, and as he thereafter showed at our conferences, that he had obtained permission, on behalf of the Italian Society for Neurosurgery, to conduct a research project over the following years, analyzing and publishing the results for selected clinical cases of the Study Group of the Italian Neurosurgical Society [7–9, 11].

Due to his personality and his great expertise in neurological surgery, including functional neurosurgery for pain, Parkinson's disease, and epilepsy, I invited Dr. Pagni to become one of our first academic members after I established the Euroacademy of Multidisciplinary Neurotraumatology (EMN) in 1994. Three years later, in 1977, I was able to convince the WFNS Board to found the a new ad hoc Committee in the world Federation of Neurosurgical Societies, dedicated exclusively to neurosurgical neurorehabilitation and reconstructive neurosurgery. Last but not least in following an international demand I instituted the World Academy for Multidisciplinary Neurotraumatology (AMN) in following EMN in 2003 (Fig. 2). Dr. Pagni used to participate in all these scientific activities and enjoyed presenting new results, together with his team, in Europe, Asia, and the Far East, even when recovering from a heart attack and during his retirement. He was always enthusiastically engaged and discussed intensively his personal views and experiences, especially in the field of EMCS for Parkinson's disease and intractable pain, as well as deep brain and spinal cord stimulation for patients in a permanent vegetative state and those in a minimally conscious state (MCS), when he met with our Academicians and friends Drs. T. Tsubokawa, Y. Katayama, and T. Yamamoto from Nihon University, Tokyo; Dr. T. Kanno and his team at Fujita Health University, Toyoake, Japan; and Dr. Jin Woo Chang, from Yonsei University College of Medicine, Seoul, Korea.

Dr. Pagni had and was known for his broad vision and creative curiosity in neuroscience, and so he was invited by Dr. Katayama and me to join our then new International Society for Reconstructive Neurosurgery (ISRN). Dr. Katayama established the ISRN together with me to economically support our WFNS Neurorehabilitation Committee when Dr. Katayama was in charge of the committee that organized the biannual international conference after the XII WFNS World Congress of Neurosurgery, in Sydney, Australia, on 19–23 September, 2001; this was when Professor Madjid Samii (Germany) was the World President. On the occasion of that congress, Dr. Pagni, his wife Sandra, and I spent a very nice time together, enjoying a classical opera performance, a memorable sailing experience at sunset on Sydney Harbour, and a dinner of excellent fresh seafood accompanied by excellent Australian white wine (Figs. 3 and 4).

In the following years Dr Pagni and I met each other on many occasions. In March 2005, we met when two meetings took place simultaneously in Nagoya, Japan. Dr. T. Kanno, WFNS Neurorehabilitation Committee member and founding member of the AMN, was the Congress President, who had invited us to participate in the 3rd World Congress of the AMN, in conjunction with the 6th International Congress of Minimally Invasive Neurosurgery (MIN) [10, 12]. As usual,

Fig. 3 WFNS World Congress, Sydney, Australia, 20 September, 2001. C.A. Pagni enjoying a fresh fish dish and typical Australian white wine at a harborside restaurant

Fig. 4 Sandra Pagni, Sydney, 20 September, 2001

Dr. Pagni took an active part in the discussions, demonstrating his great personal experience and knowledge. One day, during the lunch break, to relax, we went for a sightseeing walk to Nagoya Castle, a walking distance of about 40 minutes. Together we climbed up the many steps of the tower; this was not sot easy for him as it was not so long since his cardiac attack! From the top he very much enjoyed the view of the fine landscape with a green park and white flowers around, and on the way down he admired some Japanese antiques in the reconstructed rooms. On the way back, bright-eyed, he told me about one of his favorite passions: fishing on the Mediterranean coast, where he came from. From time to time we slowed down or stopped for a minute so that he could recover his breath—it was the first time that I observed he needed a short rest. It seemed to be not a good omen—for either of us.

Later the same year, Dr. Jin Woo Chang, Congress President, had invited us for the 4th Scientific Meeting of the WFNS Neurorehabilitation Committee and the 1st Congress of the ISRN, both of which were held in Seoul on 1–3 September, 2005. This was the time when Dr. Pagni introduced his ambitious Italian colleague, Dr. Massimiliano Visocchi, from the Institute of Neurosurgery, Catholic University Medical School, Rome, Italy, to become an ISRN member and AMN Academician. Dr. Pagni knew him well from his scientific contributions, also on functional neurosurgery, from the Italian Society of Neurosurgery. Dr. Visocchi presented part of his research on spinal cord stimulation and cerebral hemodynamics [11]. At this congress Dr. Pagni agreed to organize the next but one conference in Rome, together with Dr. Visocchi. However, this year, when Dr. Visocchi finally organized our congresses as the president in Cerveteri, we lament the loss of our dear friend Carlo Alberto Pagni, who passed away on March 1, 2009.

Dr. Jin Woo Chang, now Head, Division of Stereotactic and Functional Neurosurgery, Yonsei University College of Medicine, Seoul, Korea, remained as the chairman of both the WFNS Committee and the ISRN until 2013, when Dr. Wai. S. Poon, Prince of Wales University, Hong Kong, took the chair of the WFNS, and Dr. Maximilian Mehdorn, Medical Faculty Christian-Albrechts-University, Kiel, Germany, took the ISRN chair. In 2015 Maximilian Mehdorn was nominated chair once again of both the WFNS Committee and the ISRN. Last, but not least, in Cerveteri near Rome, Dr. Visocchi was nominated ISRN chair elect. I wished to show, in this short review, the roots of the friendship between Dr. Pagni, Dr. Visocchi, and me in the WFNS Committee, ISNR, and AMN.

Carlo Alberto Pagni: Music and Mind

Carlo Alberto Pagni was an excellent pianist. As mentioned before, his favorite composer was Ludwig van Beethoven. Dr. Pagni is quoted in *Remembrance*, by Marco Pagni Frette (Milano 22 April, 2012), and in *Breviario Beethoveniano* (Milano Maggio, 2005) [13]. Dr. Pagni quotes various authors on music:

> "La musica è una legge morale. Essa è l'essenza dell'ordine ed eleva tutto ciò che è buono, giusto e bello, di cui essa è la forma invisibile." Platone, Engkish PLATO. "Vergogna ai ciechi che presero Beethoven per sordo!"
>
> Wilhelm von Lenz: "Der Menschheit Würde ist in eure Hand gegeben, bewahret sie! Sie sinkt mith euch! Mit euch wird sie sich heben!"
>
> Friedrich Schiller: "L'unica cosa importante al mondo è la musica: la musica, i libri e un paio di quadri. Fonderò una comunità in cui non ci si sposerà, a meno che per caso non ci si innamori di una sinfonia di Beethoven"
>
> Virginia Woolf: "Un musicista sordo! Possiamo noi immaginare un pittore cieco? Ma noi conosciamo però il veggente divenuto cieco. Il musicista sordo somiglia ora a Tiresia che, cieco alla visione del mondo dei fenomeni, contempla con l'occhio dell'anima il centro da cui muovono tutti i fenomeni. Non disturbato dai frastuoni della vita Beethoven rimane solo, intento alle sue armonie interiori, e da quelle profondità parla ancora a quel mondo che non ha più nulla da dirgli".

It had always been our wish to listen to the music of Richard Wagner together and to experience one of his operas at La Scala in Milan. And so, after Dr. Pagni's retirement, when he and his wife had moved to Milan, they invited us to the performance of *Lohengrin* on 28 January, 2007. It was conducted by Daniele Gatti, and directed by Nikolaus Lehnhoff, and Klaus Florian Vogt sang a memorable Lohengrin, accompanied by Solveig Kringelborn (Elsa), Linda Watson (Ortrud), and Jurgen Linn (Telramund). After the unforgettable opera event Sandra and Carlo Alberto Pagni invited us to their new home for dinner, where Carlo Alberto had prepared perfectly, as always, his Italian noodles, followed by a fresh fish dish that came with excellent wines. Dr. Pagni was a gourmet and wine connoisseur—he really appreciated Barolo wines!

As mentioned before, one of his favorite passions was fishing. Dr. Pagni adored staying with his three grandchildren in his summer house in Levantc, Liguria—fishing, swimming, boating (Fig. 5), and afterwards, cooking the fish for them, as his daughter, Benedetta Pagni Frette, has just recently confirmed. In the summertime she said the children spent almost 2 months with Mrs Pagni and Dr. Pagni, exploring the coast he knew so well (Fig. 6) from the time that his parents bought their summer house in Liguria in the late 1940s. Now his grandchildren Valeria, Valentina, and Tommaso, are between 18 and 12 years of age. Dr. Pagni adored fresh fish dishes, but he also cooked Italian noodles and enjoyed Barolo wines. Before his children were born, he was a passionate glider and competed in some Italian championships. All his life he was an excellent, passionate skier, and in wintertime the family went skiing . He also went mountaineering and hiking (Fig. 7). These sporting activities were a natural source of regeneration for him, where he relaxed from his neurosurgical duties and freed his mind for creative thinking about new challenges and new projects.

Fig. 5 Carlo Alberto Pagni with Valeria, 6 months old (the first child of his son); Levanto, August 1996. (Fotos by permission in writing, copyright by the daughter Pagni Frette Benedetta (bpagni@sperling.it)

Fig. 6 Carlo Alberto Pagni with Tommaso (now 12, son of his daughter Benedetta) in summer. (Fotos by permission in writing, copyright by the daughter Pagni Frette Benedetta (bpagni@sperling.it)

Fig. 7 Carlo Alberto Pagni in the Brenta Dolomites –1985. (Fotos by permission in writing, copyright by the daughter Pagni Frette Benedetta (bpagni@sperling.it)

He loved both his libraries as part of his life: his library at home, with numerous antique and valuable books, and his medical library in his office. When he retired he asked me what to do with all the medical books and medical journals in his comprehensive library that he did not want just to sell. He knew about my friendship with Professor Madjid Samii, from Hanover, and my professorship in functional neurorehabilitation in neurosurgery and restoration of brain and spinal cord lesions at the International Neuroscience Institute (INI), Hanover, at the medical faculty since 2003. He then followed my proposal to give his medical library of more than 2000 books, including his own publications, as a donation to Professor Madjid Samii for his recently opened new INI, to be held in The Professor Carlo Alberto Pagni Library. Gratefully, Professor Samii himself came to help in transferring the library from Turin to his INI, which, at this time, had become a scientific Institute at Otto-von-Guericke-University, Magdeburg, Germany. What a nice generous memento!

Last, but not least, we have to remember Carlo Alberto Pagni for his passion and talent in playing chess, and, above all, playing correspondence chess in his later years. The Ken Whyld Chess Association (KWA) wrote in their obituary:"In March 2009 we the KWA received the bad news from Italy, Professor Carlo Alberto Pagni had unexpectedly departed this earth". Ralf Binnewirtz et al. wrote, on March 8, 2009: "Our Italian KWA member Professor Carlo Alberto Pagni had completely unexpectedly passed away in Milan on Monday, 2nd of March. Nobody could know that after his 78th birthday – we still congratulated him at the end of February – only such a short time was left to him." (courtesy KWA).

CV of Carlo Alberto Pagni (for KWA, from his own point of view)

"Born in La Spezia 13 February 1931. Medical Doctor, Professor of Neurology and Neurosurgery. Resident at the Clinic for Nervous and Mental Diseases (Neurosurgical section) of the Turin University 1955. Assistant Professor: Clinic for Nervous and Mental Disease of the Cagliari University and Neurosurgical Department of Cagliari 1956–1958; Neurosurgical Clinic University of Milano 1959–1979. Professor of Neurosurgery and Chairman of the Neurosurgical Clinic of the University of Torino 1980–2003. In 1947 he began to play chess; Champion of La Spezia in 1950. Owing to study and professional duties he abandoned playing chess. In 1970 he began to play by correspondence chess and became Master of the Associazione Scacchistica Italiana Giocatori per Corrispondenza when participating in many tournaments. Won in the 28th Italian Championship in 1977–1978, scored 7–9. In the IV Coppa ICCF 1977–1987, he scored 16 out of more than 3000 chess tournaments."

Fig. 8 Professor Pagni giving his lecture in Venice at the Ken Whyld Foundation and Association, May 2008

Acknowledgments Figures 1, 2, 3, and 4, copyright K. von Wild 2015. Figures. 5, 6, and 7 and family remarks, courtesy of Benedetta Pagni Frette, December 2015. Figure 8 and quotation from KWA obituary, courtesy of Dr. Ralf J. Binnewirtz, author and KWF&A-Webmaster (http://www.kwabc.org/index.php/in-memoriam-members/carlo-alberto-pagni, 2012, 08, 15). Special thanks to my good friend Federico Hernández Meyer, Magister Pharm., Av Andalucia 78, 18198 Huétor Vega, Spain, who kindly revised this paper for publication.

Conflict of Interest Statement I certify that there is no actual or potential conflict of interest in relation to this article.

Dr. Klaus von Wild, Münster 22.12.2015

References

1. von Wild K (Hrsg.) (1993) Spektrum der Neurorehabilitation. Frührehabilitation, Rehabilitation von Kindern und Jugendlichen. W. Zuckschwerdt Verlag, München/Bern/Wien/New York
2. Thompson K, Pohlmann-Eden B, Campbe LA, Abel H (2015) Pharmacological treatments for preventing epilepsy following traumatic head injury. Cochrane Database Syst Rev. 8:CD009900. doi: 10.1002/14651858.CD009900.pub2
3. Schramm J (2014) Epilepsy surgery and the evolution of clinical and translational science. Neurosurgery 61(Suppl 1):54–65. doi:10.1227/NEU.0000000000000399
4. Pagni CA (1990) Posttraumatic epilepsy, incidence and prophylaxis. Acta Neurochir Suppl 50:38–47
5. Pagni CA, Lo Rosso GM, Benna P, Paglia G, Naddeo M (1993) Posttraumatic epilepsy. In: von Wild K (ed) Spektrum der Neurorehabilitation, vol 1, Fortschritte in der Neurotraumatologie und klinischen Neuropsychologie. W. Zuckschwerdt Verlag, München/Bern/Wien/New York, pp 71–78
6. Pagni CA, Canavero S, Bonicalzi V, Nurisso C (2001) The important role of pain in neurorehabilitation. The neurosurgeon's approach. In: von Wild KRH in cooperation with Lipovsek M, Mendelow AD, Truelle JL (eds) Functional rehabilitation in neurosurgery and neurotraumatology, Acta Neurochir, Suppl 79. Springer Publisher, Wien/New York, pp 67–74
7. Pagni CA, Zeme S, Zenga F (2003) Neurosurgical clinic, university of Turin, Turin, further experience with extradural motor cortex stimulation for treatment of advanced Parkinson's disease. Report of 3 new cases. J Neurosurg Sci 47(4):189–193
8. Pagni CA, Zegna F (2005) Posttraumatic epilepsy with special emphasis on prophylaxis and prevention. In: von Wild KRH (ed) Re-engineering of the damaged brain and spinal cord – evidence-based neurorehabilitation, Suppl 93. Springer Publisher, Wien/New York, pp 27–34
9. Pagni CA, Altibrandi MG, Bentivoglio A, Caruso G, Cioni B, Fiorella C, Inso Extradural la A, Lavano A, Maina R, Mazzone P, Signorelli CD, Sturiale C, Valzania F, Zeme S, Zenga F (2005) Extradural motor cortex stimulation (EMCS) for Parkinson's disease. History and first results by the study group of the Italian neurosurgical society. In: von Wild KRH (ed) Re-engineering of the damaged brain and spinal cord – evidence-based, Neurorehabilitation Acta Neurochir, Suppl 93. Springer Publisher, Wien/New York, pp 113–119
10. Kanno T, Kato Y (eds) (2006) Minimally invasive neurosurgery and multidisciplinary neurotraumatology. Springer-Publisher, Tokyo
11. Visocchi M (2006) Spinal cord stimulation and cerebral hemodynamics. In: Chang JW, Katayama Y, Yamamoto T (eds) Advances in functional and reparative neurosurgery, Acta Neurochir, Suppl 99. Springer Publisher, Wien/New York, pp 111–116
12. Pagni CA et al (2008) Results by motor cortex stimulation in treatment of focal dystonia, Parkinson's disease and post-ictal spasticity. The experience of the Italian Study Group of the Italian Neurosurgical Society. Acta Neurochir Suppl 101:13–21
13. Brevario Beethoveniano Milano Maggio. 2005. www.lvbeethoven.it

Part I

Restorative Neurosurgery

Spinal Cord Stimulation: An Alternative Concept of Rehabilitation?

Antonella Giugno, Carlo Gulì, Luigi Basile, Francesca Graziano, Rosario Maugeri, Massimiliano Visocchi, and Domenico Gerardo Iacopino

Abstract

Background

Chronic low back and leg pain is a disabling condition, affecting, in most cases, older patients with congenital or acquired spinal stenosis or patients with failed back surgery syndrome. Spinal cord stimulation has been introduced as an effective therapeutic option for those patients who have previously been operated without significant clinical benefits, or for all those patients who are ineligible for traditional surgery.

Methods

We report our experience with ten patients treated with spinal cord stimulation plus medication and physical therapy between November 2014 and September 2015. Inclusion criteria were: previous surgical treatments for lumbar stenosis and metameric instability and persistent or ingravescent disabling low back and leg pain, with a mean duration of symptoms of at least 18 months. A visual analog scale (VAS) was employed for back and leg pain, and the Oswestry Disability Index (ODI) score was determined, and findings were analyzed after 6 months.

Findings

No intra- or postoperative complication was recorded. The mean VAS score for back pain decreased from 7.5 to 2.9, while leg VAS decreased from 8.2 to 3.0. Analysis of ODI values showed evident improvement in daily life activities, ranging from a median value of 75.7% to 32.7% after the stimulation.

Conclusion

Spinal cord stimulation has a recognized impact on the pain and on the quality of life of patients with failed back surgery syndrome.

Keywords Spinal cord stimulation • Chronic low back pain • Failed back surgery syndrome

A. Giugno, MD (✉) • C. Gulì, MD • L. Basile, MD
F. Graziano, MD • R. Maugeri, MD, PhD
D.G. Iacopino, MD, PhD
Department of Experimental Biomedicine and Clinical Neurosciences, School of Medicine, Neurosurgical Clinic, University of Palermo, Via del Vespro 129, Palermo 90100, Italy
e-mail: alfagamma82@libero.it

M. Visocchi, MD, PhD
Institute of Neurosurgery, Medical School, Catholic University of Rome, Rome, Italy

Introduction

Chronic low back and leg pain is a disabling condition, affecting, in most cases, patients aged between 50 and 80 years old with congenital or acquired spinal stenosis [3]. About 30 % of patients who undergo spinal surgery are affected by chronic pain from failed back surgery syndrome (FBSS), characterized by neuropathic leg pain with radicular distribution [7]. The strong impact of such a condition on quality of life, regarding both interpersonal relationships and professional activities, is well documented. Conservative management strategies include medications, physical treatments, and/or peripheral nerve blocks. Spinal cord stimulation (SCS) was introduced 45 years ago as an effective therapeutic option for those patients previously operated without resolution, or for all those patients who are refractory to conservative treatment or who are ineligible for traditional surgery [8]. A large study, the Randomized Controlled Multicenter Trial of the Effectiveness of Spinal Cord Stimulation (PROCESS), supports the role of stimulation in improving the quality of life after 6 months, with further amelioration after 24 months, also considering the advantages in terms of economic costs [7]. The PROMISE study, a multicenter, prospective, randomized, open-label, parallel-group study, proved the efficacy of this technique also in those patients with back pain predominant over leg pain. It is widely accepted that spinal stimulation plus conventional medical management (CMM) versus CMM alone, in

M. Visocchi et al. (eds.), *Trends in Reconstructive Neurosurgery*, Acta Neurochirurgica Supplement 124,
DOI 10.1007/978-3-319-39546-3_3, © Springer International Publishing Switzerland 2017

refractory patients, has a strong positive impact on recovery of function, and on the economic and the social profile. The aim of our retrospective study was to evaluate the effects on pain and on quality of life in a clinical series of ten patients treated at our Unit of Neurosurgery by spinal cord stimulation plus medication and physical therapy. A visual analog scale (VAS) was employed for back and leg pain, and the Oswestry Disability Index (ODI) score was determined, with analysis done after 6 months.

Materials and Methods

We report our experience with a group of ten patients treated with SCS plus medication and physical therapy between November 2014 and September 2015. Inclusion criteria were the following: persistent low back and leg pain for at least 6 months, previous lumbar surgery (decompressive or arthrodesis), and loss of response to pharmacological treatments. The patients were: six females and four males with a median age of 59.3 years (range, 48–73 years). All patients had undergone previous surgical treatments for lumbar stenosis and metameric instability and had persistent or ingravescent disabling back and leg pain with a mean duration of symptoms of 18 months. Pretreatment evaluations of pain and of quality of life were performed. Eight patients had prevalent bilateral leg pain while the other two were affected by low back and leg pain of the same intensity. The mean pretreatment VAS scores were 7.5/10 for low back pain and 8.2/10 for leg pain. The ODI score was also determined to study several aspects of daily life. The median pretreatment value was 75.7 %. All patients presented with at least a severe disability, with limitations in taking care of themselves. The demographic values of this SCS group are reported in Table 1.

Description of the Procedure

Preoperative computed tomography (CT) scans of the dorsal vertebral tract were performed to individuate eventual canal anomalies such as stenosis and scoliosis. All patients underwent a trial of SCS for about 2 weeks in order to evaluate toleration of the system. The trial treatments were performed under local anesthesia. An octopolar epidural electrode (Medtronic, Memphis, TN, USA®) was placed at the T8-L1 level. The electrode was constituted by two lines of four platinum iridium contacts in its terminal portion. During the treatment the patients referred to their subjective perceptions, guiding the surgeon in the procedure. When the system was tolerated, after about 2 weeks, implant of the permanent system was performed and the generator was placed subcutaneously in the abdomen. Patients were able to easily adjust the stimulation amplitude, to minimize possible uncomfortable paresthesias, through an external remote control device. After the procedure, all patients underwent close clinical follow-up at 1, 3, and 6 months.

Results

No postoperative neurological deficits were encountered and all patients treated with the SCS tolerated the system, noting substantial leg and back pain improvement. More consistent results were achieved in the control of leg pain. After 6 months, together with medical care and physical therapy, the mean VAS score for back pain was reduced from 7.5/10 to 2.9/10, while VAS for leg pain was reduced from 8.2/10 to 3.0/10. ODI values showed evident improvement in daily life activities, changing from a median value of 75.7% to 32.7 % after the stimulation (Table 2).

Table 1 Demographic data of patients

Sex	Age (years)	Back VAS pre	Leg VAS pre	ODI pre	Back VAS post (6 months)	Leg VAS post (6 months)	ODI post (6 months)
M	71	7/10	7/10	80 %	3/10	2/10	27 %
F	48	7/10	9/10	75 %	4/10	3/10	20 %
M	58	9/10	10/10	67 %	3/10	6/10	12 %
F	53	8/10	8/10	94 %	2/10	2/10	83 %
F	59	10/10	9/10	96 %	4/10	3/10	75 %
M	73	6/10	8/10	57 %	2/10	5/10	12 %
F	55	8/10	7/10	72 %	3/10	2/10	29 %
M	69	6/10	9/10	71 %	2/10	2/10	21 %
F	52	8/10	7/10	83 %	2/10	2/10	37 %
F	55	6/10	8/10	62 %	4/10	3/10	11 %
Mean values	**59.3**	**7.5**	**8.2**	**75.7 %**	**2.9**	**3.0**	**32.7 %**

VAS, visual analog scale, *ODI*, Oswestry Disability Index, *M* male, *F* female, *pre* preoperative, *post* postoperative

Table 2 Profiles of VAS and ODI in pre- and postoperative periods (6 months)

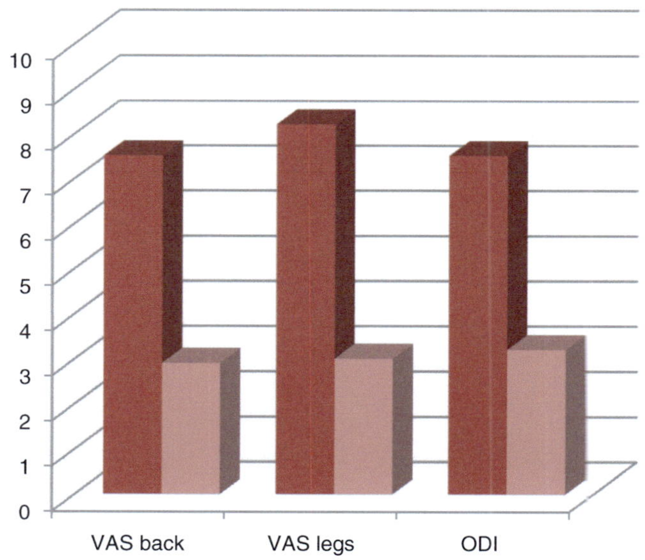

Discussion

Failed back surgery syndrome (FBSS) is a relatively common condition in patients treated with spinal surgery. It is more common in the elderly and after complex arthrodesis treatments, but it can also occur in young patients who are not responsive to traditional surgical procedures. Chronic pain leads to social and professional consequences and requires extremely high economic resources. It is well known that chronic degenerative low back and leg pain worsen the quality of life of patients, reducing their autonomy and ability to work. Indeed, such pain can be a significant obstacle in daily activities, raising the number of individuals requiring assistance, with an increase in social welfare expenses. It has been estimated by Berger et al. that the healthcare costs for patients affected by chronic pain syndromes are three times higher than the costs for patients of the same age but without pain [3].

Spinal cord stimulation was first proposed by Shealy in 1967 [8]. The technological development of SCS in recent decades has meant that it is now a widely accepted treatment for chronic intractable neuropathic pain [5]. In the past 20 years some papers have been published regarding the efficacy of SCS and comparing the cost-benefit ratios of SCS and other treatments. In 1996 Burchiel and colleagues designed a prospective multicenter study to investigate the efficacy and outcome of SCS, using a variety of clinical and psychosocial outcome measures [4]. They reported the success of the therapy, defined as pain relief in 55 % of patients at 1-year follow-up. Barolat et al. reported similar results in a group of 44 patients treated with SCS utilizing paddle electrodes and a radiofrequency (RF) stimulator [2]. Abeloos et al. evaluated the long-term efficacy and the quality of life in their series of 55 patients who underwent SCS for FBSS [1]. They reported satisfaction in 75 % of the patients after 8.3 years of follow-up, demonstrating that SCS is an effective treatment for refractory FBSS. In their retrospective study, Kamihara and colleagues evaluated the efficacy of SCS for leg pain associated with lumbar spinal stenosis (LSS) [6]. They collected 41 patients with LSS-associated leg pain who had undergone SCS implantation in the period between 2003 and 2011. The percentage of patients who showed a good response (defined as SCS continued for 1 year or longer after implantation) was 95 %, with alleviation of pain. Based on this finding, these authors proposed SCS as an LSS treatment method that could be intermediate between conservative therapy and surgery [6].

Our study evaluated the change in the quality of life in a total of ten patients with FBSS treated with SCS at the Neurosurgery Clinic at the University of Palermo over the period from November 2014 to September 2015. The maximum follow-up was 6 months. All patients had a preoperative clinical evaluation and then they were evaluated at 1, 3, and 6 months. The analysis of the results showed that SCS, associated with medical care and intensive rehabilitation, appeared to be safe and effective, with a strong positive impact on pain relief and on the quality of life of the patients.

Conclusion

Spinal cord stimulation (SCS) has a recognized impact on the quality of life of patients with FBSS. Our study confirms that SCS, in association with rehabilitation, may significantly improve functional prognosis. This study suggests that the concept of "rehabilitation" does not refer exclusively to physical treatment, but consists of a multidisciplinary approach that may include, in some cases, SCS. This allows for greater efficiency in restoring normal daily activities and also helps patients to retain their social relationships and job autonomy. To confirm these results, long-term follow-up with an extended clinical series is needed.

Conflict of Interest Statement The authors report no conflicts of interest. No funds were received in support of this work.

References

1. Abeloos L, De Witte O et al (2011) Long term outcome of patients treated with spinal cord stimulation for therapeutically refractory failed back surgery syndrome : a retrospective study. Neurochirurgie 57(3):114–119
2. Barolat G, Oakley JC et al (2001) Epidural cord stimulation with a multiple electrode paddle lead is effective in treating intractable low back pain. Neuromodulation 4(2):59–66
3. Berger A, Dukes EM, Oster G (2004) Clinical characteristics and economic costs of patients with painful neuropathic disorders. J Pain 5(3):143–149

4. Burchiel KJ, Anderson VC et al (1996) Prospective, multicenter study of spinal cord stimulation for relief of chronic back and extremely pain. Spine (Phila Pa 1976) 21(23):2786–2794
5. Devulder J, De Colvenaer L et al (1990) Spinal cord stimulation in chronic pain therapy. Clin J Pain 6:51–56
6. Kamihara M, Nakano S et al (2014) Spinal cord stimulation for treatment of leg pain associated with lumbar spinal stenosis. Neuromodulation 17:340–345
7. Kumar K, North RB, Taylor RS et al (2005) Spinal cord stimulation versus conventional medical management: a prospective, randomized, controlled, multicenter, study of patients with Failed Back Surgery Syndrome (PROCESS study). Neuromodulation 8: 213–218
8. Shealy CN, Mortimer JT et al (1967) Electrical inhibition of pain by stimulation of the dorsal column: preliminary clinical report. Anesth Analg 46:489–491

Recovery from Chronic Diseases of Consciousness: State of the Art in Neuromodulation for Persistent Vegetative State and Minimally Conscious State

Pier Paolo Mattogno, Giuseppe Barbagallo, Gerardo Iacopino, Gaetano Pappalardo, Giuseppe LA Rocca, Francesco Signorelli, Jun Zhong, and Massimiliano Visocchi

Abstract Chronic diseases of consciousness (CDC) can still be considered a challenging frontier for modern medicine, probably because of their not completely understood physiopathological mechanisms. Following encouraging evidence on cerebral hemodynamics, some authors have hypothesized a role for neuromodulation in the treatment of CDC patients. In the past 40 years, spinal cord stimulation (SCS) and deep brain stimulation (DBS) have been used experimentally for the treatment of patients in a severe altered state of consciousness, with some interesting but not conclusive results. The present review summarizes the data currently available in the literature on this particular and debated topic. On these grounds, further clinical studies are needed to better understand the altered dynamics of neuronal network circuits in CDC patients as a step towards novel therapeutic strategies.

Keywords Disorders of consciousness • Neuromodulation • SCS • DBS • Minimally conscious state • Vegetative state

P.P. Mattogno, MD (✉)
Neurosurgery, Catholic University Medical School, Rome, Italy

Neurosurgery – A, Gemelli Hospital,
Largo A. Gemelli, 8, Rome 00168, Italy
e-mail: pierpaolo.mattogno@gmail.com

G. Barbagallo
Division of Neurosurgery, Department of Neurosciences, Policlinico "G. Rodolico" University Hospital, Catania, Italy

G. Iacopino
Department of Experimental Biomedicine and Clinical Neurosciences, School of Medicine, Neurosurgical Clinic, University of Palermo, Palermo, Italy

G. Pappalardo • G. LA Rocca • F. Signorelli
Neurosurgery, Catholic University Medical School, Rome, Italy

J. Zhong
Department of Neurosurgery, XinHua Hospital, Shanghai JiaoTong University School of Medicine, Shanghai 200092, China

M. Visocchi, MD
Institute of Neurosurgery, Medical School, Catholic University of Rome, Rome, Italy

Introduction

At present, chronic diseases of consciousness (CDC) represent an issue for national healthcare assistance systems [27]. The causes of CDC are numerous and include different pathological noxae: traumatic brain injury (TBI) surely represents the most common event causing CDC, especially in the younger population, while hypoxic damage, cerebrovascular ischemic or hemorrhagic injury, infection of the central nervous system, toxins, poisoning, degenerative diseases, and tumors, as well as congenital or developmental disorders, are other possible related causes [25, 26]. The increased number of CDC patients is probably due to improvements in worldwide standards of emergency care, providing rapid and quite efficient assistance to those suffering serious accidents, in which in the past most cases were fatal. CDCs can be still considered as a challenging frontier for neuroscience research, probably because the physiopathological mechanisms behind them are still not completely understood [3]. Following encouraging evidence on cerebral hemodynamics in conditions such as ISCHAEMIA some authors have hypothesized a role for neuromodulation in the treatment of CDC patients. The aim of this review was to compare the results of different studies of the management of CDC with spinal cord stimulation (SCS) and deep brain stimulation (DBS) in order to provide a framework for the future development of neurosurgical stimulation.

Vegetative State

The vegetative state (VS) was first described in 1972 by Jennett and Plum [12], and was subsequently defined, by the Multi-Society Task Force on PVS [26], as "a clinical condition of *complete* unawareness of the self and the environment, accompanied by sleep-wake cycles, with either complete or partial preservation of hypothalamic and brainstem autonomic

M. Visocchi et al. (eds.), *Trends in Reconstructive Neurosurgery*, Acta Neurochirurgica Supplement 124,
DOI 10.1007/978-3-319-39546-3_4, © Springer International Publishing Switzerland 2017

function". The persistence of VS for over 12 months after a TBI, according to international guidelines; persistence for over 3 months after non-traumatic causes, according to guidelines from the United States; or persistence for 6 months, according to guidelines from the United Kingdom, is considered to be 'permanent' VS (PVS) [25, 31].

Minimally Conscious State

In 2002, Giacino et al., analyzing a group of VS patients, identified some particular clinical features, introducing the concept of the *minimally conscious state* (MCS), characterized by "inconsistent but clearly discernible behavioral evidence of consciousness", and well distinguished from coma and VS. Clinical criteria used to identify MCS are: the following of commands, intelligible verbalization, discernible yes–no signals (regardless of accuracy), and specific responses to selective environmental stimuli, in order to capture nonreflexive behaviors reliably triggered in the presence of a specific stimulus, with infrequent occurrence in its absence [10]. Some authors have proposed two MCS subcategories [3]: MCS+, patients following commands, and having intelligible verbalization and gestural or verbal yes/no responses; and MCS-, patients demonstrating pursuit eye movement, ability to localize (and orient) noxious stimuli, and appropriate movements or affective behaviors. This possibility of exploring covert behaviors, combined with recent advances in neural correlates, such as functional magnetic resonance imaging (fMRI), is expected to yield further insight into the phenomenology of consciousness applied to clinical phenotypes [7].

Diagnosis of CDC

Although the clinical criteria of CDC are well established, a careful assessment of the patient's level of awareness is necessary in order to make a differential diagnosis of VS or MCS. Accurate examinations have to be performed to distinguish purely reflex responses to stimuli from responses that require awareness; nevertheless, this distinction is not always possible [1]. It is crucial to determine correctly the resting brain function. Some recent studies suggest the use of electrophysiological examinations in order to establish the resting brain function in CDC patients [36]. Particularly, auditory brainstem responses (ABRs) are generally used for the evaluation of brainstem function, somatosensory evoked potentials (SEPs) are used for studying thalamocortical pathways, continuous electroencephalogram (EEG) frequency analysis is used to determine the relationship between the brainstem and the cerebral cortex, and pain-related P250 analysis is used to evaluate higher brain functions [35–37].

Treatment

As a matter of fact, there is still no standard treatment for CDC, VS, or MCS, either medical or surgical, that has reached the level of evidence-based treatment. Some medical treatments have been tested, such as dopaminergic agents (levodopa and amantadine), zolpidem (a nonbenzodiazepine-specific, indirect GABA agonist), or intrathecal baclofen (ITB): they have shown, in some cases, beneficial effects in CDC patients, but evidence on this topic is still very limited [9]. Recently, growing interest has been focused on the application of neuromodulation for CDC, based on evidence regarding the possibility of modifying the locoregional cerebral blood flow through an electrical stimulus, applied to cerebral structures, creating functional changes in those brain circuits that seem to be involved in consciousness and the comatose state. The application of these techniques to several cerebral low-perfusion syndromes seems to be a promising new trend for the treatment of cerebrovascular diseases in order to allow 'functional' cerebral revascularization [33].

Materials and Methods

Searches were performed in the Medline databases (until November 2015) using the following key words: 'vegetative state', 'minimally conscious', 'disorders of consciousness', 'coma', 'effect', 'therapy', and 'recovery', along with 'neuromodulation' 'spinal cord stimulation', 'SCS', 'dorsal column stimulation', 'cervical stimulation', 'deep brain stimulation', and 'DBS'. The electronic search was complemented by cross-checking the references of all relevant articles. Only papers dealing with subjects with a diagnosis of VS and MCS were included in the research. In order to avoid duplications, special attention was paid to identify those patients who had been described in more than one publication (for example, follow-up of already published studies).

Results and Discussion

Spinal Cord Stimulation for CDC

Spinal cord stimulation (SCS) is a neuromodulation technique that has been experimentally applied in challenging neurosurgical diseases, such as cerebral ischemia and vasospasm, both in animal and human models, with encouraging results, although its effective use has been hampered by a lack of understanding of its mechanism(s) of action [33]. On these grounds, several studies have hypothesized the application of SCS to CDC patients, with interesting and surprising results [37].

On performing a review of the actual literature, we found that 318 CDC patients were treated with SCS from 1988 to 2012 (Table 1); a clinical response was reported in 166 patients (52.2 %), with improved neurological function and arousal. Patients were studied using clinical evaluation, neuroimaging (magnetic resonance imaging [MRI], computed tomography [CT]), metabolite assessment, and neurophysiological and brain metabolism investigations. Reported causes of CDC were mainly trauma, anoxia, or stroke. In all the studies the SCS device was implanted into the epidural space between C2 and C4, stimulating the spinal cord in a cyclic mode, without reaching the motor threshold. Clinical improvement latency was quite variable, ranging from 3 days to 1 year after initiation of the stimulation. A better clinical response to SCS was evidenced in patients treated sooner, especially in younger ones. Radiological findings in all patients showed no marked cerebral atrophy and no diffuse low-density areas involving the thalamus or brainstem. Cerebral blood flow (CBF), when reported, was assessed to be >20–25 ml/100 g/min in patients who demonstrated clinical improvement. Increases of catecholamines and decreases of superoxide free radicals were reported after SCS. Yamamoto et al. reported that SCS increased CBF diffusely in the entire brain in MCS patients, with an increase of more than 22 % during the stimulation period compared with before stimulation (*P* 0.0001, paired *t*-test). As reported, SCS can induce muscle twitches of the upper extremities, with evidence of functional recovery. On these grounds, some authors speculate that both cervical and thoracolumbar stimulation may be useful for the functional recovery of the upper and lower extremities [34–37].

Deep Brain Stimulation for CDC

Deep brain stimulation (DBS) is a validated neuromodulation technique traditionally considered to be the "gold standard" for severe cerebral motor symptoms refractory to optimal drug trials. The mechanism of action is the chronic disruption of abnormal neural synchrony between affected brain regions, inhibiting neural activity and moderating abnormal brain function related to disease symptoms [2, 6, 22]. Targets of stimulation have been numerous: some examples are: the subthalamic nucleus for Parkinson's Disease, the ventralis intermedius (VIM) nucleus of the thalamus for posttraumatic tremor, and the ventroposterolateral nucleus of the thalamus for dystonia. Nowadays, DBS is proposed as a way to alleviate extrapyramidal motor disorders, and research is on-going to explore or validate further indications, such as depression, obsessive compulsive disorder, pain, obesity, anorexia, and epilepsy [4, 6, 18].

In recent years, encouraged by these experiences, the use of DBS has been extended to patients with severe CDC [17]. However, this application entails a number of challenges: the physiopathological mechanisms of CDC involve multiple and often combined events, such as trauma, hemorrhage, ischemia, and anoxia, and so, prospectively, patients liable to benefit from DBS for severe, chronic, or sometimes long-term disorders of consciousness have a wide range of individual phenotypes.

On examining the pertinent literature, we found that a small number of studies explored the effect of DBS in CDC patients (Table 2). A total of 58 CDC patients were treated using DBS from 1968 to 2015; 29 of them (52 %) demonstrated clinical and instrumental signs of arousal from VS. In 1968 McLardy et al. described the first DBS implantation, in a vegetative 19-year-old male, implanted about 3 months after severe head injury. The only reported effects were slight midbrain contacts, such as left orientation of the head and movements of the left hand; slight EEG modifications were also observed [23].

One year later, Hassler et al. published a DBS case of a 26-year-old male in VS implanted about 5 months after head injury. He was stimulated bilaterally, electrodes were placed in the anterior thalamus (left) and in the pallidum (right). Stimulation of the right pallidum was associated with sporadic contralateral arm movements, and bilateral acute stimulations provoked stronger effects plus head movements following the eyes. Convulsive seizures sometimes occurred during high thalamic stimulations (up to 100 Hz). During the 19-day follow-up, the clinical condition improved: purposeful gaze and mimicking; arm, leg, and head movements; and unintelligible vocalization were evidenced. EEG frequencies increased during acute thalamic stimulation [11]. Sturm reported a DBS implantation in a 68-year-old male in VS, implanted about 1 month after the rupture of a basilar artery aneurysm and surgical clipping. Bilateral stimulation of the thalamus was performed for 4 weeks. The patient improved, with verbal contact, sitting in a chair, oral feeding, and sleep-waking periods, and general status improved (body temperature, resolved pulmonary infection); EEG activity shifted to a less monotonous activity with theta and delta waves [30]. Tsubokawa et al. first speculated that DBS may facilitate the recovery of consciousness in patients with chronic CDC, studying a group of eight VS patients (four traumas, three strokes, one anoxia) implanted 6 months after the primary cerebral injury. The electrodes were placed in the nucleus cuneiformis of the mesencephalic reticular formation (two patients) and in the centromedian parafascicular nucleus complex (CMPf; six patients). Clinical assessment was based on the institutional prolonged coma score (PCS), ranging from 1 (alive with spontaneous respiration) to 10 (verbal response). Prior to DBS, the eight patients had PCS values ranging from 2 to 4. Open eyes with dilated pupils, open mouth and meaningless vocalization, slight increase in blood pressure, and EEG arousal patterns were described in all these patients after implantation. Increased CBF (10–72 %), metabolic rate of oxygen (10–40 %), and glucose uptake

Table 1 SCS in CDC patients

Author	Etiology	Patients	Site of lesion	Stimulation	Time to surgery	Clinical improvement latency	Evaluation parameters	Responsive *n* (%)
Kanno et al. (1988) [13]	Trauma/ischemia	10	Cerebral cortex	Cervical SCS	3–6 months	3–14 days	Clinical, EEG, rCBF, dynamic CT, xenon-enhanced CT, catecholamine	9 (90 %)
Matsui et al. (1989) [21]	Trauma/ischemia/tumors	8	Cerebral cortex	Cervical SCS	3–19 months	1–2 months	Clinical	2 (25 %)
Momose et al. (1989) [24]	Trauma/ischemia	1	Cerebral cortex	Cervical SCS	3 months	1 week	PET, 18 FDG, rCBF	1 (100 %)
Kanno et al. (1989) [14]	Trauma/ischemia	6	Diffuse	Cervical SCS	3 months	n.a.	Clinical/CT/ABR/MRI/EEG	4 (66.6 %)
Yokoyama et al. (1990) [38]	Trauma/CVD/hypoxia/tumor	23	Cerebral cortex	Cervical SCS	3–78 months	1–42 months	Clinical, EEG	8 (34.7 %)
Kuwata (1993) [16]	Trauma/vascular/meningitis/tumors	15	Cerebral cortex	Cervical SCS	1–27 months	2 months	EEG, ABR, SEP, neurotrasmitter	4 (26.6 %)
Fujii et al. (1998) [8]	Hypoxia	12	N/A	Cervical SCS	1 month	3 months	MRI, CBF, xenon-enhanced CT, ABR, SEP, clinical	7 (58.3 %)
Liu et al. (2008) [20]	Ischemia	20	Cerebral cortex	Cervical SCS	4.4–95 months	71–287 days	SPECT, neurotransmitter	9 (45 %)
Liu et al. (2009) [19]	Ischemia/trauma	12	Cerebral cortex	Cervical SCS	3–7 months	1–107 days	SPECT, Clinical	6 (50 %)
Kanno et al. (2009) [15]	Ischemia/trauma	201	Diffuse	Cervical SCS	3–12 months	n.a.	Clinical, CT, MRI, SPECT	109 (54.2 %)
Yamamoto et al. (2012) [36]	Trauma/vascular/inflammatory	10	Diffuse	Cervical SCS	3–53 months	6–7 months	Clinical, SPECT	7 (70 %)
Total		318						166 (52.2 %)

CDC chronic diseases of consciousness, *SCS* spinal cord stimulation, *EEG* electroencephalogram, *SPECT* single-photon emission computed tomography, *PET* positron emission tomography, *rCBF* regional cerebral blood flow, *CBF* cerebral blood flow, *ABR* auditory brainstem response, *n.a.* not available, *CVD* cardiovascular disease, *SEP* somatosensory evoked potential, *FDG* fluoro deoxy glucose, *MRI* magnetic resonance imaging

Table 2 DBS in CDC patients

Author	Etiology	Patients	Hemisphere	Electrode location	Time from trauma	Follow-up	Evaluation parameters	Responsive (%)
McLardy et al. (1968) [23]	Trauma	1	Unilateral left	MRF	8 m	1 m	Clinical, EEG	1 (100 %)
Hassler et al. (1969) [11]	Trauma	1	Bilateral	Anterior thalamus (l), pallidus (r)	n.a.	5 m	Clinical, EEG	1 (100 %)
Sturm et al. (1979) [30]	Vascular	1	Bilateral	Lamella medialis (r), nucleus reticularis (l)	1 m	1 m	Clinical, EEG	1 (100 %)
Tsubokawa et al. (1990) [32]	Trauma,stroke, anoxia	8	n.a.	MRF (two cases) CMPf (six cases)	6 m	12 m	Clinical, PET, EEG, CBF	4 (50 %)
Cohadon and Richer (1984–1993) [5]	Trauma	25	Unilateral right	n.a.	6–15 m	2 m	Clinical	13 (52 %)
Schiff et al. (2007) [29]	Trauma	1	Bilateral	Anterior thalamus	6 y	6 m	Clinical, EEG	1 (100 %)
Yamamoto et al. (2003–2013) [35–37]	Trauma, vascular, anoxia	21	n.a.	MRF (2 cases) CMPf (19 cases)	4–8 m	12 m	Clinical, EEG, CBF, PET, ABR, SEP, pain-related P250	8 (38.1 %)
Total		58						29 (50 %)

DBS deep brain stimulation, *MRF* mesencephalic reticular formation, *EEG* electroencephalogram *CMPf* centromedian parafascicular nucleus complex, *PET* positron emission tomography, *CBF* cerebral blood flow, *ABR* auditory brainstem response, *SEP* somatosensory evoked potential, *l* left, *r* right, *n.a.* not available, *m* months, *y* years

(200–300 %), measured by positron emission tomography (PET) imaging, were also reported. At 12-month follow-up, treatment was rated excellent in three of the eight patients, slightly effective in one, and ineffective in four [32]. Cohadon et al., in a series of dedicated papers from 1984 to 1993, collected 25 VS patients who had undergone DBS implantation 6–15 months after TBI. Electrodes were placed in the centromedian nucleus of the right thalamus in all patients. At the 2-month follow-up, 13 patients (52 %) demonstrated significant improvements in global functional score; between 2 weeks and 2 months after DBS was started all of them became communicative [5]. In a case report by Schiff et al. in 2007, DBS of the central thalamus was shown to lead to behavioral improvements in an MCS patient 6.5 years following injury, especially in arousal state, limb control, and verbalization, due probably to the preserved innervation of the frontal cortex and basal ganglia by thalamic association nuclei [28]. The Tokyo Neurosurgical Group, led by Yamamoto, studied the possibility of recognizing predictive parameters of responsiveness to neuromodulation procedures in CDC patients, in order to quantify the "resting brain" after severe brain damage. Electrophysiological criteria (electroencephalography, ABRs, SEPs, and pain-related evoked potentials) were proposed in order to identify CDC patients eligible for DBS implantation. These authors identified a total of 21 eligible CDC patients. Electrodes were placed in the midbrain reticular formation in 2 patients, and in the CMPf in 19 patients during a period of 4–8 months following injury. After 12-month follow-up, 8 of the 21 patients had emerged from a VS and were able to follow verbal instructions [35–37].

Conclusions

In the light of this review, it seems that neuromodulation techniques may be a challenging therapeutic option for severe consciousness disorders post-injury, and the techniques must be weighed up in the context of the clinical course over time, from resuscitation to rehabilitation. However, in the brain injury setting, it absolutely vital to identify favorable prognostic factors and the physiopathological processes involved, in order to quantify brain damage and evaluate the possible functional responses. On these grounds, the severity of handicap in patients emerging from consciousness disorders should also be anticipated. However, caution must be applied in interpreting the effectiveness of these techniques within an early time frame following injury, since patients are likely to exhibit some spontaneous recovery [29]. Modern technical improvements in surgery, such as the direct identification of anatomo-functional targets on brain images, are important steps forward to aid in the precise placing of electrodes in stereotactic conditions; these improvements make it possible to control electrode positioning in the chosen target with significantly reduced errors. Reliable clinical evaluation, helped by instrumental quantifications of behavior assessment, such as fMRI, PET scan, and neurophysiology, are crucial for monitoring the treated patients. However, although quite suggestive of efficacy (in 50 % of CDC patients who underwent DBS and in 52.2 % of patients treated with SCS), the clinical and experimental results of neuromodulation techniques for CDC are still not strong enough to lead to an unequivocal interpretation, and further studies are needed.

Conflict of Interest Statement We declare that we have no conflicts of interest.

References

1. Bernat JL (2009) Chronic consciousness disorders. Annu Rev Med 60:381–392
2. Brown P, Mazzone P, Oliviero A, Altibrandi MG, Pilato F, Tonali PA, Di Lazzaro V (2004) Effects of stimulation of the subthalamic area on oscillatory pallidal activity in Parkinson's disease. Exp Neurol 188:480–490
3. Bruno MA, Vanhaudenhuyse A, Thibaut A, Moonen G, Laureys S (2011) From unresponsive wakefulness to minimally conscious PLUS and functional locked-in syndromes: recent advances in our understanding of disorders of consciousness. J Neurol 258:1373–1384
4. Castrioto A, Lozano AM, Poon YY, Lang AE, Fallis M, Moro E (2011) Ten-year outcome of subthalamic stimulation in Parkinson disease: a blinded evaluation. Arch Neurol 68(12):1550–1556
5. Cohadon F, Richer E (1993) Deep cerebral stimulation in patients with post-traumatic vegetative state, 25 cases. Neurochirurgie 39:281–292
6. Della Pepa GM, Fukaya C, La Rocca G, Zhong J, Visocchi M (2013) Neuromodulation of vegetative state through spinal cord stimulation: where are we now and where are we going? Stereotact Funct Neurosurg 91(5):275–287
7. Demertzi A, Soddu A, Laureys S (2013) Consciousness supporting networks. Curr Opin Neurobiol 23:239–244
8. Fujii M, Sadamitsu D, Maekawa T, Uesugi S, Ozaki S, Koizumi S et al (1998) Spinal cord stimulation therapy at an early stage for unresponsive patients with hypoxic encephalopathy. No Shinkei Geka 26:315–321
9. Georgiopoulos M, Katsakiori P, Kefalopoulou Z, Ellul J, Chroni E, Constantoyannis C (2010) Vegetative state and minimally conscious state: a review of the therapeutic interventions. Stereotact Funct Neurosurg 88:199–207
10. Giacino JT, Ashwal S, Childs N, Cranford R, Jennett B, Katz DI et al (2002) The minimally conscious state: definition and diagnostic criteria. Neurology 58:349–353
11. Hassler R, Ore GD, Dieckmann G, Bricolo A, Dolce G (1969) Behavioural and EEG arousal induced by stimulation of unspecific projection systems in a patient with post-traumatic apallic syndrome. Electroencephalogr Clin Neurophysiol 27:306–310
12. Jennett B, Plum F (1972) Persistent vegetative state after brain damage. A syndrome in search of a name. Lancet 1:734–737
13. Kanno T, Kamei Y, Yokoyama T, Shoda M, Tanji H, Nomura M (1988) Effects of neurostimulation on the reversibility of neuronal function: experience of treatment for vegetative status. No Shinkei Geka 16:157–163
14. Kanno T, Kamel Y, Yokoyama T, Shoda M, Tanji H, Nomura M (1989) Effects of dorsal column spinal cord stimulation (DCS) on reversibility of neuronal function–experience of treatment for vegetative states. Pacing Clin Electrophysiol 12:733–738
15. Kanno T, Morita I, Yamaguchi S, Yokoyama T, Kamei Y, Anil SM, Karagiozov KL (2009) Dorsal column stimulation in persistent vegetative state. Neuromodulation 12:33–38
16. Kuwata T (1993) Effects of the cervical spinal cord stimulation on persistent vegetative syndrome: experimental and clinical study. No Shinkei Geka 21:325–331
17. Lemaire JJ, Sontheimer A, Nezzar H, Pontier B, Luauté J, Roche B, Gillart T, Gabrillargues J, Rosenberg S, Sarret C, Feschet F, Vassal F, Fontaine D, Coste J (2014) Electrical modulation of neuronal networks in brain-injured patients with disorders of consciousness: a systematic review. Ann Fr Anesth Reanim 33(2):88–97
18. Lipsman N, Woodside DB, Giacobbe P, Hamani C, Carter JC, Norwood SJ, Sutandar K, Staab R, Elias G, Lyman CH, Smith GS, Lozano AM (2013) Subcallosal cingulate deep brain stimulation for treatment-refractory anorexia nervosa: a phase 1 pilot trial. Lancet 381:1361–1370
19. Liu JT, Lee JK, Tyan YS, Liu CY, Chen YH, Lin TB (2009) Neuromodulation on cervical spinal cord combined with hyperbaric oxygen in comatose patients – a preliminary report. Surg Neurol 72(Suppl 2):S28–S34
20. Liu JT, Tan WC, Liao WJ (2008) Effects of electrical cervical spinal cord stimulation on cerebral blood perfusion, cerebrospinal fluid cathecholamine lecels, and oxidative stress in comatose patients. Acta Neurochir Suppl 101:71–76
21. Matsui T, Asano T, Takakura K, Yamada R, Hosobuchi Y (1989) Beneficial effects of cervical spinal cord stimulation (cSCS) on patients with impaired consciousness: a preliminary report. Pacing Clin Electrophysiol 12:718–725
22. McIntyre CC, Hahn PJ (2010) Network perspectives on the mechanisms of deep brain stimulation. Neurobiol Dis 38:329–337
23. McLardy T, Ervin F, Mark V, Scoville W, Sweet W (1968) Attempted inset-electrodesarousal from traumatic coma: neuropathological findings. Trans Am Neurol Assoc 93:25–30
24. Momose T, Matsui T, Kosaka N, Ohtake T, Watanabe T, Nishikawa J, Abe K, Tanaka J, Takakura K, Iio M (1989) Effect of cervical spinal cord stimulation (cSCS) on cerebral glucose metabolism and blood flow in a vegetative patient assessed by positron emission tomography (PET) and single photon emission computed tomography (SPECT). Radiat Med 7:243–246
25. (1994) Multi-society task force on PVS: medical aspects of the persistent vegetative state. Part 2. N Engl J Med 330:1572–1579
26. (1994) Multi-society task force on PVS: medical aspects of the persistent vegetative state. Part 1. N Engl J Med 330:1499–1508
27. Oliveira L, Fregni F (2011) Pharmacological and electrical stimulation in chronic disorders of consciousness: new insights and future directions. Brain Inj 25:315–327
28. Schiff ND, Giacino JT, Kalmar K, Victor JD, Baker K, Gerber M, Fritz B, Eisenberg B, Biondi T, O'Connor J, Kobylarz EJ, Farris S, Machado A, McCagg C, Plum F, Fins JJ, Rezai AR (2007) Behavioural improvements with thalamic stimulation after severe traumatic brain injury. Nature 448:600–603

29. Schiff ND (2012) Moving toward a generalizable application of central thalamic deep brain stimulation for support of forebrain arousal regulation in the severely injured brain. Ann N Y Acad Sci 1265:56–68
30. Sturm V, Kuhner A, Schmitt HP, Assmus H, Stock G (1979) Chronic electrical stimu-lation of the thalamic unspecific activating system in a patient with coma due to midbrain and upper brain stem infarction. Acta Neurochir (Wien) 47:235–244
31. (2003) The vegetative state: guidance on diagnosis and management. Clin Med 3:249–254
32. Tsubokawa T, Yamamoto T, Katayama Y, Hirayama T, Maejima S, Moriya T (1990) Deep brain stimulation in a persistent vegetative state: follow-up results and criteria for selection of candidates. Brain Inj 4:315–327
33. Visocchi M, Della Pepa GM, Esposito G, Tufo T, Zhang W, Li S, Zhong J (2011) Spinal cord stimulation and cerebral hemodynamics: updated mechanism and therapeutic implications. Stereotact Funct Neurosurg 89:263–274
34. Yamamoto T, Katayama Y, Kobayashi K, Oshima H, Fukaya C, Tsubokawa T (2010) Deep brain stimulation for the treatment f vegetative state. Eur J Neurosci 32:1145–1151
35. Yamamoto T, Katayama Y, Obuchi T, Kabayashi K, Oshima H, Fukaya C (2013) DBS and SCS for vegetative state and minimally conscious state. World Neurosurg 80:S30.e1–9
36. Yamamoto T, Katayama Y, Obuchi T, Kobayashi K, Oshima H, Fukaya C (2012) Spinal cord stimulation for treatment of patients in the minimally conscious state. Neurol Med Chir (Tokyo) 52:475–481
37. Yamamoto T, Kobayashi K, Kasai M, Oshima H, Fukaya C, Katayama Y (2005) DBS therapy for the vegetative state and minimally conscious state. Acta Neurochir Suppl 93: 101–104
38. Yokoyama T, Kamei Y, Shoda M, Abe M, Onouchi K, Kasama A, Inoue T, Kanno T (1990) Treatment of vegetative status with dorsum column stimulation. No Shinkei Geka 18:39–45

Update on Mechanism and Therapeutic Implications of Spinal Cord Stimulation and Cerebral Hemodynamics: A Narrative Review

Elisabetta Peppucci, Rina Di Bonaventura, Vincenzo Esposito, Jun Zhong, Gerardo Iacopino, Giuseppe Barbagallo, and Massimiliano Visocchi

Abstract Spinal cord stimulation (SCS) is well known for its early role in the management of chronic pain, mainly failed back surgery syndrome (FBSS), spasticity, and bowel and bladder dysfunction. In more recent years, SCS has been proposed for patients suffering from refractory angina or peripheral vasculopathies in order to gain symptom relief, thus indicating some hemodynamic effect on the peripheral circulation. Taking into account this scientific observation, since the late1980s, researchers have started to investigate the potential effect of SCS on cerebral blood flow (CBF) regulation and its possible application in certain pathological settings dealing with vascular pattern dysfunction, such as ischemia, subarachnoid hemorrhage, head trauma, and brain tumors. The aim of this study was to review the scientific literature about SCS and its effect on CBF, evaluating the results both in "physiological" experimental models and clinical studies, as well as in the particular pathological conditions we have mentioned above.

E. Peppucci, MD (✉)
Neurosurgery, Catholic University Medical School, Rome, Italy

Institute of Neurosurgery, Catholic University School of Medicine, Largo A. Gemelli 8, Rome 00168, Italy
e-mail: elisabettapeppucci@gmail.com

R. Di Bonaventura, MD • M. Visocchi, MD
Insitute of Neurosurgery, Medical School, Catholic University of Rome, Rome, Italy

V. Esposito, MD
Università la Sapienza Roma, IRCCS Neuromed Via Atinense 18, Pozzilli (IS) 86077, Italy

J. Zhong, MD, PhD
Department of Neurosurgery, XinHua Hospital, Shanghai JiaoTong University School of Medicine, Shanghai 200092, China

G. Iacopino, MD
Department of Experimental Biomedicine and Clinical Neurosciences, School of Medicine, Neurosurgical Clinic, University of Palermo, Palermo, Italy

G. Barbagallo, MD
Division of Neurosurgery, Department of Neurosciences, Policlinico "G. Rodolico" University Hospital, Catania, Italy

Keywords Spinal cord stimulation • Cerebral blood flow • Stroke • Brain injury • Vasospasm • Cerebral autoregulation • Functional sympathectomy

Introduction

Spinal cord stimulation (SCS) has been shown to be safe and effective in the management of chronic pain of spinal origin and complex regional pain syndrome; thus it has been indicated in refractory neuropathic pain [67, 68]. Subsequently, the indication has been extended to the treatment of spasticity and bowel and bladder dysfunction. Advances in this therapy have led to its use in enhancing peripheral blood flow and reducing ischemic pain patterns: the application of SCS to treat angina and improve outcomes in patients suffering from peripheral vascular disease is now becoming part of the algorithmic standard of care [2]. Many experimental models have widely investigated the effects of SCS on cerebral blood flow (CBF) since the first experience [46]. Since the publication of the experimental study by Hosobuchi [21] on the relation between SCS and CBF about 30 years ago, many advances have been made in understanding SCS-mediated effects on CBF. The aim of this study was to review the scientific literature on the effects of SCS on CBF in the form of laboratory observations, neurophysiological theories, and clinical trials. We collected studies of SCS related to different kinds of pathological settings (subarachnoid hemorrage [SAH], stroke, brain injury, brain tumor) in order to point out the effect of SCS on cerebral hemodynamics and its possible future clinical applications.

Materials and Methods

We performed a systematic review of the literature on the cerebral hemodynamic effects of SCS, using the following key words in association:"spinal cord stimulation", "cerebral

DOI 10.1007/978-3-319-39546-3_5,

blood flow", "stroke", "brain injury", "vasospasm", "cerebral autoregulation", and "functional sympathectomy". We searched on Pubmed, Medline, and Scopus for articles in English published between 1985 and November 2015. We included only clinical or experimental studies in humans and animal models investigating the physiopathological mechanism and results of SCS on CBF.

Results

We found 1644 articles, of which we selected 45 according to our inclusion criteria. Of these 45 articles, 17 were experimental and clinical studies in humans (10) or animals (7) investigating the cerebral hemodynamic effects of SCS and its possible mechanism, 13 were experimental studies evaluating the effects of SCS on cerebral hemodynamics in low-perfusion syndrome (stroke, brain injury, and subarachnoid hemorrhage SAH), and 15 were clinical studies of the effects of SCS on cerebral ischemia, vasospasm, and brain tumor perfusion.

Discussion

It is estimated that each year more than 14,000 SCS implantations are performed worldwide [31]. Even though SCS is still mostly indicated for the treatment of neuropathic pain and, specifically, failed back surgery syndrome (FBSS) [24, 27], its use in peripheral cardiac and cerebral vascular diseases advanced rapidly during the late 1980s and 1990s, particularly in Europe [69]. Clinical and basic studies indicate that the beneficial effects of SCS on these diseases are mainly associated with increased blood flow or the redistribution of blood flow to the ischemic area [14, 32]. It is accepted that the stimulation has to be applied in the upper cervical spinal segments (C1-C2) in order to obtain a remarkable effect. The gain of CBF is associated with a decrease in sympathetic activity (so-called functional sympathectomy), an increase in vasomotor center activity, and the release of neurohumoral factors [70]. Table 1 reports experimental studies on hemodynamic modification induced by SCS. The first experimental observations in an animal model were conducted between 1987 and 1989 by Garcia-March [15, 16], who reported a CBF increase of more than 60% induced by SCS in goats and dogs during the first 15 min of stimulation. We reported the same results in our laboratory model of cervical SCS in rabbits in 1994 [57]. Isono et al. [23], in 1995, reported, in a cat model, a long-lasting hemodynamic effect with low-frequency (20 Hz) stimulation in the high cervical cord, with the effect not seen in dorsal column section. Since 2000, Sagher and Huang [52] and Zhong [76] have focused on the mechanism of the increase in cortical CBF in SCS mediated by sympathetic and parasympathetic pathways. The concept of "functional sympathectomy" was already hypothesized in 1991 by Myklebust [41] in monkeys and by Linderoth et al. [29, 30] in rats submitted to SCS in order to obtain improvement of peripheral vascular circulation. Linderoth et al. studied SCS-induced neurotransmitter release in the spinal dorsal horn and the periaqueductal gray substance of rats and cats, demonstrating that SCS-induced peripheral vasodilatation was abolished by bilateral sympathectomy in rats, and that peripheral sympathetic activity might be suppressed by SCS applied at the T2 level in cats. Two different papers by Patel, in 2003–2004 [44, 45], focused on the strict relation between the hemodynamic effect of SCS and sympathetic tone through pharmacological and surgical blocking system, suggesting a possible role for brainstem vasomotor centers in CBF regulation by SCS. In 2004 Zhong et al. [75] demonstrated, in a rat model, that the magnitude of the hemodynamic effect of SCS varied according to the stimulation parameters, with better results with a stimulation of 0.25-ms pulse and 50-Hz frequency. More recently the research has focused on the molecular pathways involved in the regulation of the cerebral [73] and peripheral [71, 72] circulation in SCS animal models. Wu et al. concluded that SCS activated the transient receptor potential vanilloid type 1 (TRPV1), which contains sensory fibers, and produced the release of calcitonin gene-related peptide, twhich is known to be a powerful mediator of vascular dilatation. Yang et al. further explored the role of these mediators, suggesting that cerebral but not spinal TRPV1 is involved in SCS-induced cerebral vasodilatation. Again, Wu et al. identified two new molecular pathways in the spinal cord, extracellular signal-regulated kinase and protein kinase B, which are involved in SCS-induced vasodilatation. In regard to observations in humans (Table 2), in 1985 Hosobuchi [21] was the first to plan a clinical study on the effects of SCS on cerebral hemodynamics, using a single-photon emission computed tomography (SPECT) technique in ten patients treated with SCS for chronic intractable pain. In 1981 [37] and subsequently in 1986, a clinical study published by Meglio et al. [38] confirmed the functional sympathectomy induced by SCS, confirming decreases in the heart rate in a group of 25 patients who had undergone neuromodulation for different pathologies. The same authors, in 1991 [39, 40], reported a clinical study with a larger series than Hosobuchi's one, leading to similar results. Mazzone et al. [35, 36] introduced the concept of 'redistribution of CBF': it would be influenced mainly in the anterior regions related to ascending reticular pathways through the thalamocortical projections. Clinical studies conducted in our laboratories since 1991 [58, 61, 66] have been performed by using a CO_2 autoregulation test evaluating transcranial Doppler velocimetry (TCD) patterns during progressive hypercapnia, both in baseline conditions and during SCS. We demonstrated that SCS reduced the vasodilator response to CO_2: we hypothesized that SCS and CO_2 act on the mechanism of CBF regulation in a competitive way. In 2004 Robaina et al. [47] performed the most recent clinical study of SCS, in 35 patients, with cancer (29) and pain

Table 1 Experimental studies in animal models investigating the mechanisms involved in SCS effects on CBF

Author (year) references	Animal model	Observation method	Site of stimulation	Results
Garcia-March et al. (1987, 1989) [15, 16]	Dog and goat	Electromagnetic flowmetry; ^{131}I antipyrine brain scintigraphy	Upper cervical (C2)	Increase of CBF at the common and internal carotid level of more than 60 % and an increase of CBF of more than 50 % increase in cerebral blood flow
Visocchi et al. (1994) [57]	Rabbit	Electromagnetic flowmetry; CW Doppler	Cervical	Reduction of sympathetic excitability
Isono et al. (1995) [23]	Cat	Hydrogen clearance method	Cervical	No CBF effects during SCS after lesion of dorsal column
Sagher and Huang (2000), Zhong (2006) [52, 76]	Rat	Laser Doppler flowmetry	Cervical	Changes in CBF were significantly attenuated after interruption of sympathetic or parasympathetic pathways
Patel et al. (2003) [44]	Rat	Laser Doppler flowmetry	Cervical	Sympathetic nervous system blockers (hexamethonium and prazosin) abolish the SCS-induced increase in CBF
Patel et al. 2004 [45]	Rat	^{14}C-inosine monophosphate radiotracer studies	Cervical	Spinal transection strongly attenuated the effect of SCS on CBF, otherwise no similar effect in cases of superior cervical ganglion resection (effects of SCS involve central influences rather than cervical sympathetic outflow)
Zhong et al. (2004) [75]	Rat	Laser Doppler flowmetry	Cervical	Magnitude of the response varied in a dose-dependent fashion with the stimulation amplitude
Yang et al. (2008) [73]	Rat	Laser Doppler flowmetry	Cervical	TRPV1 agonists block CBF effects; cerebral but not spinal TRPV1 is involved in cSCS-induced cerebral vasodilatation

SCS spinal cord stimulation, *CBF* cerebral blood flow, *TRPV1* transient receptor potential vanilloid type 1, *cSCS* cervical spinal cord stimulation, *CW* cervical spinal cord stimulation, *Doppler* continuous-wave Doppler

syndromes (6), treated with cervical SCS (voltage 1–3 V, pulse 200 ms, frequency 80–100 Hz). These authors used a semiquantitative method, TCD, for the middle cerebral artery, and a quantitative method, color Doppler, for the common carotid artery, and showed an increase in CBF induced by SCS. It is quite clear that the scientific literature published in the past 30 years (summarized in Tables 1 and 2) supports the idea that the autonomic nervous system plays the most important role in the cerebral hemodynamic modification induced by cervical SCS [54], even though more recent studies focus attention on the importance of some chemical mediators in determining this well-known effect of SCS.

From the first experimental observation concerning the role of SCS in the regulation of CBF, researchers soon started to think about the possible application of the same stimulation in pathological models that could resemble so-called low-perfusion syndromes. Thus, several experimental studies on animals came along, each one of which was designed to mimic a pathological status such as SAH and/or related vasospam [50], or cerebral ischemia and brain injury, searching in this particular condition for some possible blood flow changes after SCS application. A large number of papers about this matter have been published since the late 1980s (Table 3). The first to describe some evidence in this field was Matsui, in 1989 [34], who reported the effects of cervical SCS on an experimental stroke model. Matsui induced a middle cerebral artery occlusion (MCAO) in a group of 31 cats, dividing them into three subgroups, one of which underwent MCAO and subsequent cervical SCS. Data on infarct size were checked in the three group underwent middle cerebral artery occlusion; among them just one group underwnt spinal cord stimualtion after the occlusion, using a computer system model, and compared. The analysis showed a prolonged survival rate within 24 h after ischemia and prevention of infarct size progression in the SCS group, demonstrating that cervical SCS prevents the progression of brain infarction. In 1991 Gonzales-Darder [19] discovered the role of SCS in reducing brain edema in an experimental stroke model induced by

Table 2 Clinical studies investigating the mechanisms involved in SCS effects on CBF

Author (year) references	Observation method	Site of stimulation	Results
Hosobuchi (1985) [21]	SPECT (^{133}Xe)	Cervical/thoracic	Indomethacin blocks CBF effects; atropine dose not affected CBF effects
Meglio et al. (1991) [39, 40]	TCD; SPECT (^{133}Xe)	Cervical/thoracic	Decrease in cerebral vascular resistance and increase in flow velocity observed in 63 % cervical SCS and 29.4 % thoracic SCS
Visocchi et al. (1992) [58]	TCD; OBF system of Langhans	Cervical/thoracic	SCS changes CBF and OBF at the same time with the same sign (same autoregulatory mechanisms?)
Visocchi et al. (1996) [61]	TCD	Cervical	SCS and CO_2 interact with mechanisms of regulation of CBF in a competitive way
Mazzone et al. (1995, 1996) [35, 36]	TCD; SPECT (^{133}Xe)	Cervical/thoracic	Increase of CBF induced by SCS mainly in the anterior region
Robaina et al. (2004) [47]	TCD for middle cerebral artery velocity and color Doppler for common carotid artery blood flow	Cervical	Increase in flow velocity in middle cerebral artery and common carotid artery blood flow

CW Doppler Continuous-wave Doppler, *NE* norepinephrine, *OBF* ocular blood flow, *SPECT* Single Photon Emission Computed Tomography, *TCD* Trans-cranial doppler

bilateral carotid artery occlusion in rats. That study suggested that the stimulation could have the most relevant role just during the reperfusion period, because the hemodynamic effect was described only when SCS was applied 60 min before ischemia or at the beginning of the reperfusion period. The basilar mechanism mediating such an effect of SCS could be related to a global increase in CBF, protecting the brain against ischemia, or it could be related to the activation of vascular regulation systems. A similar experience was published by Broseta et al. [3] in 1994, in 45 rabbits with different stroke models (bilateral carotid ligation, unilateral microcoagulation of the middle cerebral artery [MCA], and microcoagulation of the vertebral artery). More recently, Sagher et al. [51] used an MCA occlusion model in rats and observed that SCS reduced stroke extension; they proposed SCS as feasible for the treatment and prevention of stroke. Visocchi et al. [63] carried on these studies, building an experimental model combining head trauma and ischemic injury in 20 rabbits. They reported some results indicating the 'preventive' effect of SCS on the secondary damage induced by trauma in an already established vascular insufficiency setting. Since 2001 various authors have carried out experimental studies on SCS in cerebral vasospasm and SAH models, focusing attention on this possible therapeutic subset. Goksel et al. [18] investigated the mechanism of the SCS-induced increase in CBF and its relation with physiological vasomotor mediators, using an NO synthase inhibitor in an animal model of SAH . They concluded that the effect of SCS on CBF could be attenuated but not completely suppressed by NG-nitro-L-arginine methyl ester (L-NAME). Gurelik et al. [20] were the only ones to propose a study in which the hemodynamic effect of SCS in a vasospasm model was measured indirectly, recording motor evoked potentials and changes occurring in the latency and amplitudes of the signal. The studies published by Karadag et al. [26] and Lee et al. [28] reported similar animal models (rabbits and rats) enhancing the remarkable effect of SCS on vasospasm in the anterior and posterior circulation. Visocchi et al. [60, 62, 65] pointed out the protective role of SCS in SAH regarding the prevention of 'early' vasospasm. TCD provides an easy and reliable indication of corresponding CBF modifications both in homeostatic conditions (CBF augmentation with increase in TCD velocities along with a reduction in the resistive index) and during vasospasm (decrease in CBF with increase in TCD velocities, as well as an increase in the resistive index). These pioneer observations were followed by Ebel et al. [12, 13], who evaluated the effect of cervical SCS on CBF in rats with SAH, demonstrating an enhancement of cerebral and cerebellar blood flow (Table 4).

The fruitful research on the hemodynamic effects of cervical SCS has been applied in clinical studies since 1989, as the preliminary observations in experimental models of SCS in low-perfusion syndromes had led to great interest in their possible applications in humans. The clinical studies published in the literature deal with patients I a persistent vegetative-state, and those with cerebral ischemia, cerebral vasospasm within SAH [43, 55, 74], and brain tumors.

It was Kanno et al. [25] who first reported treating patients in a vegetative state with cervical SCS; good clinical improvement was reported in 8 of their 23 patients, although there was no clear correlation between the modification of

Table 3 Experimental studies on animals checking the effect of cervical SCS on cerebral low-perfusion syndromes

Author (year) references	Experimental model	Animal	Observation method	Results
Matsui and Hosobuchi (1989) [34]	Experimental stroke (middle cerebral artery occlusion)	Cat	TTC method; measurement by computer technique (PDP-11/23)	Reduction in death rate within 24 h after MCAO; prevention of progression of brain infarction
Gonzales-Darder and Canadas-Rodriguez (1991) [19]	Experimental stroke with bilateral carotid artery occlusion	Rat	Microgravimetric technique	Reduction in ischemic brain edema
Broseta et al. (1994) [3]	Experimental ischemic infarction in three models (middle cerebral artery microcoagulation; bilateral carotid artery occlusion; vertebral artery occlusion)	Rabbit	LDF	Improved CBF in right emisphere and posterior fossa (mean of 27 % and 32 %, respectively)
Visocchi et al. (2001) [63]	Experimental stroke and brain injury in conjunction (bilateral carotid artery occlusion plus "mechanical injury"over the dura mater after bone removal; right emispheric craniectomy)	Rabbit	MRI	No ischemic damage far from the "traumatic injury site"
Gőksel et al. (2001) [18]	Experimental SAH and cerebral vasospasm (intracisternal saline injection and administration of L-NAME)	Rabbit	TDU	SCS induced vasodilation in all SAH animals even after the administration of L-NAME
Sagher et al. (2003) [51]	Experimental stroke-middle cerebral artery occlusion	Rat	LDF; radiotracer clearance	Reduction in stroke volume
Visocchi et al. (1996, 2001) [60, 62]	Experimental vasospam in SAH	Rabbit	Angiography; electromagnetic flowmetry	Prevention of vasospasm
Ebel et al. (2001) [12, 13]	Experimental SAH	Rat	(99 m)Tc-HMPAO investigation	Improved cerebral and cerebellar blood flow
Gurelik et al. (2005) [20]	Experimental vasospasm	Rat	MEP evaluation	Improved changes in MEP latency and amplitude in "vasospasm group"
Karadag et al. (2005) [26]	Experimental vasospasm in SAH	Rabbit	LDF	Increasedcortical cerebral blood flow
Lee et al. (2008) [28]	Experimental vasospasm in SAH	Rat	Studies of the BA diameter and LDF	Increase in the diameter of BA and increase in cortical blood flow

TTC triphenyltetrazolium chloride solution, *SAH* subarachnoid hemorrhage, *MEP* motor-evoked potential, *LDF* Laser Doppler flowmetry, *TDU* transcranial Doppler ultrasonography, *L-NAME* L-Nitro-L-arginine methyl ester, *(99 m)Tc-HMPAO* flow tracer, *BA* basilar artery *MRI* magnetic resonance imaging, *MCAO* middle cerebral artery occlusion, *PDP* 11/23 Programmed Data Processor model 11

CBF recording and the clinical outcome. A similar conclusion was reached by Matsui et al. [34] in a series of patients with impaired consciousness due to trauma, cerebral hemorrhage, and vasospasm. A clinical study by Liu et al. [33] performed in 2008 in 20 comatose patients suggested that SCS increased cerebral blood perfusion, attenuated oxidative stress, and increased biogenic amines. As can be gathered from these studies, comatose patients represent a complex clinical entity because of the variability of the pathological basis, the extensive damage, and the remarkable pre-existing functional alterations: from this assumption the correct interpretation of the real effect of SCS can be difficult.

Clinical studies of the possible application of SCS in patients with cerebral ischemia have appeared in the literature since the early1990s. Hosobuchi [22] first used cervical SCS specifically for three patients with cerebral ischemia caused by basivertebral or bilateral carotid occlusive disease, describing an improvement of the ischemic symptoms associated with an increase in CBF. Visocchi et al. [59] reported, in 1994, the case of a 64-year-old patient with spastic hemiparesis and dysphasia following a left parietotemporal ischemic stroke who was treated with cervical SCS, with improvement shown in clinico-neurophysiological findings and CBF velocity. In this case, as these authors have already discussed extensively, the

Table 4 Clinical studies on the effect of cervical SCS in cerebral ischemia, vasospasm, and tumor perfusion in humans

Author (year) references	Clinical condition	No. of patients	Effect recording	Observation
Kanno et al. (1989) [25]	Impaired consciousness (persistent vegetative states)	23	SPECT	Increase in regional CBF; increase in the metabolism of catecholamines in CSF; improvement in EEG and clinical condition
Matsui and Hosobuchi (1989) [34]	Impaired consciousness (persistent vegetative states)	8	SPECT	Clinical improvement (no correlation of clinical outcome with SPECT CBF investigation)
Hosobuchi (1991) [22]	Cerebral ischemia	3	SPECT	Increase in CBF
Visocchi et al. (1994) [59]	Cerebral ischemia	1	NIRS; surface polyelectromyography; TCD	Clinical-neurophysiological improvement
Broseta et al. (1994) [3]	Cerebral low-perfusion syndrome	10	SPECT	Clinical improvement, no MRI changes; increase in CBF
Takanashi and Shinonaga (2000) [53]	Cerebral vasospasm (SAH patients; Hunt Hess 2–3, Fisher 3)	10	xenon computed tomography and cerebral angiography	Increase in CBF, clinical improvement
Visocchi et al. (2001) [64]	Ischemic/hemorrhagic brain damage	18	TCD; NIRS; SPECT	Clinical improvement, no absolute correlation increase, reduction, no changes in cerebral blood flow, or either increase/reduction of cerebral blood flow
Clavo et al. (2007, 2004, 2003, 2002) [4–7]	Brain tumor perfusion (advanced gliomas and head and neck tumors)	64	TCD, polarographic probes	Increased tumor blood flow and oxygenation
Robaina and Clavo (2007) [48]	Cerebral ischemia	49	SPECT; TCD; Color doppler, PET	Significant increases in velocity in middle cerebral artery flow and common carotid artery flow and glucose metabolism
De Andrés et al. (2007) [9]	Ischemic brain damage	1	Doppler, functional MRI	Clinical improvement, increase in CBF
Liu et al. (2008) [33]	Impaired consciousness (persistent vegetative states)	20	SPECT	Clinical improvement, increase in CBF; increase in DA and NE in CSF
Clavo et al. (2012) [8]	High-grade glioma	26	TCD, SPECT, Polarographic probes	Increase in locoregional blood flow, oxygenation, and glucose metabolism, improved chemotherapy delivery, oxygen effect during radiotherapy

CBF cerebral blood flow, *CSF* cerebrospinal fluid, *EEG* electroencephalogram, *DA* dopamine, *MRI* Magnetic Resonance Imaging, *NE* norepinephrine, *NIRS* Near Infrared Spectroscopy, *PET* Positron Emission Tomography, *SPECT* Single Photon Emission Computed Tomography, *TCD* trans cranial doppler

evidence of improvement of dysphasia suggested the primary role of CBF changes as an effect of SCS. Later the same authors published a personal series of 18 patients with stroke in whom, through TCD, single-photon emission computed tomography (SPECT), and Near Infrared Spectroscopy (NIRS) they demonstrated an increase in regional CBF in the areas surrounding the stroke area during SCS in half of the patients [64].

Broseta et al. [3] reported similar results in various cerebral low-perfusion syndrome cases in which the patients achieved a significant but not constant increase of superior cognitive abilities in skilled acts, probably due to an 'increase in blood flow in the penumbral perilesional area' revealed by SPECT. In 2007 De Andres et al. [9] and Robaina [48] reported a series of patients with cerebral ischemia treated with SCS. The latter study demonstrated, in 49 post-stroke patients treated with cervical SCS, a considerable improvement in CBF and brain metabolism, in the form of statistically significant increases in systolic and diastolic velocity in the MCA, increases in common carotid artery (CCA) confirm other edits in paragraph blood flow, and increases in glucose metabolism.

The only clinical study of SCS for cerebral vasospasm was the one published by Takanashi and Shinonaga [53] in 2000. They reported a series of ten SAH patients with secured cerebral aneurysm (ranging from Hunt Hess grades 2 to 4, Fisher grade 3) implanted with percutaneous epidural cervical leads and stimulated continuously from the 5th day post-bleeding for 10–15 days, in association with standard medical treatment. They observed a significant increase of CBF in the MCA territory. Four of the ten patients showed angiographic vasospasm, but none developed severe clinical sequelae: the overall outcome was good or excellent in seven of the ten patients. No serious adverse effect was attributed to SCS. The main limitations of this study were the small number of patients and the severe clinical status (Hunt Hess grade 3 or 4) on admission; such severity could represent a confounding element in attributing morbidity and mortality to hemorrhage itself and to vasospasm. No definitive proof of clinical efficacy can originate from this study, but surely it could represent the first attempt to evaluate SCS as an adjunctive therapy for cerebral vasospasm following SAH. A different field in the possible application of SCS is represented by high-grade tumors, in that SCS can increase blood flow to the lesion and thus increase the radiochemosensitivity. In the past 10 years Clavo and colleagues [4–7, 49] published several scientific works on the application of SCS in patients with high-grade gliomas and head and neck tumors, investigating its ability to increase locoregional blood flow and thus the delivery of chemotherapy and oxygen. In a more recent paper, in 2012 [8]: Clavo reported a series of 26 patients with high-grade gliomas in whom he assessed pre- and post-SCS CBF, tumor blood flow, tumor partial pressure of oxygen (pO_2), and glucose metabolism, using (18) fludeoxyglucose positron emission tomography (FDG-PET). He demonstrated how SCS can modify the tumor microenvironment, improving its radiochemotherapy sensitivity. These clinical studies demonstrate that SCS can enhance locoregional blood flow, increasing the delivery of chemotherapeutic drugs and of oxygen, with improvement of tumor radiosensitivity . Even though they are still preliminary, these findings open up an exciting field of research in the clinical application of SCS, in that it can influence radiochemotherapy sensitivity and, thus, the prognosis of tumor patients [4, 49].

Conclusions

SCS has been widely used for the the treatment of chronic pain [42], such as that described in patients with FBSS, complex regional pain syndrome (CRPS), peripheral neuropathy, phantom limb pain, angina [1, 11], and ischemic limb pain [10, 24]. Since the late 1980s researchers have started to think about the possible application of SCS in the regulation of CBF, checking its hemodynamic effects in animal models and in humans, both under “physiological conditions” and “in particular settings such as low-perfusion syndrome”.

Multiple data have been published concerning the effect of neurostimulation in increasing local or global CBF in physiological conditions and also in low-perfusion syndrome; although it is quite clear that SCS applied at the cervical level has some effects on CBF [17], it is still necessary to continue with the research in order to acquire some evidence that can justify the use of this particular treatment in clinical practice, and to clarify the proper way to do it.

Regarding what it is in the literature nowadays, SCS seems to have shown some true effect in “protecting” the brain from the progression of damage in cerebrovascular disease models (animals), as demonstrated for cerebral edema, stroke, and vasospasm, whether or not the latter is related to SAH; however, the latest clinical studies on this topic have not demonstrated any clear useful effect in clinical practice; it is important to emphasize that, according to the literature, the main role of SCS remains for brain “protection”; there is still no strong data available that can sustain the “therapeutic” use of SCS in “low-perfusion syndromes”, and this is probably the main aspect that should therefore be the focus of our attention in future years.

Possible therapeutic perspectives in the future will also deal with the observations that SCS can modify blood flow circulation within the “target area” in high-grade malignant brain tumors; the first data published suggest that SCS might represent an “adjuvant” weapon in the therapeutic approaches following surgery, working in conjunction with standard chemo- and radiotherapy [4, 49].

Spinal cord stimulation has been established over the past years as a very promising tool in the hands of the neurosurgeon, in regard to the interesting results obtained in the regulation of cerebral hemodynamics, both in animal models and in humans [56, 76]. It can also be seen that SCS, due to this effect, can modify in some way the course of some well-known detrimental pathological events in particular diseases, such as ischemia/stroke, SAH, and vasospasm, and can also have a significant role in brain tumor management, to increase the sensitivity of pathological tissue to radiochemotherapy.

However, statistical evidence is still lacking, and this negatively affects the possible application of SCS in daily clinical practice; the collection of statistical evidence is the way researchers should continue on this pathway, with more clinical studies and randomized control trials: to give SCS the possible wide application it could have in functional neurosurgery and, more widely speaking, in the management of some of the most relevant brain diseases.

As far as we know, worldwide, there is only one ongoing trial aimed at evaluating the global hemodynamic effects and metabolic effects of SCS on chronic stroke; this trial is being conducted in Italy at the Neurosciences Institute "Neuromed" in Pozzilli (IS). The Chief of the Institute is Professor Vincenzo Esposito and the Principal Investigator is Massimiliano Visocchi.

Conflict of Interest Statement Conflict of interest statement: the authors declare no conflict of interest and they have not received any financial support in conjunction with the generation of this manuscript

References

1. Augustinsson LE, Eliasson T, Mannheimer C (1995) Spinal cord stimulation in severe angina pectoris. Stereotact Funct Neurosurg 65:136–141
2. Augustinsson LE, Linderoth B, Mannheimer C, Eliasson T (1995) Spinal cord stimulation in cardiovascular disease. Neurosurg Clin N Am 6(1):157–165
3. Broseta J, Garcia-Mark G, Sancez-Ledesma MJ, Goncalves J (1994) High cervical spinal cord electrical stimulation in brain low-perfusion syndromes: experimental basis and preliminary clinical report. Stereotact Funct Neurosurg 62:171–178
4. Clavo B, Robaina F, Morera J, Ruiz-Egea E, Pérez JL, Macías D, Caramés MA, Catalá L, Hernández MA, Günderoth M (2002) Increase of brain tumor oxygenation during cervical spinal cord stimulation. Report of three cases. J Neurosurg 96:94–100
5. Clavo B, Robaina F, Catala L, Valcarcel B, Morea J, Carames MA, Ruiz-Egea E, Paneo F, Lloret M, Hernandez MA (2003) Increase locoregional blood flow in brain tumors after cervical SCS. J Neurosurg 98:1263–1270
6. Clavo B, Robaina F, Catalá L, Pérez JL, Lloret M, Caramés MA, Morera J, López L, Suárez G, Macías D, Rivero J, Hernández MA (2004) Effect of cervical spinal cord stimulation on regional blood flow and oxygenation in advanced head and neck tumours. Ann Oncol 15:802–807
7. Clavo B, Robaina F, Catalá L, Lloret M, Pi-nar B, Caramés MA, Ruiz A, Cabezón A, González G, Lara P, Ruiz-Egea E, Hernández MA (2007) Cerebral blood flow increase in cancer patients by applying cervical spinal cord stimulation. Neurocirugia (Astur) 1(8):28–32
8. Clavo B, Robaina F, Valcarcel B, Catala L, Perez JL, Cabezon A, Jorge IJ, Fiuza D, Hernandez MA, Jover R, Carreras JL (2012) Modification of loco-regional microenvironment in brain tumors by spinal cord stimulation. Implications for radio-chemotherapy. J Neurooncol 106(1):177–184
9. De Andrés J, Tatay J, Revert A, Valía JC, Villanueva V (2007) The beneficial effect of spinal cord stimulation in a patient with severe cerebral ischemia and upper extremity ischemic pain. Pain Pract 7:135–142
10. Deer TR, Mekhail N, Provenzano D, Pope J, Krames E, Leong M, Levy RM, Abejon D, Buchser E, Burton A, Buvanendran A, Candido K, Caraway D, Cousins M, DeJongste M, Diwan S, Eldabe S, Gatzinsky K, Foreman RD, Hayek S, Kim P, Kinfe T, Kloth D, Kumar K, Rizvi S, Lad SP, Liem L, Linderoth B, Mackey S, McDowell G, McRoberts P, Poree L, Prager J, Raso L, Rauck R, Russo M, Simpson B, Slavin K, Staats P, Stanton-Hicks M, Verrills P, Wellington J, Williams K, North R, Committee NAC (2014) The appropriate use of neurostimulation of the spinal cord and peripheral nervous system for the treatment of chronic pain and ischemic diseases: the Neuromodulation Appropriateness Consensus Committee. Neuromodulation 17(6):515–550
11. De Vries J, De Jongste MJ, Spincemaille G, Staal MJ (2007) Spinal cord stimulation for ischemic heart disease and peripheral vascular disease. Adv Tech Stand Neurosurg 32:63–89
12. Ebel H, Semmelmann G, Schomaker K, Balogh A, Volz M, Funke J, Schicha H, Klug N (2001) Effects of high cervical SCS on regional CBF after induced SAH in rats. Acta Neurochir Suppl 77:225–227
13. Ebel H, Semmelmann G, Schomaker K, Balogh A, Volz M, Funke J, Schicha H, Klug N (2001) High cervical spinal cord stimulation increases regional cerebral blood flow after induced SAH in rats. Minim Invasive Neurosurg 44:167–171
14. Erdek MA, Staats PS (2003) Spinal cord stimulation for angina pectoris and peripheral vascular disease. Anesthesiol Clin North America 21:797–804
15. Garcia-March G, Sanchez-Ledesna MJ, Broseta J (1987) Effect of high cervical spinal cord stimulation on carotid and cerebral blood flow. An experimental study. Proc 8th European Congress of Neurosurgery, Barcelona, p 106
16. Garcia-March G, Sánchez-Ledesma MJ, Anaya J, Broseta J (1989) Cerebral and carotid haemodynamic changes following cervical spinal cord stimulation. An experimental study. Acta Neurochir Suppl (Wien) 46:102–104
17. Goellner E, Slavin KV (2009) Cervical spinal cord stimulation may prevent cerebral vasospasm by modulating sympathetic activity of the superior cervical ganglion at lower cervical spinal level. Med Hypotheses 73:410–413
18. Göksel HM, Karadag O, Turaçlar U, Taş F, Oztoprak I (2001) Nitric oxide synthase inhibition attenuates vasoactive response to spinal cord stimulation in an experimental cerebral vasospasm model. Acta Neurochir (Wien) 143(4):383–390
19. Gonzalez-Darder J, Canadas-Rodriguez D (1991) Effects of cervical spinal cord stimulation in experimental ischemic oedema. Neurol Res 13:229–232
20. Gurelik M, Kayabas M, Karadag O, Goksel HM, Akyuz A, Topaktas S (2005) Cervical spinal cord stimulation improves neurological dysfunction induced by cerebral vasospasm. Neuroscience 134:827–832
21. Hosobuchi Y (1985) Electrical stimulation of the cervical spinal cord increases cerebral blood flow in humans. Appl Neurophysiol 48:372–376
22. Hosobuchi Y (1991) Treatment of cerebral ischemia with electrical stimulation of the cervical spinal cord. Pacing Clin Electrophysiol 14:122–126
23. Isono M, Kaga A, Fujiki M, Mori T, Hori S (1995) Effect of spinal cord stimulation on cerebral blood flow in cats. Stereotact Funct Neurosurg 64:40–46
24. Jeon YH (2012) Spinal cord stimulation in pain management: a review. Korean J Pain 25(3):143–150
25. Kanno T, Yoshifumi K, Yokoyama T, Shoda M, Tanji H, Nomura M (1989) Effects of dorsal column spinal cord stimulation on reversibility of neuronal function. Experience of treatment for vegetative states. Pacing Clin Electrophysiol 12:733–742
26. Karadağ Ő, Eroğlu E, Gürelik M, Göksel HM, Kiliç E, Gültürk S (2005) Cervical spinal cord stimulation increases cerebral cortical blood flow in an experimental cerebral vasospasm model. Acta Neurochir (Wien) 147(1):79–84
27. Kumar K, Caraway DL, Rizvi S, Bishop S (2014) Current challenges in spinal cord stimulation. Neuromodulation 17(Suppl 1): 22–35

28. Lee JY, Huang DL, Keep R, Sagher O (2008) Effect of electrical stimulation of the cervical spinal cord on blood flow following subarachnoid hemorrhage. J Neurosurg 109:1148–1154
29. Linderoth B, Gherardini G, Ren BB, Lundeberg T (1995) Preemptive spinal cord stimulation reduces ischemia in an animal model of vasospasm. Neurosurgery 37:266–271
30. Linderoth B, Gherardini G, Ren B, Lundeberg T (1995) Severe peripheral ischemia after vasospasm may be prevented by spinal cord stimulation. A preliminary report of study in a free-flap animal model. Acta Neurochir Suppl 64:101–105
31. Linderoth B, Foreman RD (2006) Mechanisms of spinal cord stimulation in painful syndromes: role of animal models. Pain Med 7:S14–S26
32. Linderoth B, Foreman RD (1999) Physiology of spinal cord stimulation: review and update. Neuromodulation 2:150–164
33. Liu JT, Tan WC, Liao WJ (2008) Effects of electrical cervical spinal cord stimulation on cerebral blood perfusion, cerebrospinal fluid catecholamine levels, and oxidative stress in comatose patients. Acta Neurochir Suppl 101:71–76
34. Matsui T, Hosobuchi Y (1989) The effects of cervical spinal cord stimulation on experimental stroke. Pacing Clin Electrophysiol 12:726–732
35. Mazzone P, Pisani R, Nobili F, Arrigo A, Rosadini G (1995) Assessment of regional cerebral blood flow during spinal cord stimulation in humans. Stereotact Funct Neurosurg 64:197–201
36. Mazzone P, Rodriguez G, Arrigo A, Nobili F, Rosadini G (1996) Cerebral haemodynamic changes induced by spinal cord stimulation in man. Ital J Neurol Sci 17:55–57
37. Meglio M, Cioni B, Dal Lago A, De Santis M, Pola P, Serricchio M (1981) Pain control and improvement of peripheral blood flow following epidural spinal cord stimulation. Case report. J Neurosurg 54:821–823
38. Meglio M, Cioni B, Rossi GF, Sandric S, Santarelli P (1986) Spinal cord stimulation affects the central mechanisms of regulation of heart rate. Appl Neurophysiol 49:139–146
39. Meglio M, Cioni B, Visocchi M, Nobili F, Rodriguez G, Rosadini G, Chiappini F, Sandric S (1991) Spinal cord stimulation and cerebral haemodynamics. Acta Neurochir (Wien) 111:43–48
40. Meglio M, Cioni B, Visocchi M (1991) Cerebral haemodynamics during spinal cord stimulation. PACE 14:127–130
41. Myklebust JB, Cusick JF, Boerboom LE, Prieto TE, Khan TA (1995) Vascular effects of spinal cord stimulation in the monkey. Stereotact Funct Neurosurg 64:32–39
42. Nagamachi S, Fujita S, Nishii R, Futami S, Wakamatsu H, Yano T, Kodama T, Tamura S, Kunitake A, Uno T, Takasaki M (2006) Alteration of regional cerebral blood flow in patients with chronic pain – evaluation before and after epidural spinal cord stimulation. Ann Nucl Med 20:303–310
43. Naredi S, Lambert G, Edén E, Zäll S, Runnerstam M, Rydenhag B, Friberg P (2000) Increased sympathetic nervous activity in patients with nontraumatic subarachnoid hemorrhage. Stroke 31:901–906
44. Patel S, Huang DL, Sagher O (2003) Sympathetic mechanisms in cerebral blood flow alterations induced by spinal cord stimulation. J Neurosurg 99:754–761
45. Patel S, Huang DL, Sagher O (2004) Evidence for a central pathway in the cerebrovascular effects of spinal cord stimulation. Neurosurgery 55:201–206
46. Reis DJ (1984) Central neural control of cerebral circulation and metabolism, vol 2. Raven Press, New York, pp 91–119
47. Robaina F, Clavo B, Catalá L, Caramés MÁ, Morera J (2004) Blood flow increase by cervical spinal cord stimulation in middle cerebral and common carotid arteries. Neuromodulation 7(1):26–31
48. Robaina F, Clavo B (2007) Spinal cord stimulation in the treatment of post-stroke patients: current state and future directions. Acta Neurochir Suppl 97(Pt 1):277–282
49. Robaina F, Clavo B (2007) The role of spinal cord stimulation in the management of patients with brain tumors. Acta Neurochir Suppl 97:445–543
50. Slavin KV, Vannemreddy PS, Goellner E, Alaraj AM, Aydin S, Eboli P, Mlinarevich N, Watson KS, Walters LE, Amin-Hanjani S, Deveshwar R, Aletich V, Charbel FT (2011) Use of cervical spinal cord stimulation in treatment and prevention of arterial vasospasm after aneurysmal subarachnoid hemorrhage. Technical details. Neuroradiol J 24(1):131–135
51. Sagher O, Huang DL, Keep RF (2003) Spinal cord stimulation reducing infarct volume in a model of focal cerebral ischemia in rats. J Neurosurg 99:131–137
52. Sagher O, Huang DL (2000) Effects of the cervical spinal cord stimulation on cerebral blood flow in the rat. J Neurosurg 93:71–76
53. Takanashi Y, Shinonaga M (2000) Spinal cord stimulation for cerebral vasospasm as prophylaxis. Neurol Med Chir (Tokyo) 40:352–356
54. TerLaan M, van Dijk JM, Elting JW, Fidler V, Staal MJ (2010) The influence of transcutaneous electrical neurostimulation (TENS) on human cerebral blood flow velocities. Acta Neurochir (Wien) 152:1367–1373
55. Treggiari MM, Romand JA, Martin JB, Reverdin A, Rüfenacht DA, de Tribolet N (2003) Cervical sympathetic block to reverse delayed ischemic neurological deficits after aneurysmal subarachnoid hemorrhage. Stroke 34:961–967
56. Upadhyaya CD, Sagher O (2007) Cervical spinal cord stimulation in cerebral ischemia. Acta Neurochir Suppl 97:267–275
57. Visocchi M, Cioni B, Vergari S, Marano G, Pentimalli L, Meglio M (1994) Spinal cord stimulation and cerebral blood flow: an experimental study. Stereotact Funct Neurosurg 62:186–190
58. Visocchi M, Bucci MG, Cioni B, Manni G, Meglio M, Puca A, Quaranta L (1992) Cerebral and ophthalmic haemodynamics during SCS. I INS Congress Rome, Bologna, Monduzzi, pp 137–143
59. Visocchi M, Cioni B, Pentimalli L, Meglio M (1994) Increase of cerebral blood flow and improvement of brain motor control following spinal cord stimulation in ischemic spastic hemiparesis. Stereotact Funct Neurosurg 62:103–107
60. Visocchi M, Meglio M, Argiolas L, Pentimalli L, Cioni B (1996) Spinal cord stimulation prevents cerebral early vasospasm. An experimental study. III INS Congress, Orlando, Mar 6–10, pp 56–57
61. Visocchi M, Cioni B, Meglio M (1996) Spinal cord stimulation impairs cerebral autoregulation. Acta Med Rom 28:173–178
62. Visocchi M, Di Rocco C, Meglio M (2001) Protective effect of spinal cord stimulation on experimental early cerebral vasospasm. Conclusive results. Stereotact Funct Neurosurg 76:269–275
63. Visocchi M, Tartaglione T, Romani R, Meglio M (2001) Spinal cord stimulation prevents the effects of combined ischemic and traumatic brain injury. An MR study. Stereotact Funct Neurosurg 76:276–281
64. Visocchi M, Giordano A, Calcagni M, Cioni B, Di Rocco C, Meglio M (2001) Spinal cord stimulation and cerebral blood flow in stroke: personal experience. Stereotact Funct Neurosurg 76:262–268
65. Visocchi M, Argiolas L, Meglio M, Cioni B, Dal Basso P, Rollo M, Cabezas Cuevas D (2001) Spinal cord stimulation and early experimental cerebral spasm: the functional monitoring and the preventing effect. Acta Neurochir (Wien) 143:177–185
66. Visocchi M (2006) Spinal cord stimulation and cerebral haemodynamics. Acta Neurochir Suppl 99:111–116
67. Walsh KM, Machado AG, Krishnaney AA (2015) Spinal cord stimulation: a review of the safety literature and proposal for perioperative evaluation and management. Spine J 15:1864–1869
68. Wolter T (2014) Spinal cord stimulation for neuropathic pain: current perspectives. J Pain Res 7:651–663

69. Deer TR (2009) Spinal cord stimulation for the treatment of angina and peripheral vascular disease. Curr Pain Headache Rep 13(1): 18–23
70. Wu M, Linderoth B, Foreman RD (2008) Putative mechanisms behind effects of spinal cord stimulation on vascular diseases: a review of experimental studies. Auton Neurosci 138(1–2):9–23
71. Wu M, Komori N, Qin C, Farber JP, Linderoth B, Foreman RD (2007) Roles of peripheral terminals of transient receptor potential vanilloid-1 containing sensory fibers in spinal cord stimulation-induced peripheral vasodilation. Brain Res 1156:80–92
72. Wu M, Komori N, Qin C, Farber JP, Linderoth B, Foreman RD (2008) Extracellular signal regulated kinase and protein kinase B pathways involved in spinal cord stimulation-induced vasodilation. Brain Res 1207:73–83
73. Yang X, Farber JP, Wu M, Foreman RD, Qin C (2008) Roles of dorsal column pathway and transient receptor potential vanilloid type 1 in augmentation of cerebral blood flow by upper cervical spinal cord stimulation in rats. Neuroscience 152:950–958
74. Yin D, Slavin KV (2015) A hypothesis on possible neurochemical mechanisms of action of cervical spinal cord stimulation in prevention and treatment of cerebral arterial vasospasm after aneurysmal subarachnoid hemorrhage. Med Hypotheses 85(3):355–358
75. Zhong J, Huang DL, Sagher O (2004) Parameters influencing augmentation of cerebral blood flow by cervical spinal cord stimulation. Acta Neurochir (Wien) 146:1227–1234
76. Zhong J (2006) Effects of spinal cord stimulation on cerebrovascular flow: role of sympathetic and parasympathetic innervations. Neurosci Bull 22:261–266

Spinal Cord Stimulation for Vegetative State and Minimally Conscious State: Changes in Consciousness Level and Motor Function

Takamitsu Yamamoto, Mitsuru Watanabe, Toshiki Obuchi, Kazutaka Kobayashi, Hideki Oshima, Chikashi Fukaya, and Atsuo Yoshino

Abstract Twenty-one vegetative state (VS) patients and 10 minimally conscious state (MCS) patients were treated by spinal cord stimulation (SCS) following an electrophysiological evaluation 3 months or more after the onset of brain injury.

A flexible four-contact cylindrical electrode was inserted into the epidural space of the cervical vertebrae, and placed at cervical levels C2–C4. Five-hertz stimulation was applied for 5 min every 30 min during the daytime at an intensity that produced muscle twitches of the upper extremities.

Both the fifth wave in the auditory brainstem response (ABR) and N20 in the somatosensory evoked potential (SEP) were detected in 8 of the 21 VS patients and 9 of the 10 MCS patients. Of the 3 VS patients and 7 MCS patients who recovered following SCS therapy, all showed a preserved fifth wave in the ABR and N20 in the SEP, and all had received SCS therapy within 9 months after the onset of brain injury. Although the 3 patients who recovered from VS remained in a bedridden state, all 7 patients who recovered from MCS were able to emerge from the bedridden state within 12 months after the start of SCS.

Five-hertz cervical SCS caused increased cerebral blood flow (CBF) and induced muscle twitches of the upper extremities, and MCS patients showed a remarkable recovery of consciousness and motor function in the upper extremities compared with the lower extremities. This SCS method could be a new neuromodulation and neurorehabilitation technique, and MCS patients may be good candidates for SCS therapy.

Keywords Minimally conscious state • Spinal cord stimulation • Vegetative state • Auditory brainstem response • Somatosensory evoked potential

T. Yamamoto, MD, PhD (✉) • M. Watanabe • T. Obuchi
K. Kobayashi • H. Oshima • C. Fukaya • A. Yoshino
Divisions of Applied System Neuroscience, Department of Neurological Surgery, Nihon University School of Medicine, 30-1 Ohyaguchi Kamimachi, Itabashi-ku, Tokyo 173-8610, Japan
e-mail: yamamoto.takamitsu@nihon-u.ac.jp

Introduction

In 1972, Jennett and Plum first reported the concept of a persistent vegetative state (PVS) [9]. Later, in 1994, The Multi-Society Task Force on PVS proposed a definition of PVS [21, 22] as a clinical condition of complete unawareness of the self and the environment, accompanied by sleep-wake cycles, with either complete or partial preservation of hypothalamic and brainstem autonomic function. In 2002, Giacino et al. proposed the concept of the minimally conscious state (MCS) [5], which is characterized by inconsistent but clearly discernible behavioral evidence of consciousness and can be distinguished from coma and VS by the presence of specific behavioral features not found in either of these conditions.

Results of many medical and surgical interventions for the treatment of VS and MCS have been reported, but the estimations of resting brain function in VS and MCS patients have generally been unclear [19, 21, 22]. We have reported that VS and MCS can be distinguished neurologically, but that the resting brain function in VS and MCS patients differs markedly among patients, and that estimations of resting brain function in VS and MCS patients are essential in discussions about the effects of treatment [24–26]. Therefore, we previously carried out an electrophysiological evaluation to clarify the resting brain function in VS patients and found that deep brain stimulation (DBS) of the thalamic centre médian parafascicular (CM-pf) complex was useful for the treatment of VS patients when candidates were selected on the basis of an electrophysiological evaluation [26].

In this study, we report the long-term follow-up results of 5-Hz cervical SCS for the treatment of VS and MCS patients, in conjunction with the results of electrophysiological evaluation, which included assessments of auditory brainstem response (ABR) and somatosensory evoked potentials (SEPs). We examined the effects of 5-Hz cervical SCS for the recovery of motor function as well as consciousness in VS and MCS patients.

DOI 10.1007/978-3-319-39546-3_6,

Patients and Methods

Chronic SCS in VS and MCS Patients

Electrophysiological evaluations, including assessments of ABR and SEP, were carried out more than 3 months (90 days) after the onset of brain injury and the initial coma. in 21 VS and 10 MCS patients whose families desired and agreed to SCS therapy; all patients were treated with cervical SCS. The ages of the 21 VS patients ranged from 16 to 73 years (mean 47 ± 18.4) while those of the 10 MCS patients ranged from 16 to 67 years (mean 32 ± 15.9). The causes of the initial coma in the VS patients were head injury (4 patients), cerebrovascular accident (13 patients), and anoxic brain insult (4 patients); in the MCS patients, the causes were head injury (6 patients), encephalomyelitis (1 patient), and cerebrovascular accident (3 patients).

Electrophysiological Evaluation

We used Neuropack 2208 (Nihon Kohden, Tokyo, Japan) for the recording of ABR and SEP. For the ABR recording, needle electrodes were placed on the earlobes, the top of the head (Cz at the International 10–20 system), and forehead (ground). The bandpass was set from 10 Hz to 3 KHz. Binaural click stimuli were presented through earphones at 90 dB HL (hearing level) at a rate of 10/s. In each trial, 2048 responses were recorded. For the SEP recording, a four-channel montage was used: (1) Erb's point: the non-cephalic reference; (2) the spinous process of the C5 vertebra; (3) the contralateral scalp (C3′ or C4′); and (4) the ipsilateral scalp (C4′ or C3′). Square-wave pulses with a duration of 0.1 ms and a frequency of 5/s were delivered at a sufficient intensity. The bandpass was set from 0.5 Hz to 3 KHz. For the ABR and SEP recordings, at least two traces were recorded and superimposed to check their reproducibility. On the basis of the results of our previous work, we examined whether or not the fifth wave of the ABR and the N20 of SEP could be recorded, even if with a prolonged latency or on just one side [25–27].

SCS Method for VS and MCS Patients

Each patient was placed in the prone position, and the 18-gauge Tuohy needle included in the electrode package was inserted into the midline epidural space at the cervical-thoracic junction under radiographic control. The spinal epidural space was confirmed from the change of resistance observed in practice with an air injection through the Tuohy needle. Through the Tuohy needle, a flexible, four-contact cylinder electrode with a stylet (3487A PISCES-Quad; Medtronic, Minneapolis, USA was inserted into the epidural space of the cervical vertebrae, and placed at cervical levels C2-C4. The stylet and Tuohy needle were then removed, and the stimulation electrode was connected with an extension cable to an implantable pulse generator (7427 Synergy; Medtronic, Minneapolis, USA), which was implanted under the anterior chest wall (Fig. 1). Stimulation was applied automatically by the implanted pulse generator for 5 min every 30 min during the daytime, at an intensity that produced apparent muscle twitches of the upper extremities. We used 5 Hz for cervical SCS, considering that such a stimulation can activate the ascending reticular activating system indirectly, and not only increase cerebral blood flow (CBF) and cerebral metabolism but also induce muscle twitches, which can be a useful method of functional neurorehabilitation for VS and MCS patients [27].

The families of all the patients provided their written informed consent for this procedure. This study was approved by the Committee for Clinical Trials and Research on Humans of our university and conformed with the principles outlined in the Declaration of Helsinki.

Measurement of Regional Cerebral Blood Flow Before and During SCS

Employing single-photon emission computed tomography (SPECT), we could measure the CBF of the whole brain before and during SCS in eight of the ten MCS patients who were treated with SCS. SPECT was performed using a Prism 2000XP gamma camera system (Shimazu, Kyoto, Japan). Using an ethyl cysteinate dimer, quantitative regional CBF images were converted from qualitative axial SPECT images by the application of Patlak plot graphical analysis with radionuclide angiography and Lassen's linearization. We compared CBF before and during SCS.

Statistical Analysis

Changes in the regional (r) CBF of the whole brain induced by SCS in MCS patients were compared using the paired *t*-test.

Results

Electrophysiological Evaluation of VS and MCS Patients

As VS candidates for SCS therapy, we generally try to select patients who showed both the fifth wave in the ABR and

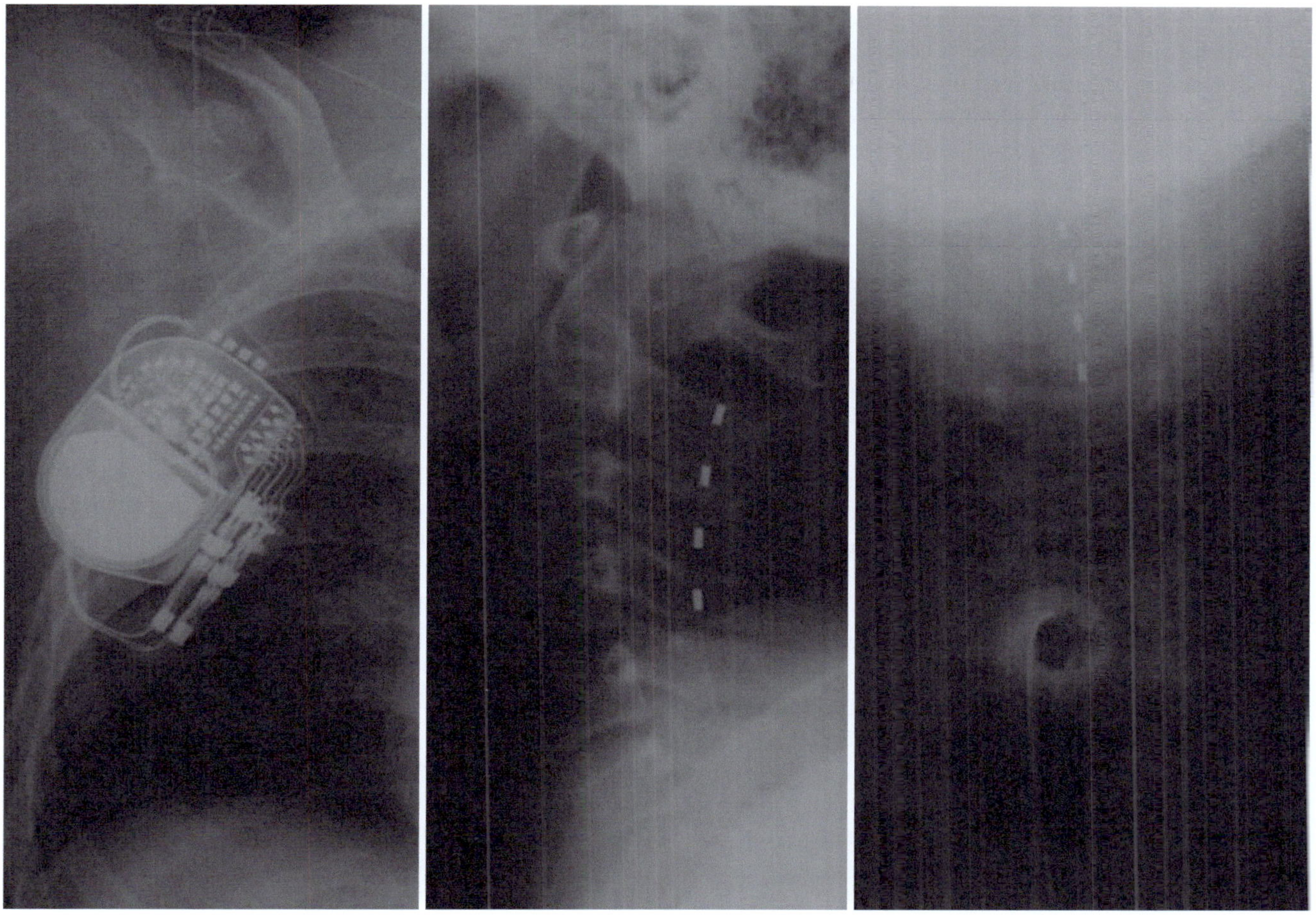

Fig. 1 Cervical spinal cord stimulation (SCS). Radiography images show the location of the stimulation electrode and implantable pulse generator (IPG). *Left*, IPG under the anterior chest wall; *Middle*, lateral view of cervical X-ray; *Right*, antero-posterior (A-P) view of cervical X-ray

N20 in SEP by electrophysiological evaluation [24–27]. However, it has not been so easy to find VS patients in whom both the fifth wave in ABR and N20 in SEP are detectable. Both measures were recorded in only 8 of the 21 VS patients treated with SCS in the present study. In contrast, we detected both the fifth wave in the ABR and N20 in SEP in 9 of our 10 MCS patients.

Phenomena Induced by 5-Hz Cervical SCS in VS and MCS

SCS did not induce strong arousal responses such as those that were observed by the deep brain stimulation (DBS) of the CM-pf complex and mesencephalic reticular formation in our previous reports [23, 25, 26]. However, 5-Hz cervical SCS induced muscle twitches of the bilateral upper extremities; these were not induced by DBS. While 5-Hz cervical SCS induced muscle twitches, 25-Hz cervical SCS induced muscle contraction of the upper extremities. Thus, we considered that 5-Hz cervical SCS is more suitable for neurorehabilitation and neuromodulation than higher-frequency cervical SCS.

In eight MCS patients, the average CBF without SCS was 38.8 ± 5.1 ml/100 g/min, while that during SCS was 47.51 ± 7.8 ml/100 g/min. SCS increased CBF diffusely in the brain, except at the lesion site, and the average CBF of the whole brain increased by 22.2 % during SCS compared with the CBF before SCS ($p<0.0001$, paired t-test).

Long-Term Effect of SCS in VS Patients

Among the 21 VS patients treated by SCS, only 8 patients showed both the fifth wave in the ABR and N20 in SEP before SCS. Of these 8 patients, only 3 patients recovered from VS and became able to communicate through speech or other responses. The causes of brain injury in the 3 VS patients who recovered were head injury (1 case) and cerebrovascular accident (2 cases). These 3 VS patients were treated by SCS at 3, 5 and 8 months after the brain injury. The 5 other VS patients showing both the fifth wave in the ABR and N20 in SEP, and who failed to recover from VS, were treated by SCS at 13, 14, 20, 36, and 51 months after the initial brain injury and coma. In the 3 patients who recovered from VS, the head injury patient recovered at 14 months, and the two cerebrovascular accident patients recovered at 8 and 12 months after the initial brain injury and coma. All 3 patients who recovered from VS remained in a bedridden state with severe disability, determined on the basis of the Glasgow Outcome Scale [10].

Long-Term Effect of SCS in MCS Patients

Giacino et al. [5] proposed criteria for determining emergence from MCS, characterized by reliable and consistent demonstration of one or both of the following: (1) functional interactive communication and (2) functional use of two different objects. On the basis of these proposed criteria, it was determined that seven of our ten patients emerged from MCS following SCS therapy.

In nine of the ten MCS patients, both the fifth wave in the ABR and N20 in SEP were detectable before SCS. All seven patients who recovered from MCS following SCS were among this group. The cause of brain injury in these seven patients was head injury (6 cases) and cerebrovascular accident (1 case), and the patients were treated by SCS at 3 months (4 patients), 8 months (1 patient), and 9 months (2 patients) after the initial brain injury. Among these seven patients, one showed improvement to only moderate disability at 1 year after the start of SCS therapy, as determined using the Glasgow Outcome Scale [10], and became able to walk by himself. Although the other six patients required the use of a wheelchair, four were able to use the wheelchair by themselves with some assistance. The other two patients were unable to handle the wheelchair by themselves even with some assistance. The two MCS patients who showed the fifth wave in the ABR and N20 in SEP but did not recover from MCS were treated with SCS at 11 and 12 months after the initial brain injury.

Discussion

Electrophysiological Evaluation and Timing of SCS

We have reported that electrophysiological evaluation is useful for estimating the resting brain function of VS and MCS patients, and we have recorded ABR, SEP, pain-related P250, and used continuous electroencephalogram (EEG) frequency analysis for the treatment of VS with DBS [24, 27]. In the present study, we performed only ABR and SEP for electrophysiological evaluation, because SCS for VS patients was applied mainly in our satellite hospital where pain-related P250 and continuous EEG frequency analysis are typically not recorded before SCS treatment. In general, ABR is used for the evaluation of brainstem function and SEP for the evaluation of thalamocortical function in the brain.

In this study, 3 of the 21 VS patients and 7 of the 10 MCS patients showed recovery of consciousness. All patients who recovered from VS or MCS showed the fifth wave in the ABR and N20 in SEP before SCS therapy. Based on these results, we consider that at least the fifth wave in the ABR and N20 in SEP must be detected by electrophysiological evaluation in order for SCS therapy to be effective in treating VS or MCS.

This study indicated that the timing for starting SCS is also very important. Three patients recovered from VS with SCS treatment, and all of these patients were started on SCS within 8 months after the onset of the initial brain injury and the comatose state. In addition, seven patients recovered from MCS with SCS treatment, and all of these seven patients underwent SCS within 9 months after the onset of the initial brain injury and comatose state. The other VS or MCS patients who showed the fifth wave in the ABR and N20 in SEP, but were treated with SCS 9 months or more after the onset of the initial brain injury and comatose state, did not recover from VS or MCS. Thus, we think that not only electrophysiological evaluation but also the timing of the start of SCS is critical for the treatment of VS and MCS. Although the neurological findings in VS and MCS are quite different, the effective timing for the start of SCS was not very different between the two groups.

Anatomical findings on computed tomography (CT) or magnetic resonance (MR) images are also useful for evaluating severe brain damage, such as bilateral diffuse injury of the thalamus or cerebral cortex; however, it is usually difficult to evaluate resting brain function correctly in the more mildly damaged brain [26]. CBF is usually lower in VS and MCS patients than in normal subjects, and some reports indicate the importance of CBF for the selection of candidates for SCS [11]; however, we must recognize that chronic stage CBF does not always represent the severity of acute brain anoxia.

Long-Term SCS for VS

The Multi-Society Task Force on PVS indicated that recovery of consciousness in post-traumatic VS is unlikely after 12 months, and that recovery from nontraumatic VS after 3 months is exceedingly rare [21, 22]. In our study, three patients recovered from VS following cervical SCS; one head injury patient recovered at 14 months, and two cerebrovascular accident patients recovered at 8 and 12 months after the initial brain injury. Based on the report from The Multi-Society Task Force on PVS, the recovery from VS in our three patients would be classified as rare if the recovery had occurred spontaneously. However, we could not demonstrate definite evidence that the SCS was useful for the recovery of VS patients in this study. Otherwise, we have reported that DBS of the CM-pf complex is useful for the treatment of VS patients when candidates are selected on the basis of an electrophysiological evaluation [25, 26, 28]. If the option of treating VS patients with SCS therapy exists, the fifth wave in the ABR and N20 in SEP should be verified as detectable, and the SCS should be started within 8 months after the initial brain injury.

Chronic DBS of the mesencephalic reticular formation and the thalamic CM-pf complex for the treatment of VS was first reported by Tsubokawa et al. [23, 24] and Chohadon and Richer [2]. Unilateral DBS of the intact side of the thalamic CM-pf complex can induce a very strong arousal response in VS patients [23, 25, 26, 28]. It seems important to increase the arousal level in VS patients for them to recover from the VS [6, 8, 16, 19, 20]. The fact that SCS does not induce an arousal response, although SCS can increase CBF, may be a clear instance of its difference from DBS therapy for VS patients.

Long-Term SCS for MCS

In contrast to VS patients, MCS patients retain some consciousness. In 2007, Schiff et al. [19] reported good results for bilateral DBS of the anterior intralaminar thalamic nuclei and adjacent paralaminar regions of the thalamic association nuclei, following a 6-month double-blind alternating crossover study in the treatment of MCS patients. Considering the persistence of a bedridden state before recovery from MCS, we selected 5-Hz cervical SCS rather than DBS. Five-hertz cervical SCS can increase CBF in the entire brain and induce muscle twitches; these cannot be induced by DBS [27, 28].

In this study, nine of the ten MCS patients showed the fifth wave in the ABR and N20 in SEP. Seven of these nine patients recovered from MCS and consistently demonstrated functional interactive communication. These findings indicate that electrophysiological evaluation with ABR and SEP is quite useful for predicting the effectiveness of SCS for MCS patients. Although SCS treatment for VS patients showed minimal improvements, we were able to confirm that cervical SCS therapy is truly effective for MCS patients.

In 1985, Hosobuchi reported that cervical SCS could significantly increase CBF in the hemisphere ipsilateral to the induced paresthesia, while thoracic SCS showed no effect on CBF [7]. The first use of SCS was for the treatment of pain or spasticity [1, 15, 17, 18]. SCS was not used for the treatment of VS and prolonged coma until 1989 [4]. Subsequent SCS studies for the treatment of VS also used high-frequency SCS [3, 12–14]. In contrast to those studies, we first reported that 5-Hz cervical SCS not only can increase CBF in the whole brain but can also induce muscle twitches of the upper extremities, which may constitute a new step toward rehabilitation [27]. We selected 5-Hz for SCS, since 5-Hz cervical SCS induced muscle twitches most clearly and strongly compared with other frequencies.

Recovery of Motor Function After 5-Hz Cervical SCS in MCS Patients

It is important that all seven patients who recovered from MCS after 5-Hz cervical SCS therapy showed good recovery of motor function in their upper extremities compared with their lower extremities. Five-hertz cervical SCS induced muscle twitches only in the upper extremities and not in the lower extremities. These results indicate that induced muscle twitches may be an important factor for the recovery of motor function. We chose stimulation for 5 min every 30 min to prevent the muscle fatigue that could been induced by continuous stimulation.

We placed the stimulation electrode at the center and dorsal part of the epidural space of the cervical spinal cord to facilitate chronic stimulation of the cervical dorsal column by the implanted SCS electrode. Regarding the mechanism of muscle twitches induced by dorsal column stimulation, there are several possible explanations: (1) current spreading to the anterior root, (2) direct stimulation of the corticospinal tract, (3) current spreading to the dorsal root, and (4) antidromical stimulation of sensory fibers from the dorsal horn to the dorsal funiculus (Fig. 2).

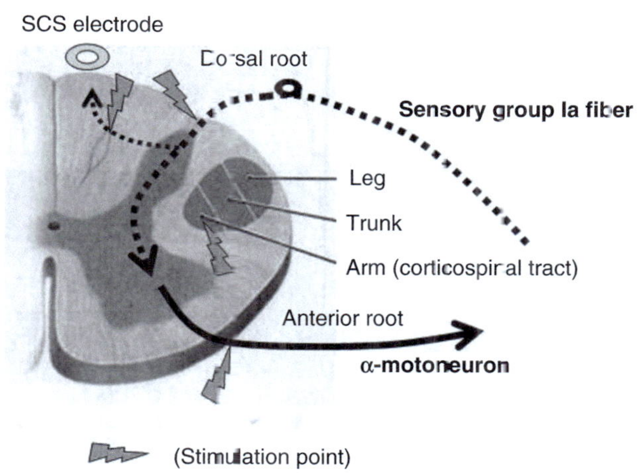

Fig. 2 Possible mechanism of muscle twitches induced by dorsal column stimulation. We speculate that dorsal column stimulation activates: (1) sensory fibers from the dorsal horn antidromically or (2) the dorsal root directly, and this stimulation activates α-motoneurons via interneurons

We speculate that dorsal column stimulation activates either: (1) sensory fibers from the dorsal horn antidromically or (2) the dorsal root directly, and this stimulation activates the α-motoneurons via interneurons. Five-hertz stimulation can synchronize the firing of α-motoneurons, and induce muscle twitches. We speculate that such repeated and induced muscle twitches are useful in both preventing the muscular atrophy and joint contracture that arises from disuse, and in encouraging the recovery of motor function. We also speculate that stimulation of the lower thoracic level, which induced muscle twitches of the lower extremities, may be useful for the recovery of the motor function of the lower extremities, just as cervical stimulation seems to be useful for the upper extremities. Further studies are required to investigate this hypothesis. Such neuromodulation may prove to be useful as a new neurorehabilitation technique.

Acknowledgments This work was supported by the MEXT-Supported Program for Strategic Research Foundation at Private Universities, and a Grant-in-Aid for Scientific Research (C-15K10375) from the Ministry of Education, Culture, Sports, Science and Technology (MEXT) of Japan.

Conflict of Interest Statement The authors confirm that this article content involves no conflicts of interest.

References

1. Campos RJ, Dimitrijevic MM, Faganel J, Sharkey PC (1981) Clinical evaluation of the effect of spinal cord stimulation on motor performance in patients with upper motor neuron lesions. Appl Neurophysiol 44:141–151
2. Chohadon F, Richer E (1993) Stimulation cerebrale profonde chez des patients etat vegetatif post traumatique. Neurochirurgia 39:281–292
3. Fujii M, Sadamitsu D, Maekawa T, Uesugi S, Ozaki S, Koizumi S et al (1998) Spinal cord stimulation therapy at an early stage for unresponsive patients with hypoxic encephalopathy. No Shinkei Geka 26:315–321 (Jpn, with Eng abstract)
4. Funahashi K, Komai N, Ogura M, Kuwata T, Nakai M, Tsuji N (1989) Effects and indications of spinal cord stimulation on the vegetative syndrome. No Shinkei Geka 17:917–923 (Jpn, with Eng abstract)
5. Giacino JT, Ashwal S, Childs N, Cranford R, Jennett B, Katz DI et al (2002) The minimally conscious state: definition and diagnostic criteria. Neurology 58:349–353
6. Hassler R, Ore GD, Dieckmann G, Bricolo A, Dolce G (1969) Behavioural and EEG arousal induced by stimulation of unspecific projection systems in a patient with post-traumatic apallic syndrome. Electroencephalogr Clin Neurophysiol 27:306–310
7. Hosobuchi Y (1985) Electrical stimulation of the cervical spinal cord increase cerebral blood flow in humans. Appl Neurophysiol 48:372–376
8. Jasper HH, Naquet R, King LE (1995) Thalamocortical recruiting responses in sensory receiving areas in the cat. Electroencephalogr Clin Neurophysiol 7:99–114
9. Jennett B, Plum F (1972) Persistent vegetative state after brain damage. A syndrome in search of a name. Lancet 1(7753):734–737
10. Jennett B, Bond M (1975) Assessment of outcome after severe brain damage. Lancet 1(7950):480–484
11. Kanno T, Morita I, Yamaguchi S, Yokoyama T, Kamei Y, Anil SM, Karagiozov KL (2009) Dorsal column stimulation in persistent vegetative state. Neuromodulation 12:33–38
12. Kanno T, Kamel Y, Yokoyama T, Shoda M, Tanji H, Nomura M (1989) Effects of dorsal column spinal cord stimulation (DCS) on reversibility of neuronal function – experience of treatment for vegetative states. Pacing Clin Electrophysiol 12:733–738
13. Liu JT, Tan WC, Liao WJ (2008) Effects of electrical cervical spinal cord stimulation on cerebral blood perfusion, cerebrospinal fluid cathecholamine lecels, and oxidative stress in comatose patients. Acta Neurochir Suppl 101:71–76
14. Matsui T, Asano T, Takakura K, Yamada R, Hosobuchi Y (1989) Beneficial effects of cervical spinal cord stimulation (cSCS) on patients with impaired consciousness: a preliminary report. Pacing Clin Electrophysiol 12:718–725
15. Megrio M, Cioni B, Rossi GF (1989) Spinal cord stimulation in management of chronic pain. A 9-year experience. J Neurosurg 70:519–524
16. Moruzzi G, Magoun HW (1948) Brain stem reticular formation and activation of the EEG. Electroencephalogr Clin Neurophysiol 1:455–473
17. Nakamura S, Tsubokawa T (1985) Evaluation of spinal cord stimulation for postapoplectic spastic hemiplegia. Neurosurgery 17:253–259
18. Sánchez-Ledesma MJ, García-March G, Diaz-Cascajo P, Gómez-Moreta J, Broseta J (1989) Spinal cord stimulation in deafferentation pain. Stereotact Funct Neurosurg 53:40–45
19. Schiff ND, Giacino JT, Kalmar K, Victor JD, Baker K, Gerver M et al (2007) Behavioural improvements with thalamic stimulation after severe traumatic brain injury. Nature 448:600–603
20. Strum V, Kuhner A, Schmitt P, Assmus H, Stock G (1979) Chronic electrical stimulation of the thalamic unspecific activating system in a patient with coma due to midbrain and upper brain stem infarction. Acta Neurochir 47:235–244
21. The Multi-Society Task Force on PVS (1994) Medical aspects of the persistent vegetative state. (First of two parts). N Engl J Med 330:1499–1508
22. The Multi-Society Task Force on PVS (1994) Medical aspects of the persistent vegetative state. (Second of two parts). N Engl J Med 330:1572–1579
23. Tsubokawa T, Yamamoto Y, Katayama Y, Hirayama T, Maejima S, Moriya T (1990) Deep brain stimulation in a persistent vegetative state: follow-up results and criteria for selection of candidates. Brain Inj 4:315–327
24. Tsubokawa T, Yamamoto Y, Katayama Y (1990) Prediction of the outcome of prolonged coma caused by brain damage. Brain Inj 4:329–337
25. Yamamoto T, Katayama Y (2005) Deep brain stimulation therapy for the vegetative state. Neuropsychol Rehabil 15:406–413
26. Yamamoto T, Katayama Y, Kobayashi K, Oshima H, Fukaya C, Tsubokawa T (2010) Deep brain stimulation for the treatment f vegetative state. Eur J Neurosci 32:1145–1151
27. Yamamoto T, Katayana Y, Obuchi T, Kobayashi K, Oshima H, Fukaya C (2012) Spinal cord stimulation for treatment of patients in the minimally conscious state. Neurol Med Chir (Tokyo) 52:475–481
28. Yamamoto T, Katayama Y, Obuchi T, Kabayashi K, Oshima H, Fukaya C (2013) DBS and SCS for vegetative state and minimally conscious state. World Neurosurg 80:S30.e1–9

Adverse Effects and Surgical Complications in Pediatric Patients Undergoing Vagal Nerve Stimulation for Drug-Resistant Epilepsy

A. Trezza, A. Landi, D. Grioni, D. Pirillo, L. Fiori, C. Giussani, and E.P. Sganzerla

Abstract Vagal nerve stimulation (VNS) is an effective treatment for drug-resistant epilepsy that is not suitable for resective surgery, both in adults and in children. Few reports describe the adverse effects and complications of VNS. The aim of our study was to present a series of 33 pediatric patients who underwent VNS for drug-resistant epilepsy and to discuss the adverse effects and complications through a review of the literature.

The adverse effects of VNS are usually transient and are dependent on stimulation of the vagus and its efferent fibers; surgical complications of the procedure may be challenging and patients sometimes require further surgery; generally these complications affect VNS efficacy; in addition, hardware complications also have to be taken into account.

In our experience and according to the literature, adverse effects and surgical and hardware complications are uncommon and can usually be managed definitely. Careful selection of patients, particularly from a respiratory and cardiac point of view, has to be done before surgery to limit the incidence of some adverse effects.

Keywords Vagal nerve stimulation • Complications • Pediatric age • Drug-resistant epilepsy

A. Trezza, MD (✉) • A. Landi, MD • D. Pirillo, MD
L. Fiori, MD • C. Giussani, MD, PhD • E.P. Sganzerla, MD
Neurosurgery, Department of Surgery and Translational Medicine, Milan Center for Neuroscience, San Gerardo Hospital, University of Milano Bicocca, via Pergolesi 33 20900, Monza, Italy
e-mail: andrea.trezza@gmail.com

D. Grioni, MD
Unit of Pediatric Neurophysiology, Child Neuropsychiatric Clinic, San Gerardo Hospital, University of Milano Bicocca, via Pergolesi 33 20900, Monza, Italy

Introduction

The worldwide prevalence of epilepsy in the pediatric population is estimated to be 0.5–1 % [5]; an effective treatment is particularly important in children because uncontrolled seizures can generate significant effects on normal cognitive development and quality of life [5, 21]. In addition, one-third of epileptic patients may develop drug-resistant epilepsy requiring surgery [15, 16].

Whenever resective surgery was not indicated, vagal nerve stimulation (VNS) has been suggested as an alternative and it has been widely used for treatment of epilepsy for at least 15 years. Positive effects of VNS in children have been widely described throughout the literature [21].

However, there are few reports concerning the possible adverse effects of stimulation of the vagal nerve and the surgical and hardware complications of VNS.

Voice alterations, hoarseness, cough, arrhythmias, obstructive sleep apnea, dysphagia, nausea/vomiting, headache, and Horner's syndrome have been reported to be possible adverse effects of vagal stimulation, while surgical complications are reported to depend upon carotid artery or jugular vein damage, infections, or hardware failure [4, 8, 12, 14, 17, 24, 25].

The aim of our study was to review the adverse effects and complications in our consecutive series of 33 pediatric patients implanted with VNS for intractable epilepsy and to compare our data with the literature.

Materials and Methods

Patient Population

Since 2006, 33 pediatric patients (18 male and 15 female) have received VNS for refractory epilepsy not eligible for resective surgery at our Institution. Their age range was between 8 months and 17 years. Mean age at implant was 8 years and mean follow-up time was 3. Two years. On average, VNS was performed 7 years after the diagnosis of refractory epilepsy.

Each patient underwent careful presurgical evaluation that included history, physical and neurological examination, magnetic resonance imaging (MRI) and functional MRI imaging, electroencephalogram (EEG) and video-EEG, and cardiac and respiratory investigations. All patients underwent VNS as the primary surgical treatment after the failure of multiple antiepileptic drugs (AEDs). Resective surgery was excluded upon anatomo-functional considerations; in detail, these were: the radiological absence of a well-defined epileptic lesion, the unfeasibility of identifying an epileptogenic zone, and features of complex and diffuse anatomical modifications or genetic alterations.

Epilepsy was due to the following pathologies: tuberous sclerosis complex (TSC, 4 patients), genetic disease (14 patients), malformations of cortical development (MCD, 7 patients), channelopathies (4 patients), and miscellaneous pathologies (4 patients). Epilepsy was characterized by spasms in most of the patients. Three cases of refractory status epilepticus (SE) were treated acutely with VNS, with a dramatic reduction in the frequency and intensity of seizures without recurrence of further episodes of SE.

Surgical Technique

A single neck incision, on the left side, was used in the first cases; later we preferred to perform a second incision under the clavicle to be more comfortable in positioning the electrode and the implantable pulse generator (IPG). Cyberonics VNS Pulse model 102 or a DemiPulse model 103 was implanted (Cyberonics, Houston, TX, USA).

The patient is placed under general anesthesia in a supine position with a roll beneath the scapulae to provide mild neck extension. The head is lightly rotated towards the right and fixed to the operative table. To implant the electrode, a 3- to 4-cm left paramedian neck incision is made and the vagal nerve is reached deep and medial to the internal jugular vein through blunt dissection. Under the operative microscope dissection of the nerve is performed for at least 4 cm and the electrode coils are positioned along the vagal trunk. A 5-cm subclavian incision is made and a pocket is created to implant the IPG above the muscular fascia. The connecting cable is tunneled and a tension-relief loop is made and fixed to the sternocleidomastoid muscle. The IPG itself is fixed to the muscular fascia to prevent future migration of the device. Once the cable is connected to the generator, a lead test is carried out to measure impedances. Antibiotics are administered preoperatively. If no complications occur, the child is usually discharged after 24–48 h and referred to the children's neuropsychiatric clinic 2 weeks later to start the stimulation.

Results

During follow-up the patients were evaluated neurologically and with repeated EEGs. AEDs were reduced in all patients. The mean current output of the device is 1.3 mA (range 1–1.75 mA) and mean duty cycle is 13 % (range 10–35 %).

Modified McHugh's score modified McHugh's score was used to evaluate VNS efficacy. In our series we had 14 patients in class I (47 %), 10 patients in class II (33 %), 2 patients in class III (7 %), and 4 patients in class V (13 %). Three patients died of different causes: sudden unexpected death in epileptic patient (SUDEP; 1 patient), prolonged and intractable SE (1 patient), and systemic complications due to the underlying syndrome (1 patient).

The overall complication rate was 18.2 % and the adverse effects rate was 9.09 %; the details are summarized in Table 1.

Out of the 33 patients, 4 (12.1 %) experienced breakage of the connecting cable as a consequence of falls or a direct hit.

One patient (3.03 %) developed a keloid at the level of the thoracic incision.

In one patient an intraoperative lesion of the thoracic duct occurred, with consequent lymphorrhea and the need for surgical closure of the duct. No further postoperative clinical troubles have been observed in this patient.

Table 1 Summary of adverse effects and complications in our series of 33 pediatric patients

No. of patients	Incidence (%)	Event description	Solution
4	12.1	Accidental electrode rupture	Replacement (IPG + electrode)
2	6	Cough	VNS parameters adjustment
1	3.03	Thoracic duct lesion	Surgical closure of the duct
1	3.03	Nausea/vomiting	VNS parameters adjustment
1	3.03	Keloid	Conservative management by plastic surgeon

IPG implantable pulse generator, *VNS* vagal nerve stimulation
None of the patients experienced irreversible deficits or clinical problems

So far, we have not observed any infection, either postoperatively or during long-term follow-up.

As far as adverse effects are concerned, we experienced one case of cough and one case of nausea/vomiting that progressively resolved with adjustment of the stimulation parameters. In particular, the nausea and vomiting appeared in a patient in whom the stimulation was started immediately after the implant because of a challenging clinical picture, characterized by frequent falls. Given this adverse effect, the stimulation was stopped and gradually restarted after 10 days.

Discussion

Vagal nerve stimulation is nowadays considered a safe surgical and neuromodulatory technique. Positive results in seizure control and quality of life improvement have been observed in children [1, 8, 19, 21, 23]. In the literature, the rates of surgical complications and adverse effects are reported to be between 2.5 and 16.8 % [1, 12, 14, 24, 26].

However, as far as hardware complications are concerned, the literature reports a rate ranging from 0 to 21 % [1, 6, 8, 11–14, 24, 26].

After our analysis of the abovementioned studies, it was clear that the most common and relevant complications concerning VNS implantations are: infections, vocal cord palsy, and hardware-related inconveniences.

In these studies the infection rate was reported to be between 0 and 10.9 %. In most cases, infections could be considered as a minor complication, only requiring oral and/or local antibiotics. Sometimes infections become a major event and explantation of the device is needed. The risk of infection is undoubtedly due to the implantation of the prosthetic device. One comparable neurosurgical procedure is deep brain stimulation (DBS), where the literature reports infection rates between 2.5 and 8.5 %, values that are similar to those reported for VNS [12]. In our series we have not had any case of infection so far. This could be explained by several factors; in particular: the careful draping of the sterile field, the surgeon's behavior (reduction of operative time, changing gloves, use of disposable materials, etc), the administration of intravenous antibiotics (usually cefazolin) 1 h before surgery, and the care of the wound during hospitalization and at home, after accurate training of the care-givers. An increased infection rate seems to be a consequence of the number of pulse generator replacements, for battery exhaustion [12] and the rate could be reduced by the introduction of new stimulator models s with longer battery life or rechargeable generators, as employed for DBS devices.

As far as vocal cord palsy is concerned, an incidence between 1 and 5.6 % emerged from the aforementioned studies. This kind of complication appears to be a direct consequence of the manipulation of the vagus nerve, or it may be due to recurrent laryngeal nerve injury; usually, it is a transient adverse effect. A similar incidence of vocal cord palsy was also reported following a second surgery, performed for the replacement or explantation of the electrode; commonly, in these cases, the palsy may last longer [12, 27]. The reason could be that fibrosis and scar near the previous lead insertion would make dissection more troublesome. However, it seems evident that vocal cord palsy is a complication that affects adults more than children [26].

Left vocal cord adduction due to stimulation of the left recurrent laryngeal nerve could be the cause of laryngeal and respiratory symptoms that might develop in these patients. It has been demonstrated that stimulation of the vagus leads to an increase in respiratory rate and to a decrease in respiratory amplitude, tidal volume, and oxygen saturation. These respiratory alterations could cause sleep apnea or exacerbate it in patients with a preexisting diagnosis. Moreover, as the VNS device is usually implanted in patients who could, potentially, have abnormal central nervous system (CNS) development and who generally receive active CNS drugs, it has to be kept in consideration that these patients are at higher risk of apnea than the normal population, even in the absence of the usual risk factors, in that the stimulation of the vagus could influence respiratory drive and the opening of the upper airway during sleep [22].

Cardiac complications (arrhythmias) are uncommon, with an incidence lower than 1 % [3, 12]. The left vagus nerve has fewer cardiac efferent fibers than the right vagus nerve and this could potentially limit the arrhythmogenic effects of left vagal stimulation. Patients with a history of bradyarrhythmias and asystole are at greater risk of cardiac effects [10].

Given the risk of cardiac and respiratory complications, it is mandatory for potential VNS patients to undergo a careful preoperative screening.

Transient side effects of VNS, such as cough (6–45 %), headache, nausea/vomiting, and neck spasms are usually dependent on the stimulation parameters and occur during stimulus. These effects are reported to be reversible after parameter adjustment and, if persistent, are usually well tolerated by the patients and do not impinge on the efficacy of VNS [24].

Since many of the patients who have undergone VNS for drug-resistant epilepsy are mentally disabled and could not well describe some types of adverse symptoms, their frequency could be underestimated and, in particular, vocal cord palsy and consequent hoarseness could be more frequent, as reported in the literature [12].

A very uncommon surgical complication described in our series is a case of lesion of the thoracic duct with subsequent lymphorrhea; the duct was immediately surgically sealed,

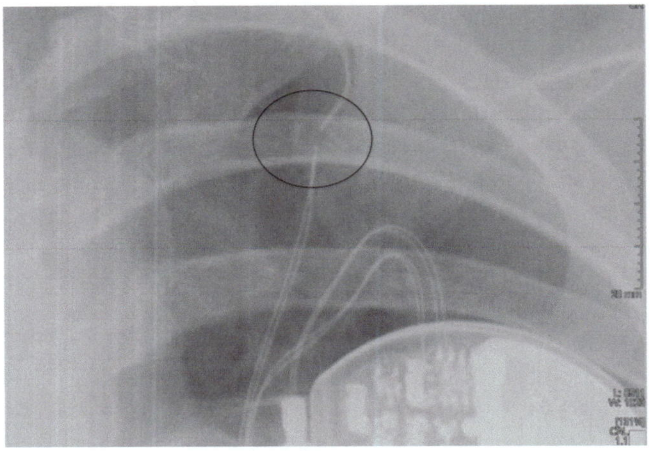

Fig. 1 Plain chest X-rays showing disconnection of the vagal nerve stimulation (VNS) lead in a patient with recurrence of preoperative symptoms and high impedance

without any systemic postoperative complication. To the best of our knowledge, this complication has never been reported in the literature. Possible reasons for the surgical accident might depend on anatomical variants of the vagal nerve and variants of its relationship with the internal jugular vein and carotid artery. In our series we observed two cases of plexiform nerve and one case of vagal nerve positioned lateral and more posterior to the internal jugular vein.

These anatomical variants required a more challenging and extensive dissection of the nerve trunk, to adequately isolate the nerve; in particular in the caudal portion, where it is more likely that the thoracic duct, which runs posterior and medial to the internal jugular vein, below the clavicle, will be found and damaged, .

Other complications, such as lesion of the jugular vein or carotid artery, neck hematoma, dysesthesia due to sensory nerve damage, keloid development, and Horner's syndrome are reported more rarely, with an incidence of less than 2 % [12].

Hardware complications most commonly consist of breakage of the connecting cable or deterioration of its isolating shield. In both such cases, replacement of the lead is required [1, 2, 6–8, 11–14, 24, 26]. The failure of VNS leads can be determined clinically, with the loss of efficacy of the stimulation, i.e., the recurrence of seizures, and failure may be highlighted by interrogation of the device, verifying the presence of very high lead impedances.

Moreover, chest and neck pain, shock-like sensations, and paresthesias could be other clinical consequences of lead failure, usually disappearing after revision [18, 26].

Dislocated leads, hardware-related infections, and device malfunction are other uncommon but reported causes of VNS failure [7, 26]. Radiological imaging of the chest and neck can help to diagnose lead dislocations or cable breaks (Fig. 1), although in most patients experiencing high impedances neither damage nor gross breakage of the lead are detectable [7]. When a history of trauma is excluded, the cause of high impedance and consequent VNS failure may be doubtful. Some authors have pointed out the possibility of micro-lesions spontaneously developing within the lead [26], while others have observed that scar tissue developing over time around the lead and the vagal nerve could be the cause of high impedance and lead failure [20].

Lead revision surgery can be challenging [7, 9, 12] and usually requires a longer operative time, due to the presence of scar tissue and fibrosis, which modify and bend the normal anatomy of the neck. In cases of extensive scars or fibrous reaction, some authors suggest leaving the damaged electrode in place and inserting the new one more cranially along the nerve [9, 12]. In our series we report four cases of traumatic lead breakage documented both radiologically (Fig. 1) and clinically, leading to a rapid increase in the intensity and frequency of seizures; high impedance along the cable was detected after system interrogation. All these four patients were implanted more than 5 years ago, with the rigid cable available at the time . Recently, a more flexible cable has been available from Cyberonics, and with the use of this cable ruptures were no longer experienced. Surgery for lead and cable replacement was performed in all four patients, with no complications, and recovery of the previous efficacy in seizure control was always obtained.

In conclusion, in our experience and according to the literature, VNS can be considered an effective and safe technique to treat refractory epilepsy not eligible for resective surgery, even in children. Complication and adverse effect rates are low; they may be successfully managed and generally they do not affect VNS efficacy.

Careful preoperative selection should be performed to determine the best candidates for VNS and to further reduce the incidence of complications. In particular, respiratory and cardiac screening should be mandatory.

Given the evidence of its efficacy and the low incidence of major complications, it seems to be possible in the future to depict a role of VNS as first-choice therapy in the treatment of drug-resistant epilepsy.

Conflict of Interest Statement The authors declare that they have no conflict of interest.

References

1. Alexopoulos AV, Kotagal P, Loddenkemper T, Hammel J, Bingaman WE (2006) Long- term results with vagus nerve stimulation in children with pharmacoresistant epilepsy. Seizure 15: 491–503
2. Ardesch JJ, Buschman HP, Wagener-Schimmel LJ, van der Aa HE, Hageman G (2007) Vagus nerve stimulation for medically refractory epilepsy: a long-term follow-up study. Seizure 16(7):579–585

3. Asconapé J, Moore D, Zipes D, Hartmann L, Duffel WH (1999) Bradycardia and asystole with the use of vagus nerve stimulation the treatment of epilepsy: a rare complication of intra- operative device testing. Epilepsia 40:1452–1454
4. Ben-Menachem E (2002) Vagus nerve stimulation for the treatment of epilepsy. Lancet Neurol 1:477–482
5. Berg AT, Zelko FA, Levy SR, Testa FM (2012) Age at onset of epilepsy, pharmacoresistance, and cognitive outcomes: a prospective cohort study. Neurology 79:1384–1391
6. Coykendall DS, Gauderer MW, Blouin BB, Morales A (2010) Vagus nerve stimulation for the management of seizures in children: an 8 year experience. J Pediatr Surg 45:1479–1483
7. Dlouhy BJ, Viljoen SV, Kung DK, Vogel TW, Granner MA, Howard MA, Kawasaki H (2012) Vagus nerve stimulation after lead revision. Neurosurg Focus 32(3):E11
8. Elliott RE, Rodgers SD, Bassani L, Morsi A, Geller EB, Carlson C, Devinsky O, Doyle WK (2011) Vagus nerve stimulation for children with treatment-resistant epilepsy: a consecutive series of 141 cases. J Neurosurg Pediatr 7:491–500
9. Espinosa J, Aiello MT, Naritoku DK (1999) Revision and removal of stimulating electrodes following long-term therapy with the vagus nerve stimulator. Surg Neurol 51:659–664
10. Hatton KW, McLarney JT, Pittman T, Fahy BG (2006) Vagal nerve stimulation: overview and implications for anesthesiologists. Anesth Analg 103:1241–1249
11. Huf RL, Mamelak A, Kneedy-Cayem K (2005) Vagus nerve stimulation therapy: 2-year prospective open-label study of 40 subjects with refractory epilepsy and low IQ who are living in long-term care facilities. Epilepsy Behav 6:417–423
12. Kahlow H, Olivecrona M (2013) Complications of vagal nerve stimulation for drug-resistant epilepsy: a single center longitudinal study of 143 patients. Seizure 22:827–833
13. Kang HC, Hwang YS, Kim DS, Kim HD (2006) Vagus nerve stimulation in pediatric intractable epilepsy: a Korean bicentric study. Acta Neurochir Suppl 99:93–96
14. Kuba R, Brazdil M, Kalina M, Prochazka T, Hovorka J, Nezadal T, Hadac J, Brozova K, Sebronova V, Komarek V, Marusic P, Oslejskova H, Zarubova J, Rektor I (2009) Vagus nerve stimulation: longitudinal follow-up of patients treated for 5 years. Seizure 18:269–274
15. Kwan P, Arzimanoglou A, Berg AT, Brodie MJ, Allen Hauser W, Mathern G, Moshé SL, Perucca E, Wiebe S, French J (2010) Definition of drug resistant epilepsy: consensus proposal by the ad hoc Task Force of the ILAE Commission on Therapeutic Strategies. Epilepsia 51:1069–1077
16. Kwan P, Schachter SC, Brodie MJ (2011) Drug-resistant epilepsy. N Engl J Med 365:919–926
17. Landy HJ, Ramsay RE, Slater J, Casiano RR, Morgan R (1993) Vagus nerve stimulation for complex partial seizures: surgical technique, safety, and efficacy. J Neurosurg 78:26–31
18. MacDonald J, Couldwell WT (2004) Revision of vagal nerve stimulator electrodes technical approach. Acta Neurochir 146:567–570
19. Morris GL, Mueller WM (1999) Long-term treatment with vagus nerve stimulation in patients with refractory epilepsy The Vagus Nerve Stimulation Study Group E01-E05. Neurology 53:1731–1735
20. Ng WH, Donner E, Go C, Abou-Hamden A, Rutka JT (2010) Revision of vagal nerve stimulation (VNS) electrodes: review and report on use of ultra-sharp monopolar tip. Childs Nerv Syst 26:1081–1084
21. Orosz I, McCormick D, Zamponi N, Varadkar S, Feucht M, Parain D, Griens R, Vallee L, Boon P, Rittey C, Jayewardene AK, Bunker M, Arzimanoglou A, Lagae L (2014) Vagus nerve stimulation for drug-resistant epilepsy: a European long-term study up to 24 months in 347 children. Epilepsia 55(10):1576–1584
22. Parhizgar F, Nugent K, Raj R (2011) Obstructive sleep apnea and respiratory complications associated with vagus nerve stimulators J Clin Sleep Med 7(4):401–407
23. Patwardhan RV, Stong B, Bebin EM, Mathisen J, Grabb PA (2000) Efficacy of vagal nerve stimulation in children with medically refractory epilepsy. Neurosurgery 47:1353–1357
24. Rychlicki F, Zamponi N, Cesaroni E, Corpaci L, Trignani R, Ducati A, Scerrati M (2006) Complications of vagal nerve stimulation for epilepsy in children. Neurosurg Rev 29:103–107
25. Smyth MD, Tubbs RS, Bebin EM, Grabb PA, Blount JP (2003) Complications of chronic vagus nerve stimulation for epilepsy in children. J Neurosurg 99:500–503
26. Spuck S, Tronnier V, Orosz I, Schönweiler R, Sepehrnia A, Nowak G, Sperner J (2010) Operative and technical complications of vagus nerve stimulator implantation. Neurosurgery 67:489–494
27. Vassilyadi M, Strawsburg RH (2003) Delayed onset of vocal cord paralysis after explantation of a vagus nerve stimulator in a child. Childs Nerv Syst 19:261–263

Vagal Nerve Stimulation for Drug-Resistant Epilepsy: Adverse Events and Outcome in a Series of Patients with Long-Term Follow-Up

R. Morace, G. Di Gennaro, P.P. Quarato, A. D'Aniello, A. Mascia, L. Grammaldo, M. De Risi, A. Sparano, F. Di Cola, M. De Angelis, and V. Esposito

Abstract

Background

Vagal nerve stimulation (VNS) is a palliative treatment option for drug-resistant epilepsy. The aim of this study was to describe the clinical and demographic features of selected patients scheduled for VNS and to evaluate the long-term efficacy of VNS in seizure control.

Materials and Methods

Between 2006 and 2013, 32 consecutive epileptic patients (14 male and 18 female) were enrolled at our Institute for VNS implantation. In all cases resective surgery had previously been excluded by the use of a noninvasive presurgical study protocol. Mean age was 32 years (range 18–50), and mean epilepsy duration 23 years (range 11–39). All subjects were followed-up for at least 2 years (mean 6 years, range 2–9) after VNS implantation. Patients were considered responders when a reduction of seizures of more than 50 % was reported.

Results

All patients had complex partial seizures, in 81 % of the patients with secondary generalization and in 56 % with drop attacks. Neurological examination revealed focal deficits in 19 % of the patients. Brain magnetic resonance imaging (MRI) was positive in 47 % of the patients. No surgical complications were observed in this series. Three patients were lost to follow-up. Twelve patients were classified as responders. Among the others, 1 patient experienced side effects (snoring and groaning during sleep) and the device was removed.

Conclusions

Our data confirm that VNS is a safe procedure and a valid palliative treatment option for drug-resistant epileptic patients not suitable for resective surgery.

Keywords Vagal nerve stimulation • Drug-resistant epilepsy • Intractable epilepsy • Neuromodulation • Palliative surgery

Introduction

Epilepsy constitutes the second most common neurological disorder in incidence and prevalence after cerebrovascular diseases [26]. Antiepileptic drugs (AEDs) are the first-line treatment that allow seizure control for most patients. However, drug-resistant epilepsy (DRE), defined as the failure of adequate trials of two tolerated, appropriately chosen and used antiepileptic drug schedules (whether given as monotherapies or in combination) to achieve sustained freedom from seizures [21], has been reported to occur in 20–40 % of cases [5, 9, 14, 20, 21]. Resective epilepsy surgery may provide seizure freedom in focal epilepsy, mainly if associated with focal lesions. Particularly for mesial temporal epilepsy, seizure freedom rates range from 60 to 90 % in hippocampal sclerosis, from 60 to 70 % in focal cortical dysplasia, and from 70 to 85 % in glioneuronal tumors or vascular malformations [27]. In most centers the selection of candidates for resective surgery is based primarily on prolonged video-electroencephalography (V-EEG), neuroimaging, and neuropsychological testing. Recent advances in neuroimaging and in digital EEG signal analysis, as well as the employment of invasive recordings both through depth

R. Morace (✉) • G. Di Gennaro • P.P. Quarato • A. D'Aniello
A. Mascia • L. Grammaldo • M. De Risi • A. Sparano • F. Di Cola
I.R.C.C.S. Neuromed, Pozzilli, Italy
e-mail: roberta.morace@yahoo.it

M. De Angelis
I.R.C.C.S. Neuromed, Pozzilli, Italy

Department of Neuroscience and Reproductive and Odontostomatological Science, University of Naples "Federico II", Naples, Italy

V. Esposito
I.R.C.C.S. Neuromed, Pozzilli, Italy

Department of Neurology and Psychiatry, University of Rome "Sapienza", Rome, Italy

M. Visocchi et al. (eds.), *Trends in Reconstructive Neurosurgery*, Acta Neurochirurgica Supplement 124,
DOI 10.1007/978-3-319-39546-3_8,

and subdural electrodes, may improve the localization of the epileptogenic zone, thus increasing the number of patients eligible for resective procedures. Nevertheless, in 40–60 % of patients with DRE, surgery cannot be indicated due to generalized seizures, multifocal epilepsy, or eloquent cortex involvement [12]. Vagus nerve stimulation (VNS; VNS Therapy System; Cyberonics, Houston, TX, USA) is a neuromodulation therapy and a palliative option to reduce epileptic seizures in patients with DRE. VNS was approved by the United States Food and Drug Administration (FDA) in 1997 and by the European Medicines Agency (EMA) in 1994 as an adjunctive therapy for the treatment of intractable partial epilepsy in adults and children over 12 years of age, although many off-label indications of VNS therapy have been reported [4, 8, 11, 12, 19, 22, 25].

We report a series of 32 patients who have received VNS therapy at our Institution since 2006, focusing on the selection criteria and clinical outcome.

Materials and Methods

This is a retrospective observational analysis of all patients who received VNS therapy at I.R.C.C.S. Neuromed, Pozzilli, Italy, from October 2006 to July 2013. Data collected included demographic information, surgical and medical history, etiology, physical and neurological examinations, epilepsy characteristics and time to implant, seizure frequency (obtained from logs kept by patients or their carers), and concomitant pharmacological treatment. A Multidisciplinary Epilepsy Committee evaluated each patient in order to select potential candidates, considering clinical history, physical and neurological examinations, psychological evaluation, V-EEG monitoring, magnetic resonance imaging (MRI) (1.5/3 T), and in some cases, MRI functional studies and positron emission tomography (PET). Among the patients with DRE, selection criteria for VNS implantation included: focal, multifocal, or diffuse seizure onset unsuitable for surgical resection; recurrent seizures following resective surgery; and/or patient or family preference for conservative measures. The surgical technique for VNS device implantation has been reported previously in the pertinent literature [2, 17, 20]. All the procedures were performed, under general anesthesia, on the left vagus nerve, with the pulse generator positioned in a subcutaneous pocket on the chest. Patients usually stayed in hospital for one night after surgery. Follow-up was conducted by means of outpatient programs. The device was turned on approximately 2 weeks after the surgery. Adjustments of VNS duty cycle or AEDs were subsequently performed when required, with variable schedules. Seizure outcome was reported as the responder rate (percentage of patients with ≥50 % seizure reduction). Information about seizure frequency, use of the external magnet to switch-off the pulse generator, and side effects were obtained from the logs kept by the patients/family members or by means of phone enquiries.

Results

Between October 2006 and July 2013, 32 patients underwent surgery for VNS implantation at our Epilepsy Unit. All the procedures were primary implantations. Table 1 summarizes patients' demographic and clinical data. Epileptic syndromes (Table 2) included partial multifocal epilepsy in 10 patients (31 %), partial (secondarily generalized) cryptogenic epilepsy in 9 cases (29 %), epileptic encephalopathy in 6 cases (19 %), bitemporal epilepsy in 2 cases (6 %), and Lennox Gastaut syndrome in the remaining 2 (6 %). Three patients had partial focal epilepsy (9 %); in two of them resective surgery was excluded because the epileptogenic zone overlapped with the eloquent cortex, while the remaining 1 patient did not give informed consent for the surgical procedure. Regarding seizure type, all patients had complex partial seizures, secondarily generalized in 26 cases (81 %) and with drop attacks in 18 (56 %). MRI was performed in 100 % of the patients, with normal results in 59 % and showing encephalomalacia in 13 %, cortical dysplasia in 13 %, atrophy in 6 %, bilateral periventricular heterotopia in 6 %, and

Table 1 Demographic data and preoperative status of patients

Patient population		
Male	14 (44 %)	
Female	18 (56 %)	
Drug resistance	32 (100 %)	
Age at onset (years)	Mean 10 (range 1–23)	
Age at surgery (years)	Mean 32 (range 18–50)	
Epilepsy duration (years)	*Mean 23 (range 11–39)*	
Abnormal neurological examination (motor deficits)		(19 %)
Mental retardation		(57 %)
Seizure type	Complex partial	6 (19 %)
	Complex partial + generalized	26 (81 %)
	Drop attack	18 (56 %)
High seizure frequency (monthly/daily)	32 (100 %)	
No. of antiepileptic drugs (AEDs)	2	4 (13 %)
	3	15 (47 %)
	≥4	13 (40 %)

Table 2 Epileptic syndromes in patients treated with vagal nerve stimulation (VNS)

Epileptic syndrome	No. of patients (%)
Epileptic encephalopathy	6 (19 %)
Partial multifocal epilepsy	10 (31 %)
Partial (secondarily generalized) cryptogenic epilepsy	9 (29 %)
Bitemporal epilepsy	2 (6 %)
Lennox Gastaut syndrome	2 (6 %)
Partial symptomatic epilepsy	2 (6 %)
(Left) temporal lobe epilepsy	1 (3 %)

hydrocephalus in only 1 patient. Four patients (13 %) had a history of traumatic brain injury and in 6 (18 %) perinatal hypoxia had been reported. Seizure frequency was daily in 15 patients, weekly in 15, and monthly in the remaining 2 patients. Twenty-eight patients (87 %) were receiving three or more AEDs.

The mean duration of follow-up was 6 years (range 2–9). Twelve patients (41 %) were considered responders (percentage of seizure reduction ≥50 %). Changes in device setting and the duty cycle were performed in 15 patients (47 %). In 1 patient the VNS device was removed after 3 months because he had experienced sleep snoring. No surgical complications were observed in this series.

Discussion

VNS is described as an effective palliative treatment option for DRE [4, 8, 19, 25]. A major indication for VNS is symptomatic epilepsy with multifocal independent foci. Several etiologies are implicated in multifocal ictal foci, such as neuronal migration disorders, cerebral palsy, traumatic brain injury, infections, phacomatosis, and metabolic syndromes [11, 12, 28]. Cryptogenic generalized epilepsy with diffuse epileptogenic abnormalities (e.g., Lennox-Gastaut syndrome) is another indication for VNS therapy. These patients are suitable candidates both for corpus callosotomy (CC) and VNS. In 2013, Lancman et al. published an interesting meta-analysis demonstrating that VNS offered rates comparable to CC in patients with tonic, generalized tonic-clonic, and complex partial and myoclonic seizures [22]. Other cases suitable for VNS can also be considered: refractory idiopathic generalized epilepsy (IGE) [24] and failed resective epilepsy surgery [3]. As widely described, resective surgery offers a satisfactory seizure outcome in patients with focal symptomatic epilepsy and is the recommended treatment option in such patients [10, 27]. Furthermore, advances in neuroimaging and the employment of intracranial recordings, as well as the recent developments in digital EEG analysis, allow a better identification of the epileptogenic zone, increasing the number of candidates for resective surgery. However, if the epileptogenic focus arises from/overlaps with the eloquent cortex, open cranial surgery should be excluded. In addition, some patients may refuse the risks associated with intracranial procedures.

In line with the current literature, VNS in our series was performed in 29 patients (91 %) due to multifocal epilepsy, and in 2 patients (6 %) in whom resective surgery could not be performed, since the epileptogenic zone involved the eloquent cortex. Only 1 patient (3 %), with left temporal lobe epilepsy that was suitable for resective surgery, underwent VNS device implantation because he refused "open" cranial surgery.

During a mean follow-up of 6 years (range 2–9) we observed 41 % of responders in our series (percentage with seizure reduction ≥50 %). DeGiorgio et al. reported a responder rate of 35 % [7]; in the E01-E05 study, 440 patients were followed-up for 3 years, with responder rates of 36.8 % at 1 year, 43.2 % at 2 years, and 42.7 % at 3 years [1, 18, 23]. The largest retrospective study, with 436 patients, demonstrated a mean seizure reduction rate of 55.8 % in nearly 5 years of mean follow-up duration, and also found that the mean seizure reduction at 10 years was 75.5 % in 65 consecutive patients [12, 13]. In their review of more than 1800 patients, Connor et al. [5] reported a mean seizure reduction of 42.8 %, with 50.9 % of patients being responders. Other studies have been performed to explore changes in health-related quality of life in patients treated by VNS [6, 15]. Patients were evaluated with the Quality of Life in Epilepsy-10 (QOLIE-10), and demonstrated significant improvements in outcomes such as energy level, memory difficulties, social aspects, mental effects, and fear of seizures [6]. VNS implantation and therapy was associated with a persistent and positive improvement in subjective quality of life [15].

Differently from results in the current literature [12, 16], we did not experience any surgical complications (e g., neck or arm pain, wound infection, hemorrhage, pneumothorax) and VNS was well tolerated in the majority of patients. The device was removed in only one patient due to sleep snoring.

Conclusion

Our results confirm that VNS is a safe and effective treatment option for reducing seizure frequency in many patients with multifocal DRE. Patients with focal symptomatic epilepsy may be treated with VNS if the results of presurgical evaluation contraindicate resective procedures.

Conflict of Interest Statement We declare that we have no conflict of interest.

References

1. A randomized controlled trial of chronic vagus nerve stimulation for treatment of medically intractable seizures. The Vagus Nerve Stimulation Study Group. Neurology 45:224–230 (1995)
2. Amar AP, Apuzzo ML, Liu CY (2004) Vagus nerve stimulation therapy after failed cranial surgery for intractable epilepsy: results from the vagus nerve stimulation therapy patient outcome registry. Neurosurgery 55:1086–1093
3. Amar AP, Heck CN, Levy ML, Smith T, DeGiorgio CM, Oviedo S et al (1998) An institutional experience with cervical vagus nerve trunk stimulation for medically refractory epilepsy: rationale, technique, and outcome. Neurosurgery 43(6):1265–1276; discussion 1276–1280
4. Blount JP, Tubbs RS, Kankirawatana P, Kiel S, Knowlton R, Grabb PA, Bebin M (2006) Vagus nerve stimulation in children less than 5 years old. Childs Nerv Syst 22(9):1167–1169
5. Connor DE Jr, Nixon M, Nanda A, Guthikonda B (2012) Vagal nerve stimulation for the treatment of medically refractory epilepsy: a review of the current literature. Neurosurg Focus 32(3):E 12
6. Cramer JA (2001) Exploration of changes in health-related quality of life after 3 months of vagus nerve stimulation. Epilepsy Behav 2:460–465
7. DeGiorgio CM, Schachter SC, Handforth A, Salinsky M, Thompson J, Uthman B, Reed R, Collins S, Tecoma E, Morris GL, Vaughn B, Naritoku DK, Henry T, Labar D, Gilmartin R, Labiner D, Osorio I, Ristanovic R, Jones J, Murphy J, Ney G, Wheless J, Lewis P, Heck C (2000) Prospective long-term study of vagus nerve stimulation for the treatment of refractory seizures. Epilepsia 41:1195–1200
8. De Herdt V, Boon P, Ceulemans B, Hauman H, Lagae L, Legros B, Sadzot B, Van Bogaert P, van Rijckevorsel K, Verhelst H, Vonck K (2007) Vagus nerve stimulation for refractory epilepsy: a Belgian multi-center study. Eur J Paediatr Neurol 11(5):261–269
9. Devinsky O, Pacia S (1993) Epilepsy surgery. Neurol Clin 11(4):951–971
10. Duchowny MS, Harvey AS, Sperling MR, Williamson PD (2007) Indications and criteria for surgical intervention. In: Engel J Jr, Pedley TA (eds) Epilepsy: a comprehensive textbook, 2nd edn. Lippincott Williams & Wilkins, Philadelphia, pp 1751–1759
11. Elliott RE, Rodgers SD, Bassani L, Morsi A, Geller EB, Carlson C, Devinsky O, Doyle WK (2011) Vagus nerve stimulation for children with treatment-resistant epilepsy: a consecutive series of 141 cases. J Neurosurg Pediatr 7(5):491–500
12. Elliott RE, Morsi A, Kalhorn SP, Marcus J, Sellin J, Kang M, Silverberg A, Rivera E, Geller E, Carlson C, Devinsky O, Doyle WK (2011) Vagus nerve stimulation in 436 consecutive patients with treatment-resistant epilepsy: long-term outcomes and predictors of response. Epilepsy Behav 20(1):57–63
13. Elliott RE, Morsi A, Tanweer O, Grobelny B, Geller E, Carlson C, Devinsky O, Doyle WK (2011) Efficacy of vagus nerve stimulation over time: review of 65 consecutive patients with treatment-resistant epilepsy treated with VNS >10 years. Epilepsy Behav 20:478–483
14. Engel J Jr (1996) Surgery for seizures. N Engl J Med 334(10):647–652
15. Ergene E, Behr PK, Shih JJ (2001) Quality-of-life assessment in patients treated with vagus nerve stimulation. Epilepsy Behav 2:284–287
16. Galbarriatu L, Pomposo I, Aurrecoechea J, Marinas A, Agúndez M, Gómez JC, Acera MA, Martínez MJ, Valle E, Maestro I, Mateos B, Cabrera A, Fernández J, Iturri F, Garamendi I (2015) Vagus nerve stimulation therapy for treatment-resistant epilepsy: a 15-year experience at a single institution. Clin Neurol Neurosurg 137:89–93
17. Garcia-March G, Sanchez-Ledesma MJ, Broseta J (2008) Vagus nerve stimulation for the treatment of refractory epilepsy. State of the art. Neurocirugia (Astur) 19(5):416–426
18. Handforth A, DeGiorgio CM, Schachter SC, Uthman BM, Naritoku DK, Tecoma ES, Henry TR, Collins SD, Vaughn BV, Gilmartin RC, Labar DR, Morris GL, Salinsky MC, Osorio I, Ristanovic RK, Labiner DM, Jones JC, Murphy JV, Ney GC, Wheless JW (1998) Vagus nerve stimulation therapy for partial-onset seizures: a randomized active-control trial. Neurology 51:48–55
19. Kostov H, Larsson PG, Roste GK (2007) Is vagus nerve stimulation a treatment option for patients with drug-resistant idiopathic generalized epilepsy? Acta Neurol Scand Suppl 187:55–58
20. Kwan P, Brodie MJ (2000) Early identification of refractory epilepsy. N Engl J Med 342(5):314–319
21. Kwan P, Arzimanoglou A, Berg AT, Brodie MJ, Hauser WA, Mathern G, Moshé SL, Perucca E, Wiebe S, French J (2010) Definition of drug resistant epilepsy: consensus proposal by the ad hoc Task Force of the ILAE Commission on Therapeutic Strategies. Epilepsia 51(6):1069–1077
22. Lancman G, Virk M, Shao H, Mazumdar M, Greenfield JP, Weinstein S, Schwartz TH (2013) Vagus nerve stimulation vs. corpus callosotomy in the treatment of Lennox–Gastaut syndrome: a meta-analysis. Seizure 22(1):3–8
23. Morris GL, Mueller WM (1999) Long-term treatment with vagus nerve stimulation in patients with refractory epilepsy. The Vagus Nerve Stimulation Study Group E0I-E05. Neurology 53: 1731–1735
24. Ng M, Devinsky O (2004) Vagus nerve stimulation for refractory idiopathic generalised epilepsy. Seizure 13:176–178
25. Parain D, Penniello MJ, Berquen P, Delangre T, Billard C, Murphy JV (2001) Vagal nerve stimulation in tuberous sclerosis complex patients. Pediatr Neurol 25(3):213–216
26. Porter R (1991) Epilepsy: prevalence, classification, diagnosis, and prognosis. In: Apuzzo MLJ (ed) Neurosurgical aspects of epilepsy. AANS, Park Ridge, pp 17–26
27. Tellez-Zenteno JF, Dhar R, Wiebe S (2005) Long-term seizure outcomes following epilepsy surgery: a systematic review and meta-analysis. Brain 128:1188–1198
28. Yamamoto T (2015) Vagus nerve stimulation therapy: indications, programing, and outcomes. Neurol Med Chir (Tokyo) 55(5): 407–415

Part II

Reconstructive Neurosurgery: Spine

Restoring Neurological Physiology: The Innovative Role of High-Energy MR-Guided Focused Ultrasound (HIMRgFUS). Preliminary Data from a New Method of Lesioning Surgery

Antonella Giugno, Rosario Maugeri, Francesca Graziano, Cesare Gagliardo, Angelo Franzini, Carlo Catalano, Massimo Midiri, and Domenico Gerardo Iacopino

Abstract

Background

Tremor is a disabling condition, common to several neurodegenerative diseases. Lesioning procedures and deep brain stimulation, respectively, of the ventralis intermedius nucleus for intentional tremor, and of the subthalamic nucleus for parkinsonian resting tremor, have been introduced in clinical practice for patients refractory to medical treatment. The combination of high-energy focused ultrasound (HIFUS) with sophisticated magnetic resonance (MR) instrumentation, together with accurate knowledge of the stereotactic brain coordinates, represents a revolution in neuromodulation.

Methods

At the Neurosurgical Clinic and the Radiology Department of the University of Palermo,, two patients affected by severe and refractory forms of intentional tremor were treated by MRI-guided FUS (MRgFUS) with a unique 1.5 T MR scanner prototype that uses FUS. This apparatus is the only one of its type in the world."

Findings

This is the first Italian experience, and the second in Europe, of treatment with MRI-gFUS for intentional tremor. But this is the very first experience in which a 1.5 T MRI apparatus was used. In both patients, the treatment completely abolished the tremor on the treated side, with results being excellent and stable after 7 and 5 months, respectively; no side effects were encountered.

Conclusion

MRgFUS, recently introduced in clinical practice, and widely used at several clinical centers, has been shown to be a valid therapeutic alternative in the treatment of tremor in several neurodegenerative diseases. It is virtually safe, noninvasive, and very efficacious. We report this technique in which a 1.5 T MR scanner was used. Further investigations with long-term follow up and larger clinical series are needed.

Keywords Parkinson's disease • Essential tremor • Functional neurosurgery • Lesioning surgery • MRI-guided focused ultrasound (MRgFUS)

A. Giugno, MD • R. Maugeri, MD PhD (✉) • F. Graziano, MD
D.G. Iacopino, MD, PhD
Department of Experimental Biomedicine and Clinical Neurosciences, School of Medicine, Neurosurgical Clinic, University of Palermo, Via del Vespro 129, Palermo 90100, Italy
e-mail: rosario.maugeri1977@gmail.com

C. Gagliardo, MD • M. Midiri, MD, PhD
Section of Radiological Sciences, DIBIMED, University of Palermo, Palermo 90127, Italy

A. Franzini, MD, PhD
Unit of Functional and Stereotactic Surgery, Fondazione IRCCS Istituto Neurologico "Carlo Besta", Milan, Italy

C. Catalano, MD, PhD
Department of Radiological, Oncological and Anatomopathological Sciences, Radiology, Sapienza, University of Rome, Rome, Italy

Introduction

"The unhappy sufferer has considered it an evil, from the domination of which he had no prospect of escape". This is how James Parkinson, in his monograph, *Essay on the Shaking Palsy* (1817) [14], depicted what came to be known as Parkinson's disease (PD). He showed deep participative feelings in his reports, joining his clinical observations to the personal experiences of affected patients, together with all their sufferings. This is the first description fully reporting all the characteristic features of PD. Since 1817, neurophysiologists have tried to explain how a dysfunction could generate pathology. The ideal goal has been to modify the neural circuits to restore normal function, and this aim guides functional neurosurgery. Tremor is the most common movement disorder-

M. Visocchi et al. (eds.), *Trends in Reconstructive Neurosurgery*, Acta Neurochirurgica Supplement 124,
DOI 10.1007/978-3-319-39546-3_9, © Springer International Publishing Switzerland 2017

related symptom. Resting tremor with an oscillation frequency of 3–5 Hz usually indicates the onset of hyperkinetic forms of PD, while in essential tremor (ET), patients suffer postural or kinetic tremor that shows an oscillation frequency between 4 and 12 Hz. Since the 1960s, lesioning procedures and, later, deep brain stimulation (DBS), respectively, of the ventralis intermedius nucleus (VIM) for intentional tremor, and of the subthalamic nucleus (STN) for parkinsonian resting tremor, have been introduced in clinical practice for patients refractory to medical treatment. Nevertheless, there is a need for a mathematically safe and precise technique, and not all patients can benefit from or are suitable for invasive surgical procedures. In this view, ultrasound delivered under magnetic resonance imaging (MRI) guidance, has been used. The combination of high-energy focused ultrasound (HIFU) under sophisticated MRI instrumentation, together with accurate knowledge of the stereotactic brain coordinates, represents a revolution in neuromodulation. Here we report our preliminary experience with two patients with intentional tremor successfully treated by this method [7].

Materials and Methods

At the Neurosurgical Clinic and the Radiology Department of the University of Palermo, two patients affected by severe and refractory forms of intentional tremor were treated with MRI-guided high-intensity focused ultrasound (MRgFUS). Actually, the US generator is worldwide integrated in a 3 T high-field MR while at the University of Palermo the Authors have used aUS generator integrated in the unique 1.5 T MR prototype scanner. At the University of Palermo there is a unique 1.5 T MR prototype scanner that uses FUS. The US generator consists of 1024 ultrasonic transducers and is connected to a system for circulating and degassing cool water. The system is controlled by a workstation to modulate intensity, energy, coordinates, and sonications. The selection of patients is critical: prior to the treatment, the diagnosis of intentional tremor was confirmed and other neurological illness was excluded. A significant preexisting tremor was documented on the Fahn-Tolosa-Marin Tremor Rating Scale (FTMTRS). Prior to the procedure, each patient underwent brain computed tomography (CT) and MRI with contrast medium. The first procedure was necessary to obtain a skull-correction algorithm for sonication through the skull. The second procedure, obtained with 1.5-mm-thick slices, allowed precise individuation of the basal ganglia. The coordinates of the VIM were determined as being three-quarters of the length of the anterior commissure-posterior commissure (AC-PC) line and 14–15 mm lateral of the midline. On the day of the treatment a stereotactic frame Radionics (Burlington, MA 01803 USA) was positioned, as low as possible, after meticulous shaving of the scalp, in order to avoid air bubble formation during the sonication procedure. The patient was pharmacologically relaxed and vital parameters were continuously monitored by the anesthesiologist. Then, the frame was connected to the MRI table together with the ultrasonic transducers. An initial pretreatment scan was acquired to adjust the coordinates of the transducer to the real position of the VIM. After the planning phase, a series of sonications of increasing energy were performed. Each sonication took about 10–15 s and was followed by a "rest phase" to avoid overheating of the system. At 50–53 °C transitory effects (mock lesions) were visible. In the meantime, continuous neurological evaluations were performed to observe the first neurological response and to note any possible deficit. At 57–58 °C, tissue coagulation and denaturing of proteins occurred within seconds and the lesion was visible, appearing as a small hyperintense point on the VIM. In the meantime, it was possible to slightly shift the transducer 1 mm in the desired direction, to precisely shape the lesion itself and to ameliorate the clinical effects. In the following 24 h the lesion grew in size because of edema, and a 48-h control MRI was used to evaluate the final results.

Results

In both patients, the treatment completely abolished the tremor on the treated side, and no side effects were encountered. The results were actually excellent and stable after 7 and 5 months, respectively. The formation of the lesions is shown in Figs. 1 and 2, and the sonication parameters are reported in Table 1.

Discussion

This is the first Italian experience [7], and the second in Europe, of treatment with MRI-gFUS for intentional tremor. But this is the very first experience in which a 1.5 T MRI apparatus was used. Tremor is a disabling condition, common to several neurodegenerative diseases. Electrophysiological studies support the peculiar role of the VIM in the pathogenesis of intractable tremor [5]. Pharmacological treatment is often inadequate: about 30–50 % of patients affected do not respond [1]. In the 1990s, DBS almost completely replaced lesioning procedures in the surgery of movement disorders, making lesioning procedures almost obsolete in the past two decades. Electrical stimulation is, in fact, adjustable and reversible. Of note, Dwarakanath et al [3]. analyzed a series of patients affected by tremor of different origins. They observed excellent clinical outcomes after lesioning, which could be considered as a strong alternative to DBS

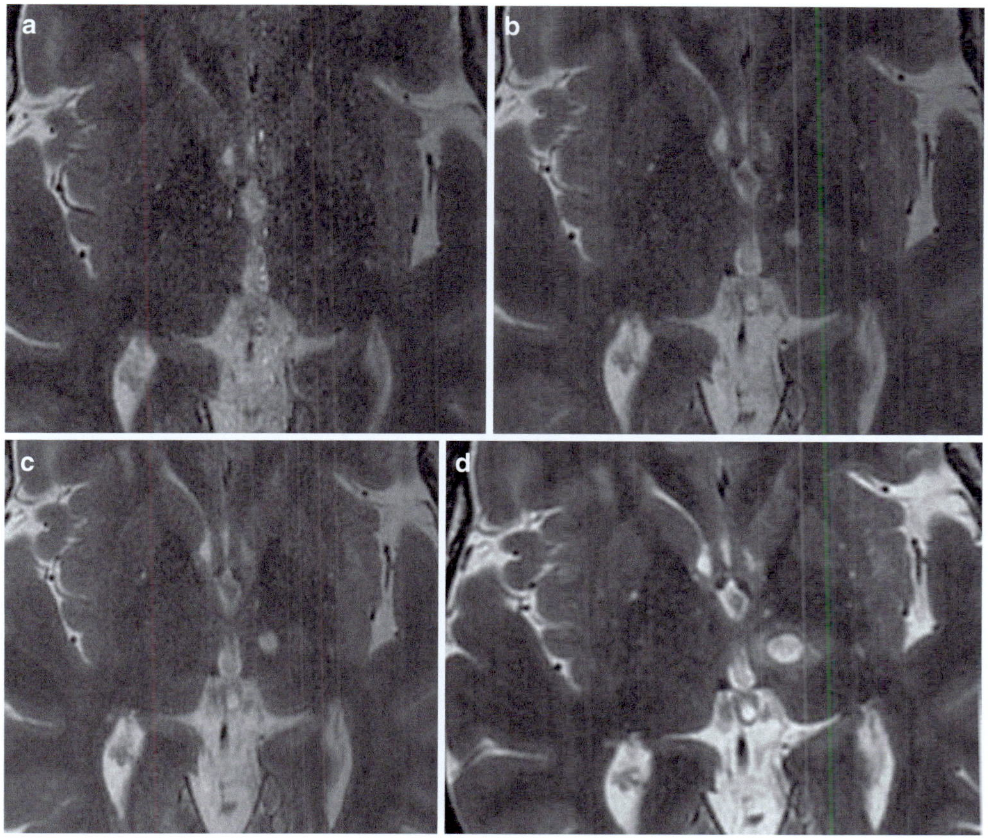

Fig. 1 Patient 1 (**a**) scan immediately before sonication; (**b**) sonication N° 21; (**c**) sonication N° 26; (**d**) 48-h follow-upscan

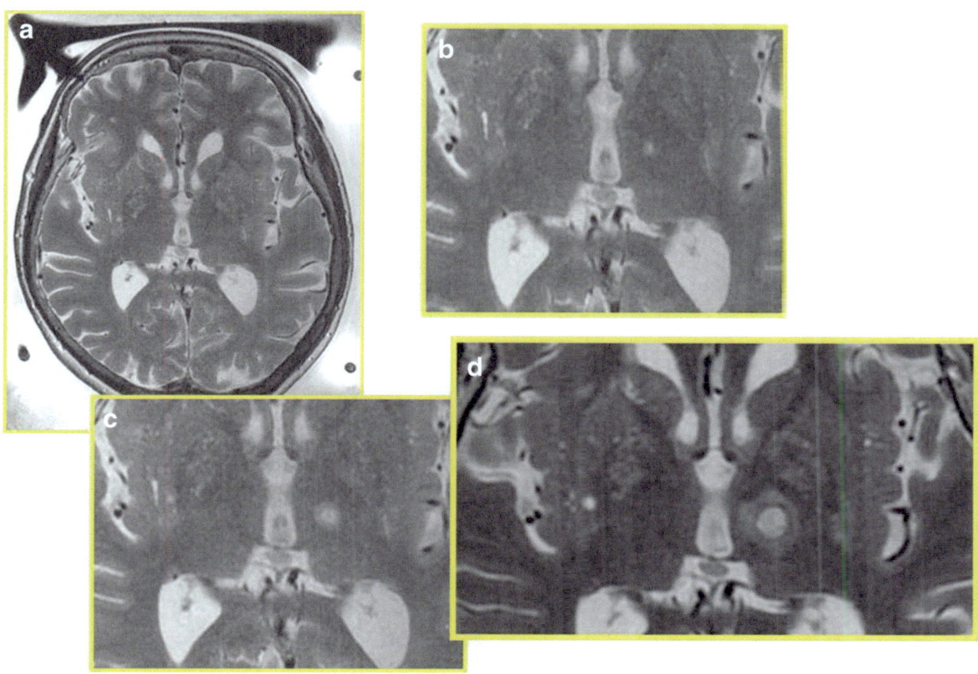

Fig. 2 Patient 2 (**a**) scan immediately before sonication; (**b**) sonication N°23; (**c**) sonication N°28; (**d**) 48-h follow-up scan

Table 1 Details of first and second treatments

	Left thalamotomy (VIM) Patient 1	Left thalamotomy (VIM) Patient 2
No. of sonications (s)	26 (10–13 s)	28 (10–17 s)
Amplitude (W) range	200–1100	150–1100
Amplitude (W) range measured	192–1028	151–1042
Energy (J) range	2000–14,300	1500–18,698
Energy (J) range measured	1903–13,218	1496–17,094
Temperature (°C) range	43–62	42–60
Duration	2 h 34 min	2 h 45 min

VIM ventralis intermedius nucleus

in prevalently unilateral tremor and in those cases in which DBS is not feasible. Nowadays DBS is the gold standard surgical procedure for treating both ET and PD-related tremor, because it is safer and allows physicians to modulate energy and related effects over a long period. The indications for lesioning and stimulation are similar. In particular, the selection of candidates enrolls patients who are refractory to the best medical treatment, with disabling tremor that, in PD, must be the predominant neurological manifestation, with better surgical outcomes if the tremor is unilateral or asymmetrical. After the correct selection of the patient, surgical treatment, either lesioning or electric stimulation, shows excellent rates of therapeutic success. The most important questions are which is the best and safest modality to recognize the target, how to reach it with the highest precision, and to how to achieve the best feedback, maybe in real time, of the creation and development of the lesion. So the ideal treatment should be mathematically precise and safe; however, neither lesioning nor DBS is mathematically precise and safe, even with sophisticated modern technology. Focused ultrasound, without the need for craniotomy, but with extremely high stereotactic precision, gives the best answers to our question. The clinical utility of FUS has been the subject of investigation since 1938, when Raimar Pohlman showed specific "therapeutic effects" of acoustic waves on human tissues [13]. The physical principle of this technique is based on thermal energy, together with the injurious mechanical effects of cavitation due to the sonication itself. The thermal energy is derived from the absorption of US passing through biological tissues; the higher the delivered energy, the higher is the local temperature. In the 1940s, the ablative effects of FUS on the animal brain were demonstrated by Lynn and Putnam [11]. Other preclinical studies with HIFUS were performed by Briquard and Langevin in 1972 [1]. During the successive decades, the advent of MR, together with galloping technological progress, led to the evolution of MRgFUS, used with ablative purposes. The term "focused" US is used because of the submillimetrical precision in delivering acoustic energy to a specific anatomical site. The main obstacle to the use of US in treating tremors is represented by the cranial bone, which can interfere with the US field and absorb the energy. The HIFU technique has gained a role of growing importance in the treatment ofseveral medical conditions [6], as well as in neurosurgery. Actually, tremors related to ET or PD and chronic thalamicpain are the only clinically accepted applications for HIFU treatment. With HIFU, the energy used is able to generate heating-related lesions in biological tissues; furthermore, this heating effect is accompanied by nonthermal effects such as cavitation and acoustic streaming. The phenomenon of cavitation consists of the formation of bubbles inside fluid tissues due to the mechanical effects of the US. The oscillations of the bubbles themselves cause cellular membrane destruction. FUS has several advantages compared with classical surgery and even radiosurgery; it does not require scalp incisions and burr holes, and there is no blood loss and no infective risk. Also, this is a high-precision treatment, with the boundary zone between the necrotic treated tissue and the remaining tissue measuring less than 0.1 mm, being more accurate than stereotactic radiotherapy. Furthermore, unlike the use of the Gammaknife or cyberknife, FUS is easily repeatable when necessary, since there are no cumulative toxic effects. The FUS-induced necrosis is immediately visible and is documented by intraoperative MRI sequences that visualize the lesion as a hyperintensity on T2-weighted images. Finally, symptom relief is almost immediate. The first experimental observations of the effects of transcranial high-intensity FUS were conducted at the University of Zurich, Switzerland, with cadaveric head preparations. Since that time, only a few centers have been able to perform such procedures, and there is only a very small number of clinical series, with limited periods of clinical follow-up. At present, the best results with such procedures are achieved in patients with ET. A detailed review of the available literature showed that, of the few clinical series, only four of them had already been concluded. The most important clinical study has been reported by Elias and colleagues [4], with 15 cases of unilateral FUS thalamotomy of the VIM nucleus in patients with refractory ET. After a 12-month follow-up, the patients showed a substantial reduction, of 75 %, in tremor in the arm opposite to the treated thalamus, documented by the FTMTRS. Also, a reduction of of about 85 % in disability was observed, assessed according to the quality of life subsection of the FTMTRS, with a strong improvement in the quality of life scores, which changed from from 11 % to 37 %, assessed by the quality of life in the ET Questionnaire. Four patients showed paresthesias as adverse effect of the treatment. Chang and colleagues [2], in a 6-month follow-up of 11 medication-refractory ET patients who had completed MRgFUS treatment, reported immediate and sustained neurological relief with a consistent reduction of tremor in 8 of the 11 patients; in the other 3 patients, the optimal temperature was not reached during the sonication process. Lipsman et al. [10] treated four patients with ET with FUS, observing, at 3 months, a stable reduction of tremor in 81.3 % of the patients. Clinical results obtained with 3-T MRgFUS are promising, despite the short follow-up interval thus far. In view of these data, Jung and colleagues [9] reported the radiological outcomes of 11 ET patients treated by MRgFUS; 8 of them had completed the whole treatment: after a monolateral VIM nucleus MRgFUS thalamotomy, at a 3-month follow-up interval, an increasing lesion volume was shown, due to perilesional edema, with a subsequent reduction of 92 % in 3 months.

Preliminary data on the first eight patients treated for PD-related tremor were presented at the Third International Symposium of the Focused Ultrasound Surgery Foundation [8]. These patients suffered from Levo-Dihydroxyphenylalanine (L-DOPA)-refractory tremor with a predominant side, and in all patients the thalamic fasciculus was targeted. Three months after treatment, a mean improvement of 57.1 %, on the Unified PD Rating Scale, was reported. No adverse symptoms occurred.

MRgFUS seems to be a very interesting technique that is suitable for treating ET and PD-related tremors. We have reported a unique experience by the use of HIFU apparatus coupled with a 1.5 T MR machine. This kind of machines are widely used in several clinical centers in Italy, while the 3 T MR are not allowed for clinical use but only for research purposes. Compared with 3-T equipment, 1.5-T equipment is safer and cheaper, and the quality of images is almost the same, with satisfactory visualization of the basal ganglia. Neurobehavioral studies have shown adverse responses in animals under exposures higher than 1.5 T [15]. On the other hand, magnetic fields lower than 1 T are not related to genotoxic risks [12]. Possible cancerogenic effects caused by higher magnetic fields are being debated; a recent study [16] reported significant time- and dose-dependent induction of micronuclei in mice exposed to 2, 3, or 4.7 T.

Conclusion

MRgFUS, recently introduced in clinical practice, has been shown to be a valid therapeutic alternative in the treatment of tremor; it is safe, noninvasive, and very efficacious. In fact the first clinical reports show promising positive results, with immediate and almost complete benefits that are stable in the medium term. The efficacy seems to be about the same as that of DBS, but without its risks. Our preliminary experience demonstrates that a 1.5-T machine may have the same clinical reliability as a 3-T machine. Further investigations are needed, and this procedure could be, potentially, the treatment of choice in the future for several neurodegenerative diseases.

Conflict of Interest Statement The Authors report no conflicts of interest. No funds were received in support of this work.

References

1. Briquard P. Paul Langevin. Ultrasonics. 1972 Sep;10(5):213-4. PubMed PMID:4577356.
2. Chang WS, Hung HH et al (2015) Unilateral magnetic resonance guided focused ultrasound thalamotomy for essential tremor practices and clinicoradiological outcomes. J Neurol Neurosurg Psychiatry 86(3):257–264
3. Dwarakanath S, Zafar A, Yadav R, Arivazhagan A, Netravathi M, Sampath S, Pal PK. Does lesioning surgery have a role in the management of multietiological tremor in the era of Deep Brain Stimulation? Clin Neurol Neurosurg. 2014 Oct;125:131-6. doi: 10.1016/j.clineuro.2014.07.016. Epub 2014 Jul 29. PubMed PMID: 25128653.
4. Elias Jeffrey W, Diane H et al (2013) A pilot study of focused ultrasound thalamotomy for essential tremor. N Engl J Med 369:640–648
5. Eskandar EN, Shinobu L, Rees Cosgrove G Surgery for intractable tremor. Funct Stereotact Neurosurg http://neurosurgery.mgh.harvard.edu/Functional/IntractableTremor.htm
6. Jolesz FA, Mc Dannold NJ (2014) Magnetic resonance guided ultrasound: a new technology for clinical neurosciences. Neurol Clin 32(1):253–269
7. Iacopino DG, Giugno A, Maugeri R, Gagliardo C, Franzini A, Catalano C, Midiri M. Is there still a role for lesioning in functional neurosurgery: the Italian experience of delivering focused ultrasound high energy through a 1,5 tesla MR apparatus. J Neurosurg Sci. 2015 Jun 9. [Epub ahead of print] PubMed PMID:26056169.
8. Jeanmonod D, Moser D, Magara A, Kowalski M, Buhler R, Pourtehrani P, Corey T, Vogel J (2012) Study on incisionless transcranial magnetic resonance-guided focused ultrasound treatment of Parkinson's disease: safety, accuracy and initial clinical outcomes. Focused Ultrasound 2102, Third symposium
9. Jung HH, Chang WS et al (2015) Different magnetic resonance imaging patterns after transcranial magnetic resonance-guided focused ultrasound of the ventral intermediate nucleus of the thalamus and anterior limb of the internal capsule in patients with essential tremor or obsessive-compulsive disorder. J Neurosurg 122 162–168
10. Lipsman N, Schwartz ML, Huang Y, Lee L, Sankar T, Chapman M, Hynynen K, Lozano AM (2013) MR-guided focused ultrasound thalamotomy for essential tremor: a proof-of-concept study. Lancet Neurol 12(5):462–468
11. Lynn JG, Putnam TJ (1944) Histology of cerebral lesion produced by focused ultrasound. J Pathol 20(3):637–649
12. Matthes R, McKinlay AF, Bernhardt JH, Vecchia P, Veyret B (eds) (2003) Biological effects and health consequences (0–100 kHz) exposure to static and low frequency electromagnetic fields ICNIRP. International Commission on Non-ionizing Radiation Protection (ICNIRP), Munich
13. R.Pohlman, R.Richter, E.Parow, Überdie Ausbreitung und Absorptiondes Ultraschalls immenschlichen Gewebe und seinethe rapeutische Wirkungan Ischiasund Plexusneuralgie, Dtsch Med Wochenschr 65(1939):251–256.
14. Parkinson J. An essay on the shaking palsy. 1817. J Neuropsychiatry Clin Neurosci. 2002 Spring;14(2):223-36; discussion 222. PubMed PMID: 11983801.
15. Saunders R (2005) Static magnetic fields: animal studies. Prog Biophys Mol Biol 87:225–239
16. Suzuki Y, Ikehata M, Nakamura K, Nishioka M Asanuma K, Koana T, Shimizu H (2001) Induction of micronuclei in mice exposed to static magnetic fields. Mutagenesis 16:499–501

Extraforaminal Disk Herniation Treatment with Surgical Exploration by Unilateral Intertransversarii Microsurgical Approach

Aldo Spallone, Massimiliano Visocchi, A. Nardella, and L. Lavorato

Abstract

Background

In this study we evaluated the role of a unilateral intertransversarii microsurgical approach for the treatment of extraforaminal lumbar disk herniations (ELDHs), with short-, medium-, and long-term follow-up.

Methods

We retrospectively evaluated 96 patients who had undergone surgery for ELDH between 2001 and 2012 at our Institution. All the patients had been examined before the intervention, immediately after, and at 6 weeks, 6 months, and 18 months after the intervention.

All the patients underwent surgical exploration with a unilateral intertransversarii microsurgical approach. After a midline incision was made, the paraspinal muscles were retracted laterally up to the transverse process, in order to visualize the intertransversarii ligament. Removal of this ligament allows microsurgical exposure of the extraforaminal pathology. If necessary, interlaminar microsurgical exploration is performed in the same session.

Results

At 6-week postoperative follow-up we noted a significant decrease of pain both in patients who suffered from leg pain and in patients who reported back pain. Also, sensory and motor deficits had improved considerably. At 6-month postoperative follow-up we observed a further improvement in the clinical conditions of almost all patients. At the 18-month postoperative follow-up we observed a very low incidence of relapse of neurological symptoms.

Conclusion

Our technique can be reasonably proposed because of its low morbidity, fast recovery, and short hospital stay.

Keywords Extraforaminal • Disk herniation • Neurosurgical procedure • Microsurgery • Paraspinal muscle

A. Spallone (✉)
Section of Neurosurgery, Department of Clinical Neurosciences, N.C.L. Neurological Center of Latium, Via Patrica 15, Rome 00178, Italy

Department of Biomedicine, University of Rome "Tor Vergata", Rome, Italy
e-mail: segreteria1@nclroma.it

M. Visocchi, MD
Insitute of Neurosurgery, Medical School, Catholic University of Rome, Rome, Italy

A. Nardella
Section of Neurology, Department of Clinical Neurosciences, N.C.L. Neurological Center of Latium, Via Patrica 15, Rome 00178, Italy

Section of Neurophysiology, Department of Neurology, IRCCS Neuromed, Pozzilli, IS, Italy

L. Lavorato
Section of Neurosurgery, Department of Clinical Neurosciences N.C.L. Neurological Center of Latium, Via Patrica 15, Rome 00178, Italy

Introduction

Extraforaminal lumbar disk herniations (ELDHs) are relatively uncommon, representing approximately 5 % of all prolapsed lumbar disks requiring surgery [1, 9, 12, 19, 23].

ELDHs have been traditionally considered as a difficult-to-solve problem from both the diagnostic and the surgical points of view. Historically, these lesions were very difficult, if not impossible, to demonstrate on myelographic examination and they became identifiable only after the introduction of modern imaging technology [4, 9]. When the classical interlaminar approach is used, the surgical removal of an ELDH is very problematic, if not impossible, for two main reasons: (1) significant removal of the articular facet is required in order to expose the far laterally located disk fragment, with the consequent risk of postoperative instability [25]; and (2) there is inadequate visualization of

DOI 10.1007/978-3-319-39546-3_10,

the fragments in most instances, with a consequent high possibility of leaving disk material in situ, leading to non-negligible postoperative problems [20].

More recently some authors have proposed a direct surgical approach to the outer portion of the root foramina, where the nerve roots are disturbed by the prolapsed disk, with the aim of directly visualizing the offending lesion without putting joint stability at risk [10, 13, 21]. A key point of this approach is the exposure of the intertransversarii ligament, which covers the nerve root exit space and the outer portion of the foramen where the ELDH is located. This exposure can be achieved either by extending the classical midline unilateral fascial muscular-subperiosteal dissection further laterally, as originally described by Abdullah et al. [1], or by using a paramedian transmuscular approach targeted to the lesion location. This latter approach has been superbly described by Tessitore and de Tribolet [27].

From 1989 we had been treating ELDH with a direct intertransversarii transmuscular approach, as originally described by Recoules-Arche [24], and shortly afterwards revised by the de Tribolet group [27].

A few years later we changed our policy and started using a midline unilateral approach to the targeted intertransversarii space for microsurgical ELDH removal, rather similar to the approach described by Jane et al. [7]; we operated on over 100 cases with this technique during a 20-year time span. Here we describe the clinical-diagnostic implications and short- and medium-term results of the microsurgical technique used for treating ELDH at our Institution. Long term-evaluation of the patients is still ongoing and will be the subject of further studies.

Materials and Methods

Patients

Ninety-six patients (53 men and 43 women) who had undergone surgery for ELDH between 2001 and 2012 at our Institution were retrospectively evaluated by reviewing their medical records and outpatient charts. Their mean age was 57.1 ± 12.6 years (minimum 36, maximum 80). Data considered included signs and symptoms before and after surgery, age, level of the ELDH, operative findings, and complications associated with the surgery. All the patients had been examined before and after the intervention (Figs. 1, 2, and 3), with postoperative examinations 6 weeks and 6 and 18 months after the intervention. Physical examination focused on radicular symptoms, the Lasègue sign, reverse Lasègue sign, crossed Lasègue sign, and Wasserman sign. Pain was evaluated at 6 weeks and 6 months after surgery using a visual analog scale (VAS). VAS ratings of 0–1 were considered as no pain; 2–4, mild pain; 5–7, moderate pain; and 8–10, severe pain [8]. At the 18-month follow-up we considered pain as indicating recurrent pathology unless specifically described as "persistent" by the patient.

Surgical Technique

After a midline 6- to 8-cm-long skin incision is made, centered on the affected disk space, a curvilinear fascial incision is made, and the paraspinal muscles are retracted laterally up to the transverse processes. For this purpose, a gentle dissection laterally to the articular facets is achieved, using only bipolar coagulation in order to minimize bleeding. As a rule this maneuver requires tilting of the operating table towards the surgeon, in order to optimize the illumination and visualization of the anatomical structures to be dissected and exposed. The paraspinal muscles are retracted by using a Finochietto retractor with one hook-like blade held on the interspinous process at the midline and an extra-long two-valve (3-cm-large) retractor spreading the muscles away from the intertransversarii space. Once the superior and inferior transverse processes are exposed and the intertransversarii ligament is visualized, the operative microscope is brought into the surgical field. A "limited" drilling is performed between the superior transverse process and the isthmus. The exposure and enlargement of the outer portion of the neural foramen is allowed by the detachment of the intertransversarii ligament from the drilled transverse process in a medial-to-lateral direction, done by using a Kerrison punch. At this step, the operating table is tilted away from the surgeon and the angle of the operating microscope is changed accordingly, in order to obtain the best view of the nerve root and of the underlying ELDH. The ruptured disk is usually seen superior and medial to the nerve root. Microdissection is carefully performed, using low-power bipolar coagulation, fine suckers, and microdissectors, in order to minimize both surgical trauma and bleeding. Depending on the size and the actual location of the ELDH, the nerve root is retracted by using an angled microretractor and the fragment is removed (Fig. 3). The disk's space is entered and emptied as much as necessary. At the end of the procedure, the intervertebral foramen is explored with a long 90° instrument, in order to check for adequate decompression of the nerve root. Careful hemostasis is achieved and the wound is carefully irrigated and closed in layers. In cases where preoperative diagnostic tests could indicate interlaminar exploration, and/or if the foraminal exploration could suggest that decompression was not adequate, a standard interlaminar exposure is made and the affected nerve root is exposed at its exit from the dural sac. In such cases, drilling of the medial part of the facet is minimized as much as possible in order to prevent the risk of postoperative instability. Approximately one-sixth of the present patient cohort required interlaminar exploration also. In those patients

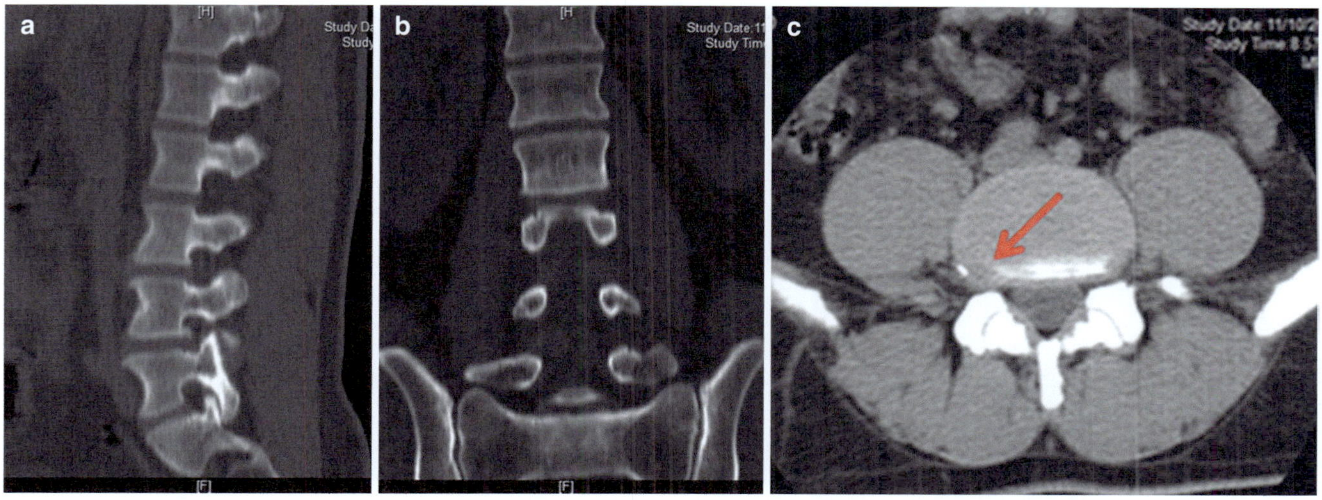

Fig. 1 Case 1. A 58-year-old man with lower back pain radiating to the right lower limb. Computed tomography (CT) study documented a herniated disk in the right lateral L4-L5 intra-and extraforaminal space. (Red arrow) Sagittal (**a**), coronal (**b**), and axial (**c**) planes show the disk pathology

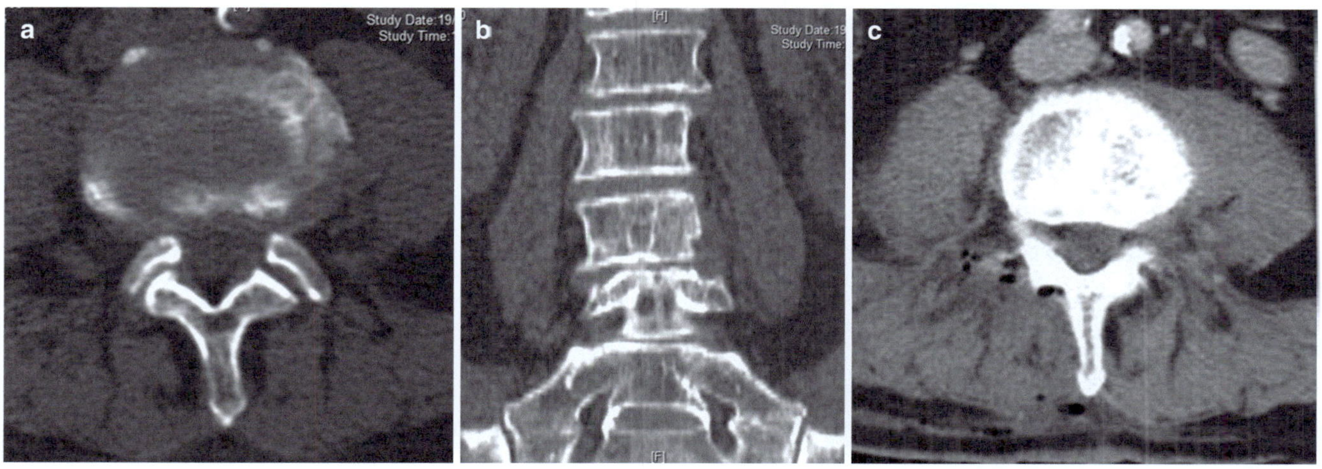

Fig. 2 Case 2. A 63-year-old woman with lower back pain radiating to the right leg over the superior-lateral surface. Axial CT study (**a**) documented the presence of a herniated disc in the left lateral L4-L5 extraforaminal space, also evident in the coronal reconstruction plane (**b**). CT after removal of the herniated disc (**c**)

in whom disk rupture was also intraspinal, laminctomy was performed in the same session and fragments were removed from both sides of the neural foramen. For L5-S1 herniations we used basically the same technique, which, however, as a rule requires a certain amount of drilling of the lateral portion of the sacrum, which varies from case to case.

The patients were allowed to stand as soon as they were discharged from the recovery room, on average 6–8 h after the end of the surgical procedure. No immobilization vest was required and the patients were asked to limit only some of their usual activities (for example, lifting weights) for 2 months only and then resume their usual activities, including sports, with the obvious exception of sporting activities requiring excessive stress load to the spine.

Results

Preoperative Findings

The ELDH was located at the level of L1-L2 in 2 patients (2.08 %), L2-L3 in 6 (6.25 %), L3-L4 in 22 (22.92 %), L4-L5 in 58 (60.42 %) L5-L6 in 3 (3.13 %), and at the level of L5-S1 in 5 patients (5.20 %). On admission 90 (93.75 %) patients suffered from leg pain, 68 had sciatic pain and 22 had femoral pain. Back pain was present in 92 patients (95.85 %), 41 of whom reported nocturnal exacerbation. VAS results are summarized in Table 1. The Lasègue sign was positive in 68 patients (70.83 %), reverse Lasègue in 22

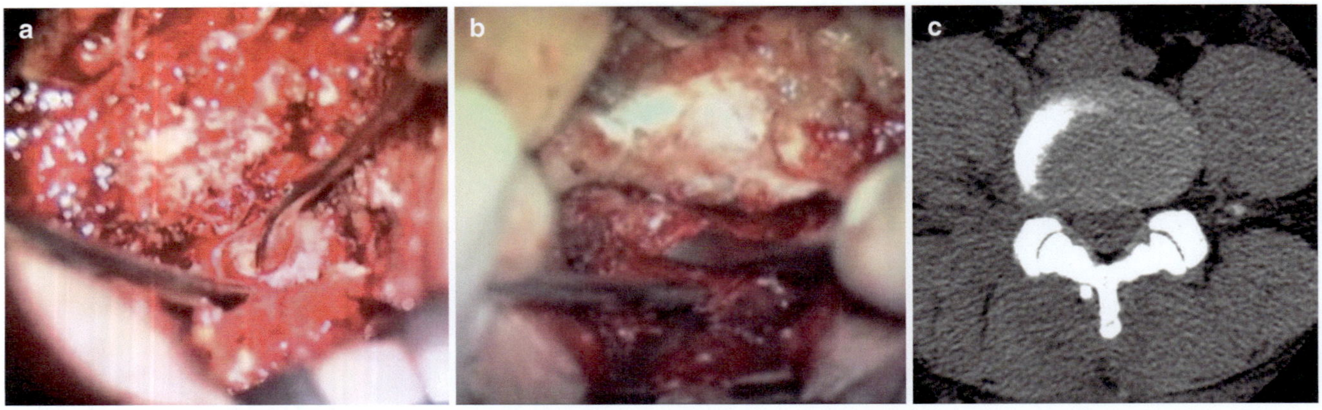

Fig. 3 Intraoperative view: *upper*, articular facet; *lower*, the lateral side with muscle retracted. *Right*, the lower transverse process; *left*, the upper transverse process. (**a**) The sucker points to the decompressed L4 left nerve root, and the dissector touches the prolapsed disk. (**b**) Following disk removal, the sucker gently retracts the decompressed nerve root. The dissector indicates the disk space. The articular facet is better shown in this picture. (**c**) CT axial plane after removal of the prolapsed disk clearly shows the foraminal decompression

Table 1 Pain occurrence in the present cohort

Localization of pain	Timing of follow-up	No pain	Mild pain	Moderate pain	Severe pain	Average VAS
Pain radiating along the leg surface	Preoperative	6	7	8	75	79.35
	Postoperative 6 weeks	85	6	3	2	8.26
	Postoperative 6 months	88	5	2	1	5.27
	Postoperative 18 months	90	5	1	1	4.14
Back pain	Preoperative	4	6	11	75	81.48
	Postoperative 6 weeks	82	6	5	3	10.06
	Postoperative 6 months	86	5	3	2	6.90
	Postoperative 18 months	89	5	2	1	3.37

Average visual analog scale (VAS) values (with levels expressed according to patient numbers), differentiated for localization of pain and timing of follow-up

(22.92 %), and crossed Lasègue in 16 (16.67 %). The Wasserman sign was found in 21 patients (21.87 %). Fifty-three patients (55.21 %) reported congruous monoradicular sensory deficits and 45 (46.86 %), motor deficits (see Tables 2 and 3). No patient had vegetative dysfunction.

Surgical Approaches

All the patients underwent surgical exploration by a unilateral intertransversarii approach. A combined intertransverse-interlaminar approach was used in 15 patients (15.63 %). An interspinous device was implanted in 12 of these patients (12.5 %). In 3 patients implantations were performed at the L5-S1 level, where a significant amount of sacral drilling was required. The remaining 9 patients were operated on with a simultaneous intra- and extraforaminal approach and required the removal of almost 50 % of the facets and/or exhibited radiological signs of segmental instability on preoperative flexion-extension X-rays.

Surgical Complications

Only one case of recurrence was observed in the present patient cohort; this occurred 3 months after surgery. The patient underwent reoperation, at which an interspinous BacJac® Pioneer Surgical Technology inc, headquartered in Marquette, Michigan. implant was implemented in order to prevent postoperative instability. Another patient was readmitted 3 months after surgery for spondylodiscitis and received antibiotic therapy, with complete benefit. In one patient a cerebrospinal fluid fistula was discovered incidentally during the intervention and managed successfully with bed rest.

Six-Week Postoperative Findings

A significant improvement of all clinical parameters occurred during the first 6 weeks after surgery in almost all patients. Particularly, pain was reduced quite significantly, with only

Table 2 Changes in clinical symptoms and signs before and after surgery, differentiated for localization of pain according to different follow-up timings

Symptoms		Preoperative	Six weeks postoperative	Six months postoperative	Eighteen months postoperative
Leg pain localization	Total leg pain distribution	90 (93.75 %)	11 (11.46 %)	8 (8.33 %)	7 (7.28 %)
	Sciatic distribution	68 (70.83 %)	8 (8.33 %)	5 (5.21 %)	5 (5.20 %)
	Femoral distribution	22 (22.92 %)	3 (3.1 %)	3 (3.1 %)	2 (2.07 %)
Back pain localization	No circadian differences	92 (95.85 %)	14 (14.58 %)	10 (10.42 %)	8 (8.34 %)
	Nocturnal pain	41 (42.71 %)	12 (12.50 %)	9 (9.38 %)	7 (7.29 %)

Values are expressed as numbers of patients and percentages

Table 3 Changes in clinical signs differentiated according to different follow-up timings

Clinical signs	Preoperative	Six weeks postoperative	Six months postoperative	Eighteen months postoperative
Lasègue sign	68 (70.83 %)	7 (7.29 %)	5 (5.21 %)	4 (4.17 %)
Reverse Lasègue sign	22 (22.92 %)	5 (5.21 %)	4 (4.17 %)	3 (3.07 %)
Crossed Lasègue sign	16 (16.67 %)	7 (7.29 %)	5 (5.21 %)	3 (3.13 %)
Wasserman sign	21 (21.87 %)	3 (3.1 %)	3 (3.1 %)	2 (2.06 %)
Sensory deficits	53 (55.21 %)	14 (14.58 %)	10 (10.12 %)	8 (8.09 %)
Motor deficits	45 (46.86 %)	15 (15.63 %)	8 (8.33 %)	6 (6.62 %)

Values are expressed as numbers of patients and percentages

11 patients (11.46 %) complaining of residual leg pain (8 of whom had sciatic pain and 3 had pain localized in the femoral area) and 14 (14.58 %) with back pain. Nocturnal pain was still complained of by 12 patients (12.50 %). The Lasègue sign was positive in 7 patients (7.29 %), reverse Lasègue was positive in 5 (5.21 %), and crossed Lasègue was positive in 7 (7.29 %). Wasserman sign was positive in 3 patients (3.1 %). Sensory deficits persisted in 14 patients (14.58 %), while weakness was noted in 15 (15.63 %).

Six-Month Postoperative Findings

A further improvement of clinical features was observed at the 6-month follow-up. Residual leg pain was present in eight patients (8.33 %), five (5.21 %) with sciatic and three (3.1 %) with femoral distribution. Ten patients (10.42 %) suffered from persistent back pain; however, they had significant improvements in VAS scores (see Table 1). Nine patients (9.38 %) reported nocturnal pain. The Lasègue sign was still positive in five patients (5.21 %), while reverse Lasègue was positive in four (4.17 %) and crossed Lasègue was positive in five (5.21). Wasserman's sign was still positive in three patients (3.1 %) Sensory deficits persisted in ten patients (10.12 %) and motor dysfunction in eight (8.33 %). The dynamics of clinical evolution in the present cohort are shown in Table 2.

Eighteen-Month Postoperative Findings

At the 18-month follow-up we observed that residual leg pain was present in seven patients (7.28); five (5.20 %) with sciatic distribution and two (2.07 %) with femoral distribution. Eight patients (8.34 %) suffered from persistent back pain. Seven patients (7.29 %) reported nocturnal pain. The Lasègue sign persisted in four patients (4.17 %), reverse Lasègue in three (3.07 %), and crossed Lasègue in three (3.13 %). Wasserman's sign was still positive in two patients (2.06 %). Sensory deficits were observed in eight patients (8.09 %) and motor dysfunction in six (6.62 %). The data for the 18-month follow-up are shown in Tables 1 and 2.

Discussion

ELDH is considered a diagnostic as well as a surgical challenge. It represents a well-known cause of unsatisfactory results following lumbar disk surgery. Although it was discovered as a clinical entity in the pre-computed tomography

(CT) era [15], it was only after the introduction of CT-scanning technology that the diagnosis of ELDH became possible on the basis of imaging technology [6].

Clinical Aspects

Some specific clinical aspects are pathognomonic of ELDH, such as the prevalence of positivity for "femoral stretching" over the Lasègue sign, much more common than in the usual clinical setting of a lumbar disk herniation, and the characteristic nocturnal pain, which forces the patient to get up at night in order to relieve the pain, as happens in the case of a lumbar spinal tumor. However, these signs are frequent but not constant ones, and to rely only on the clinical suspicion may be obviously misleading.

Radiological Diagnosis

Magnetic resonance imaging (MRI) has been recognized as the modality of choice in the evaluation of the spine. MRI findings should be interpreted with consideration of all clinical signs, symptoms, and other relevant background findings [11].

Guidelines provide the standardization of terms to more consistently describe disk herniation. We used a specified protocol for imaging (T1- and T2-weighted sequences in the sagittal plane, T2-weighted sequences in the axial plane, and T2-weighted sequences with fat saturation suppression of the signal in the sagittal plane).

In our study, when comparing MRI interpretations, radiologists and clinicians were found to be in agreement (95 %) on the presence and level of herniation, in accordance with data in the literature. However, with a relatively inexperienced eye, the hypodense appearance of the laterally placed disk fragment can sometimes be confused with the similar appearance of the root ganglia at the foraminal exit, and this is the reason why it is not so exceptional that the MRI diagnosis of ELDH can be missed, at least in the early stage of the disease.

On the other hand, the value of CT examination in the preoperative diagnosis of ELDH has been recognized since 1984. In fact, lateral disk protrusion/herniation is quite well shown on axial CT sections, due to its hyperdense appearance, which is different from the surrounding hypodense soft tissues, and the value of postoperative CT scanning for detecting surgical decompression is also quite evident for the same reason.

MRI shows potentially relevant morphological abnormalities of the end plate and marrow signal, disk abnormalities, the degree of nerve root compression and degeneration of the facet joint, and injuries of the posterior ligamentous complex of the cervical and thoracolumbar spine [14, 22].

In the present study, in cases where it was not possible to perform an MRI examination, CT examination was performed.

Other Diagnostic Modalities

Sometimes diagnosis of an ELDH can be difficult and other diagnostic modalities are required. Electromyography (EMG) examination results can be misleading. In patients in whom the clinical impact of lateral disk bulging is not clear, a CT-guided selective block of the affected nerve root can lead to the right diagnosis, since this procedure mimics the expected results of surgical decompression, at least as far as pain control is concerned. We used this procedure in five patients before performing surgery.

Surgical Approaches

Various surgical strategies have been proposed in the past for operating on extraforaminal disk herniations, including a contralateral angled-view approach [2], a variety of unilateral approaches combined with significant laminar and/or facet removal [17], total unilateral facetectomy [5], and a transmuscular approach [16, 18]. These approaches were superbly reviewed in 1995 in a fundamental paper by Epstein [3].

In our early experience we used a transmuscular approach, and noticed that muscle swelling and related local pain represented a non-negligible problem in some patients [26]. This issue appears to be solved if the transmuscular dissection is performed according to the suggestion of Tessitore and de Tribolet, with better results if an endoscopic technique is used [27]. However, this kind of exposure does not allow interlaminar exploration if necessary, and such exploration was required in one-sixth of the patients in the present study.

Unilevel facetectomy, though apparently well tolerated, accounts for a not insignificant risk of future instability, which cannot be considered marginal in the modern era of spinal neurosurgery. Moreover, the contralateral approach requires a bilateral exposure, a fact which makes its invasiveness definitely more significant than that of the more extensively used unilateral—either transmuscular or intertransversarii—approaches.

The surgical technique described in the present study not only allows the patient to stand right after the surgery but also allows the surgeon to perform simultaneous interlaminar and extraforaminal exposure, when necessary, without the need for total facetectomy. In particular, no significant complaint was reported by the patients as a possible result of the significant muscle dissection that this technique requires.

The clinical results of our series are very good, at least in the short term. We recognize that some more refinements are perhaps needed in the diagnostic techniques, and more experience in new developing mininvasive endoscopic techniques will, perhaps, restrict the indication for this surgical strategy in the near future. However, at present, our experience seems to indicate that the surgical technique described here represents a sound strategy for managing these challenging cases.

Conclusion

ELDH is a relatively uncommon condition. Clinical diagnosis remains a challenge. Apart from the different surgical strategies described in the past, our technique can be reasonably proposed because of its low morbidity, fast recovery, and short hospital stay.

The short-term results with our surgical strategy are very good. The long-term impact of this strategy confirmed that there was improvement of symptoms, with a greater effect shown in the first 6 months after the surgical treatment, and a smaller, but significant, further regression of the symptoms in the subsequent 12 months.

Conflict of Interest Statement We declare that we have no conflicts of interest.

References

1. Abdullah AF, Wolber PG, Warfield JR, Gunadi IK (1988) Surgical management of extreme lateral lumbar disc herniations: review of 138 cases. Neurosurgery 22:648–653
2. Brown LJ (1987) Definitive diagnosis of extreme lateral lumbar disc herniation. Surg Neurol 27:373–376
3. Epstein NE (1995) Evaluation of varied surgical approaches used in the management of 170 far-lateral lumbar disc herniations: indications and results. J Neurosurg 83:648–656
4. Fujisawa H, Igarashi S, Koyama T (1996) Far lateral lumbar disc herniation: clinical and radiographical features of three cases. No Shinkei Geka 24:363–367
5. Garrido E, Connaughton PN (1991) Unilateral facetectomy approach for lateral lumbar disc herniation. J Neurosurg 74:754–756
6. Godersky JC, Erickson DL, Seljeskog EL (1984) Extreme lateral disc herniation: diagnosis by computed tomographic scanning. Neurosurgery 14:549–552
7. Jane JA, Haworth CS, Broaddus WC, Lee JH, Malik J (1990) A neurosurgical approach to far-lateral disc herniation. Technical note. J Neurosurg 72:143–144
8. Jensen MP, Chen C, Brugger AM (2003) Interpretation of visual analog scale ratings and change scores: a reanalysis of two clinical trials of postoperative pain. J Pain 4:407–414
9. Kim DG, Eun JP, Park JS (2012) New diagnostic tool for far lateral lumbar disc herniation: the clinical usefulness of 3-tesla magnetic resonance myelography comparing with the discography CT. J Korean Neurosurg Soc 52:103–106
10. Kotil K1, Akcetin M, Bilge T (2007) A minimally invasive transmuscular approach to far-lateral L5-S1 level disc herniations: a prospective study. J Spinal Disord Tech 20:132–138
11. Kunihiko F, Gou K (2001) Proper use of RM imaging for evaluation of low back pain (radiologist view). Semin Muscoloskeletal Radiol 5:133–136
12. Kurobane Y, Takahashi T, Tajima T, Yamakawa H, Sakamoto T, Sawaumi A (1986) Extraforaminal disc herniation. Spine (Phila Pa 1976) 11:260–268
13. Lee S, Kang JH, Srikantha U, Jang IT, Oh SH (2014) Extraforaminal compression of the L-5 nerve root at the lumbosacral junction: clinical analysis, decompression technique, and outcome. J Neurosurg Spine 20:371–379
14. Lurie JD, Doman DM, Spratt KF, Tosteson AN, Weinstein JN (2009) Magnetic resonance imaging interpretation in patients with symptomatic spine disc herniation: comparison of clinician and radiologist readings. Spine 34:701–705
15. Macnab I (1971) Negative disc exploration: an analysys of the causes of nerve-root involvement in sixty-eight patients. J Bone Joint Surg Am 53:891–903
16. Maroon JC, Kopitnik TA, Schulhof LA, Abla A, Wilberger JE (1990) Diagnosis and microsurgical approach to far-lateral disc herniation in the lumbar spine. J Neurosurg 72:378–382
17. Marquardt G, Bruder M, Theuss S, Setzer M, Seifert V (2012) Ultra-long-term outcome of surgically treated far-lateral, extraforaminal lumbar disc herniations: a single-center series. Eur Spine J 21:660–665
18. Müller A, Reulen HJ (1998) A paramedian tangential approach to lumbosacral extraforaminal disc herniations. Neurosurgery 43:854–861
19. O'Hara LJ, Marshall RW (1997) Far lateral lumbar disc herniation. The key to the intertransverse approach. J Bone Joint Surg Br 79:943–947
20. Paolini S, Ciappetta P, Raco A, Missori P, Delfini R (2006) Combined intra-extracanal approach to lumbosacral disc herniations with bi-radicular involvement. Technical considerations from a surgical series of 15 cases. Eur Spine J 15:554–558
21. Pirris SM, Dhall S, Mummaneni PV, Kanter AS (2008) Minimally invasive approach to extraforaminal disc herniations at the lumbosacral junction using an operating microscope: case series and review of the literature. Neurosurg Focus 25 E10
22. Pizones J, Sanchez-Mariscal F, Zuniga L (2013) Prospective analysis of magnetic resonance imaging accuracy in diagnosing traumatic injuries of the posterior ligamentous complex of the thoracolumbar spine. Spine 38:745–751
23. Porchet F, Fankhauser H, de Tribolet N (1994) Extreme lateral lumbar disc herniation: clinical presentation in 178 patients. Acta Neurochir (Wien) 127:203–209
24. Recoules-Arche D (1985) Surgery of disk hernia of the lumbar vertebral canal. Neurochirurgie 31:61–64
25. Shao KN, Chen SS, Yen YS, Jen SL, Lee LS (2000) Far lateral lumbar disc herniation. Zhonghua Yi Xue Za Zhi (Taipei) 63:391–398
26. Spallone A, Gazzeri G, Floris R (1992) Extra-foraminal prolapsed lumbar disc: a possible cause of recurrent sciatica in failed low-back surgery patient. Case report. J Neurosurg Sci 36:111–115
27. Tessitore E, de Tribolet N (2004) Far-lateral lumbar disc herniation: the microsurgical transmuscular approach. Neurosurgery 54:939–942

Restoration of Thoracolumbar Spine Stability and Alignment in Elderly Patients Using Minimally Invasive Spine Surgery (MISS). A Safe and Feasible Option in Degenerative and Traumatic Spine Diseases

Giuseppe M.V. Barbagallo, Giuseppe Raudino, Massimiliano Visocchi, A. Abdulrazzaq Alobaid, A. Abdulaziz Al-Mutair, Thomas Naveen, and Francesco Certo

Abstract Minimally invasive spine surgery (MISS), including percutaneous pedicle-screw fixation (PPSF), mini-open transforaminal lumbar interbody fusion (m-open TLIF), vertebroplasty, and stentoplasty, allows the preservation of neurological function and the restoration of spine stability, while reducing associated risks and complications. This study aimed to analyze the safety and efficacy of MISS in elderly patients suffering from degenerative or traumatic thoracolumbar diseases. Forty-five patients (28 females), with a mean age of 73 years (range 65–89), suffering from osteoporotic vertebral fractures (24), degenerative spondylolisthesis (15), and lumbar canal stenosis with instability and/or de novo scoliosis (6) were included.

Twenty-one patients underwent PPSF and m-open TLIF. The remaining patients received PPSF without interbody fusion, and in six of these fenestrated screws were used for vertebral body cement augmentation.

Functional evaluation was obtained with a visual analog scale (VAS) and the Oswestry Disability Index (ODI) pre- and postoperatively. Preoperative imaging included X-rays, computed tomography (CT), and magnetic resonance imaging (MRI). Patients were followed-up with X-rays, and a CT scan was also obtained at the last follow-up. Follow-up ranged from 6 to 59 months (mean 28 months). Follow-up CT scan documented intersomatic fusion in only 14 % of patients treated with m-open TLIF. Despite the high incidence of non-union, mean VAS and ODI scores showed a significant improvement, with a reduction of mean VAS from 9 to 4 and a reduction of mean ODI from 76.33 to 38.15 %. Only three patients developed postoperative complications. No patients showed neurological deficits.

Minimally invasive spine surgery for degenerative and traumatic spinal diseases is a safe and effective treatment also in elderly patients.

G.M.V. Barbagallo, MD (✉) • G. Raudino, MD • F. Certo, MD
Department of Neurological Surgery, Policlinico "G. Rodolico" University Hospital, Via Santa Sofia 78, Catania 95123, Italy
e-mail: gbarbagallo@unict.it

M. Visocchi, MD
Insitute of Neurosurgery, Medical School, Catholic University of Rome, Rome, Italy

A.A. Alobaid, MD • A.A. Al-Mutair, MD • T. Naveen, MD
Department of Spine Surgery, Al-Razi Hospital, Kuwait, Kuwait

Introduction

Spinal canal stenosis, degenerative spondylolisthesis, and vertebral body fractures represent major causes of morbidity in patients aged ≥65 years, needing rapid and effective treatment to achieve preservation of neurological function and restoration of spinal stability [1].

A growing body of studies suggests that in the adult population minimally invasive spine surgery (MISS) is a safe and feasible treatment option for degenerative and traumatic thoracolumbar spine diseases, avoiding or reducing postsurgical complications and improving early postoperative clinical outcome [2–4]. Recently, MISS has also been proposed for deformity surgery, with the intended goal of reducing the morbidity linked to open surgery [5–7]. When open surgery is chosen, the patient's age is an important factor to consider. Elderly age, i.e., ≥ 65 years, is not always a contraindication for open surgery, but it can be an added risk factor; indeed, older patients are more prone to suffer from several medical comorbidities, including osteoporosis. And these conditions might consistently increase the risks associated with open surgery. In such a scenario, MISS might be a safe and effective alternative treatment modality [8–11]. Recent technical innovations, such as fenestrated pedicle screws [12] and expandable screws [13], have further expanded the feasibility of MISS fixations also in the elderly population.

M. Visocchi et al. (eds.), *Trends in Reconstructive Neurosurgery*, Acta Neurochirurgica Supplement 124,
DOI 10.1007/978-3-319-39546-3_11, © Springer International Publishing Switzerland 2017

This study analyses the prospectively collected clinical, functional, and radiological data of elderly patients treated with MISS for degenerative and traumatic thoracolumbar pathologies.

Patients and Methods

Forty-five patients (17 males, 28 females), with a mean age of 73 years (range 65–89), suffering from degenerative spondylolisthesis (15), lumbar canal stenosis with instability and/or de novo scoliosis (6), and traumatic osteoporotic vertebral fractures (24) were prospectively enrolled. Forty patients were treated at the Department of Neurological Surgery, Policlinico "G. Rodolico" University Hospital, Catania, Italy, and 5 patients were treated at the Department of Spine Surgery, Al-Razi Hospital, Kuwait.

Preoperatively, clinical and functional outcomes were measured with a visual analog scale (VAS) and the Oswestry Disability Index (ODI). The same outcome measures were then used at 1, 3, 6, and 12 months after the index procedure and yearly thereafter. Imaging included X-rays, with flexion/extension views, magnetic resonance imaging (MRI), and computed tomography (CT).

Because of the patients' age, thorough and accurate clinical evaluation (cardiovascular and respiratory functions, blood tests, nutritional status) was performed before surgery to rule out comorbidities. Osteoporosis was not routinely investigated before surgery, as we assumed it to be widespread among elderly people. When a likely osteoporotic condition was suggested during surgery by a reduced device grip, we opted for cement augmented or expandable screws.

In 21 patients a MISS pedicle-screw fixation with mini-open transforaminal lumbar interbody fusion (m-open TLIF) was performed (14 single-level, 5 two-level, and 2 three-level cases). Of the remaining 24 patients treated with MISS, 18 underwent percutaneous pedicle-screw fixation (PPSF) with or without vertebral body cement augmentation (Confidence Spinal Cement; DePuy Spine,) and 6 underwent vertebroplasty or kyphoplasty.

The Viper percutaneous pedicle-screw fixation system with fenestrated screws (DePuy Spine Raynham, US) or the Illico percutaneous fixation system (Alphatec Spine Carlsbad, US, US) with expandable screws (Osseoscrew; Alphatec Spine Carlsbad, US) was used.

We performed m-open TLIF on the most symptomatic side, where the most severe spinal canal stenosis or neural structure compressionwas present. Either carbon fiber cages or polyether ether ketone (PEEK) cages were used. In patients with m-open TLIF a microsurgical unilateral facetectomy and hemilaminectomy was performed; the base of the spinous process was trimmed off, and the contralateral ligamentum flavum removed. Such maneuvers, together with a contralateral surgical table inclination [14], allowed us to recalibrate and enlarge the spinal canal and decompress the neural structures (Figs. 1 and 2).

Results

Follow-up ranged from 6 to 59 months, with a mean of 28 months.

VAS and ODI scores showed significant postoperative improvements; in particular, the mean VAS score decreased from 9 to 4 and the mean ODI score decreased from 76.33 to 38.15 %.

Three of the 45 patients (6.7 %) developed the following postoperative complications: spondylodiscitis (1), screw dislocation (1) (Fig. 3), and cage dislocation (1). No patients showed postoperative neurological deficits or any age-related or device-related complications.

Solid intersomatic fusion was observed in 3 of the 21 (14 %) patients who had undergone m-open TLIF. Evaluation of intersomatic fusion was performed using multiplanar reconstructed CT scans at follow-up visits.

Discussion

Surgery in the elderly population is still a debatable issue. Boundaries between overtreatment and standard of care are often undefined and unclear; therefore, surgeons and anesthesiologists should evaluate surgical indications, as well as predict the complications, to balance the risks and benefits for patients [15]. Decision-making in the surgical management of elderly patients also has to consider the role of frailty and disability, aiming to obtain the best results with minimal risks for the delicate general homeostasis of the aging population.

Minimally invasive techniques applied to spinal surgery have gained appeal over the past decade, as they have demonstrated good clinical outcomes and low complication rates along with the well-known advantages of low invasiveness [4, 16, 17]. Currently, various spinal diseases (degenerative, neoplastic, or traumatic) may be approached using MISS techniques, with reduced blood loss, limited muscular dissection, better control of postoperative pain, and decreased hospitalization time [18]. To state the philosophy of MISS: elderly patients should be considered ideal candidates for spine surgery using less invasive techniques. Advances in MISS devices and techniques have led to an overall increase of surgical procedures in elderly patients, as demonstrated by the growing number of recently published reports on this topic [19–21].

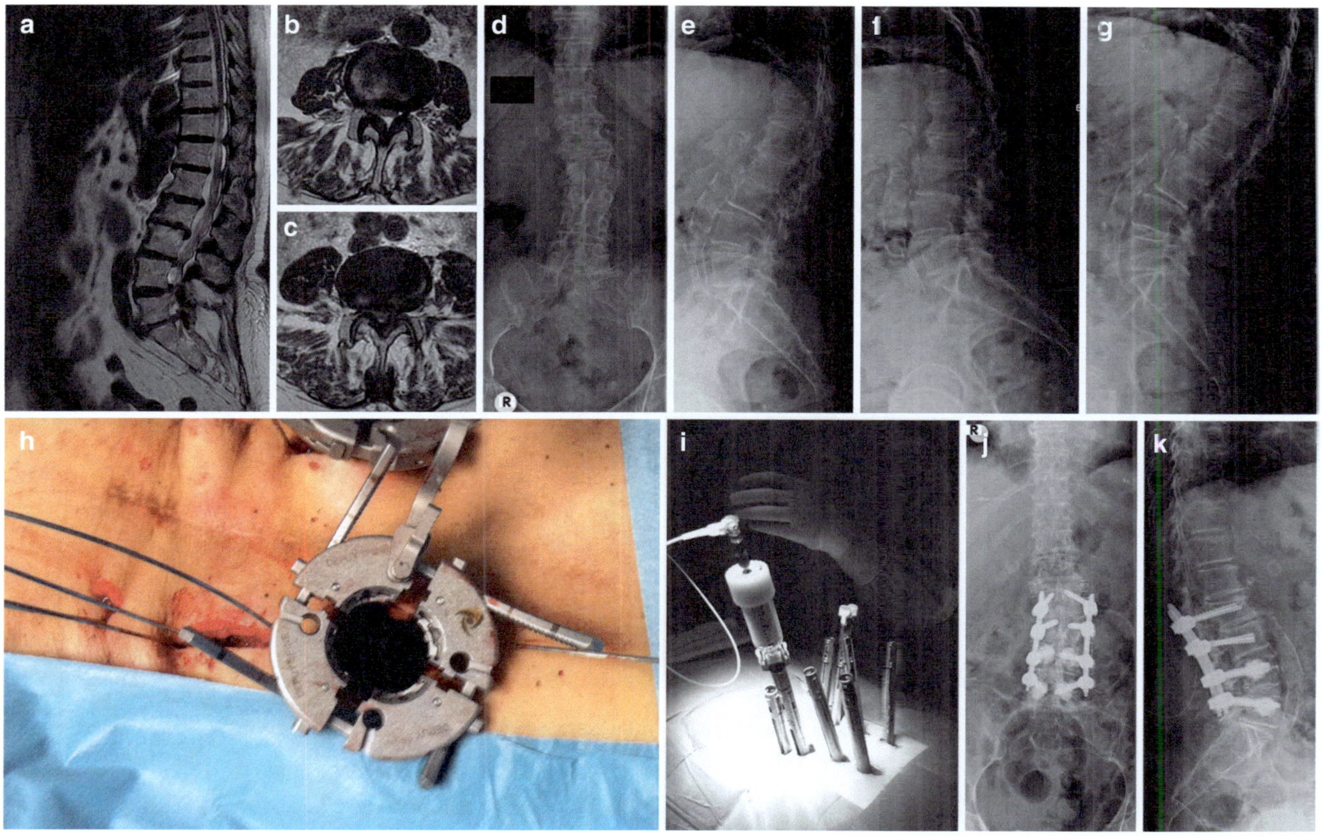

Fig. 1 Illustrative case. Minimally invasive surgical treatment of multilevel lumbar canal stenosis with L4-L5 instability is shown. A 77-year-old woman suffering from left sciatica and neurogenic claudication underwent spine magnetic resonance imaging (MRI), documenting lumbar canal stenosis (**a**) with severe bilateral foraminal stenosis at L3-L4 (**b**) and L4-L5 (**c**) levels. Anteroposterior (**d**), lateral (**e**), and flexion/extension (**f**, **g**) X-rays showed L4-L5 instability with slight L3-L4 and L4-L5 spondylolisthesis. The patient underwent L2-L5 fixation with percutaneous screws and left-sided mini-open transforaminal lumbar interbody fusion (TLIF) at L4-L5 (**h**). Cannulated screws at L4 and L5 allowed cement augmentation (**i**). Postoperative X-rays documented correct device positioning (**j**, **k**)

In 2008, Lee et al. analyzed the clinical and radiological outcomes of 27 patients, aged over 65 years old, who underwent single-level m-open TLIF as surgical treatment for degenerative spondylolisthesis [19]. They documented a high fusion rate (77.8 %), good spinal realignment (improved segmental lordosis in all patients), and a low complication rate. However, a possible inclusion bias in their series could be related to the patients' American Society of Anesthesiologists (ASA) grading, with a score ranging between 1 and 2 in all enrolled patients.

In 2011, Drazin et al. reviewed clinical data from the literature on complications and outcomes after spinal deformity surgery in patients over60 years old [7]. Interestingly, they supported the application of minimally invasive techniques in order to reduce the complication rate, encouraging future studies in this direction. A further review published in 2014 [22] highlighted the scant number of adequate clinical studies on this topic, pointing out the importance of CT scans for the thorough evaluation of fusion rates. Indeed, most of the published studies, including the previously cited one by Lee et al. [19], did not use CT scans in radiological follow-up.

A recent review of the existing literature on MISS in elderly patients [20] revealed a good clinical outcome with an acceptable complication rate, albeit that all the published studies reviewed had low-quality evidence, based on retrospective series. The authors concluded that symptomatic elderly patients may benefit from MISS techniques and they should not be excluded from surgery.

We have reported our experience with minimally invasive techniques in the surgical treatment of different spinal pathologies in a consecutive cohort of patients aged over 65 years old. Differently from Lee's series, we included patients regardless of their ASA score or surgical techniques, aiming to observe the impact of MISS in improving clinical outcome and allowing a rapid recovery. Early patient mobilization and short hospitalization have been considered primary goals, and surgical indications have been advised accordingly. We included in this series 20 patients suffering from traumatic and osteoporotic

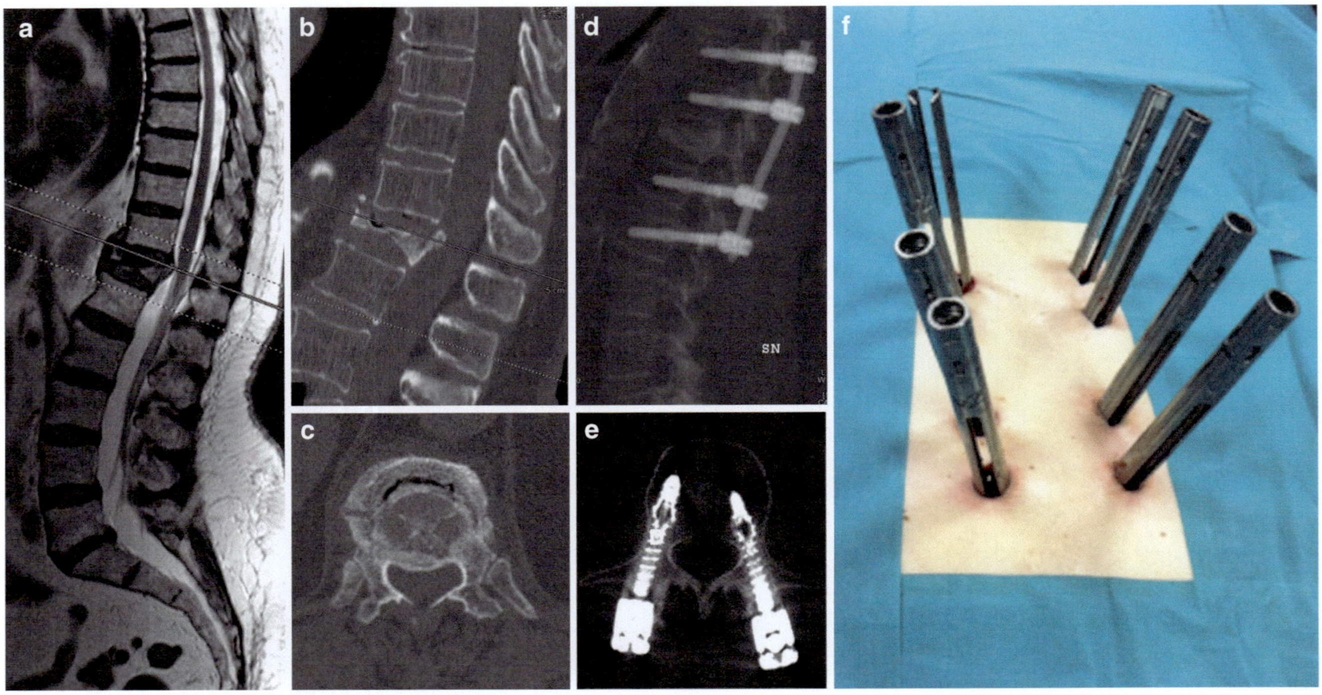

Fig. 2 Surgical treatment of Th12 burst fracture (**a–c**) in an 80-year-old man without neurological deficits is shown. Expandable screws (**d**, **e**) were inserted, using a percutaneous technique (**f**), in order to prevent screw loosening

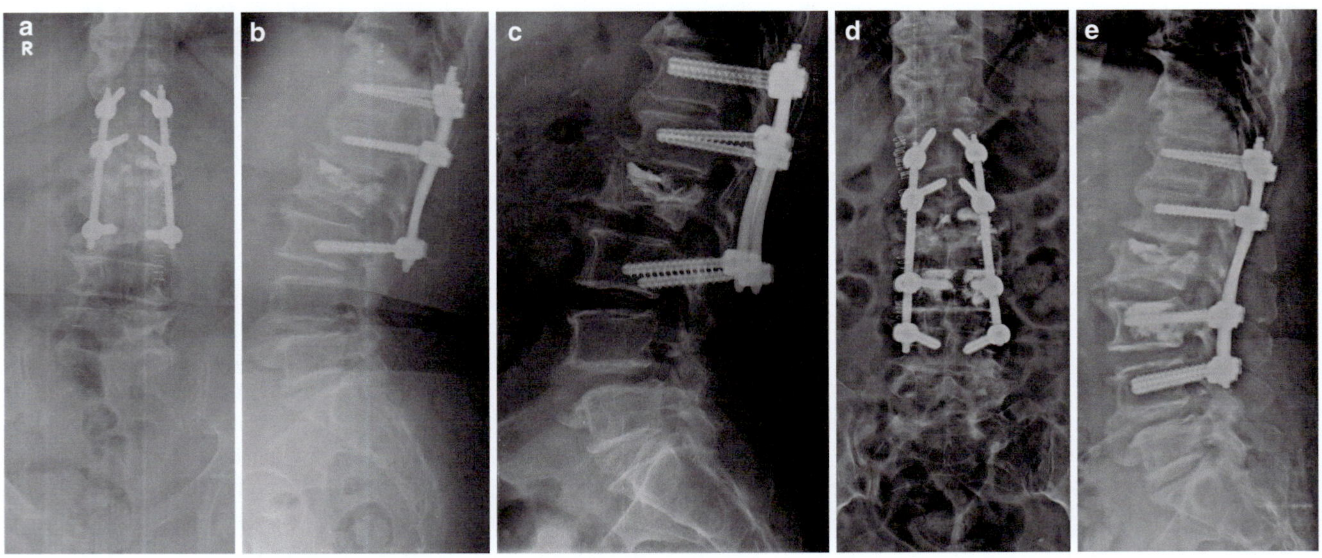

Fig. 3 Th12-L3 fixation performed in a 72-year-old woman for surgical treatment of L2 burst fracture, previously treated elsewhere with percutaneous cement injection (**a**, **b**). X-rays performed 3 weeks after the procedure documented the pull-out of L3 screws (**c**). Replacement and augmentation of the L3 screws was performed, together with caudal extension of instrumentation (**d**)

vertebral fractures. In this subgroup, minimally invasive techniques (PPSF with or without cement augmentation or expandable screws, or simple vertebro/kyphoplasty) ensured rapid recovery from symptoms, with unequivocal clinical advantages, as documented by a reduction of the VAS score from 9.3 to 3.2 (considering only patients with vertebral fractures). In this subgroup, we observed only one case of screw displacement, requiring revision surgery. Moreover, we combined PPSF and cement augmentation of the fractured vertebral body in 5 of these 20 patients, following the positive clinical experience with such a technique reported by Gu et al. [21]. Short-segment percutaneous fixation has been proposed as a viable alternative to open posterolateral fusion in patients with thoracolumbar burst fractures [23, 24]; in our series, short fixation with expandable or fenestrated screws for cement augmentation and short screws in the pedicles of fractured vertebrae were used in 8 of 24 patients, whereas 10 patients underwent percutaneous fixation with screws placed two levels above and two levels below the fractured vertebra.

The surgery-related complication rate was definitely low in our cohort. The application of our previously reported surgical technique [4] for m-open TLIF increased the safety of the procedure, reducing the risks of dural and/or neural lesions. Interestingly, the incidence of inhadvertent dural lesions during MISS in an elderly population was reported as 5 % in the review by Shamji et al. [20], whereas in our experience it was 0 %.

Despite the clinical parameters in our series showing a good or excellent outcome, our fusion rate, assessed by CT scan, in the subgroup of patients treated with m-open TLIF was low (14 %). Nonunion after MISS has been suggested as a possible weakness of such a procedure, particularly in patients with low-quality bone [8, 19]. However, in our series, pseudoarthrosis was never associated with poor clinical outcome and no device-related complications, such as dislocations, were observed. We think that this low fusion rate may also be due to the technique used to assess fusion. Indeed, X-rays may overestimate the fusion rate in patients with interbody devices, whereas CT scans offer a more accurate evaluation of intersomatic fusion.

On analyzing our results, we felt that complications could be reduced and predicted with a thorough presurgical evaluation, including assessment of osteoporosis and comprehensive medical management. Although we did not perform routine evaluation of osteoporosis and relied on intraoperative data only, we acknowledge that in elderly patients a preoperative osteoporosis assessment is advisable.

Finally, the patient's age was not a contraindication for instrumented MISS in our series, but tailored clinical evaluation and appropriate choice of a suitable device guided our indication, to achieve the best clinical results. Pseudoarthrosis did not correlate with poor clinical or functional outcome. When patients are carefully selected, MISS may be the best surgical strategy to obtain pain relief and improvement of myeloradicular signs and symptoms, as well as spinal stability.

Clinical trials with long-term follow-up should be designed to demonstrate the real impact of MISS in the management of degenerative or traumatic spinal diseases in elderly patients.

Conflict of Interest Statement We declare that we have no conflicts of interest.

References

1. Whitehouse MK, Jeyaretna DS, Wright A, Whitfield PC (2015) Neurosurgical care in the elderly: increasing demands necessitate future healthcare planning. World Neurosurg. doi: 10.1016/j.wneu.2015.10.099 [Epub ahead of print]
2. Barbagallo GM, Certo F, Sciacca G, Albanese V (2013) Bilateral tubular minimally invasive approach for decompression, reduction and fixation in lumbosacral lythic spondylolisthesis. Neurosurg Focus 35(2 Suppl):Video 9
3. Barbagallo GM, Piccini M, Alobaid A, Al-Mutair A, Albanese V, Certo F (2014) Bilateral tubular minimally invasive surgery for low-dysplastic lumbosacral lytic spondylolisthesis (LDLLS) analysis of a series focusing on postoperative sagittal balance and review of the literature. Eur Spine J 23 Suppl 6:705–713
4. Barbagallo GM, Certo F, Visocchi M, Sciacca G, Piccini M, Albanese V (2015) Multilevel mini-open TLIFs and percutaneous pedicle screw fixation: description of a simple technical nuance used to increase intraoperative safety and improve workflow. Tips and tricks and review of the literature. Neurosurg Rev 38:343–354
5. Deukmedjian AR, Ahmadian A, Bach K, Zouzias A, Uribe JS (2013) Minimally invasive lateral approach for adult degenerative scoliosis: lessons learned. Neurosurg Focus 35(2):E4
6. Park P, Wang MY, Lafage V, Nguyen S, Ziewacz J, Okonkwo DO, Uribe JS, Eastlack RK, Anand N, Haque R, Fessler RG, Kanter AS, Deviren V, La Marca F, Smith JS, Shaffrey CI, Mundis GM Jr, Mummaneni PV, International Spine Study Group (2015) Comparison of two minimally invasive surgery strategies to treat adult spinal deformity. J Neurosurg Spine 22(4):374–380
7. Drazin D, Shirzadi A, Rosner J, Eboli P, Safee M, Baron EM, Liu JC, Acosta FL Jr (2011) Complications and outcomes after spinal deformity surgery in the elderly: review of the existing literature and future directions. Neurosurg Focus 31(4):E3
8. Rosen DS, O'Toole JE, Eichholz KM, Hrubes M, Huo D, Sandhu FA, Fessler RG (2007) Minimally invasive lumbar spinal decompression in the elderly: outcomes of 50 patients aged 75 years and older. Neurosurgery 60(3):503–510
9. Acosta FL Jr, McClendon J Jr, O'Shaughnessy BA, Koller H, Neal CJ, Meier O, Ames CP, Koski TR, Ondra SL (2011) Morbidity and mortality after spinal deformity surgery in patients 75 years and older: complications and predictive factors. J Neurosurg Spine 15:667–674
10. Lee JH, Chun HJ, Yi HJ, Bak KH, Ko Y, Lee YK (2012) Perioperative risk factors related to lumbar spine fusion surgery in Korean geriatric patients. J Korean Neurosurg Soc 51:350–358
11. Daubs MD, Lenke LG, Cheh G, Stobbs G, Bridwell KH (2007) Adult spinal deformity surgery. Complications and outcomes in patients over Age 60. Spine 32(20):2230–2244
12. Pesenti S, Blondel B, Peltier E, Adetchessi T, Dufour H, Fuentes S (2014) Percutaneous cement-augmented screws fixation in the fractures of the aging spine: is it the solution? Biomed Res Int 2014:610675
13. Vishnubhotla S, McGarry WB, Mahar AT, Gelb DE (2011) A titanium expandable pedicle screw improves initial pullout strength as compared with standard pedicle screws. Spine J 11(8):777–781
14. Schonauer C, Bocchetti A, Barbagallo G, Albanese V, Moraci A (2004) Positioning on surgical table. Eur Spine J. 13 Suppl 1:S50–55
15. Murthy S, Hepner DL, Cooper Z, Bader AM, Neuman MD (2015) Controversies in anaesthesia for noncardiac surgery in older adults. Br J Anaesth 115(Suppl 2):15–25
16. Holly LT, Schwender JD, Rouben DP, Foley KT (2006) Minimally invasive transforaminal lumbar interbody fusion: indications, technique, and complications. Neurosurg Focus 20(3):E6

17. Wang J, Zhou Y, Zhang ZF, Li CQ, Zheng WJ, Liu J (2010) Comparison of one-level minimally invasive and open transforaminal lumbar interbody fusion in degenerative and isthmic spondylolisthesis grades 1 and 2. Eur Spine J 19: 1780–1784
18. Khan NR, Clark AJ, Lee SL, Venable GT, Rossi NB, Foley KT (2015) Surgical outcomes for minimally invasive vs open transforaminal lumbar interbody fusion: an updated systematic review and meta-analysis. Neurosurgery 77(6):847–874
19. Lee DY, Jung TG, Lee SH (2008) Single-level instrumented mini-open transforaminal lumbar interbody fusion in elderly patients. J Neurosurg Spine 9(2):137–144
20. Shamji MF, Goldstein CL, Wang M, Uribe JS, Fehlings MG (2015) Minimally invasive spinal surgery in the elderly: does It make sense? Neurosurgery 77(Suppl 4):S108–415
21. Gu Y, Zhang F, Jiang X, Jia L, McGuire R (2013) Minimally invasive pedicle screw fixation combined with percutaneous vertebroplasty in the surgical treatment of thoracolumbar osteoporosis fracture. J Neurosurg Spine 18:634–640
22. Bach K, Ahmadian A, Deukmedjian A, Uribe JS (2014) Minimally invasive surgical techniques in adult degenerative spinal deformity: a systematic review. Clin Orthop Relat Res 472(6):1749–1761
23. Lee JK, Jang JW, Kim TW, Kim TS, Kim SH, Moon SJ (2013) Percutaneous short-segment pedicle screw placement without fusion in the treatment of thoracolumbar burst fractures: is it effective?: comparative study with open short-segment pedicle screw fixation with posterolateral fusion. Acta Neurochir 155:2305–2312
24. Van Middendorp JJ, Barbagallo G, Schuetz M, Hosman AJF (2012) Design and rationale of a Prospective, Observational European Multicenter study on the efficacy of acute surgical decompression after traumatic Spinal Cord Injury: the SCI-POEM study. Spinal Cord. 50(9):686–694

Atlantoaxial Joint Distraction with a New Expandable Device for the Treatment of Basilar Invagination with Preservation of the C2 Nerve Root: A Cadaveric Anatomical Study

Filippo Maria Polli, Sokol Trungu, Massimo Miscusi, Stefano Forcato, Massimiliano Visocchi, and Antonino Raco

Abstract

Background

Atlantoaxial joint distraction has been advocated for the decompression of the brain stem in patients affected by basilar invagination, avoiding direct transoral decompression. This technique requires C2 ganglion resection and it is often impossible to perform due to the peculiar bony anatomy. We describe a cadaveric anatomical study supporting the feasibility of C1-C2 distraction performed with an expandable device, allowing easier insertion of the tool and preservation of the C2 nerve root.

Methods

In five adult cadaveric specimens, posterior atlantoaxial surgical exposure was performed and an expandable system was inserted within the C1-C2 joint. The expansion of the device, leading to active distraction of the joint space, together with all the surgical steps of the technique was recorded with anatomical pictures and the final results were checked with a computed tomography (CT) scan.

Results

Insertion of the device was easily performed in all cases without anatomical conflict with the C2 ganglion; CT scans confirmed the distraction of the C1-C2 joint.

Conclusion

This cadaveric anatomical study confirms the feasibility of the introduction of an expandable and flexible device within the C1-C2 joint, allowing it's distraction and preservation of the C2 ganglion.

F.M. Polli, MD, PhD (✉) • M. Miscusi, MD, PhD
Department of Medico-Surgical Sciences and Biotechnologies, "Sapienza" – University of Rome, Rome, Italy
e-mail: filippomaria.polli@gmail.com

S. Trungu, MD • S. Forcato, MD • A. Raco, MD, PhD.
Department of Neuroscience, Mental Health and Sensory Organs, "Sapienza" – University of Rome, Rome, Italy

M. Visocchi, MD
Institute of Neurosurgery, Medical School, Catholic University of Rome, Rome, Italy

Keywords Basilar invagination • Atlantoaxial dislocation • Atlantoaxial distraction • Expandable device

Introduction

In 1998, Atul Goel introduced a a new classification of congenital basilar invagination (BI) based on the presence or absence of Chiari malformation [4]. Type A, previously well defined by Von Torklus and Gehle as ' a prolapse of the cervical spine into the base of the skull" [13], exhibits a fixed atlantoaxial dislocation with the tip of the odontoid process above the Chamberlain line, the McRae foramen magnum line, and the Wackenheim clival line. In type B, the clivus, basiocciput, and the craniocervical junction are located rostrally, with the tip of the odontoid process lying above the Chamberlain line but below the McRae and Wackenheim lines. In this type there is no atlantoaxial dislocation. Nowadays the standard surgical treatment of BI is still transoral decompression followed by posterior occipito-cervical or atlantoaxial fixation surgery [1, 2, 10].

In 2004, Goel published a case series of patients with BI type A treated with atlantoaxial joint distraction and direct lateral mass fixation [5]. The technique, already described in 2002 [6], allows the indirect decompression of the brainstem by pulling the axis downward. It also requires section of the C2 ganglion in order to better expose the atlantoaxial facet and to distract C1-C2 articulation by inserting titanium or polyether ether ketone (PEEK) spacers into the joints. The quality of life (QoL) of patients who underwent C2 root section has been investigated [3, 8] confirming what Goel affirmed about his personal experience [7]: patients undergoing C2 ganglion resection seem to have a better prognosis than those who do not, in terms of postoperative neuropathic disturbances. Beyond the need for choosing whether or not to resect the root, the narrowness and the sagittal inclination

M. Visocchi et al. (eds.), *Trends in Reconstructive Neurosurgery*, Acta Neurochirurgica Supplement 124,
DOI 10.1007/978-3-319-39546-3_12,

of the C1-C2 joint in patients with BI frequently make the insertion of a cage hardly feasible.

Methods

In the Laboratory of Anatomy of Nantes University (France), five adult cadaveric specimens were surgically dissected in order to evaluate the possibility of distracting the atlantoaxial joint with an expandable device, without resecting the C2 ganglion and minimizing its stretching. The cadaveric specimens were placed prone, and the suboccipital region and the upper cervical spine were exposed through a longitudinal midline skin incision. After the sectioning of the paraspinal muscles, the posterior arch of C1, the C2 lamina and C1-C2 joints were exposed. C2 nerve roots were identified and gently moved cranially, with a dissector exposing the joint (Fig. 1a). Protecting the root with a blunt dissector, we created a cylindrical cavity in the joint bilaterally with a 4-mm diamond ball, sparing the anterior cortex (Fig. 1b).

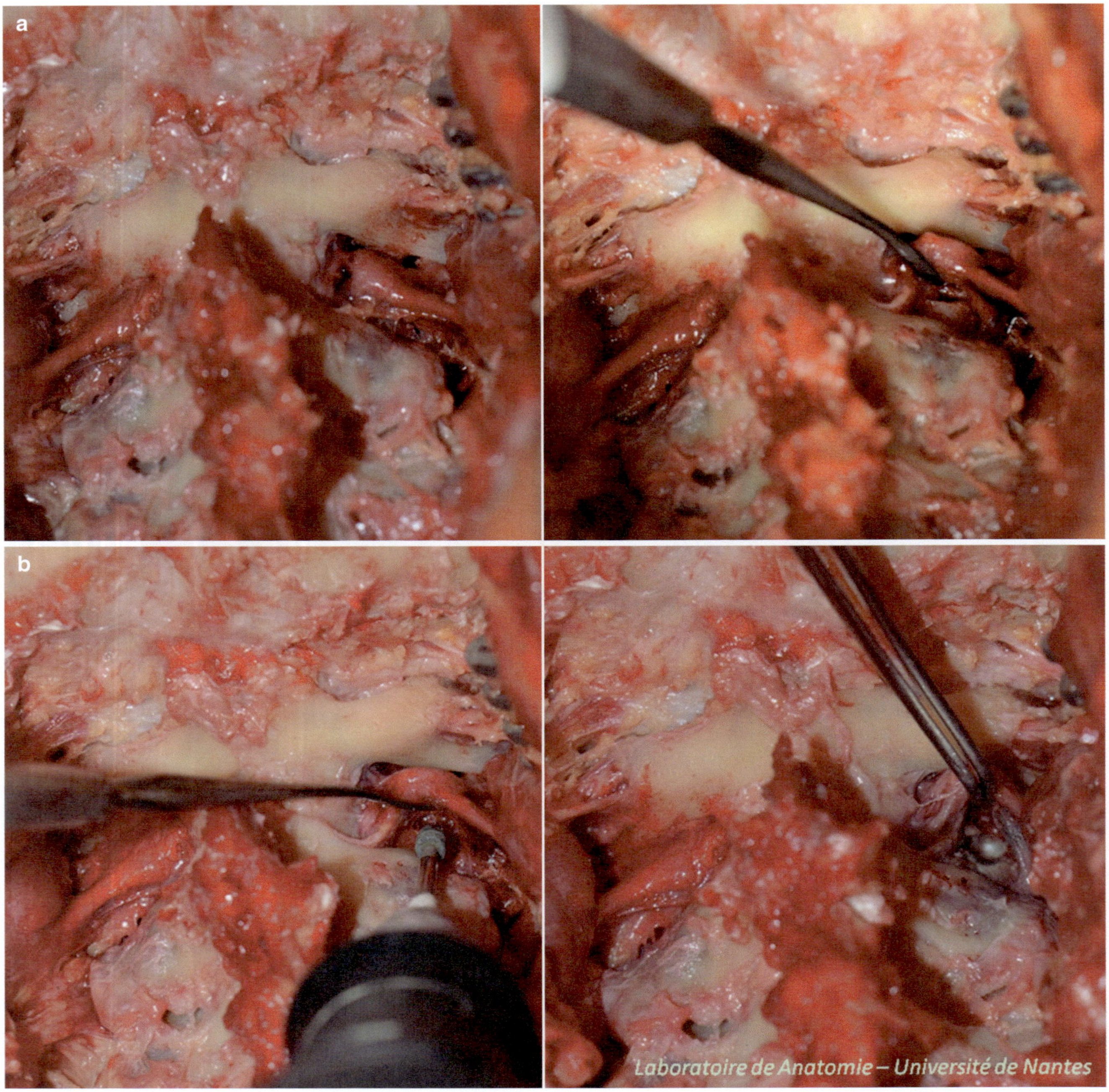

Fig. 1 (**a**) Cadaveric specimen, craniocervical junction, posterior view. Atlantoaxial joints and C2 nerve roots are exposed bilaterally. (**b**) The drilling of the C1-C2 joint to create an appropriate cavity for the expandable device, protecting the C2 nerve root with a blunt dissector

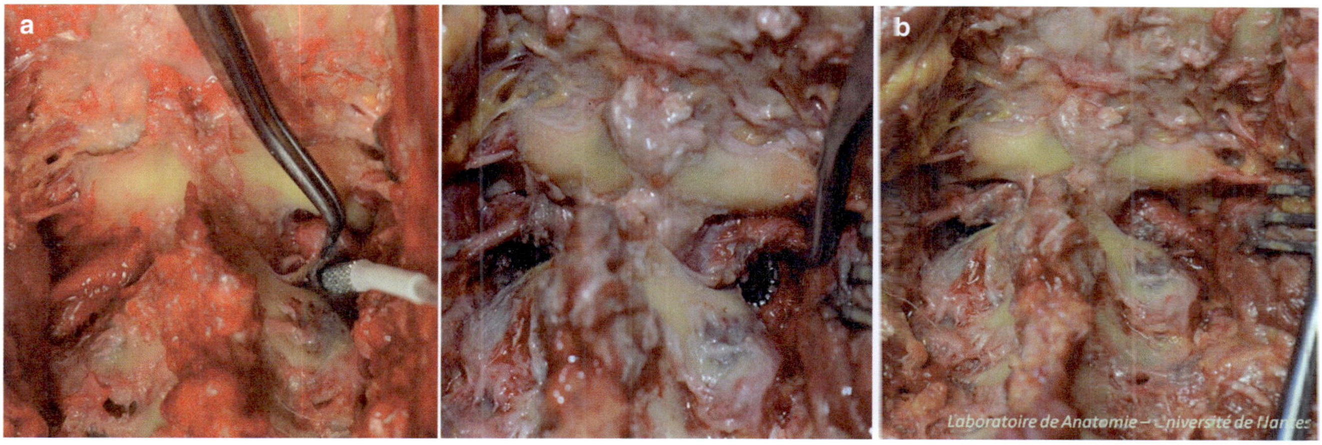

Fig. 2 (**a**) The insertion of the device is feasible with preservation of the C2 ganglion. (**b**) At the end of the procedure there was no conflict between the device and the C2 nerve root

A Vertebral Body Stent (VBS®; DePuy Synthes Spine, Zuchwil – Switzerland), a balloon-expandable metal stent mounted on a balloon-catheter for kyphoplasty, was shortened in order to fit into the atlantoaxial joint. The modified device was inserted within the joint cavity—with the C2 ganglion being preserved with a dissector—and expanded along with dilatation of the balloon. Once the distraction was visually accomplished, the balloon was pulled out, and the correct positioning of the stent within the joints bilaterally was checked visually (Fig. 2a).

Results

In all five cadaveric specimens, the device was inserted without difficulty and with a gentle mobilization of the C2 root, avoiding any stretching of the nerve. Visual inspection of the atlantoaxial joint revealed no conflict between the C2 nerve root and the device (Fig. 2b).

Control CT scans showed the correct positioning of the devices bilaterally and the symmetrical distraction of the C1-C2 joint (Fig. 3a).

Discussion

Basilar invagination is a relatively rare developmental anomaly of the craniovertebral junction, in which the odontoid abnormally prolapses into the foramen magnum. It can be associated with other osseous anomalies of the craniovertebral junction, including atlanto-occipital assimilation, incomplete C1 ring, and hypoplasia of the atlas, basiocciput, and occipital condyles. It is also associated with other malformations, including Chiari malformation, hydrocephalus, and syringomyelia. Patients often present with neurologic deficits and need surgical treatment to prevent progression [11].

Despite recent improvements in the surgical technique [14], the transoral approach for odontoid process removal still carries important surgical risks and entails the need for performing surgery in two steps [9, 12].

In order to avoid the risks of transoral surgery and to accomplish brainstem decompression in a single surgical stage, in 2004 Goel [5] proposed distraction of the atlantoaxial joint for the treatment of type A BI, using a fixed atlantoaxial dislocation with the tip of the odontoid process above the Chamberlain line, the McRae foramen magnum line, and the Wackenheim clival line. This innovative surgical strategy allows the reduction of the atlantoaxial dislocation, pulling the axis downwards with its dens, decompressing the brainstem and improving clinical symptoms, thus avoiding the complications related to the transoral approach.

The introduction of two spacers/cages in the C1-C2 joint requires the section of the C2 ganglions and of their periradicular venous plexuses, followed by a C1-C2 lateral mass fixation [5].

The postoperative QoL of patients who underwent C2 root section was investigated recently in a small group of patients. In that study the C2 nerve root resection was associated with increased occipital numbness but had no effect on the patients' QoL [3]. Interestingly, this C-2 neurectomy group showed no cases of new-onset postoperative C-2 neuralgia, in contrast to the reporting of new-onset C-2 neuralgia with C-2 ganglion preservation in a growing number of articles in the literature. Moreover, in 80 % of patients in a different series who had preoperative occipital neuralgia, this neuralgia was relieved following C1–2 instrumented arthrodesis with C-2 neurectomy in all of these patients [8].

However, ganglion resection requires a fine microsurgical technique, and the extensive bleeding from the adjacent

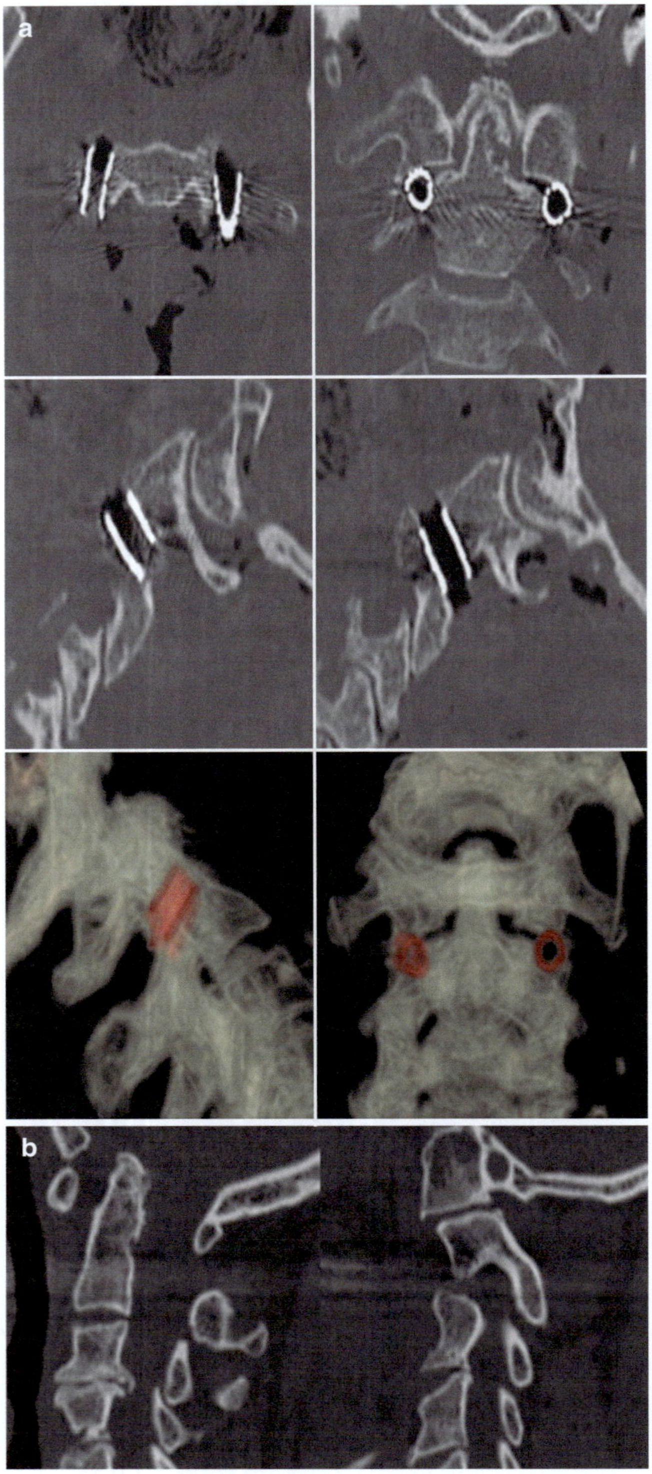

Fig. 3 (**a**) Computed tomography (CT) scan of the anatomical specimen. Axial, coronal, and sagittal planes and maximum intensity projection (device in *red*) images showing the correct positioning of the device and the symmetrical distraction of the atlantoaxial joint bilaterally. (**b**) Cervical computed tomography (CT) scan, sagittal plane, showing the unfavorable sagittal inclination of the C1-C2 joint in a patient affected by basilar invagination

venous plexus can make the exposure difficult. Furthermore, although relatively rare, as described by Goel, there are anomalies in the vertebral artery course, with possible inadvertent laceration occurring due to the close anatomical relationship of this artery with the ganglion [7]. Considering the anatomical malformation involving the craniocervical junction in patients affected by BI, the insertion of a device in the C1-C2 joint is often difficult or impossible due to the narrowness and the sagittal inclination of the joint (Fig. 3b).

The vertebral body stenting device (VBS®; DePuy Synthes Spine, Zuchwil – Switzerland) we used is flexible, it minimizes nerve root manipulation and, at the same time, it maximizes joint distraction. Although not tested and developed for the purpose described here, this device exhibits characteristics that can overcome the bony anatomical limitations of the introduction of a cage, while preserving the C2 roots. C1-C2 distraction can also increase the C2 foraminal space, thus possibly reducing the risk of postoperative suboccupital neuralgia due to C2 nerve root-C1 lateral mass screw conflict.

Conclusions

This cadaveric anatomical study confirms the feasibility of the introduction of an expandable and flexible device within the C1-C2 joints, allowing easier distraction of the joints—compared with a cage—while preserving the C2 nerve roots. Further mechanical tests are needed to analyze the stiffness to axial load parameters of the VBS® and other expandable devices before proceeding with in vivo clinical studies.

Conflict of Interest Statement We declare that we have no conflicts of interest.

References

1. Bonney G, Williams JP (1985) Trans-oral approach to the upper cervical spine. A report of 16 cases. J Bone Joint Surg Br 67:691–698
2. Crockard HA (1988) Anterior approaches to lesions of the upper cervical spine. Clin Neurosurg 34:389–416
3. Dewan MC, Godil SS, Mendenhall SK et al (2014) C2 nerve root transection during C1 lateral mass screw fixation: does it affect functionality and quality of life? Neurosurgery 74(5):475–480; discussion 480–1
4. Goel A, Bhatjiwale M, Desai KI (1998) Basilar invagination: a study based on 190 surgically treated patients. J Neurosurg 88:962–968
5. Goel A (2004) Treatment of basilar invagination by atlantoaxial joint distraction and direct lateral mass fixation. J Neurosurg Spine 1(3):281–286
6. Goel A, Desai KI, Muzumdar DP (2002) Atlantoaxial fixation using plate and screw method: a report of 160 treated patients. Neurosurgery 51:1351–1357

7. Goel A (2015) C2 ganglion resection for lateral mass fixation techniques. J Craniovertebr Junction Spine 6(1):10–11
8. Hamilton DK, Smith JS, Sansur CA et al (2011) C-2 neurectomy during atlantoaxial instrumented fusion in the elderly: patient satisfaction and surgical outcome. J Neurosurg Spine 15(1):3–8
9. Marda M, Pandia MP, Rath GP et al (2013) Postoperative pulmonary complications in patients undergoing transoral odontoidectomy and posterior fixation for craniovertebral junction anomalies. J Anaesthesiol Clin Pharmacol 29:200–204
10. Menezes AH (1995) Primary craniovertebral anomalies and hind brain herniation syndrome (Chiari I): data base analysis. Pediatr Neurosurg 23:260–269
11. Smith JS, Shaffrey CI, Abel MF et al (2010) Basilar invagination. Neurosurgery 66(3 Suppl):39–47
12. Tubbs RS, Demerdash A, Rizk E et al (2016) Complications of transoral and transnasal odontoidectomy: a comprehensive review. Childs Nerv Syst 32(1):55–59
13. Von Torklus D, Gehle W (1972) The Upper cervical spine. Regional anatomy, pathology and traumatology. A systematic radiological atlas and textbook. Grune&Stratton, New York, pp 1–98
14. Visocchi M, Di Martino A, Maugeri R et al (2015) Videoassisted anterior surgical approaches to the craniocervical junction: rationale and clinical results. Eur Spine J 24(12):2713–2723

Reconstruction of Vertebral Body After Radiofrequency Ablation and Augmentation in Dorsolumbar Metastatic Vertebral Fracture: Analysis of Clinical and Radiological Outcome in a Clinical Series of 18 Patients

Rosario Maugeri, Francesca Graziano, Luigi Basile, Carlo Gulì, Antonella Giugno, Giuseppe Roberto Giammalva, Massimiliano Visocchi, and Domenico Gerardo Iacopino

Abstract

Background

Painful spinal metastases usually occur in malignant neoplastic disease. Treatment for bone metastases has been largely conservative, and it includes the use of high doses of analgesics, radiotherapy, chemotherapy, hormone therapy, and bisphosphonates; however, results are sometimes transient and ineffective. In the presence of neurological involvement a surgical strategy should be considered. Recently, percutaneous procedures such as radiofrequency ablation, vertebroplasty, and kyphoplasty have been introduced as palliative techniques to treat painful vertebral metastases [3, 11, 25].

Methods

In our study we combined the use of radiofrequency ablation with vertebroplasty in the treatment of dorsolumbar metastatic vertebral fractures in order to examine the relationship between restoration of the vertebral structure and decrease in pain. From January 2014 to March 2015 we retrospectively analyzed 18 patients with malignant vertebral lesions who underwent radiofrequency ablation with vertebroplasty followed by cementoplasty, with posterior transpedicle fixation on levels near the lesions. The parameters examined were: demographics, pain relief, and the distribution of polymethylmethacrylate (PMMA) determined by the mean Saliou filling score; all complications were recorded.

Findings

The mean age of the patients was 55.72 years (range 34–69); average operative time was 60.4 min (range, 51–72). The average pain index score (visual analog score; VAS) decreased significantly from 8.05 at baseline to 3.0 ($p < 0.05$) after 6 months. The Saliou filling score revealed a distribution of PMMA in the vertebral body that was satisfactory (12–18) in eight patients, mediocre (6–12) in seven patients, and inadequate (0–6) in the remaining three patients. In two vertebrae, minimal asymptomatic cement leakage occurred in the lateral recess without neurological damage. No pulmonary embolism and no visceral or neural damage was recorded.

Conclusion

Radiofrequency ablation combined with vertebroplasty seems to achieve rapid and lasting improvement in clinical symptoms in patients with malignant vertebral lesions. There was wide diffusion of PMMA in the vertebral body, with a mean cement volume of 4.5 ml.

Keywords Radiofrequency ablation • Vertebroplasty • Minimally invasive spinal surgery • Spinal metastases • Vertebral fractures

Introduction

The spine is the most frequent location of skeletal metastases, which occur in up to 40 % of patients with cancer. The thoracic spine is affected in up to 70 % of cases, followed by the lumbar and cervical areas [27]. In the United States alone, more than 350 000 cases of spinal metastasis are reported each year—due to prostate, breast, kidney, lung, and thyroid cancers [30]. For patients with limited life expectancy arising from their underlying disease, high surgical complication rates with subsequent decrease in quality of life are most unacceptable; hence, minimally invasive techniques for the treatment of spine tumors have been explored with great enthusiasm [10, 27]. Recently, percutaneous procedures such as radiofrequency ablation (RFA; or

R. Maugeri, MD, PhD • F. Graziano, MD (✉)
L. Basile, MD • C. Gulì, MD • A. Giugno, MD
G.R. Giammalva, MS • D.G. Iacopino, MD, PhD
Department of Experimental Biomedicine and Clinical Neurosciences, School of Medicine, Neurosurgical Clinic, University of Palermo, Via del Vespro 129, Palermo 90100, Italy
e-mail: franeurosurgery@libero.it

M. Visocchi, MD, PhD
Institute of Neurosurgery, Medical School, Catholic University of Rome, Rome, Italy

M. Visocchi et al. (eds.), *Trends in Reconstructive Neurosurgery*, Acta Neurochirurgica Supplement 124,
DOI 10.1007/978-3-319-39546-3_13, © Springer International Publishing Switzerland 2017

coblation), vertebroplasty, and kyphoplasty have been introduced as palliative techniques to treat painful vertebral metastases [16, 17, 20, 21]. Radiofrequency ablation of tumorous masses has been proven, in numerous studies, to be an effective therapeutic option for the treatment of neoplasms of the liver, kidney, lung, and bone. It has also been increasingly used as pain therapy for unresectable spine tumors, either alone or in combination with vertebroplasty [6, 7]. The aim of performing RFA before cementoplasty in spinal metastasis is to destroy tumor tissue by favoring central necrosis and vessel thrombosis and thereby creating a cavity inside the vertebral body, allowing a low-pressure injection of high-viscosity cement, thus reducing the risk of cement leakage [18, 19, 25]. Besides this, the addition of polymethylmethacrylate (PMMA) may help to stabilize vertebral bodies [1, 20]. There are few reports regarding RFA combined with cementoplasty for painful bone metastases [14, 21, 26, 29, 30]. The aim of the present retrospective study was to review the efficacy and safety of RFA followed by percutaneous vertebroplasty in a clinical series of thoracolumbar vertebral metastatic fractures [30].

Materials and Methods

Study Design We retrospectively studied 18 patients (7 female and 11 male; mean age 55.72 years; age range 34–69) with painful osteolytic dorsolumbar vertebral body metastases. Radiofrequency ablation combined with vertebroplasty, together with short pedicle percutaneous fixation of vertebrae near the metastasis, was performed under general anesthesia, at the Department of Neurosurgery, University of Palermo. The patients' backgrounds are summarized in Table 1. We treated patients with unremitting pain over the spine, in the absence of symptomatic spinal cord or root compression and refractory to conventional therapeutic options such as radiation therapy, chemotherapy, surgery, and the use of analgesics. The primary neoplasms were: kidney carcinoma ($n=2$), breast carcinoma ($n=6$), lung carcinoma ($n=7$), melanoma ($n=1$), and bladder carcinoma ($n=2$) (Table 2). Lesion levels were D8 ($n=1$), D10 ($n=2$), D11 ($n=1$), D12 ($n=3$), L1 ($n=3$), L2 ($n=3$), L3 ($n=2$), L4 ($n=2$), and L5 ($n=1$) (Table 3). All the treated lesions were osteolytic, with massive bone destruction. Combined RFA and vertebroplasty was well tolerated by all patients. The procedures were performed using fluoroscopic guidance. Inclusion and exclusion criteria are summarized in Table 4.

Surgical Procedure Prior to the procedure, written informed consent was obtained from all patients. Radiofrequency ablation and the following vertebroplasty were performed percutaneously, with the aid of a fluoroscopic device, in all patients under general anesthesia. In all patients, short pedicle fixation was performed on the vertebral levels nearest the lesions during the RFA/ vertebroplasty surgical procedure. RFA was performed with the STAR Tumor Ablation System (consisting of the SpineSTAR ablation instrument and the MetaSTAR generator; DFine, San Jose, CA, USA). The SpineSTAR is an articulated navigational bipolar radiofrequency electrode containing a pair of thermocouples positioned along the length of the electrode, 10 and 15 mm from the center of the ablation zone. There is a 3: 2 length-to-width aspect ratio for the ablation zone, with the maximum ablation zone being 3 cm × 2 cm when the proximal thermocouple reaches 50 degrees Celsius. The MetaSTAR generator continuously displays the two thermocouple readings, permitting real-time monitoring of the peripheral edge of the ablation zone. Following ablation, cement augmentation (StabiliT Vertebral Augmentation System; DFine) was performed in all cases via the same working cannula [2, 17].

Follow-Up and Patient Evaluation For the evaluation of pain relief, a ten-point verbal visual analog scale (0–10 scale, with 0 indicating no pain and 10 mm the maximum imaginable pain) with intensity description was administered prior to the procedure and then at 1 week and at 1, 3, and 6 months after the procedure. The analgesic consumption was evaluated. Opioid analgesic medication use was translated into a morphine-equivalent dose and recorded [18]. Thin-slice CT scanning was performed immediately postoperatively and then 6 months after the procedure. Intraoperative assessment was made with plain radiographs or fluoroscopic images. Cement distribution and vertebral body cement filling were assessed according to the criteria of Saliou et al. [24]: lateral and anteroposterior views of the treated vertebrae were equally divided into nine ninths, summarizing 18 equal portions. Each portion was considered filled if some cement was visible in it, resulting in a vertebral body cement filling score (Saliou filling score) ranging from 1 to 18. Filling was considered satisfactory if the score was more than 12 (2/3 of the vertebra), mediocre when the score ranged from 6 to 12, and inadequate when the score was less than 6 [4, 24].

Results

Pain was significantly decreased after treatment. Mean VAS scores at baseline, at 1 week, and at months 1, 3, and 6 were 8.05 (range 6–10), 3.5 (range 2–6), 2.8 (range 2–5), 2.6 (range 2–4), and 3.0 (range 2–4), respectively (Table 5). Compared with the mean preoperative VAS score (8.5), the VAS score was significantly decreased at all time points

Table 1 Demographic data of the patients

Patient	Age	Gender	Primary cancer	Location	Pre OP VAS	Post OP 1 week	Post OP 1 month	Post OP 3 months	Post OP 6 Months	Analgesic reduction	Saliou filling score
1	54	M	Lung	L2	7	2	2	3	4	YES	*10*
2	39	F	Breast	D10	7	2	2	3	2	NO	*12*
3	57	M	Bladder	L4	9	5	3	3	4	YES	*4*
4	63	M	Bladder	D12	7	2	2	2	4	NO	*14*
5	62	F	Breast	D12	7	5	3	3	3	NO	*9*
6	69	M	Lung	L1	8	4	5	3	3	NO	*8*
7	60	M	Lung	L3	10	3	3	3	3	YES	*12*
8	34	F	Breast	L5	8	5	4	3	3	NO	*10*
9	53	M	Lung	L1	9	3	3	3	3	YES	*5*
10	49	F	Kidney	L2	7	5	3	2	2	NO	*13*
11	57	M	Breast	D12	6	4	2	2	3	YES	*9*
12	58	M	Lung	L1	8	3	2	2	2	NO	*5*
13	67	M	Lung	D11	10	3	4	4	4	NO	*12*
14	69	F	Breast	L3	7	2	2	3	4	YES	*14*
15	58	F	Kidney	L4	8	6	3	3	2	YES	*12*
16	62	M	Breast	D10	10	4	*2*	*2*	*4*	YES **S**	*9*
17	39	F	Lung	L2	8	2	*2*	*2*	*2*	NO	*13*
18	53	M	Melanoma	D8	9	3	*4*	*2*	*2*	YES	*8*
Mean	55.72	11 M/7 F			8.05	*3.5*	*2.8*	*2.6*	*3.0*		

Table 2 Data of primary neoplasms

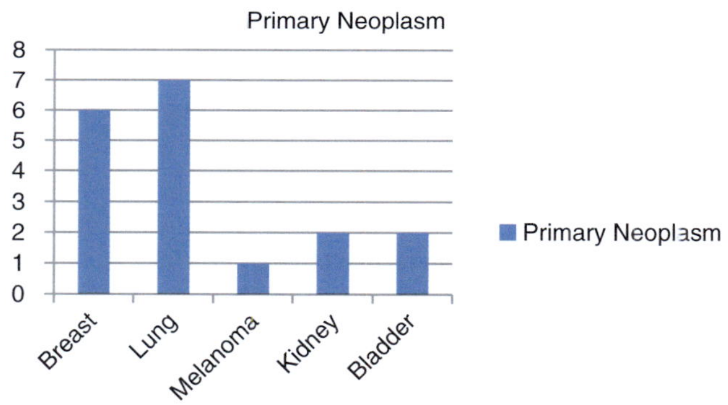

Table 3 Neoplastic vertebrae treated with radiofrequency ablation

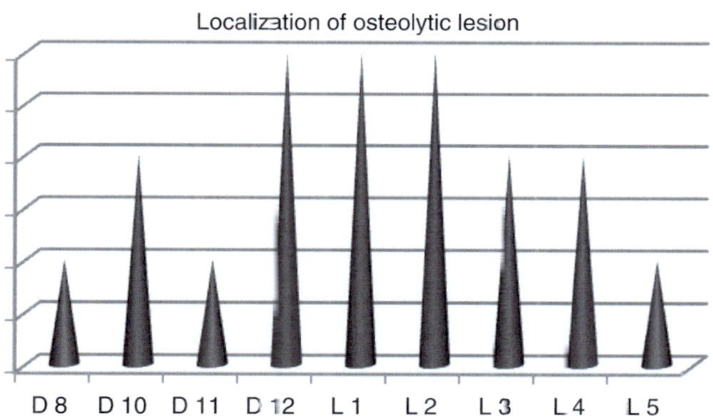

($p<0.05$) suggesting the relief of symptoms and the effectiveness of the RFA treatment combined with vertebroplasty. Analgesic reduction was achieved in most of the patients. The mean duration of the RFA procedure was 60.4 min (range, 51–72). Cement injection phases were monitored by visualizing the vertebra with fluoroscopic guidance in both anteroposterior and lateral projections. Images acquired during and immediately after the procedure demonstrated intraosseous cement distribution within the ablated cavity in all patients who were successfully coblated. Thin-slice CT scans performed immediately postoperatively revealed a satisfactory Saliou filling score (12–18) in eight patients (44.4 %), mediocre score (6–12) in seven patients (38.8 %), and inadequate score (0–6) in the other three (16.6 %), as shown in Table 1. In two vertebrae minimal cement leakage occurred in the lateral recess without any neurological damage. No pulmonary embolism and no visceral or neural damage was recorded. There was no periprocedural complication, nor was there death associated with any part of the procedure. CT scan performed 6 months after the procedure did not reveal reduction in the height of the treated metastatic vertebral body or any PMMA breakage.

Table 4 Inclusion and exclusion criteria

Inclusion criteria	Exclusion criteria
Unresecable tumors, according to Tokuhashi score	General contraindications for surgery (infection, allergy, bleeding diseases)
Karnofsky score > 60	Poor general condition (Karnofsky score < 60)
Osteolytic lesion on neurimaging	Osteoblastic tumors on neurimaging
VAS > 5	VAS < 5
Intractable pain with chemotherapy, radiation therapy, or other treatments	Spinal cord or nerve compression or intradural and intramedullary tumors

VAS visual analog scale

Table 5 Profile of the visual analog scale (VAS) in patients with back pain during follow-up

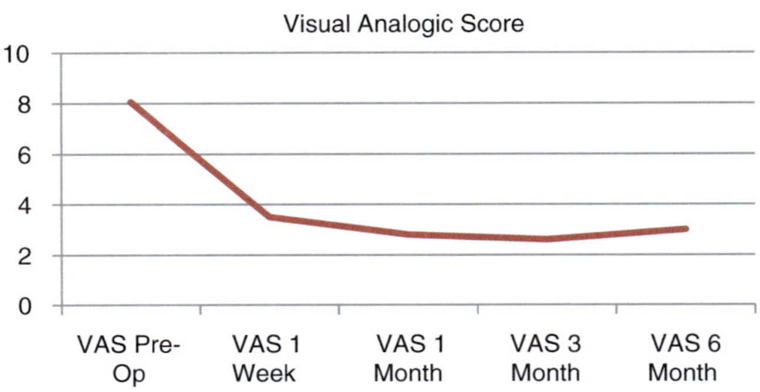

Discussion

Metastatic spine tumors, which result in destruction of the spinal column, causing mechanical instability, neurological deficit, and significant pain, affect a large number of patients each year [10]. Recently, percutaneous techniques such as RFA (or coblation), vertebroplasty, and kyphoplasty have been introduced in the armamentarium of palliative treatments for painful spinal metastases. RFA has been used successfully in many organs affected by cancer and is also receiving increasing interest as palliative treatment for vertebral metastasis [25]. However, RFA does not prevent pathologic fracture [12]. Consequently, RFA has recently been combined with vertebroplasty, with the aim being to reconstruct the mechanical stability of the vertebral body and, additionally, to remove remaining malignant cells or damaged nerve endings during cementoplasty in which the metastatic lesions are heated [30]. The aim of the present study was to evaluate whether combining RFA with vertebroplasty could restore vertebral body structure in metastatic dorsolumbar vertebral fractures, and how this may correlate with decreased pain. Our assessment of 18 patients with vertebral metastases treated with RFA with vertebroplasty revealed that this treatment had efficacy and good safety [30]. Reports on RFA, often in combination with cement injection of vertebral body metastases, are rare, and are mostly limited to case reports [14, 21, 26, 29, 30]. Dupuy et al. described a 54-year-old woman with a focal lesion in the vertebral body of L2, who was successfully treated with RFA [5]. Schaefer et al. reported a patient with an osteolytic lesion at L3 who underwent combination therapy with RFA followed by vertebroplasty in a single session [15, 26]. Gronemeyer et al., in 10 patients treated with RFA, of whom 4 were also treated with vertebroplasty, reported 74 % pain reduction at 6-month follow-up, with complete pain relief in 3 patients, without complications [13]. Goetz et al., in a total of 45 patients with osseous metastases treated with RFA, described pain relief and a reduction in analgesic consumption in all patients [12], concluding that this modality provided an effective and safe alternative method of pain palliation in patients with osteolytic metastases. Our results agree with the data in the literature and show that pain decreased significantly after treatment. Mean VAS scores at baseline, at 1 week, and at months 1, 3, and 6 were 8.05 (range 6–10), 3.5 (range 2–6), 2.8 (range 2–5), 2.6 (range 2–4), and 3.0 (range 2–4), respectively. Compared with the preoperative mean VAS (8.5), VAS was significantly decreased at all time points postoperatively ($p < 0.05$). Analgesic reduction, according to daily opioid intake during follow-up, was achieved in most patients [25]. The proposed mechanisms by which RFA decreases pain may involve: inhibition of pain transmission by destroying sensory nerve fibers in the periosteum and bone cortex; reduction of lesion volume with decreased stimulation of sensory nerve fibers; destruction of tumor cells that are producing nerve-stimulating cytokines (tumor necrosis factor-alpha [TNF-α], interleukins, etc.), and inhibition of osteoclastic activity [28]. In our study we analyzed vertebral body cement filling according to the method of Saliou, with Saliou filling scores ranging from 1 to 18. Filling was considered satisfactory if the score was more than 12 (2/3 of the vertebra), mediocre when the score ranged from 6 to 12, and inadequate when the score was less than 6. A thin-slice CT scan performed in the immediate postoperative period revealed a satisfactory Saliou filling score (12–18) in 8 patients (44.4 %), mediocre score (6–12) in 7 patients (38.8 %), and inadequate score (0–6) in the other 3 (16.6 %). As suggested by

Proschek et al., in their study of treated metastatic vertebrae, PMMA distributed inside the vertebral body provided increased biomechanical stability, and PMMA also provided analgesic effects, which were well documented [22, 23]. A review of the literature by Halpin et al. found that most studies of RFA with vertebroplasty reported significant pain reduction in 85-97 % of patients [15]. Although the mechanism of the pain reduction during vertebroplasty is unknown, possible mechanisms may include the stabilization of microfractures, the redistribution of mechanical forces, and even the cytotoxicity of the PMMA in the cement, which could destroy nerve terminals. In most series, lesions within 1 cm of the spinal cord, lesions involving the posterior wall of the vertebral body, and lesions with cortical bone destruction with involvement of soft tissue were considered to represent contraindications for treatment [28]. In our results, the particular RFA device used has an articulated electrode with two thermocouples, which allows a safer and wider ablation inside the vertebral body. The real-time monitoring procedure, with constant analysis of local temperature and near tissue impedance, reduces the risks of vascular or neural damage. We believe that creating a cavity through tissue dissolution, rather than tissue displacement alone, may aid in redirecting the cement away from the posterior aspect of the vertebral body and, hence, decrease the risk of perioperative complications. In summary, we found that combining percutaneous radiofrequency-based ablation with percutaneous cement augmentation seemed to provide a valuable addition to our surgical armamentarium for treating painful malignant spinal lesions [8, 9]. Our study has several limitations. The analysis was retrospective, implying selection bias and the lack of a control group. In addition, our vertebral sample was heterogeneous and small. Additional clinical experience and controlled prospective studies are necessary for further evaluation of the specific role of the RFA procedure combined with vertebroplasty in the management of vertebral metastatic disease.

Conclusion

The present study demonstrates that percutaneous RFA with vertebroplasty in thoracolumbar vertebral spinal metastases may be safe and effective in alleviating the pain caused by tumors and in reconstruction of the vertebral body. The articulated electrode with two thermocouples increases the extent of the vertebral area accessible to RFA because of a more controlled movement inside the vertebral body. According to our results, the combination of these two methods seems to be a promising, feasible, minimally invasive technique in the treatment of spinal metastasis.

Conflict of Interest Statement The Authors report no declarations of interest. No funds were received in support of this work.

Bibliography

1. Ahn H, Mousavi P, Chin L, Roth S, Finkelstein J, Vitken A, Whyne C (2007) The effect of pre-vertebroplasty tumor ablation using laser-induced thermotherapy on biomechanical stability and cement fill in the metastatic spine. Eur Spine J 16(8):1171–1178
2. Anchala PR, Irving WD, Hillen TJ, Friedman MV, Georgy BA, Coldwell DM, Tran ND, Vrionis FD, Brook A, Jennings JW (2014) Treatment of metastatic spinal lesions with a navigational bipolar radiofrequency ablation device: a multicenter retrospective study. Pain Physician 17(4):317–327
3. Buy X, Basile A, Bierry G, Cupelli J, Gangi A (2006) Saline-infused bipolar radiofrequency ablation of high-risk spinal and paraspinal neoplasms. AJR Am J Roentgenol 186(5 Suppl):S322–S326
4. Corcos G, Dbjay J, Mastier C, Leon S, Auperin A, De Baere T Deschamps F (2014) Cement leakage in percutaneous vertebroplasty for spinal metastases: a retrospective evaluation of incidence and risk factors. Spine (Phila Pa 1976) 39(5):E332–E338. doi:10.1097/BRS.0000000000000134
5. Dupuy DE, Hong R, Oliver B, Goldberg SN (2000) Radiofrequency ablation of spinal tumors: temperature distribution in the spinal canal. AJR Am J Roentgenol 175(5):1263–1266
6. Gazis A, Beuing O, Jöllenbeck B, Franke J, Skalej M (2012) Bipolar radio frequency ablation of spinal neoplasms in late stage cancer disease: a report of three cases. Spine (Phila Pa 1976) 37(1):E64–E68
7. Gazis AN, Beuing O, Franke J, Jöllenbeck B, Skalej M (2014) Bipolar radiofrequency ablation of spinal tumors: predictability, safety and outcome. Spine J14(4):604–608
8. Georgy BA, Wong W (2007) Plasma-mediated radiofrequency ablation assisted percutaneous cement injection for treating advanced malignant vertebral compression fractures. AJNR Am J Neuroradiol 28(4):700–705
9. Georgy BA (2009) Bone cement deposition patterns with plasma-mediated radio-frequency ablation and cement augmentation for advanced metastatic spine lesions. AJNR Am J Neuroradiol 30(6):1197–1202
10. Gerszten PC, Monaco EA (2009) Complete percutaneous treatment of vertebral body tumors causing spinal canal compromise using a transpedicular cavitation, cement augmentation, and radiosurgical technique. Neurosurg Focus 27(6):E9
11. Gevargez A, Groenemeyer DH (2008) Image-guided radiofrequency ablation (RFA) of spinal tumors. Eur J Radiol 65(2):246–252
12. Goetz MP, Callstrom MR, Charboneau JW, Farrell MA, Maus TP, Welch TJ, Wong GY, Sloan JA, Novotny PJ, Petersen IA, Beres RA, Regge D, Capanna R, Saker MB, Grönemeyer DH, Gevargez A, Ahrar K, Choti MA, de Baere TJ, Rubin J (2004) Percutaneous image-guided radiofrequency ablation of painful metastases involving bone: a multicenter study. J Clin Oncol 22(2):300–306
13. Grönemeyer DH, Schirp S, Gevargez A (2002) Image-guided radiofrequency ablation of spinal tumors: preliminary experience with an expandable array electrode. Cancer J 8(1):33–39
14. Ha KY, Kim YH, Yoo TW (2013) Intraoperative radiofrequency ablation for metastatic spine disease: report of 4 cases and review. Eur J Orthop Surg Traumatol 23(Suppl 2):S129–S134
15. Halpin RJ, Bendok BR, Sato KT, Liu JC, Patel JD, Rosen ST (2005) Combination treatment of vertebral metastases using image-guided percutaneous radiofrequency ablation and vertebroplasty: a case report. Surg Neurol 63(5):469–474; discussion 474–475
16. Hillen TJ, Anchala P, Friedman MV, Jennings JW (2014) Treatment of metastatic posterior vertebral body osseous tumors by using a targeted bipolar radiofrequency ablation device: technical note. Radiology 273(1):261–267

17. Masala S, Roselli M, Massari F, Fiori R, Ursone A, Fossile E, Laudisi A, Simonetti G (2004) Radiofrequency heat ablation and vertebroplasty in the treatment of neoplastic vertebral body fractures. Anticancer Res 24(5B):3129–3133
18. Munk PL, Murphy KJ, Gangi A, Liu DM (2011) Fire and ice: percutaneous ablative therapies and cement injection in management of metastatic disease of the spine. Semin Musculoskelet Radiol 15(2):125–134
19. Pezeshki PS, Woo J, Akens MK, Davies JE, Gofeld M, Whyne CM, Yee AJ (2014) Evaluation of a bipolar-cooled radiofrequency device for ablation of bone metastases: preclinical assessment in porcine vertebrae. Spine J14(2):361–370
20. Posteraro AF, Dupuy DE, Mayo-Smith WW (2004) Radiofrequency ablation of bony metastatic disease. Clin Radiol 59(9):803–811
21. Prologo JD, Buethe J, Mortell K, Lee E, Patel I (2013) Coblation for metastatic vertebral disease. Diagn Interv Radiol 19(6):508–515
22. Proschek D, Kurth A, Proschek P, Vogl TJ, Mack MG (2009) Prospective pilot-study of combined bipolar radiofrequency ablation and application of bone cement in bone metastases. Anticancer Res 29(7):2787–2792
23. Proschek D, Tonak M, Zangos S, Mack MG, Kurth AA (2012) Radiofrequency ablation in experimental bone metastases using a controlled and navigated ablation device. J Bone Oncol 1(2):63–66
24. Saliou G, Kocheida EM, Lehmann P et al (2010) Percutaneous vertebro-plasty for pain management in malignant fractures of the spine with epidural involvement. Radiology 254:882–890
25. Sandri A, Carbognin G, Regis D, Gaspari D, Calciolari C, Girardi V, Mansueto G, Bartolozzi P (2010) Combined radiofrequency and kyphoplasty in painful osteolytic metastases to vertebral bodies. Radiol Med 115(2):261–271
26. Schaefer O, Lohrmann C, Herling M, Uhrmeister P, Langer M (2002) Combined radiofrequency thermal ablation and percutaneous cementoplasty treatment of a pathologic fracture. J Vasc Interv Radiol 13(10):1047–1050
27. Singh K, Samartzis D, Vaccaro AR, Andersson GB, An HS, Heller JG (2006) Current concepts in the management of metastatic spinal disease. The role of minimally-invasive approaches. J Bone Joint Surg Br 88(4):434–442
28. Thanos L, Mylona S, Galani P, Tzavoulis D, Kalioras V, Tanteles S, Pomoni M (2008) Radiofrequency ablation of osseous metastases for the palliation of pain. Skeletal Radiol 37(3):189–194
29. Toyota N, Naito A, Kakizawa H, Hieda M, Hirai N, Tachikake T, Kimura T, Fukuda H, Ito K (2005) Radiofrequency ablation therapy combined with cementoplasty for painful bone metastases: initial experience. Cardiovasc Intervent Radiol 28(5):578–583
30. Zheng L, Chen Z, Sun M, Zeng H, Zuo D, Hua Y, Cai Z (2014) A preliminary study of the safety and efficacy of radiofrequency ablation with percutaneous kyphoplasty for thoracolumbar vertebral metastatic tumor treatment. Med Sci Monit 20:556–563
31. Tokuhashi Y, Uei H, Oshima M, Ajiro Y (2014) eCollection. Scoring system for prediction of metastatic spine tumor prognosis. World J Orthop 5(3):262–271. doi: 10.5312/wjo.v5.i3.262
32. Schag CC, Heinrich RL, Ganz PA (1984) Karnofsky performance status revisited: reliability, validity, and guidelines. J Clin Oncol 2:187–193. PMID: 6699671

Part III

Reconstructive Neurosurgery: Brain

Endoscopic Reconstruction of CSF Pathways in Ventricular Tumors

Piero Andrea Oppido

Abstract Neuroendoscopy is now considered to be a minimally invasive surgical approach for expanding lesions bulging into the ventricle, and it is also considered to be a relevant tool for performing biopsy procedures, fenestration of cystic walls, or for performing tumor removal in selected cases. Furthermore, the use of neuroimaging and the accurate follow-up of brain tumor patients have allowed the documentation of tumoral and pseudotumoral cystic areas that cause the obstruction of cerebrospinal fluid (CSF) pathways. Neuroendoscopic procedures enable the fenestration of cystic lesions, in addition to enabling third ventriculostomy or septostomy to restore CSF pathways. We analyze our experience regarding 77 patients affected by brain tumors arising from the wall of the third or lateral ventricle. In all cases hydrocephalus or obstruction of CSF flow was present. With an endoscopic technique, septostomy, cystostomy, endoscopic third ventriculostomy (ETV), and tumor resection were performed to control intracranial hypertension.

ETV was performed in 53 patients with noncommunicating hydrocephalus. In 4 patients with low-grade astrocytoma ETV was definitely the only surgical treatment. In 12 cystic tumors, cystostomy and marsupialization into the ventricle solved a relevant mass effect with clinical intracranial hypertension syndrome. In 10 patients, neuroendoscopic relief of CSF pathways was possible by performing septostomy with the implantation of an Ommaya reservoir or one-catheter shunt. In 5 colloid cysts and 2 cystic craniopharyngiomas, removal was possible by restoring CSF flow without other procedures. After intracranial hypertension control, in 13 malignant gliomas and 5 leptomeningeal metastases, the patients' quality of life improved sufficiently to provide for tumor adjuvant therapy.

In this series, endoscopy, due to its minimally invasive characteristics and reduced complications, was found to be safe and effective, without any relevant postoperative morbidity, gained by avoiding major surgical approaches.

Based on these results and on the increasing number of series described in the literature, we believe that endoscopic techniques should be considered a selected approach for treating CSF obstructions caused by para-intraventricular tumors. The result of using neuroendoscopy is the reconstruction of CSF pathways that bypass the tumor occlusion. This surgical procedure is not only limited to the relief of noncommunicating hydrocephalus, but it is also useful for tumor removal or biopsies and the evacuation of cystic lesions. In patients affected by malignant tumors, neuroendoscopy can be performed to control intracranial hypertension before the patients start adjuvant chemotherapy or radiotherapy.

Keywords Intraventricular tumor • Endoscopic biopsy • Endoscopic third ventriculostomy • Hydrocephalus

Introduction

Because of the central location of intra- and paraventricular tumors, the commonly employed open surgical approaches have relatively high potential morbidity and mortality [7, 21]. Notably, the increased use of neuroimaging and the accurate follow-up of brain tumor patients have now frequently allowed the documentation of tumoral and pseudotumoral cystic areas that cause the obstruction of cerebrospinal fluid (CSF) pathways [8, 18]; these tumors are often associated with dilated ventricles and intracranial hypertension [19, 20]. For this reason, even today, microsurgical removal is considered the best therapeutic option in selected cases. However, due to the deep location of intra- and paraventricular tumors, this removal remains challenging and is fraught

P.A. Oppido, MD, PhD
Neurosurgical Department, Regina Elena National Cancer Institute, Via Elio Chianesi 53, Rome 00144, Italy
e-mail: piero.oppido@ifo.gov.it

M. Visocchi et al. (eds.), *Trends in Reconstructive Neurosurgery*, Acta Neurochirurgica Supplement 124,
DOI 10.1007/978-3-319-39546-3_14,

with potential complications, which may be functional and cognitive or even life-threatening. Neuroendoscopy is now considered a minimally invasive surgical approach for expanding lesions bulging into the ventricle, as well as being considered a relevant tool for performing biopsy procedures, discontinuation of cystic walls, or performing tumor removal in selected cases [2, 4, 5, 14]. Furthermore, neuroendoscopic procedures can be used to reconstruct alternative CSF pathways and treat intracranial hypertension by the fenestration of cystic lesions, in addition to being used for endoscopic third ventriculostomy (ETV) or septostomy [6, 12, 13]. In addition to the use of endoscopic procedures for diagnostic sampling, patients with primary cystic tumors (colloid cyst, epidermoid cyst, cystic craniopharyngioma, Rathke's cleft cyst) are optimal candidates for endoscopic procedures, given the ease of cyst aspiration and cyst ablation or resection. In selected cases, partial or total tumor removal is possible to improve CSF circulation, the success of which is dependent upon the tumor consistency, a feature that may be difficult to predict preoperatively [9, 17]. Thus, many patients will benefit by treatment of the tumor manifestations of noncommunicating and compartmentalized hydrocephalus through the use of endoscopic procedures. These procedures can also be coupled with the accurate placement of catheters for intracavitary therapeutic purposes [10] or sequential aspiration. Thus, in patients in whom the disease will be treated primarily by nonsurgical means (primary central nervous system [CNS] germ cell tumors, primary CNS lymphoma, disseminated metastatic disease, malignant glioma, tectal glioma) endoscopic CSF pathway relief can offer a distinct benefit by avoiding a more extensive intracranial procedure.

Material and Methods

From 2002 to 2014, 77 patients with solid or solid cystic tumors arising from the wall of the ventricles underwent neuroendoscopic procedures. These patients ranged in age from 8 to 79 years (median 55 years.). There were 42 males and 35 females, with seven children. All patients were symptomatic: 23 (30 %) patients presented with a classical intracranial hypertension syndrome. The others presented with focal neurological signs, as well as ataxic gait, cognitive disorders, and headache with papilledema. The patients had a median presurgical Karnofsky performance score (KPS) of 55 (range 30–70). At preoperative magnetic resonance imaging (MRI), tumor sites were: 31 (40 %) in the third ventricle, 27 (35 %) in the lateral ventricle, 2 in the fourth ventricle, 4 in the Sylvian aqueduct, and 8 in the brain stem, while 5 showed leptomeningeal diffusion. Ventricular dilation with hydrocephalus or obstruction of CSF flow was present in all cases. Depending on the location of the tumor and the ventricle size, unilateral access (mainly right side) was performed in all cases. The endoscope's trajectory was planned in only 5 cases, with the help of a neuronavigation system (BRAINLab system; Feldkirchen, Germany). Septostomy, cystostomy, ETV, or tumor resection was performed with rigid (25 %) or flexible (75 %) endoscopes to control intracranial hypertension. If needed to obtain a diagnosis for further oncological treatment, a tumor biopsy was performed during the same endoscopic CSF relief procedure. We used a thulium (Tm) diode pumped solid state (DPSS) laser (Revolix LISA Laser Products, Katlenburg, Germany) for the shrinkage or tissue desiccation of thick tissue and for the hemostasis of highly vascularized tumors [11, 15].

Results

The CSF obstruction was associated with tumors of different histologies (Table 1). In 34 patients, new diagnosis tumor endoscopic biopsy was performed, combined with procedures to reconstruct CSF pathways. After diagnosis, 10 patients were operated on for microsurgical removal. From 2009, in 24 procedures the Tm laser was used for the ablation and cutting of tumor tissue, hard synechiae, and cyst membranes. With Tm laser septostomy, cyst fenestration, tumor resection, and foramen of Monro and aqueduct opening were feasible to promote the reestablishment of ventricular anatomy and hydrodynamic patterns. In 5 patients, total removal of colloid cysts was possible, and in 2 patients with cystic craniopharyngiomas subtotal removal was possible by restoring CSF flow without other procedures. In 53 patients with noncommunicating hydrocephalus, ETV was performed successfully. In 10 of these patients, the hydrocephalus was due to posterior cranial fossa tumor, while in 5 it was secondary to radionecrosis. In 12 cystic tumors cystostomy and marsupialization into the ventricle solved a relevant mass effect with clinical intracranial hypertension syndrome. In 10 patients the relief of CSF pathways was possible by performing septostomy associated with the implantation of an Ommaya reservoir or one-catheter ventriculoperitoneal (VP) shunt. In 2 patients with epidermoid cysts with postoperative entrapped fourth ventricle after aqueductoplasty and ETV, reconstruction of CSF flow was obtained (Fig. 1a, b).

Complications

No mortality or morbidity due to the procedures was present. Mild or severe bleeding was successfully controlled with the Tm Laser. Following ETV failure in six cases, a one-catheter VP shunt was implanted. In two patients with

Table 1 Histopathological diagnoses

Diagnosis	No of patients
Glioma; low grade	9
Glioma; high grade	13
Tectal glioma	4
Malignant teratoma	2
Colloid cyst	5
Radionecrosis	7
Craniopharyngioma	7
PNET	5
Lymphoma	4
Metastases	9
Leptomeningeal metastases	5
Epidermoid cyst	2
Nonspecific tumor	5

PNET primitive neuroectodermal tumor

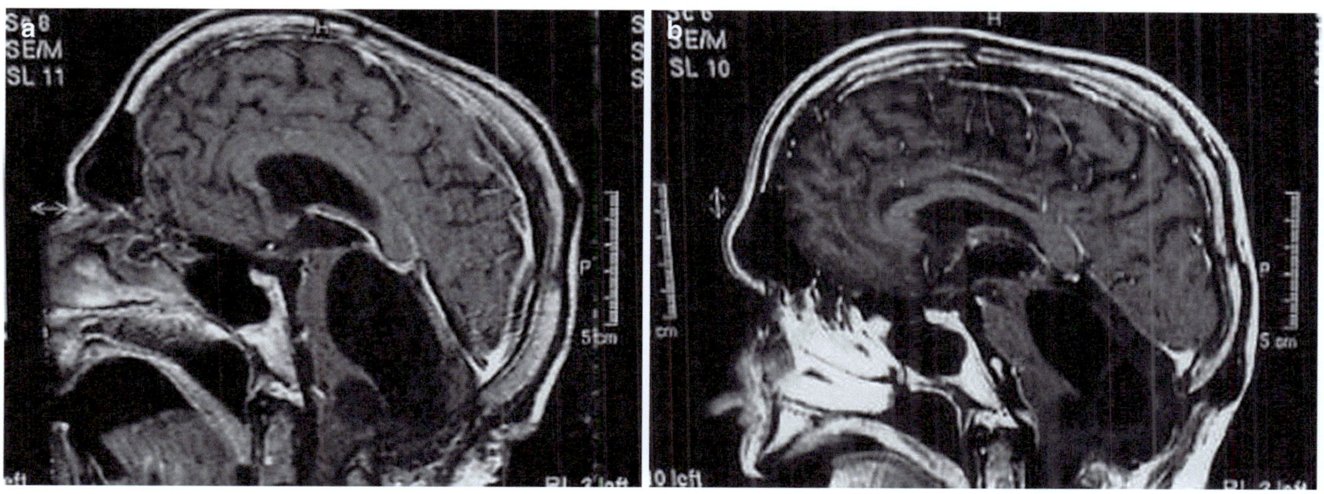

Fig. 1 (**a**) Pre-endoscopic magnetic resonance imaging (MRI), showing the entrapped fourth ventricle as a complication of microsurgical removal of an epidermoid cyst; (**b**) post-endoscopic MRI, showing reduction of the cystic fourth ventricle after aqueductoplasty and endoscopic third ventriculostomy (ETV)

entrapped ventricle the fenestration was redone, while two cystic tumors were treated microsurgically.

Follow-Up

The reconstruction of CSF pathways increased the median KPS > 80 (range 60–100). The follow-up ranged from 1 to 10 years, with periodic clinical and MRI examinations. In 4 tectal low-grade astrocytomas ETV was definitely the only surgical treatment. After intracranial hypertension control was achieved, in 13 malignant gliomas, 5 leptomeningeal metastases, and other malignant tumors, specific chemotherapy and/or radiotherapy was administered, improving quality of life and overall survival. Patients with tumors responding to therapy or radionecrosis showed a longer overall survival.

Discussion

Ventricular tumors represent a heterogeneous group in terms of histology and therapy, but they often present a common clinical history and common radiological aspects [14, 19]. The most frequent type of clinical presentation is a syndrome arising from intracranial hypertension. In the present series this syndrome was present in 30 % of our patients, accompanied by ventricle dilation due to blockage of the

CSF pathways with hydrocephalus or entrapped ventricle. In these cases the patients' clinical features ruled out the possibility of performing other therapies as an alternative to surgery, which is elective both for diagnosis and for relieving intracranial hypertension. Some tumors (i.e., lymphoma or germinoma) are radiosensitive and their surgical removal is excluded [3].

Intraventricular endoscopic surgery has an integral and expanding role in the management of patients with brain tumors. Established applications exist for tumor biopsy, concordant CSF diversion, tumor cyst decompression, and colloid cyst removal. This surgery also offers the possibility of employing ETV for the treatment of associated hydrocephalus, instead of VP shunting, a procedure that may play a role in the dissemination of some tumors, such as pineoblastomas and germ cell tumors, into the peritoneum [14].

In accordance with the literature [1, 16], in the present series endoscopy was found to be safe and effective—without any relevant postoperative morbidity—for reconstructing CSF pathways and restoring CSF flow, and for providing ETV, septostomy, tumor resection, or other endoscopic procedures. In patients affected by malignant tumors, neuroendoscopy can be performed to control intracranial hypertension before the patients start adjuvant chemotherapy or radiotherapy. Because of the tumor site these patients are frequently admitted with intracranial hypertension and a low KPS, which creates a challenge for any kind of treatment. After endoscopic procedures the reduction of intracranial pressure and improvement of the KPS is made possible, depending on the histology, adjuvant therapy, or microsurgery. Furthermore, in tectal glioma or radionecrosis ETV can be employed as the only surgical procedure without any other therapy.

Based on these results and on the increasing number of series described in the literature, we believe that endoscopic techniques should be considered as a selected approach to treat CSF obstructions caused by para-intraventricular tumors. This surgical procedure is not only limited to the relief of noncommunicating hydrocephalus, but is also useful for tumor removal or biopsies and the evacuation of cystic lesions.

Conflict of Interest Statement The authors declare that they have no conflicts of interest.

References

1. Bouras T et al (2013) Complications of endoscopic third ventriculostomy. World Neurosurg 79(2):S22.e9–S22.e12
2. Cappabianca P, Cinalli G, Gangemi M (2008) Application of neuroendoscopy to intraventricular lesions. Neurosurgery 62(suppl 2): SHC575–SHC598
3. Ferreri AJM, Dell'Oro S, Foppoli M (2006) MATILDE regimen followed by radiotherapy is an active strategy against primary CNS lymphomas. Neurology 66:1435–1438
4. Fiorindi A, Longatti P (2008) A restricted neuroendoscopic approach for pathological diagnosis of intraventricular and paraventricular tumours. Acta Neurochir (Wien) 150(12): 1235–1239
5. Gangemi M, Mascari C, Maiuri F (2007) Long-term outcome of endoscopic third ventriculostomy in obstructive hydrocephalus. Minim Invasive Neurosurg 50:265–269
6. Hellwig D, Grotenhuis JA, Tirakoroi W et al (2005) Endoscopic third ventriculostomy for obstructive hydrocephalus. Neurosurg Rev 28:1–38
7. Johnson RR, Baehring J, Piepmeier J (2003) Surgery for third ventricular tumors. Neurosurg Quart 13:207–225
8. Koeller KK, Sandberg GD (2002) Cerebral intraventricular neoplasms: radiologic-pathologic correlation. RG 22:1473–1505
9. Longatti P, Godano U, Gangemi M, Delitala A, Morace E, Genitori L, Alafaci C, Benvenuti L, Brunori A, Cereda C, Cipri S, Fiorindi A, Giordano F, Mascari C, Oppido PA, Perin A, Tripodi M, Italian neuroendoscopy group (2006) Cooperative study by the Italian neuroendoscopy group on the treatment of 61 colloid cysts. Childs Nerv Syst 22(10):1263–1267, Epub 2006 Apr 29. Erratum in: Childs Nerv Syst. 2006 22(10):1375
10. Liu W, Fang Y, Cai B, Xu J, You C, Zhang H (2012) Intracystic bleomycin for cystic craniopharyngiomas in children (Abriged Republication of Cochrane Systematic Review). Neurosurgery 71(5):909–915. doi:10.1227/NEU.0b013e31826d5c31
11. Ludwig HC, Kruschat T, Knobloch T, Teichmann HO, Rostasy K, Rohde V (2007) First experience with a 2.0 micron near infrared laser system for neuroendoscopy. Neurosurg Rev 30:195–201
12. Najjar MW, Azzam NI, Baghdadi TS, Turkmani AH, Skaf G (2010) Endoscopy in the management of intra-ventricular lesions: preliminary experience in the Middle East. Clin Neurol Neurosurg 112(1):17–22
13. O'Brien DF, Hayhurst C, Pizer B et al (2006) Outcomes in patients undergoing single trajectory endoscopic third ventriculostomy and endoscopic biopsy for midline tumors presenting with obstructive hydrocephalus. J Neurosurg 105(Suppl Pediatrics):219–226
14. Oppido PA, Fiorindi A, Benvenuti L, Cattani F, Cipri S, Gangemi M, Godano U, Longatti P, Mascari C, Morace E, Tosatto L (2011) Neuroendoscopic biopsy of ventricular tumors: a multicentric experience. Neurosurg Focus 30(4):E2. doi:10.3171/2011.1.FOCUS10326
15. Passacantilli E, Antonelli M, D'Amico A, Delfinis CP, Anichini G, Lenzi J et al (2012) Neurosurgical applicationsof the 2 μm thulium laser: histological evaluation of meningiomas in comparison to bipolar forceps and ultrasonic aspirator. Photomed Laser Surg 30(5):286–292
16. Shono T, Natori Y, Morioka T et al (2007) Results of a long-term follow-up after neuroendoscopic biopsy procedure and third ventriculostomy in patients with intracranial germinomas. J Neurosurg 107(3 Suppl Pediatrics):193–198
17. Souweidane MM, Luther N (2006) Endoscopic resection of solid intraventricular brain tumors. J Neurosurg 105:271–278
18. Tirakotai W, Hellwig D, Bertalanffy H et al (2007) The role of neuroendoscopy in the management of solid-cystic intra and periventricular tumours. Childs Nerv Syst 23:653–658
19. Waldron JS, Tihan T (2003) Epidemiology and pathology of intraventricular tumors. Neurosurg Clin N Am 14:469–482
20. Yamini B, Refai D, Rubin CM (2004) Initial endoscopic management of pineal region tumors and associated hydrocephalus: clinical series and literature review. J Neurosurg 100(Suppl Pediatrics): 437–441
21. Yasargil MG, Abdulrauf SI (2008) Surgery of intraventricular tumors. Neurosurgery 62(6 SHC Suppl 3):SHC 1029–SHC 1041

Minipterional Craniotomy for Treatment of Unruptured Middle Cerebral Artery Aneurysms. A Single-Center Comparative Analysis with Standard Pterional Approach as Regard to Safety and Efficacy of Aneurysm Clipping and the Advantages of Reconstruction

Carmelo Lucio Sturiale, Giuseppe La Rocca, Alfredo Puca, Eduardo Fernandez, Massimiliano Visocchi, Enrico Marchese, Giovanni Sabatino, and Alessio Albanese

Abstract Pterional craniotomy (PT) has long been the standard approach for the treatment of middle cerebral artery (MCA) aneurysms, even though it may cause temporalis muscle atrophy, facial nerve injury, and masticatory difficulties. Minipterional craniotomy (MPT) is an alternative approach that may provide the same surgical corridor, limiting the risk of postoperative esthetic and functional complications. From January 2011 to December 2014 we consecutively performed 68 craniotomies for surgical treatment of unruptured MCA aneurysms: 37 were standard PT and 31 were MPT. There were no significant differences in mean age, sex, and aneurysm topography between the two groups. The mean skin incision length was 14 cm in the PT group and 6 cm in the MPT group. According to the Glasgow Outcome Scale (GOS) and modified Rankin Scale (mRS), there were no significant differences in clinical outcome at discharge or follow-up between the two groups. Also, the rates of complete aneurysm exclusion were comparable. However, the number of patients complaining of masticatory disorders was higher among those treated with PT. Finally, the number of complications observed in the PT group was higher than that in the MPT group, but only the differences in mean hospitalization length and necessity for a dural patch for reconstruction were statistically significant. In conclusion, the MPT approach is a safe and effective alternative to the standard PT for the treatment of unruptured MCA aneurysms.

Keywords Pterional craniotomy • Minipterional craniotomy • Aneurysm clipping • Middle cerebral artery aneurysm • Reconstructive neurosurgery

Dr. Carmelo Lucio Sturiale and Dr. Giuseppe La Rocca are the first co-authors.
Dr. Enrico Marchese and Dr. Alessio Albanese both supervised the study as senior co-authors.

C.L. Sturiale, MD, PhD • G. La Rocca, MD (✉) • A. Puca, MD
E. Fernandez, MD • M. Visocchi, MD • E. Marchese, MD, PhD
G. Sabatino, MD • A. Albanese, MD
Institute of Neurosurgery, Medical School,
Catholic University of Rome, Rome Italy
e-mail: giularocca86@gmail.com

Introduction

Pterional craniotomy (PT) has long been the standard approach for clipping the majority of anterior circulation aneurysms [1–3]. According to Yasargil's technique, this approach provides exposure of the anterior and middle cranial fossa, the superior aspect of the posterior cranial fossa, the sellar and parasellar regions, the superior orbital fissure, and the cavernous sinus [2–4]. Nonetheless, this extremely versatile approach may show several complications, such as temporalis muscle atrophy, risk of damage to the frontal branch of the facial nerve, and masticatory difficulties. Since the first proposal by Yasargil in 1975 [1], several modifications have been developed over time. These are based on alternative dissection and reconstruction techniques, such as the interfascial and subfascial exposition of the temporalis muscle, in order to limit the risk of injury to the frontal branch of the facial nerve [5, 6], or the use of subperiosteal retrograde dissection, which has been proposed to minimize muscle atrophy [7]. Several other different surgical approaches have been described as alternatives to PT, such as the lateral supraorbital, mini supraorbital, supraorbital keyhole, sphenoid ridge keyhole, modified pterional with temporalis muscle splitting, and eyebrow approaches [8–16]. These alternative approaches have been conceived to offer surgical corridors similar to those with PT, but with shorter incisions and relative bone sparing, thus improving the cosmetic and functional results [14]. Among them, the minipterional approach (MPT) provides a surgical view comparable to that of the standard PT, and it is being increasingly used for the clipping of middle cerebral artery (MCA) aneurysms. Figueiredo et al., in 2007, first published a technical description and an anatomic assessment of this approach in a cadaveric study [12]. Accordingly, in the past 2 years we have changed the paradigm of our treatment of unruptured MCA aneurysms, systematically adopting the MPT approach. In this study we retrospectively analyzed our experience, comparing 31 cases of MPT and 37 of PT.

Material and Methods

Sixty-eight consecutive craniotomies were performed for unruptured MCA aneurysms at A. Gemelli Hospital – Catholic University School of Medicine, Rome, between January 2011 and December 2014. In particular, from January 2011 to December 2012 we treated 37 cases using a standard PT, whereas from January 2013 to December 2014 we systematically adopted an MPT approach for the treatment of 31 MCA aneurysms. Standard PT was performed according to Yasargil's technique [1]. The MPT approach, instead, followed Figueiredo's procedure: a semi-arcuate skin incision was started 1 cm above the zygomatic arch, and extended superiorly and curved gradually 1 cm above the linea temporalis superficialis (Fig. 1a) [12]. We created a single myocutaneous flap that was reflected anteriorly until the exposure of the pterion, using a subperiostal dissection in order to avoid injury to the frontalis branch of the facial nerve (Fig. 1b) [17]. A single burr hole was usually performed where the upper temporal line crosses the coronal suture (stephanion). The craniotomy proceeded superiorly along the superior temporal line, posteriorly and inferiorly curved to include the pterion within the bone flap and anteriorly directed along the sphenoid bone up to the keyhole region. After bone flap removal, the sphenoid ridge was then drilled until the meningo-orbital artery was visualized at the superior orbital fissure. This bone removal was systematically performed in order to enlarge the surgical corridor before the microsurgical dissection of the

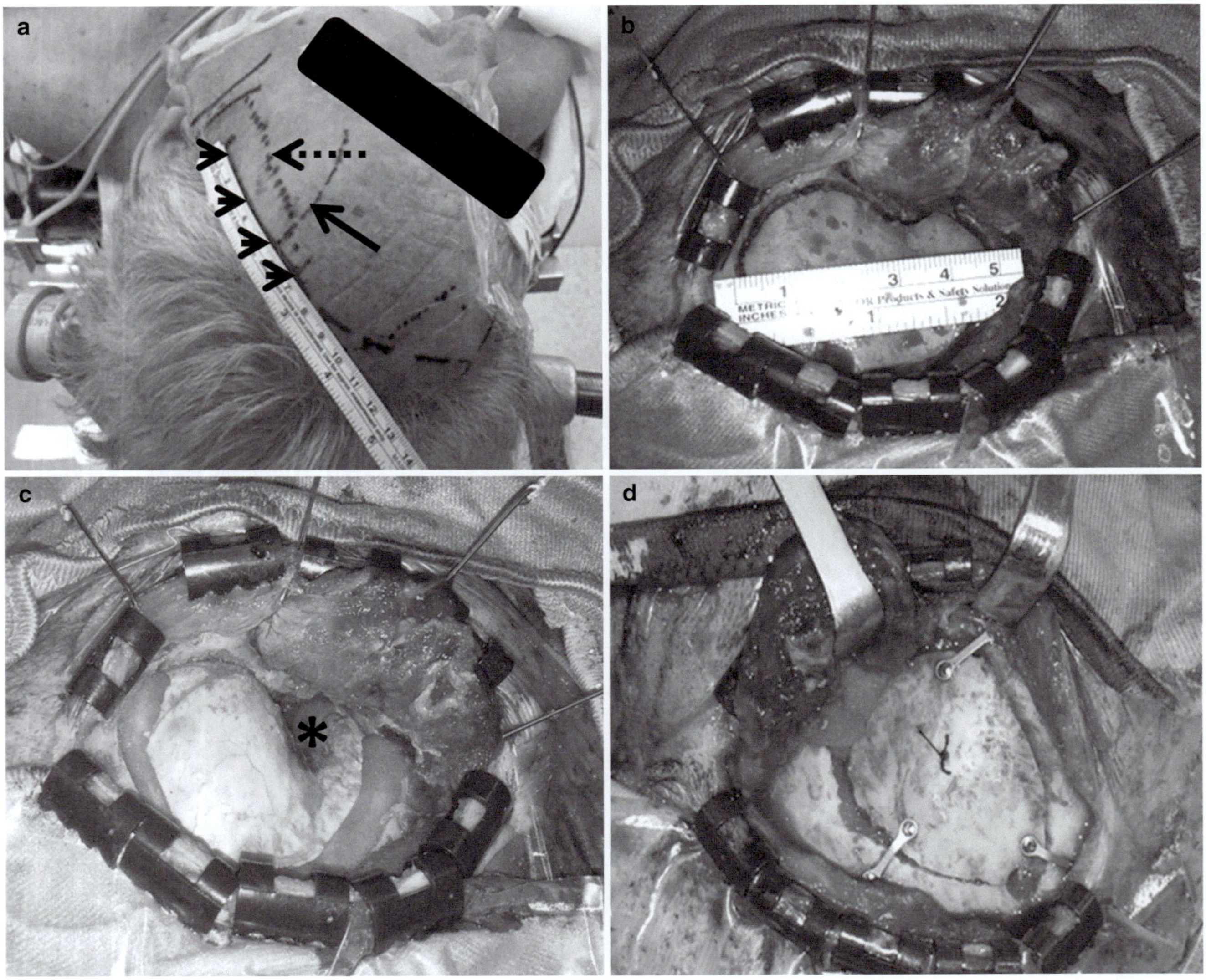

Fig. 1 Cutaneous and bony landmarks for minipterional approach. The head is fixed in a three-point Mayfield head rest, rotated about 45°, and tilted to elevate the zygoma. The hairline is marked with a dashed line (*dotted arrow*), and the hair is shaved for about 1.5 cm. The superior temporal line is then marked (*arrow*) and the skin incision is performed along a semi-arcuate trajectory (*arrowheads*) starting 1 cm above the zygomatic arch and extended 1 cm above the linea temporalis superficialis (**a**). A single burr hole is made at the stephanion and MPT of about 5 cm is performed to expose the sylvian fissure (**b**). Before dural opening, the sphenoid ridge is drilled in order to enlarge the surgical corridor before the microsurgical dissection of the sylvian fissure (**c**). The MPT craniotomy flap is synthesized with three titanium plates (**d**)

sylvian fissure was done (Fig. 1c). A semilunar dural flap was then opened, with its base directed toward the skull base. This opening provided adequate exposure of the inferolateral aspect of the frontal lobe, the proximal third of the sylvian fissure, and the superior temporal gyrus. A tailored microsurgical dissection of the sylvian fissure was then performed, starting from the proximal to distal portion, or vice-versa, depending on the fissure's vein anatomy and aneurysm localization, in order to obtain control of the parent vessel. Temporary clipping of the parent vessel was very rarely adopted and only in cases of larger aneurysms, aimed to obtain detention of the sac during manipulation and clipping. After aneurysm closure, microdoppler sonography and indocyanine green video-angiography were routinely used to confirm the complete exclusion of the sac and flow preservation in the parent vessels and perforators. At the end of the procedure, the dural flap was closed in a watertight fashion and the bone flap was replaced using three titanium plates (Fig. 1d, 2a). The temporalis muscle and galea-pericranium were reapproximated with absorbable sutures; the skin flap was usually closed with an intradermal suture in order to obtain the best esthetic result (Fig. 2b)". Outcome at discharge and follow-up was assessed by Glasgow Outcome Scale (GOS) ranking, as good recovery (GR), moderate disability (MD), severe disability (SD), vegetative state (VS), and death (D), and modified Rankin Scale (mRS) grading, from 0 to 5 points. We also submitted a questionnaire to the patients by a telephonic interview in order to assess the personal degree of cosmetic satisfaction, ranked as poor, regular, good, or excellent according to the Figueiredo et al. criteria [18], and the presence of functional disorders related to mastication, which were graded as complete restoration, pain persistence during mastication, and functional limitation. Follow-up was performed in September 2015.

Statistical Analysis

Values for quantitative variables were expressed as mean ± standard deviation and compared with each other using Student's *t*-test. Fisher's exact (two-sided) test was used to compare the categorical variables with the outcome (univariate analysis). The association between variables was considered significant when the p value was <0.05.

Results

Demographic Features

The mean age of the patients was 56.6 ± 10.6 years in the PT group, and 58.8 ± 10 years in the MPT group ($p=0.38$). There were 25 females and 12 males in the PT group, and 27 females and 4 males in the MPT group ($p=0.08$).

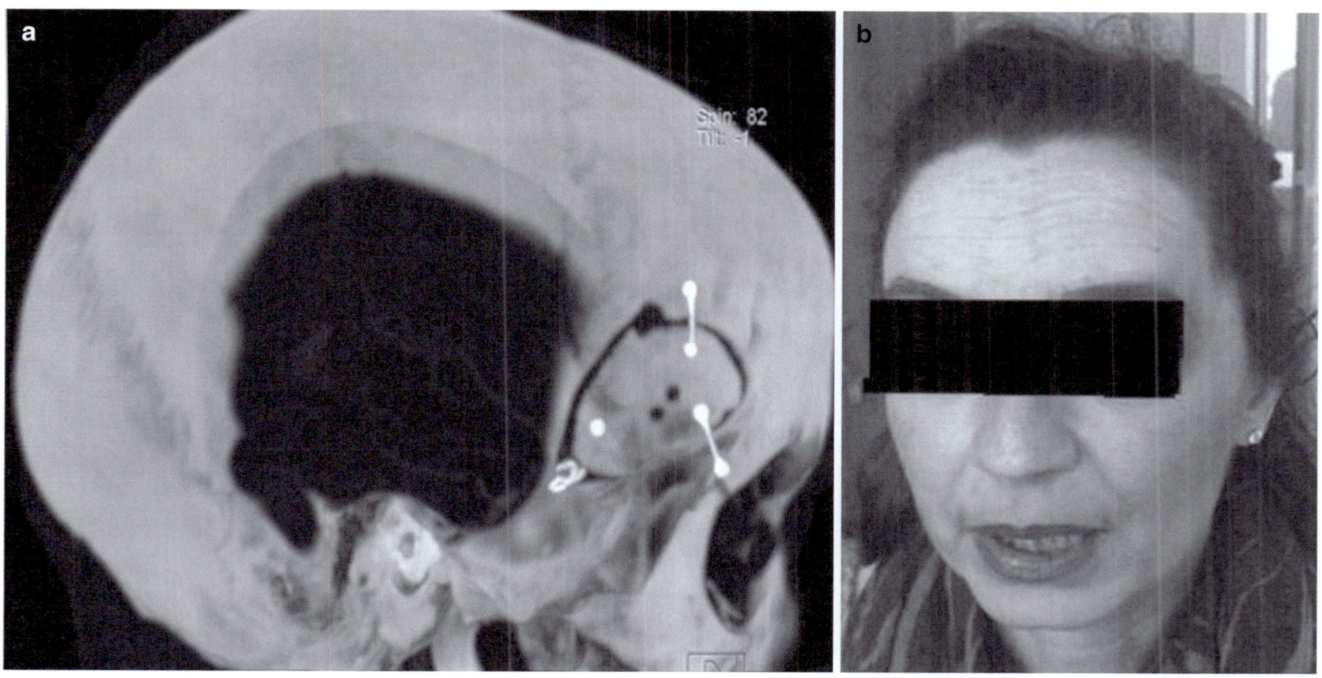

Fig. 2 Volumetric three-dimensional (3D) rendering showing the size of MPT craniotomy after osteosynthesis with titanium plates (**a**). Esthetic result after MPT approach: no alterations in the morphological profile of the patient's face are evident (**b**)

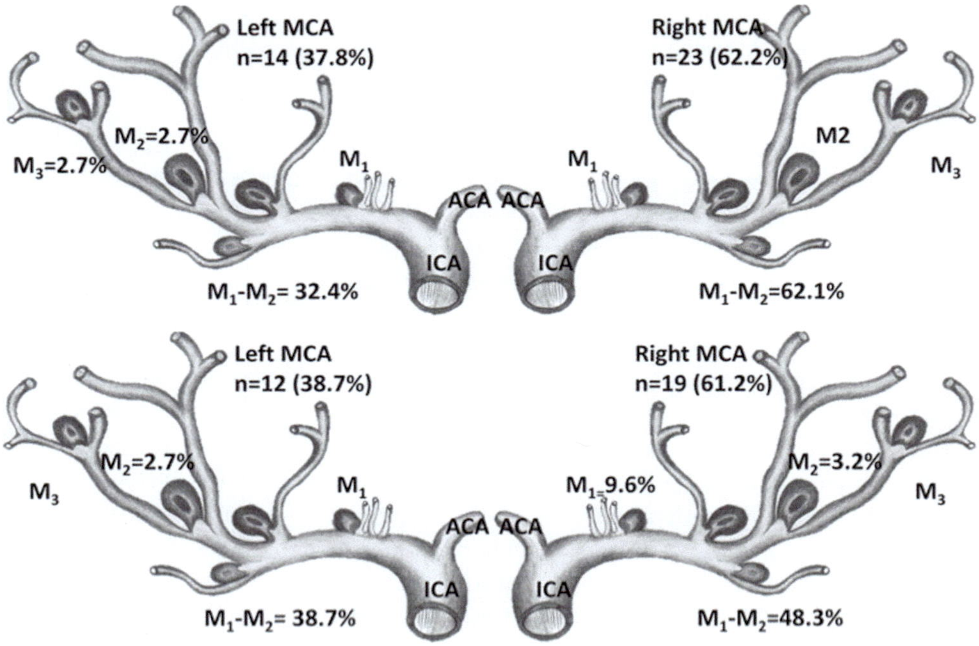

Fig. 3 *Upper figure*: topographic distribution of middle cerebral artery (MCA) aneurysm treated with standard pterional craniotomy (PT). *Lower figure*: topographic distribution of MCA aneurysms treated with minipterional craniotomy (MPT)

Topographic Distribution and Angioarchitectural Features of MCA Aneurysms

Among the 37 patients treated with standard PT, 14 (37.83 %) had an aneurysm involving the left and 23 (62.16 %) the right MCA; 17 of them had multiple aneurysms. Twelve aneurysms (32.43 %) were located at the bifurcation of the left, and 23 (62.16 %) at the bifurcation of the right MCA, 1 (2.7 %) was located at the left M2 and 1 (2.7 %) at the left M3 segment (Fig. 3 upper). Among the 31 patients treated with MPT, 12 (38.70 %) had an aneurysm involving the left and 19 (61.29 %) the right MCA; 12 of them had multiple aneurysms. Twelve aneurysms (38.70 %) involved the left MCA bifurcation, 15 (48.38 %) the right MCA bifurcation, 3 (9.67 %) the right M1, and 1 (3.22 %) the right M2 segment (Fig. 3 lower). In the PT group there were 31 small, 5 large aneurysms and 1 giant one, whereas in the MPT group, there were 30 small aneurysms and 1 large one (Table 1). The topography of the aneurysms was not significantly different between the two groups. Larger aneurysms measured 35×18 mm, whereas the smaller ones measured 3×2 mm; the mean aneurysm size was 7 mm. The mean skin incision length was 14 cm in the PT group and 6 cm in the MPT group. During the procedures with MPT, instances of compromised operative corridors requiring craniotomy extension never occurred.

Clinical Outcomes

Clinical outcomes at discharge and follow-up were assessed by the GOS and mRS. The mean length of clinical follow-up was 37.3±7.2 months for the PT group, and 20.9±7 months in the MPT group. At discharge, 35 of the 37 patients (94.59%) in the PT group showed GR, 1 (2.70%) MD, and 1 (2.70%) SD, while in the MPT group, 30 of the 31patients (96.77%) had GR, and 1 (3.22%), SD. These results were stable at follow-up. According to the mRS grading at discharge, in the PT group there were 31 patients (83.78%) with 0 points, 4 (10.81%) with 1 point, 1 (2.7%) with 3 points, and 1 (2.7%) with 4 points. At follow-up, 2 patients showed a further improvement in the quality of life, with disappearance of the previous minor symptoms, thus changing the score from 1 to 0 points. In the MPT group, in contrast, at discharge there were 28 patients (90.32%) with 0 points, 2 (6.45%) with 1 point, and 1 (3.22%) with 4 points. At follow-up, all the 3 patients who previously complained of symptoms showed an improvement; two of them, in particular, showed complete restoration from perioperative seizures and wound healing . There was no significant difference in clinical outcomes between the two groups, either at discharge or at follow-up (Table 2).

Table 1 Relationship between size and location of MCA aneurysms treated with standard PT and MPT

	M_1 segment		M_1-M_2 bifurcation		M_2 segment		M_3 segment	
	PT	MPT	PT	MPT	PT	MPT	PT	MPT
Small	–	2 R	18 R 11 L	14 R 12 L	–	1 R	1 L	–
Large	–	1 R	4 R 1 L	1 R	1 L	–	–	–
Giant	–	–	1 R	–	–	–	–	–

R Right, *L* left *MCA* middle cerebral artery, *PT* pterional craniotomy, *MTP* minipterional craniotomy

Table 2 Clinical outcomes at discharge and follow-up for PT and MPT groups

	GOS				mRS			
	Score	PT $n=37$	MPT $n=31$	*p* value	Score	PT $n=37$	MPT $n=31$	*p* value
Discharge					*0*	31	28	0.49
	GR	35	30	1.0	*1*	4	2	0.68
	MD	1	0	1.0	*2*	–	–	–
	SD	1	1	1.0	*3*	1	0	1.0
	VS	–	–	–	*4*	1	1	1.0
	D	–	–	–	*5*	–	–	–
Mean follow-up	*GR*	35	30	1.0	*0*	33	30	0.36
					1	2	0	0.49
	MD	1	0	1.0	*2*	–	–	–
	SD	1	1	1.0	*3*	1	1	1.0
	VS	–	–	–	*4*	1	0	1.0
	D	–	–	–	*5*	–	–	–

GOS Glasgow outcome scale, *mRS* modified Rankin Scale, *GR* good recovery, *MD* moderate disability, *SD* severe disability, *VS* vegetative state, *D* death (D)

Angiographic Outcome

Digital subtraction angiographic or angio-computed tomography (CT) scans were performed postoperatively in 32/37 (86.48 %) patients in the PT group, and in 29/31 (93.54 %) in the MPT group. Mean follow-up was 69.3 weeks for the PT group, and 28.5 weeks for the MPT group. With regard to patients treated with standard PT, complete aneurysm exclusion was obtained in 28/32 (87.5 %). However, in two of four aneurysms with subtotal exclusion, a residual neck was left in order to preserve the origin of a collateral branch. Therefore, the overall goal of the treatment was achieved in 30/32 cases (93.75 %). In the MPT group, in contrast, complete aneurysm exclusion was obtained in 29/29 patients (100 %).

Esthetic and Functional Outcome

Data on esthetic and functional results were also collected at the clinical follow-up. Overall personal satisfaction was high in our patients, and most of them emphasized that they were more psychologically relaxed since they had no more risk of aneurysm rupture. As regards functional disorders of mastication, a significantly higher percentage of patients treated with MPT showed complete restoration with no symptoms at follow-up. On the other hand, seven patients treated with PT still complained of pain during mastication, and seven had functional limitations (mouth opening) at follow-up, compared with only two in the MPT group. Although these differences were not statistically significant, they showed a trend (Table 3).

Table 3 Esthetic and functional comparisons between PT and MPT groups at clinical follow-up

	Score	PT group $n=37$	MPT group $n=31$	p value
Personal satisfaction	*Excellent*	30	29	0.16
	Regular	2	1	1.00
	Good	3	1	0.61
	Poor	2	0	0.49
Functional disorders	*CR*	23	29	*0.0034*
	PP	7	1	0.06
	FL	7	1	0.06

CR complete restoration, *PP* pain persistence during mastication, *FL* functional limitation

Table 4 Comparative incidence of complications in the PT and MPT groups

	PT group $n=37$	MPT group $n=31$	p value
Mean length of hospitalization (days)	9.1±4.5	6.4±2.3	*0.004*
Seizures	3	0	0.2447
CSF collection	5	1	0.2089
Hemiparesis	2	1	1.000
Infection/wound dehiscence	2	1	1.000
Hydrocephalus	1	0	1.000
Dural substitutes	8	0	*0.006*

CSF cerebrospinal fluid

Complications

Overall, we observed 13 postoperative complications in 12 patients in the PT group: 3 seizures, 5 cerebrospinal fluid (CSF) collections, 2 hemipareses, 2 wound infections, and 1 hydrocephalus. In contrast, only 3 postoperative complications were recorded in the MPT group, in 3 patients: 1 subcutaneous fluid collection, 1 hemiparesis, and 1 wound dehiscence. Moreover, in 8 patients treated with standard PT, a dural substitute patch (Liodura® (Braun, Melsungen – Germany)) was necessary for dural closure in 2 patients, and a homologous pericranium flap was necessary in 6 patients, whereas no patients who underwent the MPT approach needed a dural patch. However, the occurrence of complications did not differ significantly between the two groups. Only the length of hospitalization was significantly different in the two groups, being longer in the PT group, with a mean of 9.1±4.5 days, compared with 6.4±2.3 days in the MPT group (Table 4).

Discussion

The aim of the surgical treatment of intracranial aneurysms is the exclusion of the sac in order to eliminate the risk of bleeding. Preservation of neurological function is mandatory, in particular in cases of unruptured aneurysms [19]. Since the description of the first case of ruptured aneurysm wrapping by Norman Dott in 1933 [20], and the preliminary experience of Walter Dandy in aneurysm clipping in 1938 [21], the surgical technique of intracranial aneurysm treatment has continuously evolved. A significant improvement in surgical results was observed in 1975 when Yaşargil et al. proposed combining a standard pterional craniotomy centered on the sylvian fissure and a microsurgical technique. This keyhole approach allowed the exposure of the sylvian corridor towards the circle of Willis with a relatively small bone flap and limited exposure of the frontal and temporal lobes [22]. In addition to the limited extent of the craniotomy, the approach described by Yaşargil et al. introduced the removal of the lateral two-thirds of the lesser sphenoid wing. The surgical window so obtained allowed dissection of the sylvian fissure up to the basal cisterns with progressive CSF drainage and brain relaxation, thus reducing the retraction of the frontal lobe [3, 22, 23]. In PT according to Yaşargil's technique, the temporal muscle is usually completely dissected from the temporal fossa in order to maximize the pterional exposure. This kind of maneuver can present some drawbacks, including functional and esthetic complications such as alterations in mandibular function, chronic pain, and alterations in the facial sensory components frequently associated with temporal atrophy and injury of the frontal branch of the facial nerve and trigeminal branches [2, 12, 24–26]. The advantages of a minimally invasive procedure compared with a standard PT should be to decrease the surgical trauma and the associated pain, to provide better cosmetic results, and at the same time to reduce the risk of neurological function injury, and to reduce the length of surgical procedures and hospitalization, as well as reducing the costs. Several different techniques have been described as surgical modifications of the traditional PT craniotomy [2], but at the end they fail to provide comparable microsurgical exposure, and some do not reduce the extent of dissection of the temporal muscle [16, 27, 28]. In our experience, we found that unruptured MCA aneurysms up to 35 mm in size were effectively and safely treatable using the MPT approach. We never found that the exposure provided by the MPT limited the ability to treat these aneurysms. This finding is supported by clinical and anatomic studies, comparing the standard PT with its various modifications both in clinical series and in cadaveric dissections [12, 16–18, 29, 30]. Figueiredo et al. used computerized tracking to assess the extent of exposure in the pterional-transsylvian approach [4]. They found that dissection of the sylvian fissure distal to the anterior ascendant ramus did not provide additional exposure of the basal cisterns or circle of Willis. Therefore, in accordance with these and other findings, we conclude that MPT craniotomy is a suitable alternative to the standard PT for unruptured MCA aneurysms. In our series, MPT craniotomy was not selected for patients with aneurysmal subarachnoid hemorrhage (SAH), as the smaller bony exposure in the setting of significant brain edema could be severely limiting. The MPT craniotomy is built upon the principles of the standard PT and it is ideal for a minimally invasive approach for MCA aneurysm, with which all neurosurgeons should be confident, since it affords the patient a smaller and more cosmetically satisfactory incision. Differently from the technique described by Figueiredo et al., we did not perform an interfascial dissection of the temporalis muscle, but raised a single myocutaneous flap. One benefit of the subfascial or interfascial dissection, in addition to helping the preservation of the frontal branch of the facial nerve, is that it allows for mobilization of the temporalis muscle inferiorly to limit the bulk of muscle along the scalp flap [5, 6]. In the MPT approach, given the small size of the skin-muscle flap, we did not find this extra muscle bulk to limit the sylvian fissure dissection and aneurysm exposure. Also, we never recorded any instances in which MPT craniotomy appeared to limit the surgical exposure necessary for optimal aneurysm preparation and clipping.

Conclusion

In our study no differences in clinical outcomes were observed between the MPT and PT techniques. However, patient satisfaction was higher in the MPT group, due to the favorable cosmetic and functional results. Thus, the MPT approach appeared to be a safe and effective alternative to the standard PT for treating unruptured MCA aneurysms in all our patients.

Acknowledgments We would like to thank Francesco Cantavenera for the realization of the artwork included in this article.

Conflict of Interest Statement The authors declare no conflicts of interest and they did not receive any financial support in conjunction with the generation of this manuscript.

References

1. Yasargil MG, Fox JL (1975) The microsurgical approach to intracranial aneurysms. Surg Neurol 3(1):7–14
2. Figueiredo EG, Oliveira AMP, Plese JPP et al (2010) Perspective of the frontolateral craniotomies. Arq Neuropsiquiatr 68(3):430–432
3. Vishteh AG, Marciano FF, David CA et al (1998) The pterional approach. Oper Tech Neurosurg 1(1):39–49
4. Figueiredo EG, Deshmukh P, Zabramski JM et al (2006) The pterional-transsylvian approach: an analytical study. Neurosurgery 59(4 suppl 2):ONS263–ONS269, discussion ONS269
5. Yasargil MG, Reichman MV, Kubik S (1987) Preservation of the frontotemporal branch of the facial nerve using the interfascial temporalis flap for pterional craniotomy. Technical article. J Neurosurg 67(3):463–466
6. Coscarella E, Vishteh AG, Spetzler RF et al (2000) Subfascial and submuscular methods of temporal muscle dissection and their relationship to the frontal branch of the facial nerve. Technical note. J Neurosurg 92(5):877–880
7. Oikawa S, Mizuno M, Muraoka S et al (1996) Retrograde dissection of the temporalis muscle preventing muscle atrophy for pterional craniotomy. Technical note. J Neurosurg 84(2):297–299
8. Hernesniemi J, Ishii K, Niemelä M et al (2005) Lateral supraorbital approach as an alternative to the classical pterional approach. Acta Neurochir Suppl 94:17–21, Vienna: Springer-Verlag
9. Figueiredo EG, Deshmukh V, Nakaji P et al (2006) An anatomical evaluation of the mini-supraorbital approach and comparison with standard craniotomies. Neurosurgery 59(4 suppl 2):ONS212–ONS220, discussion ONS220
10. van Lindert E, Perneczky A, Fries G et al (1998) The supraorbital keyhole approach to supratentorial aneurysms: concept and technique. Surg Neurol 49(5):481–489, discussion 489–490
11. Nathal E, Gomez-Amador JL (2005) Anatomic and surgical basis of the sphenoid ridge keyhole approach for cerebral aneurysms. Neurosurgery 56(1 suppl):178–185, discussion 178–185
12. Figueiredo EG, Deshmukh P, Nakaji P et al (2007) The minipterional craniotomy: technical description and anatomic assessment. Neurosurgery 61(5 suppl 2):256–264, discussion 264–265

13. Czirják S, Szeifert GT (2001) Surgical experience with frontolateral keyhole craniotomy through a superciliary skin incision. Neurosurgery 48(1):145–149, discussion 149–150
14. Mitchell P, Vindlacheruvu RR, Mahmood K et al (2005) Supraorbital eyebrow minicraniotomy for anterior circulation aneurysms. Surg Neurol 63(1):47–51, discussion 51
15. Reisch R, Perneczky A (2005) Ten-year experience with the supraorbital subfrontal approach through an eyebrow skin incision. Neurosurgery 57(4 suppl):242–255, discussion 242–255
16. Harland SP, Hussein A, Gullan RW (1996) Modification of the standard pterional approach for aneurysms of the anterior circle of Willis. Br J Neurosurg 10(2):149–153, discussion 153
17. Yasargil MG (1984) Interfascial pterional (frontotemporosphenoidal) craniotomy. In: Yasargil MG (ed) Microneurosurgery, vol 1. Georg Thieme Verlag, Stuttgart, pp 215–220
18. Welling L, Figueiredo G, Wen TH et al (2015) Prospective randomized study comparing clinical, functional, and aesthetic results of minipterional and classic pterional craniotomies. J Neurosurg 122:1012–1019
19. Bederson JB, Connolly ES Jr, Batjer HH et al (2009) Guidelines for the management of aneurysmal subarachnoid hemorrhage: a statement for healthcare professionals from a special writing group of the Stroke Council, American Heart Association. Stroke 40:994–1025
20. Tan TC, Black PM (2002) Sir Victor Horsley (1857-1916): pioneer of neurological surgery. Neurosurgery 50:607–612
21. Dandy WE (ed) (1966) The Brain. WF Prior, Hagerstown
22. Yaşargil MG, Fox JL (1975) The microsurgical approach to intracranial aneurysms. Surg Neurol 3:7–14
23. Yaşargil MG, Antic J, Laciga R et al (1976) Microsurgical pterional approach to aneurysms of the basilar bifurcation. Surg Neurol 6:83–91
24. Ammirati M, Spallone A, Ma J et al (1994) Preservation of the temporal branch of the facial nerve in pterional-transzygomatic craniotomy. Acta Neurochir (Wien) 128:163–165
25. Barone CM, Jimenez DF, Boschert MT (2001) Temporalis muscle resuspension using titanium miniplates and screws: technical note. Neurosurgery 48:450–451
26. de Andrade FC Jr, de Andrade FC, de Araujo Filho CM et al (1998) Dysfuntion of the temporalis muscle after pterional craniotomy for intracranial aneurysms. Comparative, prospective and randomized study of one flap versus two flaps dieresis. Arq Neuropsiquiatr 56:200–205
27. Chehrazi BB (1992) A temporal transsylvian approach to anterior circulation aneurysms. Neurosurgery 30:957–961
28. Figueiredo EG (2008) Descrição Técnica e Avaliação Anatômica da Craniotomia Minipterional [dissertation]. Faculdade de Medicina, Universidade de São Paulo, São Paulo
29. Caplan J, Papadimitriou K, Yang W et al (2014) The minipterional craniotomy for anterior circulation aneurysms: initial experience with 72 patients. Neurosurgery 10(2):200–207
30. Kim E, Delashaw JB Jr (2011) Osteoplastic pterional craniotomy revisited. Neurosurgery 68(1 Suppl Operative):125–129

Efficacy of Microsurgical Sublabial Approach (MSA) Versus Endoscopic Endonasal Approach (EEA) for the Treatment of Pituitary Adenomas Based on Radiological and Hormonal Outcome

Pier Paolo Mattogno, Annunziato Mangiola, Giorgio Lofrese, Vera Vigo, and Carmelo Anile

Abstract

Background

The purpose of this study was to compare the surgical efficacy of the microsurgical sublabial approach (MSA) versus the endoscopic endonasal approach (EEA) for the treatment of pituitary adenomas, based on short-term (12 months) radiological and endocrinological follow-up.

Methods

One hundred and fourteen patients affected by pituitary adenoma were enrolled at our Unit between January 2007 and February 2012; 72 were treated with MSA, and 42 with EEA. The preoperative parameters considered were: type of lesion (secreting or nonsecreting), lesion size, presence of intralesional hemorrhage, lesion perimeter (nodular vs. uniform), intrasellar vs. suprasellar, involvement of cavernous sinus, and osteodural infiltration. Hormonal assays and magnetic resonance imaging (MRI) scans were performed at 12 months after the surgical procedure.

Results

Univariate analysis of the data documented a statistically significant difference in favor of MSA for the subgroups of secreting adenomas (90.9 % vs. nonsecreting 48.3 %), microadenomas (100 % vs. macroadenomas 57.1 %), adenomas without osteodural infiltration (87.5 % vs. 55.5 % with the infiltration) or those without intralesional hemorrhage (75 % vs. 45.9 % with the hemorrhage), and growth hormone (GH) adenomas (88.8 % vs. 43.7 %). Multivariate analysis confirmed the greater effectiveness of MSA for the treatment of micro-secreting adenomas.

Conclusions

Recent advances in the EEA for treating pituitary adenomas could lead to this modality replacing the microsurgical technique. In our experience the MSA allowed us to achieve better results in the treatment of microadenomas.

Keywords Pituitary adenoma • Microsurgical • Sublabial • Endoscopic • Trans-sphenoidal • Microadenoma

Introduction

Since 1907, with Schloffer's seminal report on the trans-sphenoidal approach to a sellar lesion, pituitary surgery has been in constant evolution [31]. In the 1960s – with the advent of the operative microscope – the trans-sphenoidal "route" (both sublabial and transnasal) became the gold standard to approach lesions of the sellar area, because of its remarkable reduction of morbidity and mortality compared with traditional transcranial approaches [17]. In the 1990s, the introduction of the surgical endoscope brought significant innovations to pituitary surgery [11]. Some studies in the literature have compared the most commonly used surgical techniques for approaching the sellar region, using multiparametric evaluation (costs/benefits, days of hospitalization, morbidity, extent of tumor removal, etc.) in order to define the more reliable surgical strategy [12, 29]. The purpose of our study was to compare the surgical efficacy of two procedures (the microsurgical sublabial approach [MSA] and the endoscopic endonasal approach [EEA]), based on short-term (12 months) radiological and endocrinological follow-up.

Materials and Methods

In this study, we examined a total of 114 patients affected by pituitary adenomas, who came to our attention between January 2007 and February 2012. All patients were operated on by the two senior authors (C.A. and A.M.) both with long experience in both microscopic and endoscopic pituitary surgery. Indeed,

P.P. Mattongo, MD (✉) • A. Mangiola
G. Lofrese • V. Vigo • C. Anile
Institute of Neurosurgery, Catholic University
School of Medicine, Rome, Italy
e-mail: pierpaolo.mattogno@gmail.com

M. Visocchi et al. (eds.), *Trends in Reconstructive Neurosurgery*, Acta Neurochirurgica Supplement 124,
DOI 10.1007/978-3-319-39546-3_16,

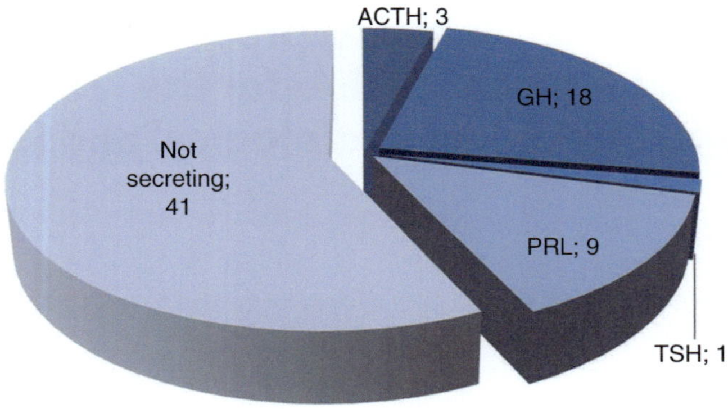

Fig. 1 The Fig. shows the endoscopic endonasal approach (EEA) subgroups in terms of hormonal secretion: there were 41 nonsecreting adenomas, and 3 adrenocorticotropic hormone (ACTH), 1 thyroid stimulating hormone (TSH), 10 prolactin (PRL), and 17 growth hormone (GH) adenomas

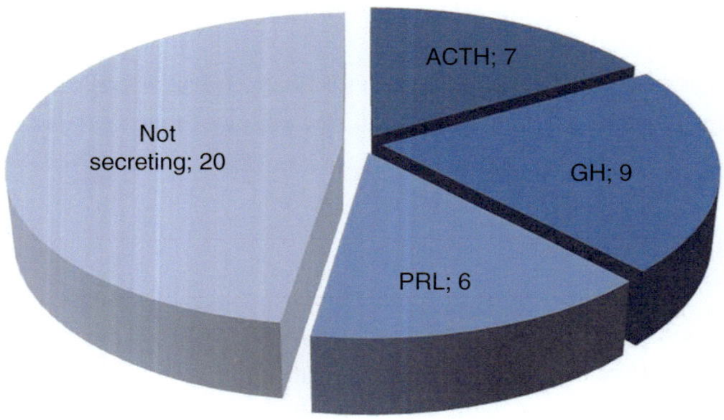

Fig. 2 The Fig. shows the microsurgical sublabial approach (MSA) subgroups, in terms of hormonal secretion: there were 20 nonsecreting adenomas, and 7 ACTH, 6 PRL, and 9 GH adenomas

their practice with microscopic pituitary surgery started in the 1980s. On the other hand, it is only since 2004 that endoscopic treatment for pituitary adenomas has been used routinely – in collaboration with ear, nose, and throat (ENT) surgeons. Informed consent was obtained from each patient enrolled in the study. The average age of the selected patients (55 men, 59 women) was 50.71 years (range 15–83). Before surgery, each patient included in the study had preoperative hormonal assays, brain magnetic resonance imaging (MRI) with gadolinium, and clinical evaluation. Of the 114 patients, 72 were treated using EEA (mean age 51.70 years; age range 21–78; 36 men, 36 women), while 42 patients were treated with MSA (mean age 48.35 years; age range 15–83; 19 men, 23 women); the patients were enrolled for each surgical technique using a randomized method. The different numbers in the two groups are due to the impossibility of assessing contemporary endocrinological and radiological follow-up 1 year after surgery in 42 patients (6 treated with EEA and 36 with MSA). Considering each group, 41 non-secreting adenomas and 31 secreting adenomas (3 adrenocorticotropic hormone [ACTH], 1 thyroid stimulating hormone [TSH], 10 prolactin [PRL], 17 growth hormone [GH]) were treated endoscopically, while 20 nonsecreting and 22 secreting adenomas (7 ACTH, 9 GH, 6 PRL) were treated using MSA (Figs. 1 and 2).

Different preoperative parameters were evaluated: type of lesion (secreting vs. nonsecreting adenoma), size (macroadenoma > 1 cm vs. microadenoma ≤ 1 cm), presence of intralesional hemorrhage, the perimeter of the lesion (nodular vs. uniform), intrasellar lesion vs. suprasellar lesion, and involvement of the cavernous sinus (no vs. yes). For the invasion of the cavernous sinus, we used the Knosp scale: grade ≥2 has been considered suggestive of invasion [22]. Osteodural invasion was evaluated during surgery. Preoperative clinical and laboratory diagnosis was confirmed by tissue histological examination (Table 1). Each of the 114 patients included in our test set were followed-up with pituitary MRI with dynamic contrast-enhancement sequences performed at 3, 6, and 12 months after surgery, and for those with secreting adenomas, a postoperative hormonal assay was performed 12 months after the surgery.

Statistical evaluation of the collected data was performed using the χ^2 test and Fisher test, in order to elaborate univariate and multivariate analyses with contingency tables.

Regarding hormonal secretion, the efficacy of the two surgical procedures was assessed in two main secreting subgroups, GH (18 EEA and 9 MSA) and PRL (9 EEA and 6 MSA). We considered patients to be healed when normalization was shown (GH: < 1 ng/ml, insulin-like growth factor 1 [IGF1] < 300 ng/ml; PRL: < 20 ng/ml) at 12-month hormonal essays, compared with the preoperative data.

Results

On preliminary analysis performed in the two groups, MSA seemed to be more effective than EEA (76.2 % vs. 47.2 %, $p<0.01$) (Table 2).

Considering the hormonal activity (secreting or nonsecreting) a significantly greater efficacy of MSA was

Table 1 Summary of all characteristics of adenomas included in the study

	EEA	MSA
Micro	7	18
Macro	65	24
Secreting	31	22
Nonsecreting	41	20
Infiltrative	36	10
Noninfiltrative	36	32
Hemorrhagic	11	2
Nonhemorrhagic	61	40
Uniform	41	35
Nodular	31	7
Intrasellar	14	23
Suprasellar	58	19
CS invasion	41	7
No CS invasion	31	35

CS Cavernous Sinus, *EEA* endoscopic endonasal approach, *MSA* microsurgical sublabial approach

evidenced in the removal of secreting adenomas: 90.9 % of patients affected by a secreting adenoma were successfully treated using the sublabial approach, vs. 48.3 % of patients operated on with endoscopy (χ^2 test, p <0.01) (Table 3). The sublabial technique was also more effective in the treatment of microadenomas, with 100 % of "sublabial" patients vs. 57.1 % of "endoscopic" patients healed (Fisher test, p <0.05) (Table 3). In noninfiltrating adenomas too, we documented better efficacy of MSA: 87.5 % sublabial vs. 55.5 % endoscopic (χ^2 test, p <0.01) (Table 3), and similar results were detected in the nonhemorrhagic adenomas, with efficacy of 45.9 % for EEA vs. 75 % for MSA (χ^2 test, p <0.01) (Table 3). Considering the main subgroups of secreting adenomas—GH and PRL—greater efficacy of MSA was documented for GH adenomas (88.8 % vs. 43.7 %, Fisher test, p < 0.05), while for the PRL adenomas the differences between the two techniques were not statistically relevant (Table 4).

In order to assess the results obtained with univariate analysis, multivariate analysis using contingency tables was performed. A statistically significant better efficacy of MSA was confirmed for microadenomas, in particular in the secreting subgroup (Fisher test, p < 0.05) (Table 5).

Discussion

The first recorded use of the trans-sphenoidal approach with successful removal of a pituitary tumor was performed in 1907 by Schloffer, through a superior nasal approach [31]. Kanavel and Hirsch suggested, in 1909, an endonasal operation via the ethmoid sinus, subsequently modified with an endonasal submucosal rhino-septal approach [17, 20]. In 1910 Cushing performed his first such operation through a sublabial incision and submucosal resection of the septum to approach the sellar region [6, 7]. However, the trans-sphenoidal approach was abandoned for the transcranial approach in the first half of the twentieth century, until the introduction of the operative microscope in the 1960s [14, 15]; nowadays, this surgical "route" represents one of the standard approaches to the sellar area. In the early 1990s the application of the endoscope for the treatment of pituitary lesions progressively gained popularity among surgeons [2, 8]. Several studies have focused on comparisons between the two different techniques (MSA and EEA) for approaching the sellar region, reporting benefits and costs for each surgical technique. Some authors have pointed out the advantage of EEA for more effective and rapid access to the sphenoid sinus compared with the microsurgical approach, with extended visualization of the sellar and parasellar areas, but it is important to consider that bleeding and air moisture could often cloud the scope and complicate the procedure [2, 19, 30]. However, MSA should permit better bleeding control as a consequence of its larger surgical corridor [32]. It is important to emphasize that the training time for the neurosurgeon seems to be longer for the EEA than for the MSA [5]. Some authors have reported that endonasal endoscopic surgery allows shorter surgical times and shorter periods in hospital [4, 18]. In addition, the endoscopic technique permits the elimination of fluoroscopy, which is mandatory in MSA for identification of the sella turcica [5]. Moreover, several authors maintain that EEA is associated with a reduced frequency of

Table 2 Total number of patients treated with EEA and MSA

	Total removal	Partial removal	p (χ^2)
EEA	34	38	
MSA	32	10	<0.01

MSA was significantly more effective than EEA ($p<0.01$)

Table 3 Univariate significance data

	EEA		MSA		
	Total removal	Partial removal	Total removal	Partial removal	p (χ^2 or Fisher test)
Secreting	15	16	20	2	$p<0.01$ (χ^2)
Microadenoma	4	3	18	0	$p<0.05$ (Fisher)
Noninfiltrative	20	16	28	4	$p<0.01$ (χ^2)
Nonhemorrhagic	28	33	30	10	$p<0.01$ (χ^2)

MSA was more effective than EEA in the treatment of the secreting, microadenoma, noninfiltrative adenoma, and nonhemorrhagic adenoma groups

Table 4 Hormonal normalization after 12 months for GH- and PRL-secreting adenomas

	EEA		MSA		
	Normalization (12-month follow-up)		Normalization (12-month follow-up)		
	Yes	No	Yes	No	p (Fisher test)
GH-secreting	7	11	8	1	$p<0.05$*
PRL-secreting	6	3	4	2	$p>0.05$

MSA was more effective than EEA for GHsecreting adenoma (*)
GH growth hormone, *PRL* prolactin

Table 5 Multivariate significance data

Micro-secreting adenomas			
	Total removal	Partial removal	p (Fisher test)
EEA	3	3	
MSA	16	0	$p<0.05$

MSA was more effective than EEA for the treatment of microadenoma, in particular for the micro-secreting adenoma subgroup

complications, although no statistically significant differences were found between the two approaches [3, 16]. Good results are reported in the use of endoscope-assisted microsurgical techniques [21, 24, 34]. However, another important point of comparison is the invasiveness of the endoscopic approach in regard to nasal structure and function: evident superiority in terms of nasal preservation is demonstrated for MSA compared with EEA [26].

In the literature, different reviews have analyzed these two surgical techniques in terms of surgical results: a systematic review by Rotenberg et al. found similar outcomes for microscopic and endoscopic techniques for pituitary tumor resection with regard to the effectiveness of tumor removal [28]; similar results were confirmed by Goudakos et al. [13]; Starke et al. reported no differences in outcome in their groups of acromegalic patients [33]. Some works, regarding pediatric and adult patients, demonstrated improved quality of the postoperative course and a reduction of surgical complications using EEA [23, 27]. Considering tumor dimensions, Atkinson et al. did not report statistically significant differences between the two surgical techniques in the treatment of microadenomas [1]. In contrast, other authors have demonstrated better efficacy of MSA in microadenoma treatment, as also documented by our results [9, 25]. Undoubtedly the microsurgical technique provides the binocular vision that allows greater care and provides better visibility during tumor removal. However, the recent introduction of three-dimensional (3D) endoscopy could overcome this difference between the techniques [10]. As a matter of fact, MSA allowed us to achieve better results in patients with microadenomas, in particular in those in the micro-secreting group, at 12-month radiological follow-up, and in the patients with GH-secreting microadenomas with the 12-month hormonal essay. We note, however, that our data could be affected either by the small patient population or by the timing of follow-up. A further evaluation based on a wider cohort will be necessary to assess our preliminary results [34].

Conclusions

EEA represents an important advance in the field of pituitary surgery. In our experience, however, MSA seemed to be more effective than EEA in the treatment of micro-secreting adenomas in regard to 12-month radiological follow-up, and MSA was more effective than EEA for GH-secreting tumors in regard to the12-month hormonal follow up. Long-term follow-up associated with a larger cohort will be necessary to confirm our results.

Conflict of Interest Statement We declare that we have no conflict of interest.

References

1. Atkinson JL, Young WF Jr, Meyer FB, Davis DH, Nippoldt TB, Erickson D, Vella A, Natt N, Abboud CF, Carpenter PC (2008) Sublabial transseptal vs transnasal combined endoscopic microsurgery in patients with Cushing disease and MRI-depicted microadenomas. Mayo Clin Proc 83(5):550–553
2. Carrau RL, Jho HD, Ko Y (1996) Transnasal–transsphenoidal endoscopic surgery of the pituitary gland. Laryngoscope 106:914–918
3. Cheng RX, Tian HL, Gao WW, Li ZQ (2011) A comparison between endoscopic trans-sphenoidal surgery and traditional trans-sphenoidal microsurgery for functioning pituitary adenomas. J Int Med Res 39(5):1985–1993
4. Cho DY, Liau WR (2002) Comparison of endonasal endoscopic surgery and sublabial microsurgery for prolactinomas. Surg Neurol 58:371–375
5. Comer BT, Young AB, Gal TJ (2011) Impact of endoscopic surgical techniques on efficiency in pituitary surgery. Otolaryngol Head Neck Surg 145(5):732–736
6. Cushing H (1932) Intracranial tumors: notes upon a series of two-thousand verified cases with surgical mortality percentages pertaining thereto. Charles C Thomas, Springfield
7. Cushing H (1912) Transsphenoidal methods of access. In: Cushing H (ed) The pituitary body and its disorders. JB Lippincott, Philadelphia, pp 296–303
8. Doglietto F, Prevedello DM, Jane JA Jr, Han J, Laws ER Jr (2005) Brief history of endoscopic transsphenoidal surgery noidal surgery—from Phillip Bozzini to the First World Congress of Endoscopic Skull Base Surgery. Neurosurg Focus 19:E3
9. Esposito V, Santoro A, Minniti G, Salvati M, Innocenzi G, Lanzetta G, Cantore G (2004) Transsphenoidal adenomectomy for GH-, PRL- and ACTH-secreting pituitary tumours: outcome analysis in a series of 125 patients. Neurol Sci 25:251–256
10. Felisati G, Lenzi R, Pipolo C, Maccari A, Messina F, Revay M, Lania A, Cardia A, Lasio G (2013) Endoscopic expanded endonasal approach: preliminary experience with the new 3D endoscope. Acta Otorhinolaryngol Ital 33(2):102–106
11. Gamea A, Fathi M, el-Guindy A (1994) The use of the rigid endoscope in trans-sphenoidal pituitary surgery. J Laryngol Otol 108(1):19–22
12. Gao Y, Zhong C, Wang Y, Xu S, Guo Y, Dai C, Zheng Y, Wang Y, Luo Q, Jiang J (2014) Endoscopic versus microscopic transsphenoidal pituitary adenoma surgery: a meta-analysis. World J Surg Oncol 11:12–94
13. Goudakos JK, Markou KD, Georgalas C (2011) Endoscopic versus microscopic trans-sphenoidal pituitary surgery: a systematic review and meta-analysis. Clin Otolaryngol 36(3):212–220
14. Hardy J, Vezina JL (1976) Transsphenoidal neurosurgery of intracranial neoplasm. In: Tompson RA, Green JR (eds) Advances in neurology. Raven, New York, pp 261–275
15. Hardy J, Wigser SM (1965) Trans-sphenoidal surgery of pituitary fossa tumors with televised radiofluoroscopic control. J Neurosurg 23:612–619
16. Higgins TS, Courtemanche C, Karakla D, Strasnick B, Singh RV, Koen JL, Han JK (2008) Analysis of transnasal endoscopic versus transseptal microscopic approach for excision of pituitary tumors. Am J Rhinol 22:649–652
17. Hirsch O (1956) Pituitary tumors - a borderland between cranial and trans-sphenoidal surgery. N Engl J Med 254:937–939
18. Jain AK, Gupta AK, Pathak A, Bhansali A, Bapuraj JR (2007) Excision of pituitary adenomas: randomized comparison of surgical modalities. Br J Neurosurg 21(4):328–331
19. Jho HD, Carrau RL (1997) Endoscopic endonasal transsphenoidal surgery: experience with 50 patients. J Neurosurg 87:44–51
20. Kanavel AB (1909) The removal of tumors of the pituitary body by an infranasal route: a proposed operation with description of the technique. JAMA 53:1704–1707
21. Kawamata T, Iseki H, Ishizaki R, Hori T (2002) Minimally invasive endoscope-assisted endonasal trans-sphenoidal microsurgery for pituitary tumors: experience with 215 cases comparing with sublabial trans-sphenoidal approach. Neurol Res 24(3):259–265
22. Knosp E, Steiner E, Kitz K, Matula C (1993) Pituitary adenomas with invasion of the cavernous sinus space: a magnetic resonance imaging classification compared with surgical findings. Neurosurgery 33(4):610–617; discussion 617–618
23. Massimi L, Rigante M, D'Angelo L, Paternoster G, Leonardi P, Paludetti G, Di Rocco C (2011) Quality of postoperative course in children: endoscopic endonasal surgery versus sublabial microsurgery. Acta Neurochir (Wien) 153(4):843–849
24. McLaughlin N, Eisenberg AA, Cohan P, Chaloner CB, Kelly DF (2013) Value of endoscopy for maximizing tumor removal in endonasal transsphenoidal pituitary adenoma surgery. J Neurosurg 118(3):613–620
25. Mortini P, Losa M, Barzaghi R, Boari N, Giovanelli M (2005) Results of transsphenoidal surgery in a large series of patients with pituitary adenoma. Neurosurgery 56:1222–1233
26. Mortini P (2014) Cons: endoscopic endonasal transsphenoidal pituitary surgery is not superior to microscopic transsphenoidal surgery for pituitary adenomas. Endocrine 47(2):415–420
27. Razak AA, Horridge M, Connolly DJ, Warren DJ, Mirza S, Muraleedharan V, Sinha S (2013) Comparison of endoscopic and microscopic trans-sphenoidal pituitary surgery: early results in a single centre. Br J Neurosurg 27(1):40–43
28. Rotenberg B, Tam S, Ryu WH, Duggal N (2010) Microscopic versus endoscopic pituitary surgery: a systematic review. Laryngoscope 120(7):1292–1297
29. Rudmik L, Starreveld YP, Vandergrift WA, Banglawala SM, Soler ZM (2015) Cost-effectiveness of the endoscopic versus microscopic approach for pituitary adenoma resection. Laryngoscope 125(1):16–24
30. Schaberg MR, Anand VK, Schwartz TH, Cobb W (2010) Microscopic versus endoscopic transnasal pituitary surgery. Curr Opin Otolaryngol Head Neck Surg 18(1):8–14
31. Schloffer H (1907) Erfolgreiche Operation eines Hypophysentumors auf nasalen Wege. Wien Klin Woch 20:621–624

32. Sheehan MT, Atkinson JLD, Kasperbauer JL, Erickson BJ, Nippoldt TB (1999) Preliminary comparison of the endoscopic transnasal vs. the sublabial transseptal approach for clinically nonfunctioning pituitary macroadenomas. Mayo Clin Proc 74:661–670
33. Starke RM, Raper DM, Payne SC, Vance ML, Oldfield EH, Jane JA Jr (2013) Endoscopic vs microsurgical transsphenoidal surgery for acromegaly: outcomes in a concurrent series of patients using modern criteria for remission. J Clin Endocrinol Metab 98(8):3190–3198
34. Yang I, Wang MB, Bergsneider M (2010) Making the transition from microsurgery to endoscopic trans-sphenoidal pituitary neurosurgery. Neurosurg Clin N Am 21(4):643–651, vi

Direct and Oblique Approaches to the Craniovertebral Junction: Nuances of Microsurgical and Endoscope-Assisted Techniques Along with a Review of the Literature

Massimiliano Visocchi, Antonino Germano', Giuseppe Umana, Armando Richiello, Giuseppe Raudino, A.M. Eldella, Gerardo Iacopino, and Giuseppe Barbagallo

Abstract

Purpose

The aim of this review is to provide an update of the technical nuances of microsurgical and endoscopic-assisted approaches to the craniovertebral junction (transnasal, transoral, and transcervical), and to report on the available clinical results in order to identify the best strategy.

Methods

A nonsystematic update of the reviews and reporting on the anatomical and clinical results of endoscopic-assisted and microsurgical approaches to the craniovertebral junction (CVJ) was performed.

Results

Pure endonasal and cervical endoscopic approaches still have some disadvantages, including their steep learning curves and their deeper surgical fields. Endoscopically assisted transoral surgery with 30° endoscopes represents an emerging option compared with standard microsurgical techniques for transoral approaches to the anterior CVJ. This approach should be considered as complementary to, rather than as an alternative to the traditional transoral-transpharyngeal approach.

Conclusions

The transoral (microsurgical or video-assisted) approach with sparing of the soft palate still remains the gold standard compared with the "pure" transnasal and transcervical approaches, due to the wider working channel provided by the former technique. The transnasal endoscopic approach alone appears to be superior when the CVJ lesion exceeds the upper limit of the inferior third of the clivus. Of particular interest is the evidence that advances in reduction techniques can avoid the ventral approach.

Keywords Craniovertebral junction • Transoral approach • Transnasal approach • Transcervical approach

M. Visocchi, MD (✉)
Institute of Neurosurgery, Medical School, Catholic University of Rome, Rome, Italy
e-mail: mvisocchi@hotmail.com

A. Germano', MD
Department of Biomedical Sciences, Neurosurgery, University of Messina, Messina, Italy

G. Umana, MD
Institute of Neurosurgery, Tor Vergata University of Rome, Rome, Italy

A. Richiello, MD
Department of Neurosurgery, S Camillo Forlanini Hospital Rome, Rome, Italy

G. Raudino, MD • G. Barbagallo, MD
Division of Neurosurgery, Department of Neurosciences, Policlinico "G. Rodolico" University Hospital, Catania, Italy

A.M. Eldella, MD
Department of Neurosurgey, Misurata Centra Hospital, Misurata, Libya

G. Iacopino, MD
Department of Experimental Biomedicine and Clinical Neurosciences, School of Medicine, Neurosurgical Clinic, University of Palermo, Palermo, Italy

Introduction

Endoscopic endonasal, transoral, and transcervical approaches have recently been developed as promising alternatives to traditional transoral microsurgery for the craniovertebral junction (CVJ), and these approaches may become more mainstream as experience with them increases (drawbacks of these newer approaches are a steeper learning curve and loss of three-dimensional visualization) [40, 41].

The transoral-transpharyngeal approach historically remains the "gold standard" for anterior approaches to the upper cervical spine when indicated according to the Menezes algorithm [15]. However, there are still technical difficulties with the operating microscope, such as the need to see and

M. Visocchi et al. (eds.), *Trends in Reconstructive Neurosurgery*, Acta Neurochirurgica Supplement 124,
DOI 10.1007/978-3-319-39546-3_17, © Springer International Publishing Switzerland 2017

work through a narrow opening in a deep cavity and the need to improve visualization; soft palate splitting and even hard palate resection along with extended maxillotomy are occasionally required. To overcome such complications, endoscopic-assisted procedures have been developed for CVJ decompression. The endoscopic approaches to the CVJ include the endoscopic endonasal approach, the endoscopic transoral approach, the robot-assisted endoscopic transoral approach, combined endoscopic transnasal and transoral approaches, and the endoscopic transcervical approach [18, 22]. The aim of the present review is to provide an update on the anatomical fundamentals of endoscopic-assisted surgery for the CVJ and to report on the available clinical results.

Anatomical Studies of Endoscopic Craniovertebral Approaches

At present, the most commonly used endoscopically assisted approaches to the CVJ include the transnasal, transoral, and transcervical routes (Table 1).

Endoscopic Transoral Approach

In 2004, de Divitiis et al. studied an endoscopic transoral-transclival intradural approach on 15 cadavers, without maxillotomy or mandibulotomy, and estimated a safe entry zone achieved endoscopically through the clivus [10].

In 2006, Balasingam et al. conducted a cadaveric anatomical study to assess the area of surgical exposure and the available liberty of action for instrument manipulation by four different surgical approaches to the extracranial periclival region: the traditional transoral route, transoral with a palate split, LeFort I osteotomy, and median labioglossomandibulotomy [4].

In 2009, Pillai et al. performed an odontoidectomy in nine specimens by a direct transoral approach; endoscope-assisted (five cases) or a combined endoscopic-microscopic procedure, evaluating the surgical working area and the surgical freedom; the authors concluded that the endoscope and image guidance allowed them to approach the ventral CVJ transorally with minimal tissue dissection, no palatal splitting, and no compromise of surgical freedom [31].

Endoscopic Endonasal Approach

The main advantages of the endoscopic endonasal approach to the ventral CVJ are minimal invasiveness, unlimited surgical access to the rostral midline CVJ, avoidance of palatal split, and less operative morbidity overall compared with the transoral approach. Thanks to a relatively inclined surgical trajectory, in a rostral-to-caudal direction, the compressive pathology of basilar invagination, including the lower clivus and odontoid tip, may be removable without removing the C1 anterior arch, thus maintaining the stability of C1–C2 [2]. In 2009, Kassam's team published the concept of the "nasopalatine line" (NPL) [12]. The NPL is a reliable predictor of the maximal length of inferior dissection, and odontoid surgery can be performed safely according to a preoperative radiological study of the potential anatomical limitations of the endonasal approach. In 2012 Aldana et al. proved that a line in the midsagittal plane, the nasoaxial line (NAxL), connecting the midpoint of the distance from the rhinion to the anterior nasal spine of the maxillary bone and the C2 vertebra, tangential to the posterior nasal spine of the palatine bone, accurately predicted the lowest limit of this approach on the cervical spine [1].

Endoscopic Transcervical Approach

In 2011, Russo et al. [35] described the microsurgical anatomy and limits of exposure of the endoscopically assisted high anterior cervical, submandibular approach to the clivus and foramen magnum; the optimal route to access pathologies located ventral to the pontomedullary region. Two extensions of the approach were studied and described: an extended anterior far-lateral clivectomy and an inferior petrosectomy, thus extending the exposure to the anterior foramen magnum and the anterior cerebellopontine region.

Comparison Studies

In a study on nine cadaver heads, in 2009, Baird et al. assessed surgical access to the craniovertebral junction using three endoscopic approaches: endonasal, transoral, and transcervical. Data suggested that the surgical goals of lower clival and odontoid decompression were achieved using the endonasal and transoral approaches, and the distance to the target area was shorter in the first approach. The transcervical approach was unable to achieve more than 1 cm of lower clival resection, thus not allowing complete odontoid resection [5]. In 2010, Seker et al. reported that the transnasal endoscopic approach provided a shorter route to the CVJ, while the transoral approach achieved a wider opening [36].

Table 1 Major findings in anatomical studies of endoscopic-assisted approaches to the craniovertebral junction (CVJ)

Author	Year	Approach	Major findings
Ammirati and Bernardo [3]	1998	Endoscopic transoral approach	Median mandibulotomy/glossotomy or the LeFort I approach with hard palate splitting of atlanto-occipital and C1–C2 joint access is not necessary
de Divitiis et al. [10]	2004		A limited clival and dural opening (20 × 15mm) allows full view of the anterolateral brainstem and cisternal spaces around it, from the spinomedullary junction to the interpeduncular cistern
Balasingam et al. [4]	2006		Both median labioglossomandibulotomy and the classic transoral route provide good exposure of the CVJ but limited exposure of the clivus, which was, instead, well visualized in its inferior third by the transoral route with a palate split. Maximal exposure of the extracranial clivus was gained by the LeFort I approach
Youssef [46]	2008		Mandibulotomy and mandibuloglossotomy decreased operative distance, while increasing exposure in the axial and sagittal planes. Palatectomy increased rostral exposure without changing the caudal or axial exposure or the operative distance
Pillai et al. [31]	2009		The use of an endoscope coupled with image guidance offers several advantages for providing access to the lower clivus and C1-C2 region
Dallan et al. [11]	2012		The combined transoral transnasal approach is the best answer to gain adequate space and optimal visualization in the rhinopharyngeal and upper clival region
Alfieri et al. [2]	2002	Endoscopic transnasal approach	First description, in an anatomical study, of the endonasal route to the craniovertebral junction, providing access from the anterior cranial fossa to the whole clivus and the upper cervical spine up to the body of C2
Messina et al. [27]	2007		Data suggest that the binostril technique provides, without any additional surgical trauma, better maneuverability of the surgical tools and the possibility to work with "three hands"
Ciporen et al. [7]	2010		The combination of supraorbital or transorbital endoscopic pathways with transnasal approaches appears to improve anatomical target visualization in the central corridor of the anterior cranial fossa
Aldana et al. [1]	2012		A line in the midsagittal plane, the nasoaxial line (NAxL), accurately predicted the lowest limit of the CVJ
Little [24]	2013		Significant increase in angular range of motion during flexion/extension and axial rotation at the C0-C1 joint after inferior-third clivectomy and intradural exposure of the foramen magnum, suggesting posterior surgical fusion
Perez-Orribo [32]	2013		Increase of range of motion mostly in flexion/extension and less in axial rotation at the C0-C1 joint after removal of the lower third of the clivus and progressive occipital condylectomy
Russo et al. [35]	2011	Endoscopic transcervical approach	The study described the microsurgical anatomy and the limits of exposure of the high anterior cervical submandibular approach to the clivus and foramen magnum, endoscopically assisted

(continued)

Table 1 (continued)

Author	Year	Approach	Major findings
Baird et al. [5]	2009	Compared approaches	Surgical goals of lower clival and odontoid decompression were achieved using the endonasal and transoral approaches. The transcervical approach was unable to achieve more than 1 cm of lower clival resection, not allowing complete odontoid resection
Seker et al. [36]	2010		Both transoral and transnasal approaches provide direct access to the CVJ, avoiding neural and brain retraction, but with a difference in level and extent of exposure. The transnasal endoscopic approach provides the shorter route to the CVJ, while the transoral exposure gains a wider opening
Visocchi et al. [41]	2014		The endoscope-assisted transoral approach allows better surgical control of the CVJ, in sagittal and transverse planes, providing a larger working channel and easier maneueverability. The transnasal approach is limited in the caudal direction down to the nasopalatine line (NPL); the transoral approach is limited in the rostral direction
Van Abel [38]	2015		According to a recent anatomical study, the lower incidence of post operative dysphagia with the endonasal approach is likely related to the lower density of neuronal elements from the pharyngeal plexus above the palatal plane
Visocchi et al. [47]	2015		The surgical palate inferior arcade (SPIA) represents the maximal extent of the superior dissection for the transoral approach. Interestingly, it can be drawn by a simple lateral head X-ray examination with open mouth. SPIA is more reliable than NAxL

However, the two approaches should be considered as complementary rather than as alternatives. When removing large lesions that extend from the upper clivus to below C2, the transnasal and transoral routes may be successfully combined. The transcervical approach has the clear clinical advantage of reducing the risk of meningitis and of cerebrospinal fluid leak; its advantages also include maintaining a sterile surgical field, a familiar approach, and an optimal surgical trajectory for pathological findings lower than C2.

In 2012, Dallan et al. [11] investigated a new robotic surgical setting, the DaVinci system, in two cadavers, comparing the traditional transoral and the combined transoral-transnasal approaches to the CVJ. They concluded that the lower the placement of the robotic arms, the easier was the dissection of the rhinopharynx, basisphenoid, and upper clivus.

Visocchi et al. [42] compared the surgical exposition angle and the working channel volume of both the transnasal and transoral approaches in a cadaver, by means of a comparative neuroradiological "real-time" study. They concluded that the transnasal approach, as widely discussed, is a viable strategy for reaching the CVJ, but that this approach has limited angular (nostrils, choanae) and linear (NPL) surgical exposure, which, in our view, makes it suitable only for certain types of diseases and prevents its systematic applicability in other conditions, such as lateral tumors and pathologies caudal to C2. However, an obvious advantage of this approach is that there is no need to cut the soft palate; this minimizes potential postoperative morbidities, such as swallowing disturbances and hypernasal speech, which have a major negative impact on the quality of life (if there is a palatine veil dysfunction). The transoral approach provides a better exposure of the CVJ, both on the sagittal plane and on the transverse plane. Finally, the combination of the two approaches must be considered as an option for accomplishing a particular surgical goal. From a purely anatomical point of view, the results of Visocchi et al. seem to suggest that, in normal anatomical conditions, the transnasal approach to the CVJ is an oblique approach, which allows only the piecemeal removal of CVJ pathology and is not recommended for large tumors and low and far laterally sited CVJ pathologies. The transnasal approach is limited in the caudal direction down to the NAxL, whereas the transoral approach is limited in the rostral direction in an anatomically normal specimen [42]. In a further study, Visocchi and colleagues have confirmed the NAxL to be a reliable preoperative predictor of the maximal extent of inferior dissection for the transnasal approach. Moreover, these authors identified the corresponding palatal line for evaluating the upper limit of the transoral approach (from the inferior dental arch up to the hard palate), which represents the maximal extent of superior dissection; they called it the surgical palate inferior arcade (SPIA), and, interestingly, it can be found by a simple lateral head X-ray examination with open mouth. The NAxL appears to vary more than the SPIA. Finally, the pros and cons of each approach have to be taken into account, as well as the choice of a combined transoral and transnasal approach [47].

Surgical Studies (Table 2)

In regard to complications associated with the endoscopic endonasal approach, Valero et al., in 2015 [29], in a comprehensive literature search of several databases indexing the English-language literature published from 1990 to November 13, 2014, reported cerebrospinal fluid (CSF) leakage in 18 % with this approach. One patient developed meningitis that was complicated by sepsis and death, resulting in a procedure-related mortality of 1.4 %. Transient velopharingeal insufficiency was seen in three patients (4.2 %) and two patients had respiratory failure in the perioperative period.

Liu et al. 2015 [26] reported the operative technique and technical nuances used in their institution. In particular they use two surgeons (neurosurgeon and otolaryngologist) with a three- to four-hand approach via binostril access. They start with a 30° angled HD (Hight Definition) 4-mm endoscope. A zero degree endoscope is preferred in cases of cranial settling in which the odontoid is located very high, above the hard palate. A pedicled nasoseptal

Table 2 Surgical results of endoscopic-assisted surgery for the craniovertebral junction

Author	Approach	No. of patients°	Patient disease	Mean age (years)	Associated posterior fusion	Complications
Frempong-Boadu et al. [17]	Endoscopic transoral approach	7	3 congenital anomalies, 1 degenerative, 1 traumatic, 1 pseudogout granulation mass, 1 neoplasm	49.3	5/7	1 death, from myocardial infarction
Kassam et al. [23]	Endoscopic transoral approach	1	1 degenerative	73	1/1	None
Husain et al. [21]	Endoscopic transoral approach	11	7 congenital anomalies, 2 trauma, 2 degenerative	27.7	11/11	2 pharyngeal wound dehiscence, 1 immediate postoperative neurological worsening, 2 posterior wall infection
Wolinsky et al. [44]	Endoscopic transcervical approach	3	3 congenital anomalies	61.6	3/3	1 intraoperative CSF leakage
Wu et al. [45]	Endoscopic endonasal approach	3	2 degenerative, 1 traumatic	44	3/3	1 intraoperative CSF leakage
McGirt et al. [28]	Endoscopic transcervical approach	4	4 Congenital anomalies	14	4/4	1 subluxation with Halo vest
Menezes [30]	Transoral approach	280	267 Congenital anomalies, 7 tumors, 6 other	16	280/280	2 pharyngeal wound dehiscence, 5 velopalatine incompetence
Perrini [33]	Transoral approach	34	34 Congenital anomalies	55	32/34	2 dural lacerations, 1 oral wound dehiscence, 2 urinary infections, 2 pulmonary embolisms, 1 pseudoarthrosis, 1 velopharyngeal dysfunction, 4 deep vein thromboses, 2 posterior wound infections, 1 chest infection
El-Sayed et al. [16]	Total Transoral approach (3) Combined endoscopic transnasal and transoral approaches (8)	11	Endoscopic: 2 tumors, 2 infections, 1 degenerative, 3 congenital anomalies; open: 3 degenerative	54	No report	Dysphagia, airway complications

(continued)

Table 2 (continued)

Author	Approach	No. of patients°	Patient disease	Mean age (years)	Associated posterior fusion	Complications
Lee et al. [25]	Endoscopic transnasal approach	4	1 degenerative, 2 congenital anomalies, 2 neoplasms	48	4/4	1 intraoperative CSF leakage
Visocchi et al. [39]	Endoscopic transoral approach	7	3 neoplasms, 1 traumatic, 1 degenerative, 2 congential anomalies	6–78	6/7	None
Salunke et al. [34]	Transoral approach	24	24 congenital anomalies	5–60	15/24	4 prolonged ventilation, 1 CSF leakage, 1 residual compression, 4 pharyngeal wound infections
Dhaliwal et al. [37]	Transoral approach	22	7 neoplasms, 7 congenital anomalies, 6 degenerative	50	19/22	1 spinal cord injury, 1 durotomy, 2 wound complications, 1 hardware failure, 3 prolonged dysphagias, 6 infections, 5 respiratory distress, 2 deep vein thromboses.
Gladi et al. [19]	Endoscopic endonasal approach	4	4 degenerative	74	2/4	None
Dasenbrock et al. [14]	Endoscopic transcervical approach	15	5 degenerative, 9 congenital anomalies	42	15/15	2 urinary tract infections, 2 upper airway swelling, 2 dysphagias, 1 asymptomatic pseudomeningocele
Choi and Crockard [8]	Transoral approach	533	95 congenital anomalies, 216 degenerative, 34 traumatic, 100 tumors, 14 infections, 20 other conditions	46.3	228/533	6 CSF leakages, 11 sepsis, 13 meningitis, 34 infections, 19 cardiovascular complications, 71 respiratory complications, 15 dysphagias, 2 hematomas, 33 velopharyngeal incompetence, 4 cranial nerve palsy, 20 fixation failures, 7 subaxial instability requiring surgery, 5 paralysis, 9 other complications
Hickman et al. [20]	Endoscopic transnasal approach	2	2 congenital anomalies	11–12	2/2	1 incomplete resection of the odontoid process, 1 minimal swallowing impairment
Morales-Valero [29]	Endoscopic transnasal approach	Review	The endoscopic endonasal approach, rather than being considered an alternative, should be considered as a complementary approach to the standard transoral-transpharyngeal route	55.8		CSF leakage; 18 % intraoperative and 4.2 % postoperative; Mortality 1.4 %, transient velopharyngeal insufficency; 2 patients respiratory failure
Gladi [19]	endoscopic transnasal	4	rheumatoid pannus and basilar invagination (BI)		2/4	None
Chaudry [9]	endoscopic transnasal	1	BI and moderate cranial settling	47	No	intermittent mild dysphagia

(continued)

Table 2 (continued)

Author	Approach	No. of patients°	Patient disease	Mean age (years)	Associated posterior fusion	Complications
Ponce-Gomez [34]	Total 5 endoscopic transnasal 7 transoral microsurgical	12	craniovertebral junction instability	18-52	12/12	In the transoral group, 2 patients had postoperative dysphonia, 1 patient presented with dysphagia, and 1 patient had intraoperative CSF leakage. The endoscopic procedure required longer surgical time, less time to extubation and oral feeding, and a shorter hospital stay, with no complications in this series
Menezes [15]	Transoral approach	800	In small children an endonasal approach may be limited by the small nares. If reduction cannot be achieved, a 540° procedure may be necessary in some cases (depending on the pathology), whereby the posterior approach and incision is temporarily closed and the patient is repositioned supine for a ventral decompression, followed by reopening of the posterior incision and posterior fixation. All patients undergo neck flexion/extension MRI of the CVJ. The patient is positioned supine with crown halo traction; an intraoperative 3D CT scan is obtained in traction. The patient is then placed prone and another 3D CT scan is obtained. The updated algorithm is shown in			Velopharyngeal insufficiency 1.8 % pharyngeal wound dehiscence 0.7 %
Burns et al. [6]	endoscopic transnasal	2	Ventral epidural abscess with osteomyelitis at the craniovertebral junction	69. 55	2/2	CSF leakage

CSF cerebrospinal, *CT* computed tomography, *MRI* magnetic resonance imaging

flap is prepared on both sides. In some cases of platybasia, it may be necessary to perform a sphenoidotomy and extend the midline incision from the floor of the sphenoid sinus down to the inferior clivus, especially if the odontoid process is located in a retroclival position.

The Menezes group (Dlouhy et al. [15]) emphasize the importance of intraoperative reduction strategies. If reduction cannot be achieved, a 540° procedure may be necessary in some cases (depending on the pathology), whereby the posterior approach and incision is temporarily closed and the patient is repositioned supine for a ventral decompression, followed by reopening of the posterior incision and posterior fixation. All patients undergo neck flexion/extension magnetic resonance imaging (MRI) of the CVJ. The patient is positioned supine with halo crown traction; an intraoperative three-dimensional computed tomography (3D CT) scan is obtained in traction. The patient is then placed prone and another 3D CT scan is obtained. The updated algorithm is shown in [13].

Conclusions

The progressive worldwide blooming of transoral procedures, thanks to the intensive care and the improvements in intraoperative neurophysiological monitoring techniques (once considered pioneering and very selective), is spreading the expertise in this field of surgery to a new population of surgeons. These techniques are performed alone or in conjunction with posterior procedures [43].

The pure endonasal and cervical endoscopic approach deserves consideration, but it still has three disadvantages, the first two being: (1) the steep learning curve and (2) the lack of 3D perception of the surgical field, which could be an operationally limiting factor. Image clarity would be diminished when endoscopes smaller than 2.7 mm are used. Standard 4-mm endoscopes give a good image quality, but 2.7-mm scopes provide better maneuverability. The third disadvantage is that there is a limited working channel, depending on the variability of the NAxL, which can make it difficult to remove huge tumors.

In our opinion, endoscopically assisted transoral surgery with 30° endoscopes represents an emerging alternative to standard microsurgical techniques for transoral approaches to the anterior CVJ. Used in conjunction with traditional microsurgery and intraoperative fluoroscopy, this endoscopically assisted transoral approach provides a safe and improved method for anterior decompression, with or without a reduced need for extensive soft palate splitting, hard palate resection, or extended maxillotomy. Virtually no surgical limitations exist for this approach, compared with the pure endonasal and transcervical approaches to the CVJ, in normal anatomical conditions.

Of note, the endoscope has an interesting role as "support" for the standard transoral microsurgical approach, since 30° angulated endoscopy strongly improves the visual but not the working channel and volume, even though soft palate splitting is often still required. In our opinion, the transoral (microsurgical or video-assisted) approach with sparing of the soft palate still remains the gold standard compared with the "pure" transnasal and transcervical approaches, due to the wider working channel provided by the former technique. The transnasal endoscopic approach alone appears to be superior when the CVJ lesion exceeds the upper limit of the inferior third of the clivus. Furthermore, the combined transnasal and transoral procedures can be tailored according to the specific pathological and radiological findings.

According to a recent anatomical study, the lower incidence of postoperative dysphagia with the endonasal approach is likely related to the lower density of neuronal elements from the pharyngeal plexus above the palatal plane [38].

However, the time to extubation and oral feeding was significantly shorter in the endonasal group in that study. Similarly, Ponce-Gómez and colleagues reported their own series of patients treated using both approaches and found comparable rates of neurological improvement after odontoidectomy, with less time to extubation and oral feeding, as well as shorter hospital stay, in the endonasal group [39].

Finally, to further validate all the endoscopic techniques, experience is required with greater numbers of patients and long-term follow-up. In our opinion, and in agreement with other authors, the endoscopic endonasal approach should be considered a complementary approach, rather than an alternative, to the standard transoral-transpharyngeal route [29].

Conflict of Interest Statement No conflict of interests does exist.

References

1. Aldana PR, Naseri I, La Corte E (2012) The naso-axial line: a new method of accurately predicting the inferior limit of the endoscopic endonasal approach to the craniovertebral junction. Neurosurgery 71(2 Suppl Operative):308–314
2. Alfieri A, Jho HD, Tschabitscher M (2002) Endoscopic endonasal approach to the ventral cranio-cervical junction: anatomical study. Acta Neurochir (Wien) 144:219–225
3. Ammirati M, Bernardo A (1998) Analytical evaluation of complex anterior approaches to the cranial base: an anatomic study. Neurosurgery 43:1398–1407
4. Balasingam V, Anderson GJ, Gross ND, Cheng CM, Noguchi A, Dogan A, McMenomey SO, Delashaw JB Jr, Andersen PE (2006) Anatomical analysis of transoral surgical approaches to the clivus. J Neurosurg 105:301–308
5. Baird CJ, Conway JE, Sciubba DM, Prevedello DM, Quin˜ones-Hinojosa A, Kassam AB (2009) Radiographic and anatomic basis of endoscopic anterior craniocervical decompression: a comparison of endonasal, transoral, and transcervical approaches. Neurosurgery 65(6 Suppl):158–163
6. Burns TC, Mindea SA, Pendharkar AV, Lapustea NB, Irime I, Nayak JV (2015) Endoscopic transnasal approach for urgent decompression of the craniocervical junction in acute skull base osteomyelitis. J Neurol Surg Rep 76(1):e37–e42. doi:10.1055/s-0034-1395492, Epub 2015 Jan 16
7. Ciporen JN, Moe KS, Ramanathan D, Lopez S, Ledesma E, Rostomily R, Sekhar LN (2010) Multiportal endoscopic ap- proaches to the central skull base: a cadaveric study. World Neurosurg 73:705–712

8. Choi D, Crockard HA (2013) Evolution of transoral surgery: three decades of change in patients, pathologies, and indications. Neurosurgery 73:296–303
9. Chaudhry NS, Ozpinar A, Bi WL, Chavakula V, Chi JH, Dunn IF (2015) Basilar invagination: case report and literature review. World Neurosurg 83(6):1180.e7–11. doi:10.1016/j.wneu.2015.02.007, Epub 2015 Feb 18
10. de Divitiis O, Conti A, Angileri FF, Cardali S, La Torre D, Tschabitscher M (2004) Endoscopic transoral-transclival approach to the brainstem and surrounding cisternal space: ana- tomic study. Neurosurgery 54:125–130
11. Dallan I, Castelnuovo P, Montevecchi F, Battaglia P, Cerchiai N, Seccia V, Vicini C (2012) Combined transoral transnasal robotic-assisted nasopharyngectomy: a cadaveric feasibility study. Eur Arch Otorhinolaryngol 269:235–239
12. de Almeida JR, Zanation AM, Snyderman CH, Carrau RL, Prevedello DM, Gardner PA, Kassam AB (2009) Defining the nasopalatine line: the limit for endonasal surgery of the spine. Laryngoscope 119:239–244
13. Dhaliwal PP, Hurlbert RJ, Sutherland GS (2012) Intraoperative magnetic resonance imaging and neuronavigation for transoral approaches to upper cervical pathology. World Neurosurg 78:164–169
14. Dasenbrock HH, Clarke MJ, Bydon A, Sciubba DM, Witham TF, Gokaslan ZL, Wolinsky JP (2012) Endoscopic image-guided transcervical odontoidectomy: outcomes of 15 patients with basilar invagination. Neurosurgery 70:351–359
15. Dlouhy BJ, Dahdaleh NS, Menezes AH (2015) Evolution of transoral approaches, endoscopic endonasal approaches, and reduction strategies for treatment of craniovertebral junction pathology: a treatment algorithm update. Neurosurg Focus 38(4), E8. doi:10.3171/2015.1.FOCUS14837
16. El-Sayed IH, Wu JC, Ames CP, Balamurali G, Mummaneni PV (2010) Combined transnasal and transoral endoscopic approaches to the craniovertebral junction. J Craniovertebr 1:44–48
17. Frempong-Boadu AK, Faunce WA, Fessler RG (2002) Endo- scopically assisted transoral-transpharyngeal approach to the craniovertebral junction. Neurosurgery 51(5 Suppl):S60–S66, Junction Spine 1:44–48
18. Funaki T, Matsushima T, Peris-Celda M, Valentine RJ, Joosdlasd W, Rhoton AL Jr (2013) Focal transnasal approach to the upper, middle, and lower clivus. Neurosurgery 73(2 Suppl Operative):155–190
19. Gladi M, Iacoangeli M, Specchia N, Re M, Dobran M, Alvaro L, Moriconi E, Scerrati M (2012) Endoscopic transnasal odontoid resection to decompress the bulbo-medullary junction: a reliable anterior minimally invasive technique without posterior fusion. Eur Spine J 21(Suppl 1):S55–S60
20. Hickman ZL, McDowell MM, Barton SM, Sussman ES, Grunstein E, Anderson RC (2013) Transnasal endoscopic approach to the pediatric craniovertebral junction and rostral cervical spine: case series and literature review. Neurosurg Focus 35:E14
21. Husain M, Rastogi M, Ojha BK, Chandra A, Jha DK (2006) Endoscopic transoral surgery for craniovertebral junction anomalies. J Neurosurg Spine 5:367–373
22. Komatsu F, Komatsu M, Di Ieva A, Tschabitscher M (2013) Endoscopic far-lateral approach to the posterolateral craniovertebral junction: an anatomical study. Neurosurg Rev 36:239–247
23. Kassam AB, Snyderman C, Gardner P, Carrau R, Spiro R (2005) The expanded endonasal approach: a fully endoscopic transnasal approach and resection of the odontoid process: technical case report. Neurosurgery 57:E213
24. Little AS, Perez-Orribo L, Rodriguez-Martinez NG, Reyes PM, Newcomb AG, Prevedello DM, Crawford NR (2013) Biome- chanical evaluation of the craniovertebral junction after inferior- third clivectomy and intradural exposure of the foramen mag- num: implications for endoscopic endonasal approaches to the cranial base. J Neurosurg Spine 18:327–332
25. Lee A, Sommer D, Reddy K, Murty N, Gunnarsson T (2010) Endoscopic transnasal approach to the craniocervical junction. Skull Base 20:199–205
26. Liu JK, Patel J, Goldstein IM, Eloy JA (2015) Endoscopic endonasal transclival transodontoid approach for ventral decompression of the craniovertebral junction: operative technique and nuances. Neurosurg Focus 38(4):E17. doi:10.3171/2015.1 FOCUS14813
27. Messina A, Bruno MC, Decq P, Coste A, Cavallo LM, de Divittis E, Cappabianca P, Tschabitscher M (2007) Pure endoscopic endonasal odontoidectomy: anatomical study. Neurosurg Rev 30:189–194
28. McGirt MJ, Attenello RJ, Sciubba DM, Gokaslan ZL, Wolinsky JP (2008) Endoscopic transcervical odontoidectomy for pediatric basilar invagination and cranial settling. J Neurosurg Pediatr 1:337–342
29. Morales-Valero SF, Serchi E, Zoli M, Mazzatenta D, Van Gompel JJ (2015) Endoscopic endonasal approach for craniovertebral junction pathology: a review of the literature. Neurosurg Focus 38(4):E15. doi:10.3171/2015.1.FOCUS14831
30. Menezes AH (2008) Surgical approaches: postoperative care and complications "transoral–transpalatopharyngeal approach to the Craniocervical Junction". Childs Nerv Syst 24:1187–1193
31. Pillai P, Baig MN, Karas CS, Ammirati M (2009) Endoscopic image-guided transoral approach to the craniovertebral junction: an anatomic study comparing surgical exposure and surgical freedom obtained with the endoscope and the operating micro- scope. Neurosurgery 64(5 Suppl 2):437–442
32. Perez-Orribo L, Little AS, Lefevre RD, Reyes PR, Newcomb AG, Prevedello DM, Roldan H, Nakaji P, Dickman CA, Crawford NR (2013) Biomechanical evaluation of the craniovertebral junction after anterior unilateral condylectomy: implications for endoscopic endonasal approaches to the cranial base. Neurosurgery 72:1021–1029
33. Perrini P, Benedetto N, Guidi E, Di Lorenzo N (2009) Transoral approach and its superior extensions to the craniovertebral junction malformations: surgical strategies and results. Neurosurgery 64(5 Suppl 2):331–342
34. Ponce-Gómez JA, Ortega-Porcayo LA, Soriano-Barón HE, Sotomayor-González A, Arriada-Mendicoa N, Gómez-Amador JL, Palma-Díaz M, Barges-Coll J (2014) Evolution from microscopic transoral to endoscopic endonasal odontoidectomy. Neurosurg Focus 37(4):E15. doi:10.3171/2014.7.FOCUS14301
35. Russo VM, Graziano F, Russo A, Albanese E, Ulm AJ (2011) High anterior cervical approach to the clivus and foramen mag- num: a microsurgical anatomy study. Neurosurgery 69(1 Suppl Operative):103–114
36. Seker A, Inoue K, Osawa S, Akakin A, Kilic T, Rhoton AL Jr (2010) Comparison of endoscopic transnasal and transoral ap- pro- aches to the craniovertebral junction. World Neurosurg 74:583–602
37. Salunke P, Sharma M, Sodhi HB, Mukherjee KK, Khandelwal NK (2011) Congenital atlantoaxial dislocation: a dynamic pro- cess and role of facets in irreducibility. J Neurosurg Spine 15:678–685
38. Van Abel KM, Mallory GW, Kasperbauer JL, Moore EJ, Price DL, O'Brien EK, Olsen KD, Krauss WE, Clarke MJ, Jentoft ME, Van Gompel JJ (2014) Transnasal odontoid resection: is there an anatomic explanation for differing swallowing outcomes? Neurosurg Focus 37(4):E16. doi:10.3171/2014.7.FOCUS14338
39. Visocchi M, Della Pepa GM, Doglietto F, Esposito G, La Rocca G, Massimi L (2011) Video-assisted microsurgical transoral approach

to the craniovertebral junction: personal experience in childhood. Childs Nerv Syst 27:825–831
40. Visocchi M (2011) Advances in videoassisted anterior surgical approach to the craniovertebral junction. Adv Tech Stand Neurosurg 37:97–110
41. Visocchi M (2014) Videoassisted anterior surgical approaches to the craniocervical junction: rationale and clinical results. Eur Spine J. doi:10.1007/s00586-015-3873-6
42. Visocchi M, La Rocca G, Della Pepa GM, Stigliano E, Costantini A, Di Nardo F, Maira G (2014) Anterior video-assisted approach to the craniovertebral junction: transnasal or transoral? A cadaver study. Acta Neurochir (Wien) 156:285–292
43. Visocchi M, Pietrini D, Tufo T, Fernandez E, Di Rocco C (2009) Pre-operative irreducible C1–C2 dislocations: intra-operative reduction and posterior fixation. The "always posterior strategy". Acta Neurochir (Wien) 151:551–559
44. Wolinsky JP, Sciubba DM, Suk I, Gokaslan ZL (2007) Endoscopic image-guided odontoidectomy for decompression of basilar invagination via a standard anterior cervical approach. Technical note. J Neurosurg Spine 6:184–191
45. Wu JC, Huang WC, Cheng H, Liang ML, Ho CY, Wong TT, Shih YH, Yen YS (2008) Endoscopic transnasal transclival odon- toidectomy: a new approach to decompression: technical case report. Neurosurgery 63(1 Suppl 1):E92–E94
46. Youssef AS, Guiot B, Black K, Sloan AE (2008) Modifications of the transoral approach to the craniovertebral junction: anatomic study and clinical correlations. Neurosurgery 62(3 Suppl 1):145–154
47. Visocchi M, Pappalardo G, Pileggi M, Signorelli F, Paludetti G, La Rocca G (2015) Experimental endoscopic angular domains of transasal and transoral routes to the craniovertebral junction. Light and Shade. SPINE 41: Number 3 (in press)

Craniovertebral Junction Transanasal and Transoral Approaches: Reconstruct the Surgical Pathways with Soft or Hard Tissue Endocopic Lines? This Is the Question

Massimiliano Visocchi, Giuseppe Barbagallo, Vincenzo Lorenzo Pascali, Pierpaolo Mattogno, Francesco Signorelli, Gerardo Iacopino, Antonino Germano', and Giuseppe La Rocca

Abstract A variety of pathological conditions may affect the clivus and the craniovertebral junction (CVJ). These include congenital disorders, chronic inflammation, neoplasms, infections, and posttraumatic conditions that could all result in CVJ compression and myelopathy Endoscopic-assisted procedures have been further developed for CVJ decompression and they have now become conventional approaches. The aims of the present study were:

(1) to compare "radiological" and "surgical" nasoaxial lines (NAxLs); (2) to introduce an analogous radiological line as a predictor of the superior extension of the transoral approach (palatine inferior dental arch line (PIA); (3) to compare the "radiological" nasopalatine line (NPL) with the "surgical" NPL (SNPL) and surgical PIA (SPIA); (4) to compare "our" SNPL with the NAxL; and (5) to find possible radiological reference points to predict, preoperatively, the maximal extent of superior dissection for the transoral approach (SPIA).

Keywords Endoscopy • Transnasal approach • Transoral approach • Craniovertebral junction

M. Visocchi, MD • M. Pierpaolo, MD
S. Francesco, MD • G. La Rocca, MD (✉)
Insitute of Neurosurgery, Medical School,
Catholic University of Rome, Rome, Italy
e-mail: giularocca86@gmail.com

G. Barbagallo
Division of Neurosurgery, Department of Neurosciences,
Policlinico "G. Rodolico" University Hospital, Catania, Italy

V.L. Pascali
Department of Public Health, Institute of Legal Medicine,
Catholic University of Rome, Rome, Italy

G. Iacopino, MD
Department of Experimental Biomedicine and Clinical
Neurosciences, School of Medicine, Neurosurgical Clinic,
University of Palermo, Palermo, Italy

A. Germano', MD
Department of Biomedical Sciences, Neurosurgery,
University of Messina, Messina, Italy

Introduction

A variety of pathological conditions may affect the clivus and the craniovertebral junction (CVJ). These include congenital disorders, chronic inflammation, neoplasms, infections, and posttraumatic conditions that could all result in CVJ compression and myelopathy. Fang and Ong, in 1962, performed transoral decompression for irreducible atlantoaxial abnormalities in the first series of patients who underwent this procedure [1]. The microsurgical ventral approach to the CVJ has since been widely described for the decompression of irreducible extradural pathology [2–5] and was popularized by Crockard to drain retropharyngeal abscesses [1, 3]. Endoscopic-assisted procedures have been further developed for CVJ decompression and they have now become conventional approaches [4, 6–8]. Kassam et al. introduced the fully endoscopic transnasal approach to the CVJ [9, 10] and emphasized a nasopalatine line (NPL) as a reliable predictor of the maximal extent of inferior dissection. According to these authors the line created by connecting the most inferior point on the nasal bone to the most posterior point on the hard palate in the midsagittal radiological plane was found to be the best way to predict the real surgical lines. Many experimental studies have been performed to test the feasibility of the endoscopic transnasal approach, as well as to compare microsurgical and endoscopic transoral approaches [6, 8, 9]. We have performed neuroradiological studies to compare transnasal and transoral surgical domains: we evaluated, in cadavers, the surgical exposition angle and the working channel volume of both the transnasal and transoral approaches, employing a procedure with open mouth, with an oral distractor [10].

A novel radiological and surgical nasoaxial line (NAxL) was conceived in order to overcome the unreliability of the NPL, this being due to the resistance of the skin of the nose [11–13] (Fig. 1). We note that, so far, no conceptually analogous radiological line has been introduced as a reliable predictor of the maximal superior extension of the transoral approach.

M. Visocchi et al. (eds.), *Trends in Reconstructive Neurosurgery*, Acta Neurochirurgica Supplement 124,
DOI 10.1007/978-3-319-39546-3_18, © Springer International Publishing Switzerland 2017

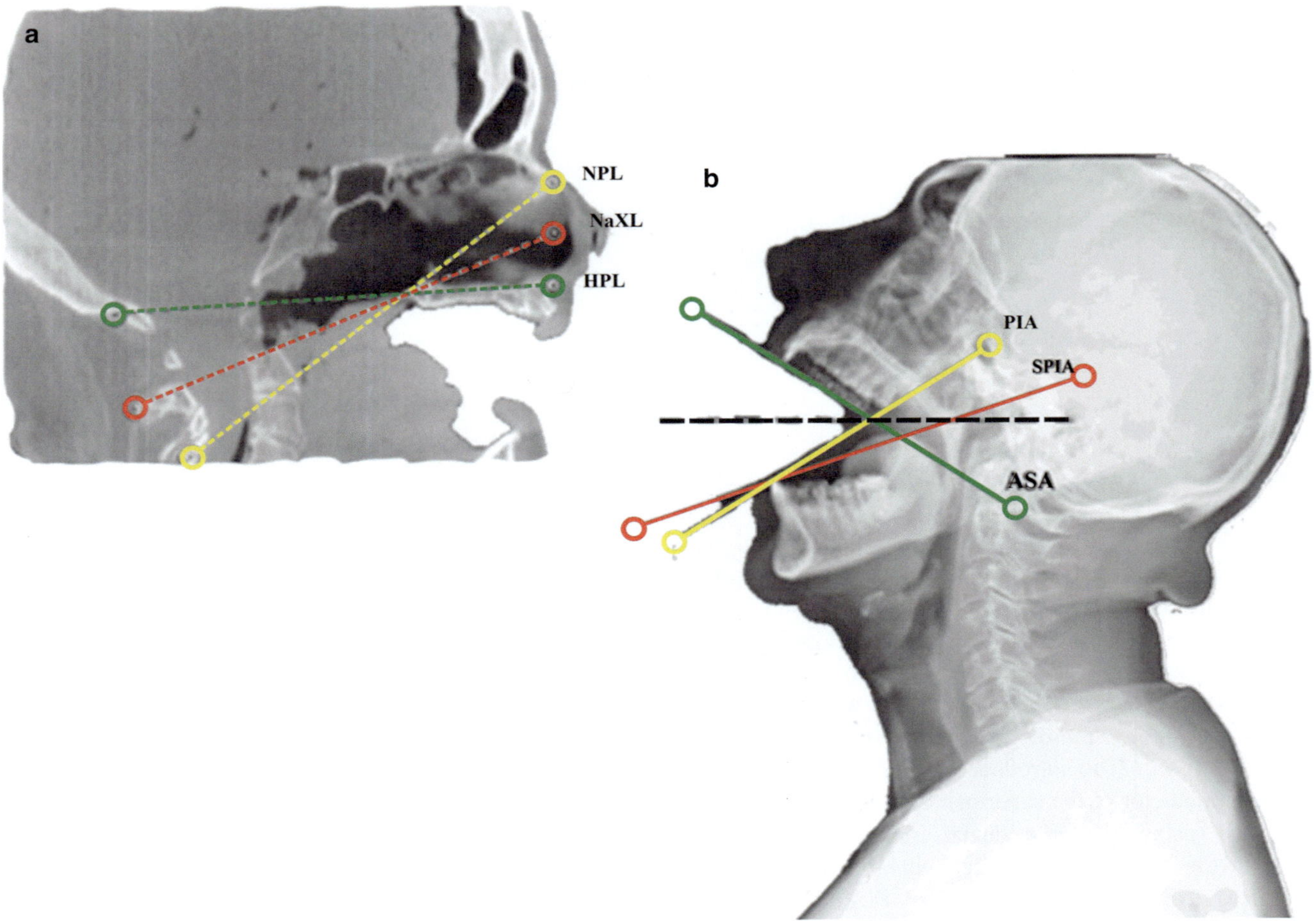

Fig. 1 (**a**) Comparison of the nasoaxial line (*red*) and nasopalatine line (*yellow*) with the actual surgical extent. The NAxL closely corresponds to the lowest limit of the endoscopic endonasal approach to the craniovertebral junction; the NPL overestimates the prediction on preoperative images. *NAxL* nasoaxial line, *NPL* nasopalatine line, *EEA* endoscopic endonasal approach, *HPL* hard palate line. (**b**) Lateral open-mouth skull X-ray with palatine inferior dental arch line (PIA; *continuous line*), atlanto superior dental arch line (ASA; *continuous line*), and surgical PIA (SPIA; *red line*). The SPIA was found to be engaged at the soft palate with the line in the midsagittal plane that crosses, at the midpoint, two more lines: the radiological PIA (RPIA) and ASA; these are defined as the line (*dotted line*) joining the superior dental arch and the anterior base of the atlas (see text)

The aims of the present study were:

1. to compare radiological and surgical NAxLs;
2. to introduce an analogous radiological line as a predictor of the superior extension of the transoral approach (palatine inferior dental arch line [PIA]);
3. to compare radiological with surgical NPL (SNPL) and surgical PIA (SPIA);
4. to compare "our" SNPL with the NAxL;
5. to find possible radiological reference points to predict preoperatively the maximal extent of superior dissection for the transoral approach (SPIA).

Material and Methods

With Ethics Committee Approval of the experimental protocol granted by the Catholic University of Rome, Italy (protocol number P663/CE/2010 approved on July 28, 2010; subsequent amendment number P437/CE 2012 approved on May 2, 2012) we studied nine fresh nonperfused cadavers—five female and five male—median age 72 years (interquartile range 33; minimum 41, maximum 94), at the CVJ Surgery Research Center in the Department of Public Health, Institute of Legal Medicine, of our University. With the cadaver in the supine position with the head slightly extended (about 25°), a Crockard transoral distractor (Crockard Transoral Instrument Set; Codman and Shurtleff, Raynham, MA, USA) was placed in the oral cavity to expose the CVJ. The C1 tubercle was identified with the finger in all the cadavers and the position of the distractor was chosen according to fluoroscopic assessment (MPX+ portable X-ray unit; Philips Healthcare, Best, The Netherlands). We considered the NPL according to the Kassam definition and conceived a new PIA line from the inferior dental arch up to the hard palate, for preoperative transoral approach planning [11]. The radiological NPL and PIA lines were evaluated by means of X-ray and computed tomography (CT) scan (GE

LightSpeed VCT 64 Slice, 1.25 mm thin; General Electric, Milwaukee, WI, USA). Subsequently two thin stainless probes mimicking the endoscopic tools (30 cm length) were inserted through the nostrils (choanae) and the oral cavity, as exposed by the Crockard distractor. The SNPL and SPIA were then radiologically evaluated and compared more as usually [10, 11]. In detail the values of the angular (°) exposure of the transoral and the transnasal approaches, in reference to the hard palate line first described by Aldane [13], were evaluated for each subject by lateral reconstructions (Fig. 1). Percentage differences (%) between the radiological and surgical NPLs were evaluated, along with the radiological and surgical NPL ratio.

The same procedure was used for determining radiological and surgical PIA values. Box plot minimum-maximum values of the NPL and PIA are reported in Tables 1a and 1b. Furthermore, we also evaluated the NAxL and compared it with the SNPL [12].

No platybasia or basilar invagination was identified radiologically, nor was jaw-opening impairment found in any of the cadavers. The collected data were statistically analyzed. A descriptive analysis of the sample was carried out by means of median, interquartile range (IQR), and range for continuous variables, and absolute and relative frequencies for qualitative variables. In order to find statistically significant differences between the two surgical approaches, we performed a Wilcoxon signed rank test. We chose to use a nonparametric test because data were not normally distributed, as demonstrated by the Shapiro-Wilk test [15]. The analysis was performed using SPSS software version 12.0 for Windows and the statistical significance level was set at $P = .05$.

Results (Tables 1a and 1b; Fig. 1)

X-ray and CT scan measurements of the CVJ were performed in all the subjects. Statistically significant differences ($P = 0.05$) were found between the radiological (minimum 33°, maximum 41°) and surgical (minimum 22°, maximum 27°) NPLs and the radiological (minimum 36°, maximum 59°) and surgical (minimum 29°, maximum 49°) PIA angle values. The results of the study are summarized in Tables 1a and 1b. In all the cadavers the angular gap between the radiological and surgical lines was wider for the transnasal than for the transoral approach. The most reliable radiological preoperative line was found to be the PIA, with a mean ratio between the radiological PIA and surgical PIA of 0.82. On the other hand, the mean ratio between the radiological and surgical NPL was found to be only 0.66; in this case the differences were statistically significant (Fig. 1).

Moreover, we found a 100 % correspondence between the NAxL and the SNPL (NAxL/SNPL = 1) and finally we were able to identify the SPIA radiologically.

The SPIA was found to be the line, in the midsagittal plane, that crosses, in the midpoint, two more lines: the radiological PIA (RPIA) and the atlanto superior dental arch line (ASA), defined as the line joining the superior dental arch and the anterior base of the atlas (Fig. 1). We defined

Table 1a Angles of transnasal and transoral radiological and surgical routes

	Mean angle (°)
RNPL	36.4°
SNPL	24°
RNPL/SNPL	0.66
NAxL/SNPL	1
RPIA	47.2°
SPIA	38.9°
RPIA/SPIA	0.82

RNPL radiological nasopalatine line, *SNPL* surgical nasopalatine line, *NAxL* nasoaxial line, *RPIA* radiological palatine inferior dental arch line, *SPIA* surgical palatine inferior dental arch line.

Table 1b Medians and statistical analysis of radiological and surgical transnasal and transoral routes

Variable	Median (IQ range)	Wilcoxon signed-rank test (*P*)
RNPL	37.45° (3.57)	$P = 0.05$
SNPL	24.75° (3.07)	
RPIA	47.60° (4.83)	$P = 0.05$
SPIA	38.25° (3.38)	

IQ interquartile, *RNPL* radiological nasopalatine line, *SNPL* surgical nasopalatine line, *RPIA* radiological palatine inferior dental arch line, *SPIA* surgical palatine inferior dental arch line the NPL and the PIA as *"hard-tissue lines"*, since they both deal with bone tissue only; we defined the NAxL (i.e., SNPL) and SPIA as *"soft-tissue lines"*, due to their relationship with soft tissues such as the skin and the soft palate.

Discussion

The transoral-transpharyngeal approach provides surgical access to the anterior clivus, C1, and C2. However, the use of microscopes, high-speed drills, self-retaining mouth retractors, flexible oral endotracheal tubes, intraoperative neuroradiological investigations, neuronavigation, and electrophysiological monitoring has made transoral procedures much safer than they were previously [4, 15, 17, 18]. The 30° endoscope has been proposed for the transoral approach to avoid full soft-palate splitting, hard-palate splitting, or extended maxillo/mandibulotomy [19, 20]. Using the endoscope, the operator is able to look in all directions by rotating the instrument. The last high-profile cadaveric study recently available in the literature is the one by Pillai et al. [15], which quantified the surgical volume gained by the endoscopic approach: the surgical area exposed over the posterior pharyngeal wall was significantly improved using an endoscope (606.5–127.4 mm^3) compared with the finding with an operating microscope (425.7–100.8 mm^3), without any compromise of surgical freedom ($P = 0.05$). The extent of the clivus exposed with the endoscope (9.5 ± 0.7 mm) without splitting the soft palate was significantly improved compared with that associated with the microscopic approach (2.0 ± 0.4 mm) ($P = .05$) [15]. Some authors have reported anatomical studies and surgical experience with the endoscopic endonasal approach [8, 14]. In 2002, Alfieri et al. [21] were the first to perform a cadaveric study of totally transnasal endoscopic odontoidectomy through a one-or two-nostril route [12, 19]. Cavallo et al. [22] confirmed the observations of Alfieri et al. in a cadaveric study as, late on Messina et al. [14] and Kassam et al., in 2005, operated the first case through a fully transnasal endoscopic resection of the odontoid [8], [14] and concluded: "The transoral approach remains the 'gold standard'", but in contrast with this, "the defect created by transnasal approach is above the level of soft palate and should not be exposed to the same degree of bacterial contamination". Messina et al. [20], in further anatomical studies, concluded that, similar to the transoral approach, the endoscopic endonasal approach provides a direct route to the surgical target, but it seems to be related to lower morbidity. De Almeida et al. [12] published, in 2009, the concept of the NPL, a line created by connecting the most inferior point on the nasal bone to the most posterior point on the hard palate in the midsagittal plane, and they concluded that the NPL was a reliable predictor of the maximal extent of inferior dissection [16]. A novel line, the NAxL, used for the best preoperative planning to determine the inferior limit of the endonasal approach to the CVJ, has been identified as the line in the midsagittal plane that starts from the midpoint of the distance from the rhinion to the anterior nasal spine of the maxillary bone and ends at the C2 vertebra, tangential to the posterior nasal spine of the palatine bone.

Cadaveric Study

In our cadaveric study we showed a novel PIA and compared the surgical domains of the NPL and PIA. Radiological examination and comparison of both the transnasal and transoral CVJ sagittal surgical domains in the same subject by means of NPL and PIA allowed us to recognize which preoperative radiological planning might be more reliable and closer to the effective surgical route allowed. The cranial settling, CVJ kyphotic deformity, and other changes would alter the utility of the two approaches. In fact, it must be pointed out that we studied only "normal" subjects. We used the *classic* NPL and a *novel* PIA, i.e., *"hard-tissue lines"*, as ideal reference points to compare the two surgical strategies. The transnasal approach is a viable strategy to reach the CVJ, but it has more limited angular (nostrils, choanae) and linear (NPL) surgical exposure, mainly related *to the stiffness of the skin of the nose*, which, in our view, makes it suitable only for certain types of diseases and prevents its systematic applicability in all other conditions, such as pathologies caudal to C2 (and obviously lateral tumors) [12, 23–26]. Moreover, although an obvious advantage of the transnasal approach is that there is no need to cut the soft palate (which minimizes potential postoperative morbidities such as swallowing disturbances and hypernasal speech, which are really limiting to quality of life if the palatine veil is lacking), the transoral approach provides better exposure of the CVJ, both on the sagittal plane and on the transverse plane, providing a larger working channel and allowing the easier handling of surgical instruments such as the endoscope [6]. With this tool, the advantage of soft palate-sparing might make the transnasal endoscopic approach less common than the transoral endoscopic-assisted one. However, we believe that the transnasal and transoral endoscopic procedures should not be considered in competition but as complementary approaches [11–13]. The present experience seems to emphasize that preoperative planning by means of *hard-tissue lines* seems to be closer to the surgical reality (i.e., *soft-tissue lines)* with the transoral approach compared with the transnasal, as demonstrated by the low radiological and surgical NPL ratio compared with the radiological and surgical PIA ratio.

Conclusions

1. The NAxL is confirmed to be a reliable preoperative predictor of the maximal extent of inferior dissection for the transnasal approach.
2. With the novel SPIA, it is possible to determine the maximal extent of superior dissection for the transoral approach with a simple lateral head X-ray examination with open mouth.
3. The NAxL/SNPL ratio appeared to vary more than the RPIA/SPIA and more than RNPL/SNPL (Table 1a).
4. There is 100 % correspondence between the NAxL and the SNPL;
5. *The SPIA was found to be the line, in the midsagittal plane, that crosses, in the midpoint, two more lines: the RPIA and the atlanto superior dental arch line (ASA), defined as the line joining the superior dental arch and the anterior base of the atlas* (Fig. 1b).

   In other words, both the soft-tissue lines vary from the hard-tissue lines, the but the NAxL varies more than the SPIA.

   The pros and cons of each approach have to be taken into account; as well, a combined transoral and transnasal approach may be chosen.

Conflict of Interest Statement The authors declare that they have not received any funds for this work from any of the following organizations: National Institutes of Health (NIH); Wellcome Trust; Howard Hughes Medical Institute (HHMI); and other foundation(s) requiring open access. Moreover, the authors declare that they have no personal or institutional financial interest in the drugs, materials, or devices described in their submissions.

References

1. Fang HSY, Ong GB (1962) Direct anterior approach to the upper cervical spine. J Bone Joint Surg Am 44A:1588–1604
2. Crockard HA (1991) Ventral approaches to the upper cervical spine. Orthopade 20:140–146
3. Symonds CPMS (1937) Compression of the spinal cord in the neighborhood of the foramen magnum with a note on the surgical approach by Julian Taylor. Brain 60:52–84
4. Visocchi M, Della Pepa GM, Doglietto F, Esposito G, La Rocca G, Massimi L (2011) Video-assisted microsurgical transoral approach to the craniovertebral junction: personal experience in childhood. Childs Nerv Syst 27:825–831
5. Visocchi M, Pietrini D, Tufo T, Fernandez E, Di Rocco C (2009) Preoperative irreducible C1-C2 dislocations: intra-operative reduction and posterior fixation. The “always posterior strategy”. Acta Neurochir (Wien) 151:551–559
6. Frempong-Boadu AK, Faunce WA, Fessler RG (2002) Endoscopically assisted transoral-transpharyngeal approach to the craniovertebral junction. Neurosurgery 51:S60–S66
7. Visocchi M, Doglietto F, Della Pepa GM et al (2011) Endoscope-assisted microsurgical transoral approach to the anterior craniovertebral junction compressive pathologies. Eur Spine J 20:1518–1525
8. Kassam AB, Snyderman C, Gardner P, Carrau R, Spiro R (2005) The expanded endonasal approach: a fully endoscopic transnasal approach and resection of the odontoid process: technical case report. Neurosurgery 57:E213; discussion E213
9. Seker A, Inoue K, Osawa S, Akakin A, Kilic T, Rhoton AL Jr (2010) Comparison of endoscopic transnasal and transoral approaches to the craniovertebral junction. World Neurosurg 74(583):602
10. Visocchi M, La Rocca G, Della Pepa G et al (2004) Anterior videoassisted apporach to the craniovertebral junction: tansnasal or transoral? Acadaveric study. Acta Neurochir (Wien) 156(2):285–292
11. Visocchi M, Pappalardo G, Pileggi M, Signorelli F, Paludetti G, La Rocca G (2015) Experimental endoscopic angular domains of transnasal and transoral routes to the craniovertebral junction light and shade. Spine 41(3)
12. De Almeida JR, Zanation AM, Snyderman CH et al (2009) Defining the nasopalatine line: the limit for endonasal surgery of the spine. Laryngoscope 119:239–244
13. Aldana PR, Naseri I, La Corte E (2012) The naso-axial line: a new method of accurately predicting the inferior limit of the endoscopic endonasal approach to the craniocervical junction. Neurosurgery 71(2 Suppl Operative):308–314
14. Messina A, Bruno MC, Decq P et al (2007) Pure endoscopic endonasal odontoidectomy: anatomical study. Neurosurg Rev 30:189–194
15. Shapiro S, WilkMB CHJA (1968) Comparative study of various tests for normality. J Am Stat Assoc 63:1343–1372
16. Pillai P, BaigMN KCS, Ammirati M (2009) Endoscopic image-guided transoral approach to the craniovertebral junction: an anatomic study comparing surgical exposure and surgical freedom obtained with the endoscope and the operating microscope. Neurosurgery 64:437–442
17. Visocchi M (2011) Advances in videoassisted anterior surgical approach to the craniovertebral junction. Adv Tech Stand Neurosurg 37:97–110
18. Visocchi M, Doglietto F, Della Pepa GM et al (2011) Endoscope-assisted microsurgical transoral approach to the anterior craniovertebral junction compressive pathologies. Eur Spine J 20:1518–1525
19. Vougioukas VI, Hubbe U, Schipper J, Spetzger U (2003) Navigated transoral approach to the cranial base and the craniocervical junction: technical note. Neurosurgery 52:247–250
20. Jho HD, Ha HG (2004) Endoscopic endonasal skull base surgery: part 3-The clivus and posterior fossa. Minim Invasive Neurosurg 47:16–23
21. Messina A, Bruno MC, Decq P et al (2007) Pure endoscopic endonasal odontoidectomy: anatomical study. Neurosurg Rev 30:189–194
22. Alfieri A, Jho HD, Tschabitscher M (2002) Endoscopic endonasal approach to the ventral cranio-cervical junction: anatomical study. Acta Neurochir (Wien) 144:219–225
23. Cavallo LM, Messina A, Cappabianca P et al (2005) Endoscopic endonasal surgery of the midline skull base: anatomical study and clinical considerations. Neurosurg Focus 19(1):E2
24. Visocchi M, Trevisi G, Iacopino DG, Tamburrini G, Caldarelli M, Barbagallo GM (2014) Odontoid process and clival regeneration with Chiari malformation worsening after transoral decompression: an unexpected and previously unreported cause of “accordion phenomenon”. Eur Spine J 24(Suppl 4):S564–S568
25. Visocchi M, Di Martino A, Maugeri R, González Valcárcel I, Grasso V, Paludetti G (2015) Videoassisted anterior surgical approaches to the craniocervical junction: rationale and clinical results. Eur Spine J 24(12):2713–2723
26. Visocchi M (2015) Transnasal and transoral approach to the clivus and the craniovertebral junction. J Neurosurg Sci

Spheno-Orbital Meningiomas: When the Endoscopic Approach Is Better

Stefano Peron, Andrea Cividini, Laura Santi, Nicola Galante, Paolo Castelnuovo, and Davide Locatelli

Abstract Spheno-orbital meningiomas were historically treated by traditional craniotomies. However, in the past few years new endoscopic treatments have been successfully performed. In this study, we analyzed different indications for craniotomy and endoscopy, and the advantages and disadvantages of these procedures.

Thirty patients with spheno-orbital meningiomas were operated on over 2 years, between 2013 and 2014. Computed tomography (CT) and magnetic resonance imaging (MRI) were performed in all patients preoperatively. Navigated surgical removal and histological confirmation, as well as follow-up examinations, including CT scan at 24 h and MRI at 3, 6, and 12 months after surgery, were performed. Twenty-three patients were treated by traditional fronto-temporal, fronto-temporo-orbital, and supraorbital craniotomies; in six cases the tumor was removed via endoscopic endonasal and lateral transorbital resection. Only one case required a combined supraorbital and endoscopic endonasal approach.

We analyzed the results of the different surgical techniques, in particular those of the endoscopic approaches.

In selected cases, the endoscopic approach to spheno-orbital meningiomas, compared with traditional approaches, may be more effective in removing tumors completely. The surgical technique is easy and the rate of complications is low.

Keywords Spheno-orbital meningiomas • Orbital tumors • Transorbital approach • Endonasal approach

S. Peron (✉) • A. Cividini • L. Santi • N. Galante • D. Locatelli
Department of Neurosurgery, Legnano Civil Hospital,
Via Papa Giovanni Paolo II, 20025 Legnano, Italy
e-mail: ste.peron@libero.it

P. Castelnuovo
Department of Otorhinolaryngology, Di Circolo Hospital,
University of Insubria, Varese, Italy

Introduction

Spheno-orbital meningiomas (SOMs) are secondary tumors of the orbit arising from the sphenoid ridge. SOMs are the most frequent meningiomas of the skull base, accounting for up to 18 % of all intracranial meningiomas [3, 5, 14, 17, 19].

These tumors may expand from the sphenoid medially into the lateral wall of the cavernous sinus, anteriorly into the orbit, and laterally into the temporal bone. SOMs are often associated with hyperostosis of the sphenoid ridge and may have a really invasive characteristic, spreading to the dura of the frontal, temporal, orbital, and sphenoidal regions [2, 7, 8, 11, 15, 16, 18, 23–25].

The most common symptoms at presentation include slowly developing unilateral exophthalmos, vision or visual field impairment, and extraocular movement palsy, as well as cosmetic deformities, such as a bony prominence in the temporal region [4, 9, 10, 17, 18, 21, 22].

SOMs are very difficult to manage, with high surgical morbidity and mortality. Sphenoidal hyperostosis represents a limit for complete resection, and the rate of recurrence is very high when compared with meningiomas in other locations [3–5, 10, 18, 20–22, 24].

For years, traditional fronto-temporal, fronto-temporo-orbital, and supraorbital craniotomies were the only approaches to treat SOMs. In the past few years, endoscopic endonasal and transorbital approaches to remove these tumors have been successfully performed.

Material and Methods

All patients underwent preoperative and postoperative ophthalmological evaluation to assess visual acuity, visual field campimetry, and extraocular movement integrity.

M. Visocchi et al. (eds.), *Trends in Reconstructive Neurosurgery*, Acta Neurochirurgica Supplement 124,
DOI 10.1007/978-3-319-39546-3_19, © Springer International Publishing Switzerland 2017

Brain magnetic resonance imaging (MRI) with gadolinium and bone computed tomography (CT) were performed in all patients before and the day after surgery.

Different surgical approaches were employed to treat the SOMs. Transcranial fronto-temporal, fronto-temporo-orbital, and supraorbital approaches, as well as endoscopic endonasal and lateral transorbital approaches, were used in different cases, with some combined approaches also used.

Illustrative Case

A 43-year-old woman presented at our Department with a 6-month history of worsening left eye proptosis.

CT scan and MRI showed a spheno-orbital meningioma with large hyperostosis of the sphenoid lesser wing involving the lateral orbital wall and extracranial compartment (Fig. 1).

Ophthalmological evaluation was negative for oculomotor deficits or visual field impairment.

A lateral transorbital endoscopic approach was performed to remove the tumor completely (Fig. 2).

No cranial nerve deficits or cosmetic deformities were observed after the surgery (Fig. 1).

The patient was discharged 3 days after the procedure.

Results

Thirty patients, 22 females (73 %) and 8 males (27 %), suffering from SOM were operated on between 2013 and 2014.

The mean age was 46 years (range, 8–82 years).

Proptosis was the most frequent sign at presentation, with 21 patients suffering from it, 13 in the right eye and 8 in the left eye. Twenty patients had visual impairment, with campimetric defects in 19 and amaurosis in 1. Oculomotor nerve deficits were found in 8 patients, with diplopia in 5. In particular, 4 patients had a deficit of the third cranial nerve; 3, a deficit of the fourth cranial nerve; and 1, a deficit of the sixth cranial nerve. One patient had trigeminal hypoesthesia in V1 and V2.

Traditional transcranial surgery was performed in 23 patients, using a fronto-temporal approach in 19, fronto-temporal orbital approach in 2, and supraorbital approach in 2.

Six patients underwent endoscopic surgery, in four cases by a lateral transorbital approach, in one case by an endonasal approach, and in one case by a combined transorbital and endonasal approach.

A combined transcranial-transorbital endoscopic approach was performed in one patient.

No approach-related mortality or morbidity, such as temporomandibular joint dysfunction or trismus, occurred after the surgery.

Four patients complained of new, but temporary, third cranial nerve paresis in the postoperative period. A new permanent fourth cranial nerve paresis occurred in one patient. No hyperpathic trigeminal sensation appeared after the procedure.

Visual function, as well as proptosis, improved in all patients, remaining stable at 2-year maximum follow-up.

Postoperative neuroradiological evaluation by MRI with gadolinium and bone CT scan on the day after surgery confirmed a gross-total resection in 24 (80 %) patients, with no recurrences after 2 years.

Six patients with residual tumor in the cavernous sinus were considered candidates for gamma-knife radiosurgery.

Discussion

Spheno-orbital meningiomas usually arise from the inner or outer parts of the sphenoid lesser wing, with intraosseous tumor growth, resulting in hyperostosis, and thin soft-tissue growth at the dura [2, 7, 8, 11, 15, 16, 18, 19, 23–25].

Bony tumor growth usually involves the lesser sphenoid wing, the orbital roof, the lateral orbital wall, the superior orbital fissure, the optic canal, and the anterior clinoid process. In cases of orbital extension the growth may occur through the natural canals, such as the optic canal and superior orbital fissure, or the lateral orbital wall [2, 7, 11, 14–16, 18, 23–25].

Soft-tissue growth can spread to extracranial compartments, including the orbital content and the infratemporal fossa with the temporalis muscle.

Dural growth is often widespread, including the basal sphenoid wing, cavernous sinus, and temporal convexity [24].

In most patients, minor symptoms, such as minimal painless proptosis and mild visual impairment, are complained of at presentation. However, cases of loss of vision, severe proptosis, and large cosmetic deformities can be observed [9, 18, 19, 23, 25].

Due to their anatomical, radiological, and morphological aspects, SOMs are considered complex tumors to remove. The involvement of bone, orbit, and neural structures makes the surgery difficult and the resection often incomplete [3–5, 8, 10, 18, 20–22, 24].

However, in cases of visual impairment, oculomotor dysfunction, and severe proptosis, tumor removal is required.

Conversely, a 'wait and see' strategy might be appropriate in patients with barely visible proptosis, incidental tumor finding, or little ocular pain.

For years, surgical removal by traditional fronto-temporal and fronto-temporal orbital craniotomy was the gold standard of treatment for SOMs.

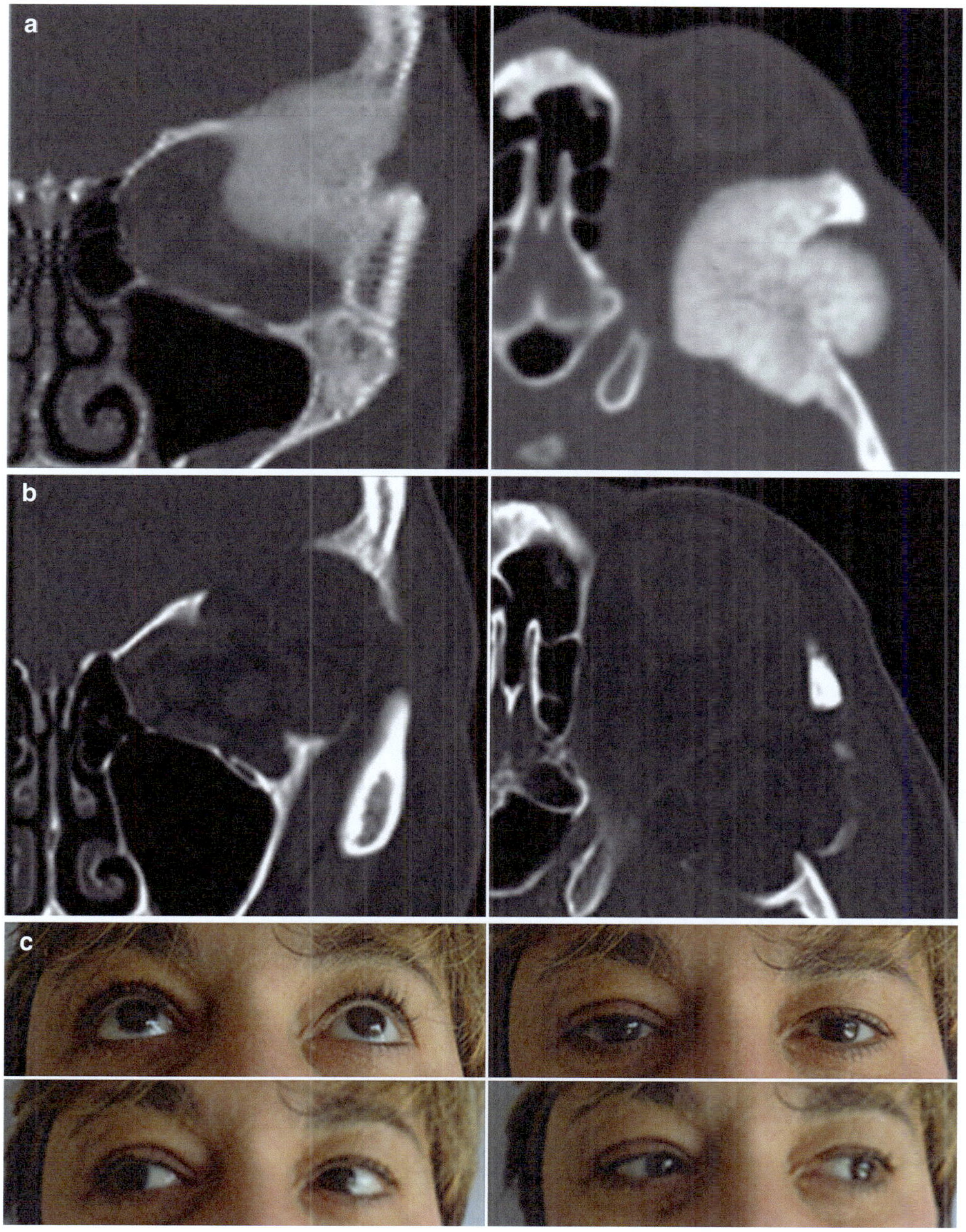

Fig. 1 Illustrative case: preoperative (**a**) and postoperative (**b**) computed tomography (CT) scans; there were no extraocular movement deficits or cosmetic deformities after surgery (**c**)

Recently, new endoscopic approaches to these tumors have been proposed and have been performed successfully and safely.

The use of endoscopes in orbital surgery was first described in 1981, but their use was limited to the biopsy excision of orbital tumors and to the removal of foreign bodies from the orbit [12, 13].

Even though the endonasal route to approach intraorbital pathologies is increasing in surgical practice, transorbital non-endonasal endoscopic approaches are still little known and used.

In using the endonasal intraorbital approach, a standard spheno-ethmoidectomy has to be performed together with a medial maxillectomy, thus exposing the lamina papyracea. After the lamina papyracea opening, free access to the medial and inferomedial walls of the orbit and, afterwards, to the periorbita, can be obtained [1, 6].

The endoscopic transnasal approach is mainly effective in cases of orbital and optic canal decompression, repair of medial and inferomedial wall fractures, and in intraconal and extraconal lesions with inferomedial location. Accordingly, SOMs located on the medial orbital wall and

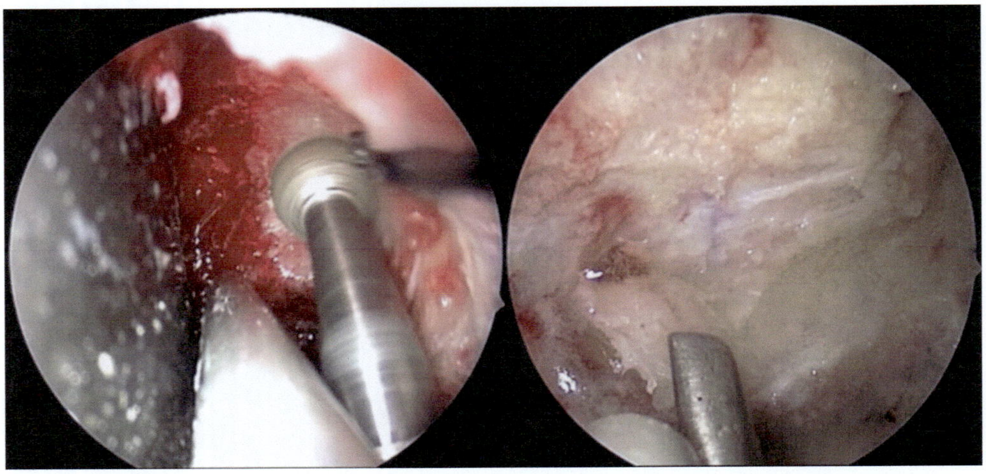

Fig. 2 Illustrative case: drilling of the tumor with spatula protecting the periorbit (*left*); removing all the pathologic bone tissue up to the dura mater of the middle cranial fossa (*right*)

the inferomedial part of the orbital floor can be removed in this way.

Surgical access to the orbit and periorbital structures through the eyelids and anterior orbital compartment can be achieved through different cutaneous and transconjunctival incisions.

A lateral transorbital approach is performed with an incision on the superior eyelid. The orbital rim is reached by dissection in a superolateral direction. Once the orbital rim is identified and exposed, a careful subperiosteal dissection has to be performed until the superior and inferior orbital fissures are reached [1, 6] (Fig. 3).

In case of SOMs with extension to the middle cranial fossa, the greater wing of the sphenoid should be removed as far as the dura mater (Fig. 2). In this case, the superomedial boundary of the approach is defined by the superior orbital fissure, while the lateral boundary is delineated by the temporalis muscle. Superiorly, the approach can be partially extended to the lesser wing of the sphenoid towards the anterior clinoid process. If necessary, the frontal bone can be partially resected and the spheno-orbital sinus can be coagulated. When you need to go intracranial, the dura mater can be opened and the anterior part of the temporal lobe reached.

In all cases, even in patients with no visual impairment, the abnormal bone has to be removed as much as possible, including the opening of the optic canal, if required. On the other hand, in cases of periorbital infiltration by the tumor, complete resection is not mandatory, considering the high risk of damaging intraconal structures in a setting where the rate of recurrence is low.

SOMs with intraorbital extension and widespread dural growth involving the anterior or middle cranial fossa, as well as those infiltrating the cavernous sinus, can be properly treated by combined endoscopic-transcranial decompression and radiosurgery on the sinus infiltration.

Finally, when a large endoscopic decompression is carried out, a reconstruction of the bone defect is useful to avoid enophthalmos and other cosmetic defects [2]. An autologous fat graft can be used for this purpose.

The transorbital superior eyelid approach can be successfully performed in patients with superiorly and laterally located extra- and intraconal lesions, as well as for lesions located in the anterior and middle cranial fossa. Actually, if the endonasal approach is preferred for SOMs involving the medial orbital wall and the inferomedial part of the orbital floor, a lateral transorbital approach is mainly indicated in cases of tumors that extend to the superior and lateral orbital wall or to the lateral part of the orbital floor, a lateral transorbital approach is mainly indicated.

This transorbital access, in combination with the transnasal route, enables the performance of a ‘multiportal’ endoscopic approach to lesions located in the anterior and middle cranial fossa [6].

We performed endoscopic procedures in seven patients, with no mortality or morbidity during or after the surgery. The extent of removal was high and the rate of permanent deficits very low.

Summing up, SOMs with large hyperostosis and growth into the orbit can be successfully removed by endoscopic approaches. In selected cases, endoscopic approaches, compared with traditional craniotomies, are more effective in removing the tumor completely, reducing proptosis and cranial nerve compression.

Endoscopic surgery is safe and quite easy to perform, with a low complication rate and reduced hospital stay, as well as an optimal neurological outcome and cosmetic result for the patient.

Conflict of Interest Statement The authors declare that they have no conflicts of interest.

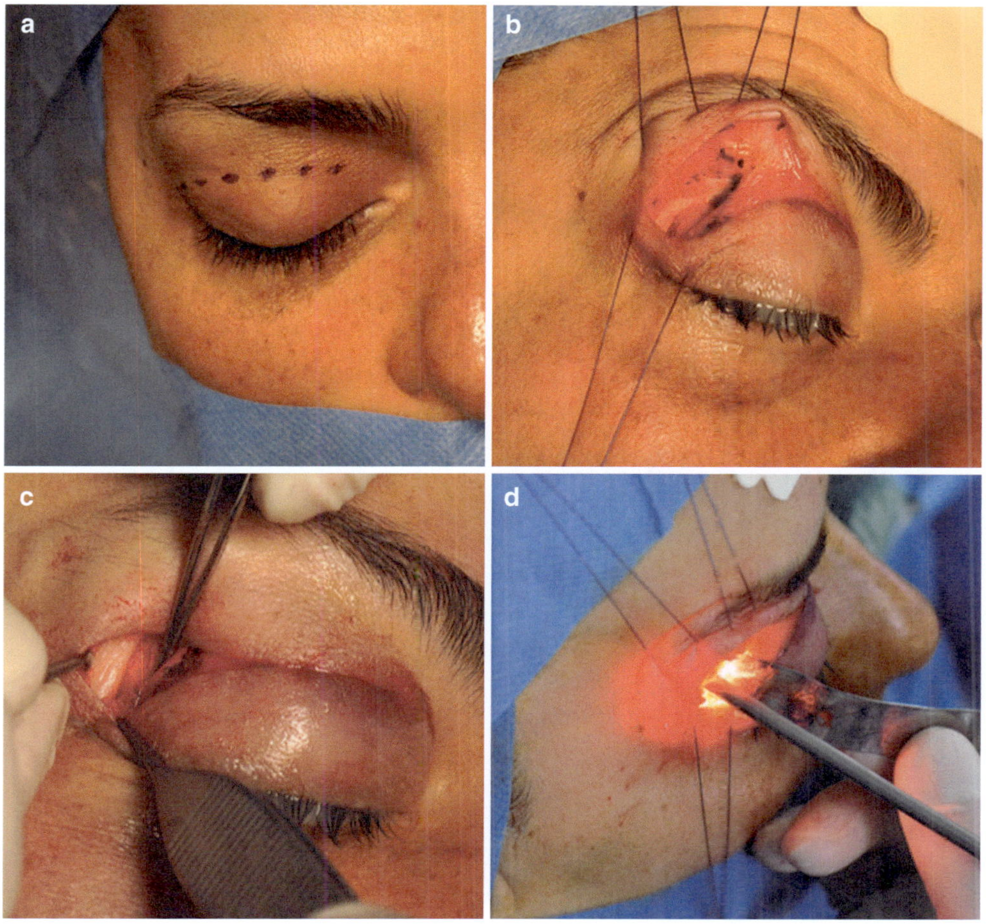

Fig. 3 Endoscopic lateral transorbital approach: superior eyelid incision (**a**), sparing the levator palpebrae aponeurosis (**b**), identifying the orbital rim (**c**), and performing a subperiosteal dissection (**d**)

References

1. Alqahtani A, Padoan G, Segnini G, Lepera D, Fortunato S, Dallan I, Pistochini A, Abdulrahman S, Abbate V, Hirt B, Castelnuovo P (2015) Transorbital transnasal endoscopic combined approach to the anterior and middle skull base: a laboratory investigation. Acta Otorhinolaryngol Ital 35(3):173–179
2. Bikmaz K, Mrak R, Al-Mefty O (2007) Management of bone-invasive, hyperostotic sphenoid wing meningiomas. J Neurosurg 107(5):905–912
3. Cannon PS, Rutherford SA, Richardson PL, King A, Leatherbarrow B (2009) The surgical management and outcomes for spheno-orbital meningiomas: a 7-year review of multi-disciplinary practice. Orbit 28(6):371–376
4. Carrizo A, Basso A (1998) Current surgical treatment for spheno-orbital meningiomas. Surg Neurol 50:574–578
5. Civit T, Freppel S (2010) Sphenoorbital meningiomas. Neurochirurgie 56(2–3):124–131
6. Dallan I, Castelnuovo P, Locatelli D, Turri-Zanoni M, AlQahtani A, Battaglia P, Hirt B, Sellari-Franceschini S (2015) Multiportal combined transorbital transnasal endoscopic approach for the management of selected skull base lesions: preliminary experience. World Neurosurg 84(1):97–107
7. De Jesús O, Toledo MM (2001) Surgical management of meningioma en plaque of the sphenoid ridge. Surg Neurol 55:265–269
8. Forster MT, Daneshvar K, Senft C, Seifert V, Marquardt G (2014) Sphenoorbital meningiomas: surgical management and outcome. Neurol Res 36(8):695–700
9. Grassi P, Strianese D, Mariniello G, Bonavolontà G (2015) Delayed recovery of visual acuity after sphenoorbital meningioma surgical removal: case report and review of the literature. Is visual acuity recovery possible after an initial decline? J Neurol Surg A Cent Eur Neurosurg 76(4):328–331
10. Honeybul S, Neil-Dwyer G, Lang DA, Evans BT, Ellison DW (2001) Sphenoid wing meningioma en plaque: a clinical review. Acta Neurochir (Wien) 143:749–757
11. Luetjens G, Krauss JK, Brandis A, Nakamura M (2011) Bilateral sphenoorbital hyperostotic meningiomas with proptosis and visual impairment: a therapeutic challenge. Report of three patients and review of the literature. Clin Neurol Neurosurg 113(10):859–863
12. Norris JL, Cleasby GW (1981) Endoscopic orbital surgery. Am J Ophthalmol 91(2):249–252
13. Norris JL, Cleasby GW, Nakanishi AS, Martin LJ (1981) Intraocular endoscopic surgery. Am J Ophthalmol 91(5):603–606
14. Oya S, Sade B, Lee JH (2011) Sphenoorbital meningioma: surgical technique and outcome. J Neurosurg 114(5):1241–1249
15. Pieper DR, Al-Mefty O, Hanada Y, Buechner D (1999) Hyperostosis associated with meningioma of the cranial base: secondary changes or tumor invasion. Neurosurgery 44:742–746
16. Pompili A, Derome PJ, Visot A, Guiot G (1982) Hyperostosing meningiomas of the sphenoid ridge–clinical features, surgical

therapy, and long-term observations: review of 49 cases. Surg Neurol 17:411–416

17. Ringel F, Cedzich C, Schramm J (2007) Microsurgical technique and results of a series of 63 spheno-orbital meningiomas. Neurosurgery 60:214–221
18. Roser F, Nakamura M, Jacobs C, Vorkapic P, Samii M (2005) Sphenoid wing meningiomas with osseous involvement. Surg Neurol 64:37–43
19. Saeed P, van Furth WR, Tanck M, Kooremans F, Freling N, Streekstra GI, Regensburg NI, van der Sprenkel JW, Peerdeman SM, van Overbeeke JJ, Mourits MP (2011) Natural history of spheno-orbital meningiomas. Acta Neurochir (Wien) 153(2):395–402
20. Saeed P, van Furth WR, Tanck M, Freling N, van der Sprenkel JW, Stalpers LJ, van Overbeeke JJ, Mourits MP (2011) Surgical treatment of sphenoorbital meningiomas. Br J Ophthalmol 95(7):996–1000
21. Sandalcioglu IE, Gasser T, Mohr C, Stolke D, Wiedemayer H (2005) Spheno-orbital meningiomas: interdisciplinary surgical approach, resectability and long-term results. J Craniomaxillofac Surg 33:260–266
22. Scarone P, Leclerq D, Heran F, Robert G (2009) Long-term results with exophthalmos in a surgical series of 30 sphenoorbital meningiomas. Clinical article. J Neurosurg 111:1069–1077
23. Schick U, Bleyen J, Bani A, Hassler W (2006) Management of meningiomas en plaque of the sphenoid wing. J Neurosurg 104:208–214
24. Shrivastava RK, Sen C, Costantino PD, Della RR (2005) Sphenoorbital meningiomas: surgical limitations and lessons learned in their long-term management. J Neurosurg 103:491–497
25. Solmaz I, Tehli O, Temiz C, Kural C, Hodaj I, Kutlay M, Gonul E, Daneyemez MK (2014) Surgical strategies for the removal of sphenoorbital meningiomas. Turk Neurosurg 24(6):859–866

Giant Basilar Artery Aneurysm Involving the Origin of Bilateral Posterior Cerebral and Superior Cerebellar Arteries: Neck Reconstruction with pCONus-Assisted Coiling

Francesco Signorelli, Carmelo Lucio Sturiale, Giuseppe La Rocca, Alessio Albanese, Francesco D'Argento, Pierpaolo Mattogno, Alfredo Puca, Massimiliano Visocchi, Enrico Marchese, and Alessandro Pedicelli

Abstract Giant aneurysms of the basilar artery are rare and are frequently associated with obstructive hydrocephalus and brainstem compression. Treatment still remains a challenge both for neurosurgeons and for interventional neuroradiologists. Cases reported in the literature are anecdotal and, overall, their outcomes are poor. We present the case of a patient with a giant aneurysm of the basilar artery tip, involving the origin of both the posterior cerebral and superior cerebellar arteries, who underwent coiling and ventriculoperitoneal shunting for associated obstructive hydrocephalus. A pCONus ® stent (Phenox; Bochum, Germany) was detached with its petals opened over the ostia of the parent vessels, with the aim being to reconstruct the neck of the aneurysm and to preserve the flow in the parent vessel. Moreover, the presence of the stent was useful to maintain the coils within the dome of the aneurysm. The pCONus is a new neurovascular device that is also useful for treating cases of complex basilar artery aneurysms when the ostia of the parent vessel origin is at the level of the aneurysm neck.

Keywords Giant aneurysm • Basilar artery • pCONus • Coiling • Reconstructive neurosurgery

Francesco Signorelli, MD, and Carmelo Lucio Sturiale, MD, PhD, are first co-authors.
Dr. Enrico Marchese and Dr. Alessandro Pedicelli both supervised the study as senior co-authors.

F. Signorelli, MD (✉) • C.L. Sturiale, MD, PhD • G. La Rocca, MD • A. Albanese, MD • P. Mattogno, MD • A. Puca, MD
M. Visocchi, MD • E. Marchese, MD, PhD
Institute of Neurosurgery, Medical School, Catholic University of Rome, Rome, Italy
e-mail: francesco.signorelli1984@gmail.com

F. D'Argento, MD • A. Pedicelli, MD
Institute of Bio-Imaging, Catholic University School of Medicine, Rome, Italy

Introduction

Giant aneurysms, overall, account for about 5 % of all aneurysms, and only a few of them involve the basilar artery. These commonly originate from the basilar tip and are frequently associated with obstructive hydrocephalus and brainstem compression [1, 2]. Although most of them continue to enlarge over time, increasing the risk of rupture, their treatment still remains a challenge both for neurosurgeons and for interventional neuroradiologists. Regardless of difficulties in the decision-making process for the management of the aneurysm, cerebrospinal fluid (CSF) shunting is often mandatory for treating the concomitant obstructive hydrocephalus [3, 4]. Recently, endoscopic third ventriculostomy (ETV) has also been proposed as an alternative option for dealing with this associated condition, both in cases of ruptured and unruptured giant basilar artery aneurysms (BAAs) [5, 6]. In this article we report the case of an elderly patient affected by a giant BAA, of 2.55 cm, bilaterally, involving the origin of the posterior cerebral artery (PCA) and the superior cerebellar artery (SCA), who underwent pCONus ® (Phenox, Bochum, Germany)-assisted coiling with the aim being to reconstruct the neck of the aneurysm and to preserve the blood flow at the ostia of the parent vessels. The patient also showed associated hydrocephalus, which was managed with a ventriculoperitoneal (VP) shunt. A few cases of pCONus-assisted coiling for complex BAAs have been reported in the literature so far, but only one of these BAAs was large and showed the origin of both the PCAs and the SCAs from the neck of the aneurysm [7]. To the best of our knowledge, this is the first case reported in the literature of a giant BAA treated with pCONus-assisted coiling (Table 1).

M. Visocchi et al. (eds.), *Trends in Reconstructive Neurosurgery*, Acta Neurochirurgica Supplement 124,
DOI 10.1007/978-3-319-39546-3_20, © Springer International Publishing Switzerland 2017

Table 1 Giant basilar artery aneurysms associated with obstructive hydrocephalus

Author	Age (years), sex	Aneurysm treatment	Shunt	Follow-up
Goetz et al. (1990) [16]	62, M	NA	VP	Died from SAH 3 weeks after VP shunt placement
Bose et al. (1983) [1]	55, F	Attempted aneurysm removal	VP	Died after craniotomy
Borrie et al. (1985) [3]	72, F	NA	VP	Improved
Wozniak et al. (1978) [17]	NA	NA	VP	Died
Kinoshita et al. (1979) [18]	52, M	NA	VP	Improved
Shimizu et al. (1985) [19]	52, M	NA	VP	Aneurysm still disappeared 3 years after VP shunt placement
Yasargil (1984) [20]	24, M	Trapping	VP	Aneurysm growth after shunt placement
Koga et al. (1983) [21]	65, M	NA	VP	Improved
Piek et al. (1983) [22]	60, F	NA	VP	Improved
Ishibashi et al. (1993) [23]	63, M	NA	VP	Improved
Drake et al. (1979) [24]	64, M	NA	VP	Died from SAH 4 weeks after VP shunt placement
Drake et al. (1979) [24]	32, F	LVA clamp occlusion	VP	Improved
Souer et al. (1979) [25]	74, F	NA	VP	Died from aneurysm growth
Vishteh et al. (1999) [26]	41, M	NA	VP	Aneurysm growth 7 years after VP shunt placement
Kim et al. (2002) [11]	58, M	LVA occlusion	VP	Died from aneurysm growth 11 months after VP shunt placement
Kim et al. 2002 [11]	30, M	Neck clipping	VP	Aneurysm growth 5 months after VP shunt placement
Hongo et al. (2001) [8]	70, F	RVA occlusion	E septostomy	Died from aneurysm rupture
Koyama et al. (1996) [27]	67, F	None	EVD	Died from aneurysm rupture
Liu et al. (2005) [28]	55, M	Transcallosal biopsy	Open TV	Improved
Oertel et al. (2008) [5]	80, M	None	ETV	Improved
Oertel et al. (2008) [5]	55, M	Coil embolization	ETV	Died from brain infarction 5 months after ETV
Oertel et al. (2008) [5]	32, F	Coil embolization	ETV	Died from BA thrombosis 1 week after ETV
Obaid et al. (2012) [6]	53, F	Coil embolization	ETV	Improved
Obaid et al. (2012) [6]	77, F	Coil embolization	ETV	Improved
Obaid et al. (2012) [6]	81, F	None	ETV	Improved
Kaptan et al. (2013) [29]	NA	None	VP	Died from brain infarction 2 weeks after VP shunt placement
Stachura et al. (2008) [30]	NA	None	ETV	Improved
Present patient, 2015	76, F	pCONus-assisted coiling	VP	Died from aneurysm rupture 9 months after VP shunt placement

EVD VP External Ventricular Drainage ventriculoperitoneal, *ETV* endoscopic third ventriculostomy, *RVA* right vertebral artery , *LVA* left vertebral artery, *BA* basilar artery, *NA* not available, *SAH* subarachnoid hemorrhage

Case Report

A 76-year-old woman was admitted to our neurosurgical department with a 2-month history of confusion, gait imbalance, and urinary incontinence. The patient underwent a brain magnetic resonance imaging (MRI) investigation, which revealed the presence of a giant BAA associated with brainstem compression and obstructive hydrocephalus (Fig. 1a, b). Digital subtraction angiography (DSA) also showed the origin of both the PCAs and SCAs from the sac at the neck of the aneurysm (Fig. 1c, d).

After we engaged in a thorough multidisciplinary discussion, surgical clipping reconstruction was excluded because of the patient's advanced age and comorbidities; traditional coiling was also not appropriate because it could not guarantee preservation of the blood flow in the parent arteries

Therefore, we decided to use a pCONus ® stent (Phenox) with its upper extremities (petals) anchored over the ostia of the parent vessels (Fig. 2a–d). The provided bridging structure allowed the dislocation of 52 coils (Target, Axium, and Versatile Range Fill Advanced Coils (Microvention), ®) to completely fill out the dome of the sac, which was excluded by the circula-

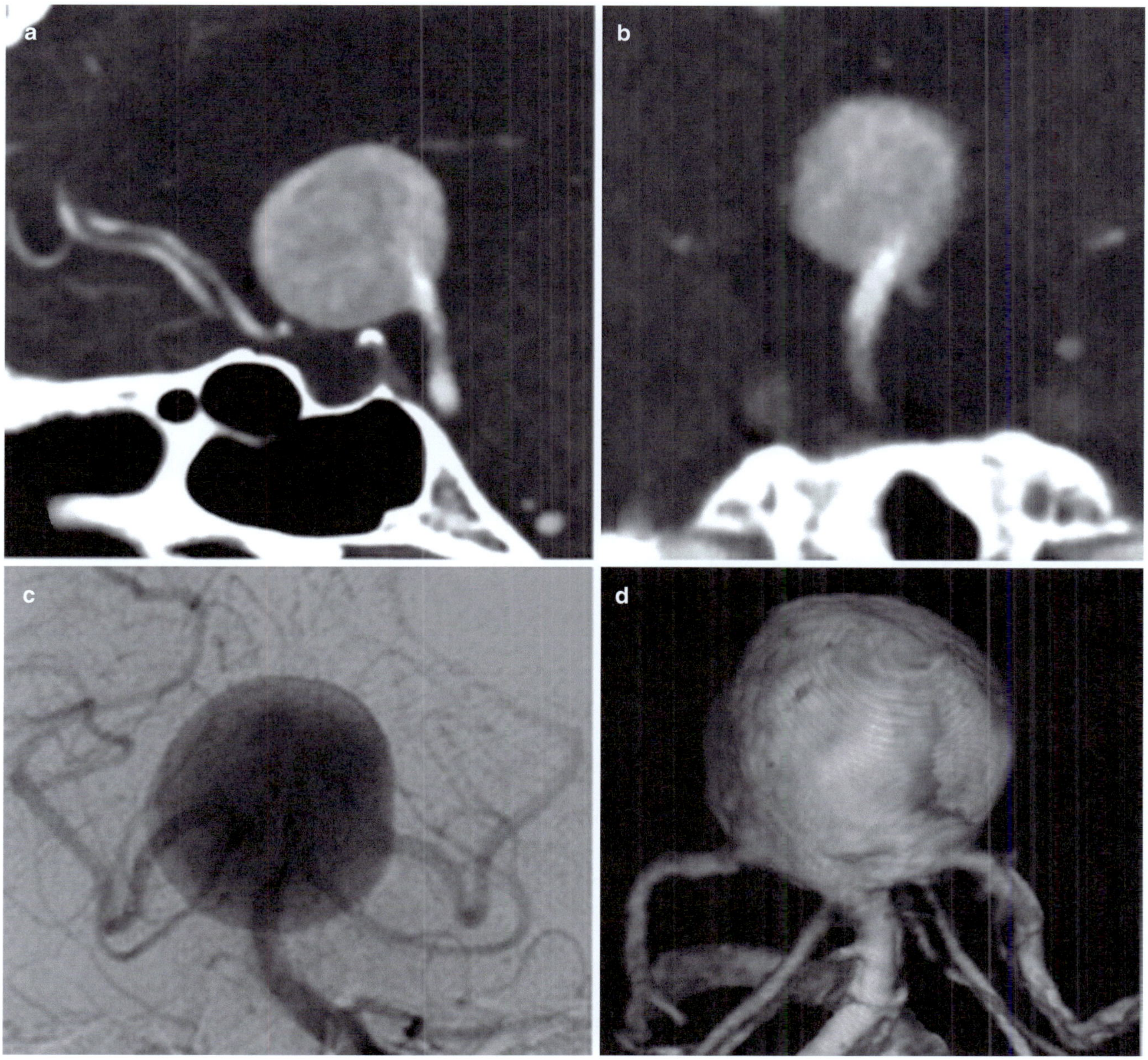

Fig. 1 Preoperative sagittal and coronal angio-computed tomography (CT) scan images showing the giant basilar artery (BA) aneurysm associated with obstructive hydrocephalus (**a**, **b**). Cerebral digital subtraction angiography (DSA) showing the origin of both the posterior cerebral and the superior cerebellar arteries from the neck of the aneurysm (**c**, **d**)

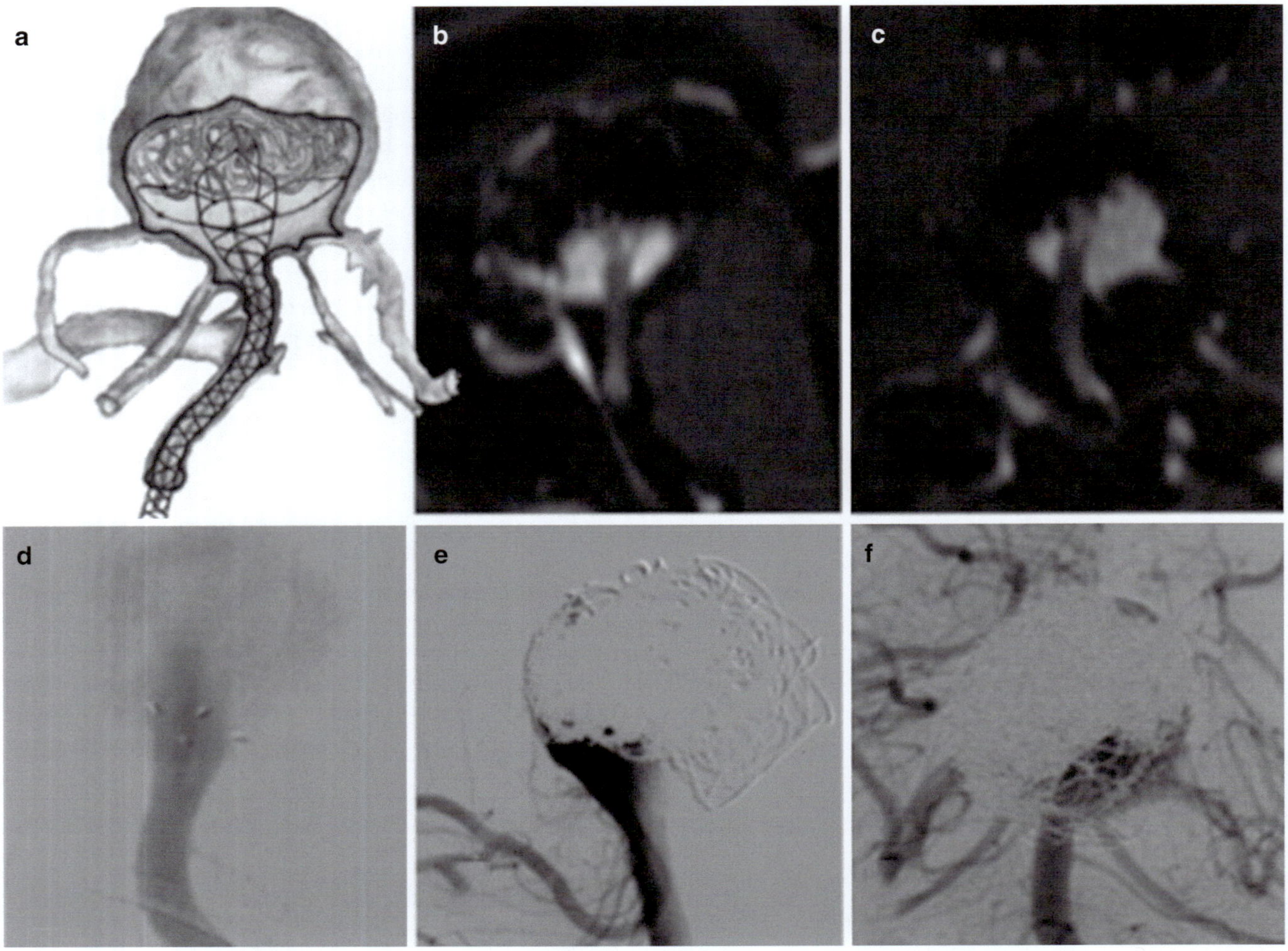

Fig. 2 Artwork showing the final positioning of the pCONus stent along with the coiling of the aneurysmatic sac (**a**). Post-procedural DSA showing the positioning of the pCONus stent, whose petals appear open over the ostia of the parent vessels; the coils have obliterated 4/5 of the aneurysmatic sac (**d–f**). Three-month follow-up magnetic resonance imaging (MRI) scan showing coil compaction and partial reperfusion of 2/5 of the sac, with no stent displacement (**b**, **c**)

tion (Fig. 2b, c and e), and at the same time, we obtained a reconstruction of the neck of the aneurysm, which preserved the blood flow in the parent vessels (Fig. 2f). The procedure was performed by the senior author (Alessandro Pedicelli).

Before undergoing the endovascular procedure, the patient underwent ventriculoperitoneal shunting with a programmable Hakim-Medos valve (Codmann; Johnson&Johnson ®), which was temporarily closed with a silk suture at the supraclavicular region until completion of the endovascular procedure. Also, the patient received double antiplatelet therapy, with aspirin 100 mg per day and clopidogrel 75 mg per day, for 7 days before the endovascular procedure; once the aneurysm was coiled, the shunt was opened by removing the silk suture.

During the postoperative period, the patient's neurological condition improved slightly, especially in regard to her cognitive status, although the gait disturbance persisted.

At 3-month follow-up the clinical condition of the patient had significantly improved, with almost complete regression of the preoperative symptoms. A brain computed tomography (CT) scan confirmed the absence of hydrocephalus, and MRI showed coil compaction, with extension of the reperfusion of the sac (Fig. 2). The patient refused further treatments.

Nine months after the procedure, the patient presented with a sudden loss of consciousness. A CT scan showed a massive subarachnoid hemorrhage (SAH) due to the rupture of the aneurysm.

Discussion

Giant BAAs are very rare and are frequently associated with brainstem compression and hydrocephalus [8]. They tend to progressively grow in size, showing an increasing risk of rupture. Therefore, whenever possible, treatment is mandatory, along with management of the associated hydrocephalus.

Nowadays coiling is the preferred treatment for unruptured aneurysms, in particular for those located at the tip of the basilar artery, because it can achieve complete exclusion of the sac in the majority of cases, while at the same time preserving the small brainstem perforators. For large aneurysms, complete coiling is able to preserve the aneurysm from the risk of rupture, and in some cases the procedure limits the pulsation, reducing the mass effect-related symptoms. As the size of the BAA gets larger, coiling by itself may sometimes worsen the symptoms related to the brainstem compression, increasing the mass effect.

Before the advent of the stent-assisted coiling technique, only saccular aneurysms with a favorable dome/neck ratio could be managed endovascularly. Nowadays new stent devices allow us to consider endovascular treatment also for irregular BAAs that do not show a favorable morphology. In the present patient, we used a pCONus stent in order to reconstruct the neck region of the aneurysm (Fig. 2a), since the ostia of both the PCAs and SCAs opened at this level. The device was introduced within the sac and the petals were unfolded over the region of the ostia, with no wall anchoring. Fifty-two Guglielmi Detachable Coil were then introduced within the aneurysm, obtaining exclusion of 4/5 of the sac (Fig. 2).

The pCONus is a new stent-like self-expanding nitinol implant with four distal petals, designed for extra-intrasaccular neck-bridging aneurysm implantation, to assist the coil occlusion of selected bifurcation aneurysms [7]. Stent placement may permanently modify the angulation of the vessel bifurcation, which may be advantageous for detaching coils and maintenance. Moreover, stent placement allows coil occlusion of the aneurysm fundus with reliable protection of the efferent vessels, as in our patient, in whom the PCA and SCA originated from the neck of the aneurysm.

Clipping of recurrent aneurysms after pCONus implantation is discouraged. In fact, after pCONus implantation, if the device position is stable, the aneurysm neck usually remains bridged by the struts of the pCONus, which would most likely interfere with the proper closure of an aneurysm clip [7]. However, coil compaction and aneurysm reperfusion are not prevented by the pCONus, while the device can also assist in re-coiling.

Aguilar-Perez et al. [7] treated a fusiform upper basilar trunk and bifurcation aneurysm (neck 12.7 mm; fundus 4.7 mm) with both PCAs and superior SCAs originating from the lateral aspect of the aneurysm. After positioning the pCONus petals within the lumen of the aneurysm, they catheterized both PCAs, and two Solitaire (Covidien) stents were deployed in a Y-technique. The resulting structure was stable and allowed a well-controlled coil occlusion of the aneurysm. In our patient, reconstruction of the neck obtained with pCONus proved to provide a sufficiently stable structure to ensure that the coils were within the dome of the aneurysm, preserving the blood flow in the parent vessels. Aguilar-Perez et al. also described a second large backward-directed aneurysm of the basilar tip (neck 15.5 mm; fundus 18 mm) incorporating both PCA origins, treated with two 5-mm pCONus devices deployed in a crossing position to protect the neck [7].

Particular consideration is necessary for the timing of a ventriculoperitoneal shunt. Acute reduction of intracranial pressure could result in an increase of the transmural pressure gradient across the aneurysm wall [8–10], causing sac growth and leading to a premature rupture of the aneurysm, as reported by several authors [11]. However, this topic of shunting has been long debated in the literature: in fact, while some authors have argued that shunting is not conclusively associated with the rebleeding of a ruptured aneurysm [12–14], some others have demonstrated an increased risk of aneurysmal rebleeding in patients undergoing ventricular drainage [15]. It is not yet clear whether an increased risk of aneurysmal bleeding occurs in unruptured aneurysms.

Recently ETV has been proposed as an alternative option for dealing with obstructive hydrocephalus associated with unruptured giant BAA. However, compared with conventional shunting, ETV carries the risk of aneurysm perforation during the procedure. Thus, ETV should be considered only in the presence of a prepontine space large enough to perform the opening of the floor of the third ventricle with the aneurysm dome pointing backward [5], and with the endoscopic technique in very experienced hands.

In our patient, we planned to implant a ventriculoperitoneal shunt before the endovascular procedure, but we implanted a temporarily closed shunt in order to avoid the risk of aneurysm rupture before coiling. Ideally, in fact, aneurysm treatment should precede the shunting procedure in order to prevent an increase of the transmural pressure gradient.

The treatment of giant BAAs with obstructive hydrocephalus still remains a challenge. Cases reported in the literature are anecdotal and overall outcomes are poor.

In a review of the pertinent literature, to the best of our knowledge, we found 27 patients harboring BAAs with hydrocephalus treated with CSF diversion (VP shunt or ETV) [1, 3, 5–8, 11, 16–30]. Thirteen of them improved after CSF diversion.

However, information regarding the management of the associated aneurysm was available for only 16 cases. In 5 of them, a CSF diversion was performed before the aneurysm was secured; 2 patients died, from aneurysm rupture and from brain infarction, respectively. In the remaining 11 cases, the patients underwent aneurysm treatment before treatment of the hydrocephalus. Seven of them had a poor outcome due to treatment complications or further aneurysm growth (Table 1).

Conclusions

Giant BAAs are rare and are often associated with obstructive hydrocephalus and brainstem compression. Treatment of the aneurysm and the associated hydrocephalus is mandatory, due to the high risk of aneurysm rupture. Shunting procedures performed without securing the aneurysm are associated with an increased risk of aneurysm bleeding. The pCONus is a new neurovascular device designed to assist coil occlusion in cases of unfavorable dome-neck ratio or in those cases in which the neck hosts the opening of the ostia of the parent vessel, allowing a feasible vessel reconstruction. Nevertheless, these procedures remain at high risk of complications, and they do not guarantee prolonged control of the risk of bleeding.

Acknowledgments We would like to thank Francesco Cantavenera for the realization of the artwork (Fig. 2a) included in this article.

Conflict of Interest Statement The authors declare no conflicts of interest and they have not received any financial support in conjunction with the generation of this manuscript.

References

1. Bose B, Northrup B, Osterholm J (1983) Giant basilar artery aneurysm presenting as a third ventricular tumor. Neurosurgery 13:699–702
2. Gelal F, Vidinli BD, Yurt A, Cakir Y, Uygur M (2002) Giant fusiform basilar artery aneurysm causing obstructive hydrocephalus. Skull Base 12:197–200
3. Borrie MJ, Campbell AJ, Caradoc-Davies TH (1985) Obstructive hydrocephalus secondary to giant cerebral aneurysm in the elderly. J Am Geriatric Soc 33:210–212
4. Morota N, Ohtsuka A, Kameyama S, Suzuki Y, Yasukawa K (1988) Obstructive hydrocephalus due to a giant aneurysm of the internal carotid bifurcation. Surg Neurol 29:227–231
5. Oertel JM, Mondorf Y, Gaab MR (2009) Endoscopic third ventriculostomy in obstructive hydrocephalus due to giant basilar artery aneurysm. J Neurosurg 110(1):14–18
6. Obaid S, Weil AG, Bojanowski MV (2012) Endoscopic third ventriculostomy in the presence of large or giant basilar artery aneurysms. Acta Neurochir 154:1845–1850
7. Aguilar-Perez M, Kurre W, Fischer S, Bazner H, Henkes H (2014) Coil occlusion of wide-neck bifurcation aneurysms assisted by a novel intra- to extra-aneurysmatic neck-bridging device (pCONus): initial experience AJNR Am J Neuroradiol 35:965–971
8. Hongo K, Morota N, Watabe T, Isobe M, Nakagawa H (2001) Giant basilar bifurcation aneurysm presenting as a third ventricular mass with unilateral obstructive hydrocephalus: case report. J Clin Neurosci 8:51–54
9. Nornes H (1973) The role of intracranial pressure in the arrest of hemorrhage in patients with ruptured intracranial aneurysm. J Neurosurg 39:226–234
10. Rosenorn J, Westergaard K, Hansen PH (1983) Mannitol-induced rebleeding from intracranial aneurysm. Case report. J Neurosurg 59:529–530
11. Kim MS, Oh CW, Han DH (2002) Growth of basilar artery aneurysm after ventriculo-peritoneal shunt. J Clin Neurosci 9(6): 696–702
12. Hasan D, Vermeulen M, Wijdicks EFM et al (1989) Management problems in acute hydrocephalus after subarachnoid hemorrhage. Stroke 20:747–753
13. Troupp H (1975) Intracranial pressure in hydrocephalus after subarachnoid hemorrhage. Zentralbl Neurochir 36:11–17
14. Voldby B, Enevoldsen EM (1982) Intracranial pressure changes following aneurysm rupture. Part 3: Recurrent hemorrhage. J Neurosurg 56:784–789
15. Pare L, Delfino R, Leblanc R (1992) The relationship of ventricular drainage to aneurysmal rebleeding. J Neurosurg 76:422–427
16. Goetz C, Seifert V, Haubitz B (1990) The foramen of Monro-blockage caused by a giant aneurysm of the basilar artery. A case report and review of the literature (in German). Neurochirurgia (Stuttg) 33:122–126
17. Wozniak M, Mazurowski W, Chrzanowski M (1978) Case of giant aneurysm of the basilar artery (in Polish). Neurol Neurochir Pol 12:345–349
18. Kinoshita K, Sunami N (1979) Occult hydrocephalus due to unruptured giant basilar bifurcation aneurysm (in Japanese). No Shinkei Geka 7:603–608
19. Shimizu Y, Namba S, Kinoshita K (1985) Spontaneous angiographical disappearance of an unruptured giant aneurysm of the basilar top demonstrated by a sequential series of angiograms. A case report (Japanese). No Shinkei Geka 13:1121–1127
20. Yasargil MG (1984) Microsurgery, vol 2. Georg Thieme Verlag/Thieme-Stratton Inc, Stuttgart/New York, pp 232–295
21. Koga H, Mori K, Kawano T, Tsutsumi K, Jinnouchi T (1983) Parinaud's syndrome in hydrocephalus due to a basilar artery aneurysm. Surg Neurol 19:548–553
22. Piek J, Lim DP, Bock WJ (1983) Obstructive hydrocephalus caused by a growing, giant aneurysm on the upper basilar artery. Surg Neurol 20:288–290
23. Ishibashi A, Yokokura Y, Kojima K, Abe T (1983) Acute obstructive hydrocephalus due to an unruptured basilar bifurcation aneurysm associated with bilateral internal carotid occlusion±a case report. Kurume Med J 40:21–25
24. Drake CG (1979) Giant intracranial aneurysms: experience with surgical treatment in 174 patients. Clin Neurosurg 26:12–95
25. Soeur M, Brihaye J, Moerman C (1979) Pseudo-tumoural aneurysm in the third ventricle: report of three cases. Acta Neurochir (Wien) 45:247–258
26. Vishteh AG, Spetzler RF (1999) Evolution of a dolichoectatic aneurysm into a giant serpentine aneurysm during long-term follow up. Case illustration. J Neurosurg 91(2):346
27. Koyama S, Kotani A, Sasaki J (1996) Giant basilar artery aneurysm with intramural hemorrhage and then disastrous hemorrhage: case report. Neurosurgery 39:174–178
28. Liu JK, Gottfried ON, Couldwell WT (2005) Thrombosed basilar apex aneurysm presenting as a third ventricular mass and hydrocephalus. Acta Neurochir (Wien) 147(4):413–416
29. Kaptan H, Koktekir E, Recber F, Akdemir G (2013) Obstructive hydrocephalus caused by giant basilar artery aneurysm. Neurosciences (Riyadh) 18(2):173–175
30. Stachura K, Libionka W, Moskała M (2008) Neuroendoscopic third ventriculostomy in the management of noncommunicating hydrocephalus secondary to giant basilar artery bifurcation aneurysm – case report and review of literature. Neurol Neurochir Pol 42(3):255–262

Part IV

Reconstructive Neurosurgery: Technical Notes

Skull Bone Defects Reconstruction with Custom-Made Titanium Graft shaped with Electron Beam Melting Technology: Preliminary Experience in a Series of Ten Patients

Natale Francaviglia, Rosario Maugeri, Antonino Odierna Contino, Francesco Meli, Vito Fiorenza, Gabriele Costantino, Roberto Giuseppe Giammalva, and Domenico Gerardo Iacopino

Abstract

Background

Cranioplasty represents a challenge in neurosurgery. Its goal is not only plastic reconstruction of the skull but also to restore and preserve cranial function, to improve cerebral hemodynamics, and to provide mechanical protection of the neural structures. The ideal material for the reconstructive procedures and the surgical timing are still controversial. Many alloplastic materials are available for performing cranioplasty and among these, titanium still represents a widely proven and accepted choice.

Methods

The aim of our study was to present our preliminary experience with a "custom-made" cranioplasty, using electron beam melting (EBM) technology, in a series of ten patients. EBM is a new sintering method for shaping titanium powder directly in three-dimensional (3D) implants.

Findings

To the best of our knowledge this is the first report of a skull reconstruction performed by this technique. In a 1-year follow-up no postoperative complications have been observed and good clinical and esthetic outcomes were achieved

Conclusion

Costs higher than those for other types of titanium mesh, a longer production process, and the greater expertise needed for this technique are compensated by the achievement of most complex skull reconstructions with a shorter operative time.

N. Francaviglia, MD • A. Odierna Contino, MD • F. Meli, MD, PhD
V. Fiorenza, MD • G. Costantino, MD
Department of Neurosurgery, ARNAS Hospital Civico, Palermo, Italy

R. Maugeri, MD, PhD (✉) • R.G. Giammalva, MD
D.G. Iacopino, MD, PhD
Department of Experimental Biomedicine and Clinical Neurosciences, School of Medicine, Neurosurgical Clinic, University of Palermo, Via del Vespro 129, Palermo 90100, Italy
e-mail: rosario.maugeri1977@gmail.com

Keywords Skull bone defects • Cranioplasty • Custom-made titanium implants • 3D computer-aided design and computer-aided manufacturing (CAD/CAM) technologies • Electron beam melting technology

Introduction

Cranioplasty is defined as the full-thickness reconstruction of calvarial bone. It restores the integrity of the skull, protects neural structures, and improves cerebral hemodynamics [11].

Decompressive craniectomy is often required in cases of traumatic brain injuries, hemorrhage, tumor removal, or other severe cerebral damage [1]. This procedure is also performed to treat growing skull fractures and congenital anomalies in pediatric patients, where the skull reconstruction is more challenging, considering the singularly rapid bone growth in children [1, 4].

Archeological evidence proves that cranioplasty dates back to 7000 B.C., making this one of the most ancient surgical procedures [1]. Since then, many different methods and grafts have been adopted throughout the history of neurosurgery, and great improvements have been achieved since the nineteenth century, with the development of metal and nonmetal synthetic allografts [1, 4]. Some authors still support the primary role of an autologous cranial bone flap in primary cranioplasty, when possible, ensuring protection by freezing the bone at −70 °C [3, 9, 12]. However, a secondary cranioplasty is sometimes needed because of bone resorption, autograft devitalization, or infections. In this and other cases, such as severe traumatic brain injuries with comminuted fractures, skull reconstruction has to be performed by the use of allografts, which allows the surgeon to reshape the cranial profile and restore its function.

Among the non-metal synthetic allografts, polymethylmethacrylate (PMMA), polyether ether ketone (PEEK), and hydroxyapatite are the most used [4]. Among the whole variety of materials that have been adopted as metal allografts, titanium

DOI 10.1007/978-3-319-39546-3_21,

Table 1 Demographic data of patients who underwent cranioplasty with EBM custom-made titanium graft

Age (years)	Sex	Localization of skull defect	Etiology
70	F	Frontal-parietal right	Failure of primary cranioplasty
32	M	Bifrontal	TBI with edema
60	F	Frontal-parietal right	Aneurysm of MCA
58	F	Frontal-parietal left	Cerebral hemorrhage
50	M	Frontal-parietal right	TBI with edema
45	F	Pterional right	Aneurysm of MCA
40	M	Frontal-parietal left	TBI with edema
55	M	Frontal-parietal left	Cerebral hemorrhage
35	F	Frontal-parietal right	Aneurysm of MCA
45	M	Bifrontal	Failure of primary cranioplasty

MCA middle cerebral artery, *TBI* traumatic brain injury

maintains its role as the only metal currently used in cranioplasty, and it has been used since 1965 [4]. Titanium can be modelled manually or by molding, also intraoperatively [6]. Even if it is hard to shape, titanium is cheaper than other alloplastic grafts, and it is radiotransparent and biocompatible [4]. It can be used as a plate covering the bone defect itself or as a mesh supporting other plastic materials or cement. With the increasing development of new synthetic materials and modelling technologies, such as three-dimensional (3D) printing, using computer-aided design and computer-aided manufacturing (CAD/CAM) technologies, cranioplasty has been improved by many alternative materials and methods when autologous bone is not suitable as a graft. Here we present our preliminary experience in a clinical series of ten patients who underwent "custom-made" cranioplasty, using novel electron beam melting (EBM) technology, together with the support of a skull computed tomography (CT) scan to model the complex-shaped titanium allograft with CAD/CAM technology, without the need for a plastic mold.

Materials and Methods

A single-center retrospective study was performed at the Department of Neurosurgery of the Hospital CIVICO in Palermo, Italy. Between September 2014 and October 2015, ten patients (five males and five females) aged from 32 to 70 years (median, 49 years) underwent cranioplasty with an EBM custom-made titanium graft after craniotomy, performed for different etiologies: traumatic brain injuries in three patients, rupture of a middle cerebral artery aneurysm in three patients, hypertensive cerebral hemorrhage in two patients, and failure of a primary cranioplasty in two patients. A total of ten cranioplasties were performed with this graft; seven of the ten patients presented a frontal-parietal skull defect; two, a bifrontal skull defect; and one patient had a pterional bone defect. The grafts were designed and built at MT Ortho, a manufacturing engineering company in Catania, Italy.

Demographic data are summarized in Table 1.

Description of the Procedure

The grafts used to perform cranioplasty were made of titanium modelled by CAD/CAM technology and produced by electron beam melting (EBM) technology. CAD/CAM EBM is a novel technique for titanium powder sintering that provides the precise shape of the titanium implant in a virtual 3D model of the patient's skull (obtained from DICOM (Digital Imaging and Communications in Medicine) CT-imaging). The technique uses a high-energy focused electron beam that melts the titanium powder in a vacuum chamber. This beam not only models a simple plate but also precisely shapes the CAD model of the prosthesis in three dimensions. A high-resolution CT scan of the entire skull is performed on every patient before the surgical procedure (Fig. 1a–c). The acquired images in DICOM format are transferred to the manufacturing company (Fig. 2a). An accurate 3D virtual image of the skull is created, via software, for each patient and the titanium plate is then precisely shaped on this, respecting the symmetry and individual bone curvature (Fig. 2b, c). The 3D virtual images of the titanium graft and skull are digitally verified with the prior CT scan, and then a 3D plastic prototype of the patient's skull defect is printed (Fig. 2d). With the EBM technique, the previously designed titanium graft is printed by the sintering of titanium powder and then fitted on the 3D plastic prototype of the skull defect to ensure the best clinical and esthetic results before surgical implantation (Fig. 2d, e). The custom-made EBM titanium grafts are provided with holes for drainage and textured surfaces to improve their integration with soft tissue. Moreover, the graft's thin and definitely shaped margins precisely follow the bone defect margins. The

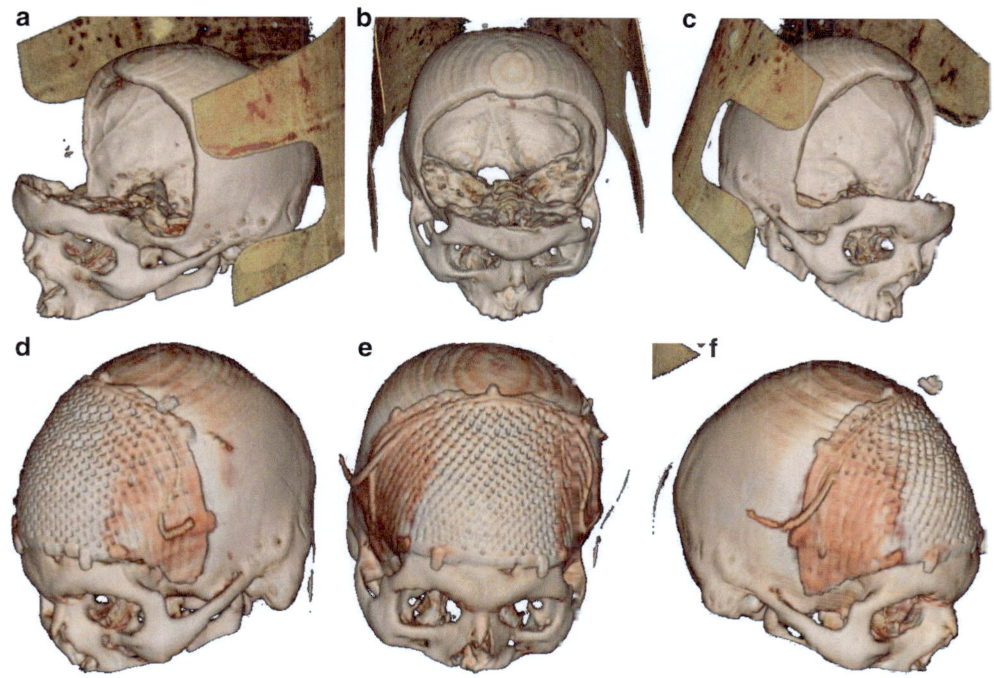

Fig. 1 Three-dimensional (3D) volume rendering computed tomography (CT) scan. (**a–c**) Preoperative skull CT scan. (**d–f**) Postoperative CT scan with implanted bifrontal electron beam modelling (EBM)-created titanium graft

fixation is performed with titanium screws directly onto the bone circumference. A postoperative CT scan is performed to evaluate the surgical result (Fig. 1d–f).

Results

All cranioplasty procedures were performed according the aforementioned technique. The mean duration of the skull reconstruction was 111 min and the range was between 78 and 184 min. A thin-slice CT scan, with multiplanar reconstruction, was performed in the immediate postoperative period and then after 6 and 12 months. No intraoperative or postoperative complications were observed with any of the ten surgical procedures. No infective or hemorrhagic complications were recorded. After a 1-year follow-up period good clinical and functional outcomes, and good esthetic results, were achieved in all the patients.

Discussion

While some authors still confirm the effectiveness of autologous bone as the graft for primary cranioplasty [3], many other clinical series support the use of PEEK, PMMA, and titanium [3, 14].

There is a deep lack of consensus on the best choice of materials and techniques for reconstructing the skull bones [1, 12]. Some of the qualities of an ideal material for cranioplasty are that it has to allow the complete and easy closure of the cranial defect, with other required qualities being radiotransparency, low infection rate [8], biocompatibility, good elastic modulus and resistance to strain, and the maintenance of its chemical and physical properties over time [1, 3]. Among alloplastic bone substitutes in cranioplasty, titanium is the most common metal still in use, with different modelling techniques being employed. (Table 2).

In critical-size skull defects, such as those wider than 25 cm^2 and those too large to heal unaided [3], titanium demonstrates chemical and physical properties that are the most similar to those of bone [3]. Titanium has also been demonstrated to be corrosion resistant [3]. Meshes and solid plates made of titanium for reconstructive purposes provide an elastic modulus that is similar to and not greater than that of cortical bone [3]. These properties make surgical titanium implants less likely to fail during traumatic injuries, compared with other alloplastic bone substitutes [3] (i.e., PMMA, PEEK, hydroxyapatite, and ceramics). Moreover, the textured and porous surfaces of the titanium implants allow these implants to incorporate soft tissues and improve integration [3].

The aim of our present study was to illustrate our preliminary experience with a "custom-made" cranioplasty, using CAD/CAM and EBM technology, in a retrospective clinical series of ten patients at the Department of Neurosurgery of the Hospital CIVICO in Palermo.

Although hydroxyapatite has greater osteoinductive power than other allograft materials, titanium ensures lower infection and rejection rates than other allografts [8–10].

In our preliminary experience, after a 1-year follow-up period (range 3–12 months), good clinical and functional

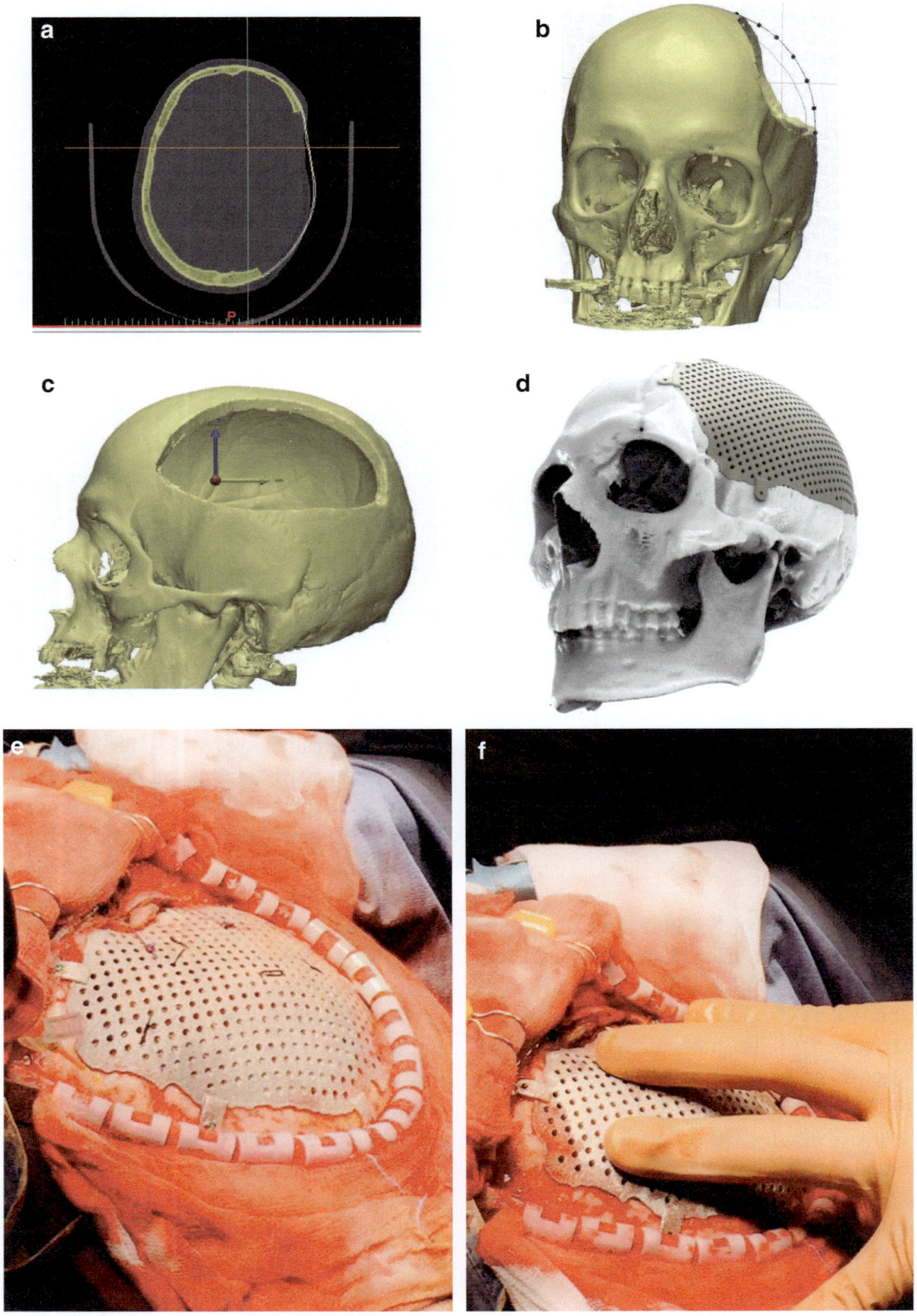

Fig. 2 (**a**) A 3D virtual model of the skull, rendered for the preventive virtual reconstruction of the bone defect surface. (**b**) Correction of the virtual bone reconstruction curvature to prevent intersections and ensure optimal strain dissipation. (**c**) Digital check of the compatibility of the designed implant with prior skull CT images. (**d**) Check of the match between the created titanium implant and 3D-printed plastic skull model of the patient. (**e, f**) Implantation of computer-aided design and computer-aided manufacturing (CAD/CAM) EBM custom-made titanium graft for cranioplasty

outcomes, as well as good esthetic results, were achieved in all the patients. No infective or hemorrhagic complications were recorded in this clinical series. To the best of our knowledge, this is the first report of a skull reconstruction performed by this technique. Considering titanium's cost, it must be emphasized that hand-made titanium allografts are less expensive than CAD/CAM titanium implants [7, 15], and the latter also require *ad hoc* edu-

Table 2 Titanium implant modelling techniques

Custom-made techniques	Modelling technology
Hand-made	Titanium meshes directly shaped by hand during surgical operation
Negative cast from bone flap	Use of bone flap to pour a negative cast on which a titanium plate is modelled by hydraulic press
Stereolithography (STL)	Titanium modelled by hydraulic press on 3D-printed plastic model of the skull acquired by DICOM CT images, with CAD/CAM technology
Selective laser sintering	Titanium sintered directly with a laser from a CAD model of the implant, with CAD/CAM technology
Electron beam melting	Titanium sintered directly with an electron beam from a CAD model of the implant, with CAD/CAM technology

CAD/CAM computer-aided design and computer-aided manufacturing, *CT* computed tomography, *DICOM* Digital Imaging and Communications in Medicine

cated technicians and a longer production process [15]. Titanium is, however, significantly cheaper than other alloplastic grafts [1, 4], and it is radiotransparent [10] and biocompatible [3]. Although the costs of titanium implants are higher, it has emerged that this CAD/CAM technique provides a shorter operative time, lower failure rates, a reduction in the number of screws used to fix the graft to the surrounding bone, and better esthetic results, increasing the patient's satisfaction [7, 13]. Postoperative complications are decreased too, reducing subcutaneous fluid collection, tilting of the mesh and its exposure, and reducing compression of the temporal muscle, with the containment of hospital costs [7]. These advantages make titanium a good choice [15]. To sum up, according to previous data, EBM technology is now emerging in the wide horizon of CAD/CAM shaping techniques as a promising method for shaping titanium grafts.

Conclusions

There is a deep lack of consensus on the best choice of materials and techniques for reconstructing the skull bones [12]. A huge variety of techniques using titanium as a metal allograft have been reported: hand-made custom implants shaped intraoperatively, titanium meshes and plates molded on stereolithographic CAD models of the skull, titanium plates shaped on the craniectomy bone flap as a template, and CAD/CAM titanium implants made by numerically controlled milling machines [2, 3, 5, 6]. The use of EBM technology to create custom-made titanium grafts allows us to precisely shape the areas of skull to be reconstructed, from calvarial bone to the most complex ones (such as pterion and temporal fossa). It also avoids the need to manually mold any mesh or titanium plate, guaranteeing the preservation of fine details and allowing one-step surgical application. A lower rate of postoperative complications and better results over a long time period are expected from the ongoing follow-up of our data.

Conflict of Interest Statement The Authors report no conflicts of interest. No funds were received in support of this work

References

1. Aydin S, Kucukyuruk B, Abuzayed B, Aydin S, Sanus GZ (2011) Cranioplasty: review of materials and techniques. J Neurosci Rural Pract 2(2):162–167
2. Bhargava D, Bartlett P, Russell J, Liddington M, Tyagi A, Chumas P (2010) Construction of titanium cranioplasty plate using craniectomy bone flap as template. Acta Neurochir (Wien) 152(1):173–176
3. Bonda DJ, Manjila S, Selman WR, Dean D (2015) The recent revolution in the design and manufacture of cranial implants: modern advancements and future directions. Neurosurgery 77:814–824
4. Harris DA, Fong AJ, Buchanan EP, Monson L, Khechoyan D, Lam S (2014) History of synthetic materials in alloplastic cranioplasty. Neurosurg Focus 36(4):E20
5. Höhne J, Brawanski A, Gassner HG, Schebesch KM (2013) Feasibility of the custom-made titanium cranioplasty CRANIOTOP(®). Surg Neurol Int 4:88
6. Lau D, McDermott MW (2015) A method for combining thin and thick malleable titanium mesh in the repair of cranial defects. Cureus 7(5):e267
7. Luo J, Liu B, Xie Z, Ding S, Zhuang Z, Lin L, Guo Y, Chen H, Yu X (2012) Comparison of manually shaped and computer-shaped titanium mesh for repairing large frontotemporoparietal skull defects after traumatic brain injury. Neurosurg Focus 33(1):E13
8. Matsuno A, Tanaka H, Iwamuro H, Takanashi S, Miyawaki S, Nakashima M, Nakaguchi H, Nagashima T (2006) Analyses of the factors influencing bone graft infection after delayed cranioplasty. Acta Neurochir (Wien) 148(5):535–540; discussion 540
9. Moreira-Gonzalez A, Jackson IT, Miyawaki T, Barakat K, DiNick V (2003) Clinical outcome in cranioplasty: critical review in long-term follow-up. J Craniofac Surg 14(2):144–153
10. Rakowski JT, Chin K, Mittal S (2012) Effects of titanium mesh implant on dosimetry during gamma knife radiosurgery. J Appl Clin Med Phys 13(5):3833
11. Sakamoto S, Eguchi K, Kiura Y, Arita K, Kurisu K (2006) CT perfusion imaging in the syndrome of the sinking skin flap before and after cranioplasty. Clin Neurol Neurosurg 108(6):583–585
12. Servadei F, Iaccarino C (2015) The therapeutic cranioplasty still needs an ideal material and surgical timing. World Neurosurg 83(2):133–135
13. Stoodley MA, Abbott JR, Simpson DA (1996) Titanium cranioplasty using 3-D computer modelling of skull defects. J Clin Neurosci 3(2):149–155
14. Thien A, King NK, Ang BT, Wang E, Ng I (2015) Comparison of polyetheretherketone and titanium cranioplasty after decompressive craniectomy. World Neurosurg 83(2):176–180
15. Williams LR, Fan KF, Bentley RP (2015) Custom-made titanium cranioplasty: early and late complications of 151 cranioplasties and review of the literature. Int J Oral Maxillofac Surg 44(5):599–608

Myelomeningocele Repair: Surgical Management Based on a 30-Year Experience

Pier Paolo Mattogno, Luca Massimi, Giampiero Tamburrini, Paolo Frassanito, Concezio Di Rocco, and Massimo Caldarelli

Abstract

Background

Myelomeningocele (MMC) is a rare but severe spinal defect resulting from a failed neurulation process. Surgical repair is a delicate procedure that needs accurate neuroanatomical knowledge and adequate surgical planning.

Materials and Methods

The authors report on the main problems of the surgical management of this type of dysraphism. The technique used for reconstruction, and the management strategies, are analyzed according to the authors' experience with more than 600 cases, in comparison with the relevant literature.

Results

Tip and tricks of more than 30 years of MMC surgical management are reported, with special mention of the timing of surgery, the management of early hydrocephalus, and the principles of the repair, from the dissection to the closure of five separate layers (arachnoid, dura, muscular fascia, subcutaneous layer, skin).

Conclusions

Accuracy and attention to the reconstruction are crucial for achieving good results and avoiding perioperative complications.

Keywords Myelomeningocele • Hydrocephalus • Spina bifida • Surgical repair

P.P. Mattogno, MD (✉) • L. Massimi • G. Tamburrini • P. Frassanito
C. Di Rocco • M. Caldarelli
Pediatric Neurosurgery – A. Gemelli Hospital, Catholic University Medical School, Largo A. Gemelli, Rome, Italy
e-mail: pierpaolo.mattogno@gmail.com

Introduction

The malformed spinal cord or primitive placode appears as a flat tongue of neural tissue with its borders merging into the contiguous malformed meningeal coverings [6]. As an effect of the failed neurulation process, both ventral and dorsal spinal roots exit from the ventral aspect of the placode, surrounded by the "junctional zone", which is a boundary between the malformed arachnoid membrane covering the placode and dystrophic skin. The presence of an intact subarachnoid space ventral to the placode confirms the lesion as a myelomeningocele (MMC), while its absence suggests a myelocele.

A first, milestone, principle of the surgical repair of the MMC is to recognize and carefully handle the neural tissues to minimize the risk of harming the residual functional nervous tissue. A second principle concerns the protection of the covering structures and their reconstruction to restore the cerebrospinal fluid (CSF) circulation and to prevent the retethering of the spinal cord and the occurrence of CSF leakage. Finally, knowledge of the correct timing for surgical repair and the prevention of perioperative complications allows us to achieve accurate management of MMC.

Timing of Surgery

According to the literature and to the authors' experience with over 600 cases treated in the past 30 years, the surgical repair of MMC has to be carried out as soon as possible after birth [11, 20]. Such a goal is usually achievable, because, thanks to prenatal diagnosis, the birth of a child with MMC is an expected and well-planned event in more than 90 % of cases. However, a certain time interval is necessary to obtain comprehensive information on the patient's clinical condition and to plan the surgical reconstruction adequately [23]. Surgery should be performed within the first 48 h of life to reduce the risk of infection (e. g., the

M. Visocchi et al. (eds.), *Trends in Reconstructive Neurosurgery*, Acta Neurochirurgica Supplement 124,
DOI 10.1007/978-3-319-39546-3_22, © Springer International Publishing Switzerland 2017

frequency of ventriculitis drops from 37 to 7 % if surgery is done within 48 h) or neurological deterioration resulting from the dehydration or the stretching of the placode [36].

In the majority of cases, fetuses with MMC are delivered at term, thus reaching pulmonary maturity and adequate weight at birth, factors which are crucial for a favorable prognosis. The delivery is obtained through a cesarean section to reduce the risk of adjunctive neurological dysfunction resulting from the possible compression and/or contamination of the placode inside the vaginal canal [8, 18, 25]. The cesarean procedure is also required because of the macrocrania that is often found in fetuses with early hydrocephalus.

At our institution, the planned cesarean section of a mother with a fetus affected by MMC is carried out early in the morning in order to have the newborn ready for the neurosurgical repair late in the morning or early in the afternoon. This has allowed us to repair the spinal defect within the first 6–12 h of life in the great majority of our patients. Only in a few cases, the repair was realized later (but always within 48 h); these were the cases without a prenatal diagnosis.

Early Management of Hydrocephalus

Hydrocephalus is so frequently associated with MMC (85–90 % of cases) that it can be considered as part of the malformation [10, 26]. Aqueduct stenosis and fourth ventricle outlet obstruction and obliteration of the posterior fossa subarachnoid spaces are responsible for such a frequent association [10, 12]. The hydrocephalus, which is symptomatic at birth in 15 % of the cases, usually worsens in the early postnatal period, with signs of raised intracranial pressure (ICP; split sutures, tense anterior fontanel, sunsetting eyes, vomiting) or even brainstem dysfunction (stridor, apneic spells, poor sucking and swallowing, high-pitched cry, nasal regurgitation, repeated coughing). The neurological focal deficits related to brainstem dysfunction are increased by the presence of the Chiari II malformation and must be quickly evidenced since they require urgent treatment.

The timing of treatment is still matter of debate. In the majority of cases (70–85 %), the treatment can be postponed for days, weeks, or even months following MMC repair. In some instances; namely, in cases of severe fetal hydrocephalus, MMC repair and hydrocephalus shunting have to be realized in the same procedure to obtain rapid relief of intracranial hypertension and to favor spinal wound healing [7, 26, 31]. However, infective complications seem to be higher in children treated at the same time for MMC and hydrocephalus than in those treated separately [5]. In our Institution, in cases of suspected CSF infection in MMC children with acute hydrocephalus, we place a temporary external ventricular drain, which is useful for managing the raised intracranial pressure and for the administration of intraventricular antibiotics in cases of confirmed CSF infection. Generally, we treat the hydrocephalus some days after the surgical repair of MMC, in order to reduce early malfunction of the shunt system.

Ventriculoperitoneal shunt (VPS) is still considered the "gold standard" treatment for MMC-related hydrocephalus. Due to the young age of the patients, the often associated infectious complications, and the possibly unfavorable anatomical conditions, endoscopic third ventriculostomy (ETV) is burdened by a high rate of failure in MMC children when performed as a primary procedure, while it offers a relatively high success rate in cases of a "secondary" procedure after VPS malfunction. The failure rate of ETV in MMC children less than 1 year old actually ranges from 50 to 100 % (mean, 75 %) [13, 27, 37]. The main reasons for this high rate are the immaturity of the CSF re-adsorption systems, which is typical of newborns and small infants [29], and the distorted anatomy that may prevent an ETV (small and/or oblique Monro foramen, malformed third ventricle, opaque floor of the third ventricle, intraventricular septations, large massa intermedia, etc.). Accordingly, ETV is usually performed in children aged more than 6–12 months or as a second-step procedure in cases of late VPS malfunction [35].

Myelomeningocele Repair

The main aim of the surgical treatment of MMC is to dissect the malformed area from the surrounding tissues and to create an adequate barrier above the spine to prevent infections and to preserve the residual motor and sensory functions [9, 11, 22, 23].

A latex-free setting in the operation room is strongly recommended for MMC children, because of the predisposition of these patients to develop latex sensitization [33]. The operation is realized under microscopic magnification with a standard microsurgical instrumentarium.

After completing the anesthetic work-up with the patient in the supine position, the newborn is positioned prone on a soft thoracic support to optimize the thoracic expansion. Pads are positioned on all pressure points to avoid pressure sores. A mild Trendelenburg position is generally used to avoid CSF escape from the spinal canal and the risk of pneumocephalus.

MMC repair is accurately planned considering the specific characteristics of the affected child, accurate inspection of the spinal defect extension and characteristics, and the radiological images, when available [11, 23, 32]. Spinal deformities such as kyphosis or kyphoscoliosis, for example, can strongly influence the procedure: these alterations can be associated, in a variable percentage of cases, with multilevel vertebral cleft or split cord malformation, which need adjunctive treatments [14, 15]. Particular care should be taken to identify the skin available for closure (Fig. 1) and

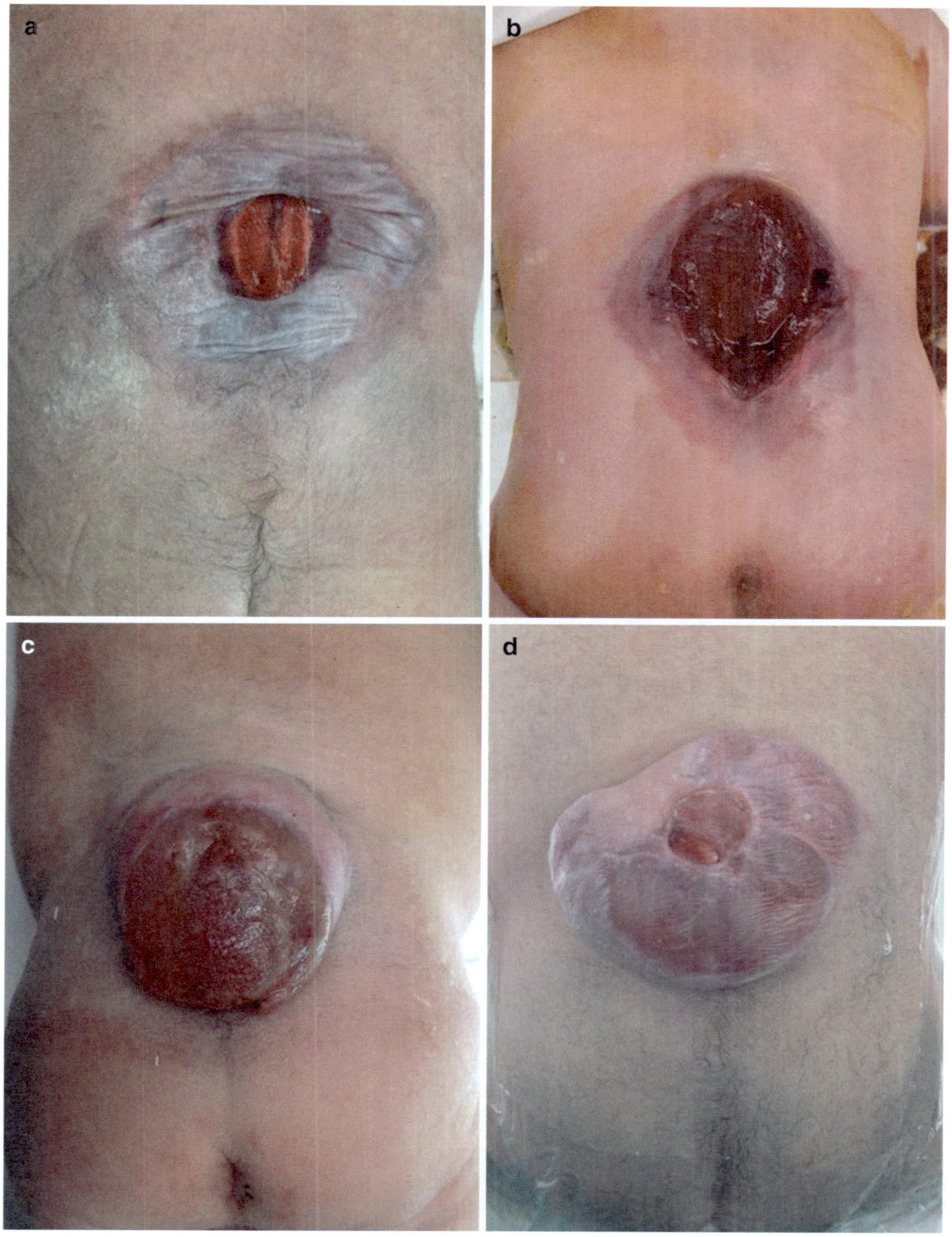

Fig. 1 Spectrum of the possible presentations of myelomeningocele (MMC). (**a**) Complete exposure of the placode surrounded by thin and dystrophic skin. This skin layer must be preserved in order to attempt a reliable closure, but achieving such a closure may be very difficult. (**b**) Complete exposure of the placode and large skin defect: challenging closure. (**c**) The placode is exposed and elevated by a filled meningocelic sac: after the dissection of the arachnoid plane and cerebrospinal fluid (CSF) escape, the placode can be easily replaced in the spinal canal and the redundant, intact skin allows a relatively easy closure. (**d**) Placode partially covered by dystrophic skin and contained in a large meningocelic sac: after the neurulation and the replacement of the placode in the spinal canal, redundant skin may result and its reduction may be required

the "junctional zone", an area around the malformed placode where the arachnoidal and dural coverings are contiguous with the surrounding dystrophic skin: it is important to recognize this region in order to avoid injury to the nerve roots during the initial incision and dissection.

The operation starts with a limited midline linear skin incision at the upper and lower limits of the malformation, in order to identify the normal elements rostral and caudal to the spinal defect. Then the incision is brought along the border between the dystrophic skin and the arachnoid that surrounds the malformation and circumferentially until the entire placode (with its incomplete arachnoid margin) is completely freed and normal dural margins are identified. The next step is the microsurgical dissection of the free placode along the junctional zone: care must be taken when manipulating the border of the placode for the presence of the adjacent dorsal root entry zone. Care has to be taken to avoid arachnoid and dermal/epidermal remnants on the placode, since they can be responsible for delayed cord tethering or dermoid/epidermoid cyst formation [4, 19, 24, 34].

Sectioning of the filum terminale is an integral part of the procedure that is needed to minimize the risk of secondary tethering [19, 24, 30]. Inspection of the inner aspect of the open dural sac is mandatory to reveal the presence of aberrant nerve roots that terminate in the dural sac: the sectioning of them is not related to worsening of the patient's neurological status. Differently, all the intradural vascular vessels must be manipulated carefully and mobilized from the arachnoidal adhesions, and their coagulation should be avoided.

Subsequently, the freed lateral edges of the placode are approximated in the midline, and their pia-arachnoid borders are sutured under microscope magnification, with a 7.0 non-absorbable monofilament, avoiding an excessively tight closure, in order to reconstruct the medullary anatomy (surgical neurulation) [9, 23]. This procedure is not performed to improve the neurological outcome, but to reduce the occurrence of symptomatic late tethering [21, 23].

The dural layer is dissected starting from rostral and caudal normal tissues and then approximated on the midline; afterwards, it is sutured with either 5.0 silk suture or monofilament (continuous suture is preferred). In cases of a narrow spinal canal or when the dural layer is missing, dural patching is necessary. In the first instance, the dural patching is required to obtain a capacious dural sac to prevent re-tethering of the placode, while in the second instance, dural patching is necessary to restore the dural layer itself. The dural patch may be realized using various dural substitutes, the most physiological being represented by autologous tissues, such as muscle or muscular fascia. Alternatively, synthetic dural substitutes and/or dural glues can be useful to obtain a watertight dural closure. A Valsalva maneuver should be performed after completing the dural suture to verify water-tightness. The thoracolumbar fascia is then dissected and sutured over the reconstructed dural sac in order to reinforce the dural closure.

Reconstruction of the superficial layers is performed undermining the skin and the subcutaneous tissue from the muscular fascia. Attention must be paid to preserve, as much as possible, the vessels that provide the blood supply of the cutaneous coverings [17, 28]. The subcutaneous layers are approximated and anchored to the underlying fascia in order to create an adequate support for the skin and to reduce the incidence of retracting scars. Initially, following closure, the skin may be blanched as a sign of tension (Fig. 2). Such initial discoloration usually improves rapidly; wound dehiscence rarely occurs. A nitroglycerine ointment has been suggested to be helpful in some cases [16].

Skin closure can be performed in a midline vertical fashion or with a horizontal or oblique suture, with skin edges having little or no tension (Fig. 2). Although rare, in cases of a big myocutaneous defect (namely, larger than

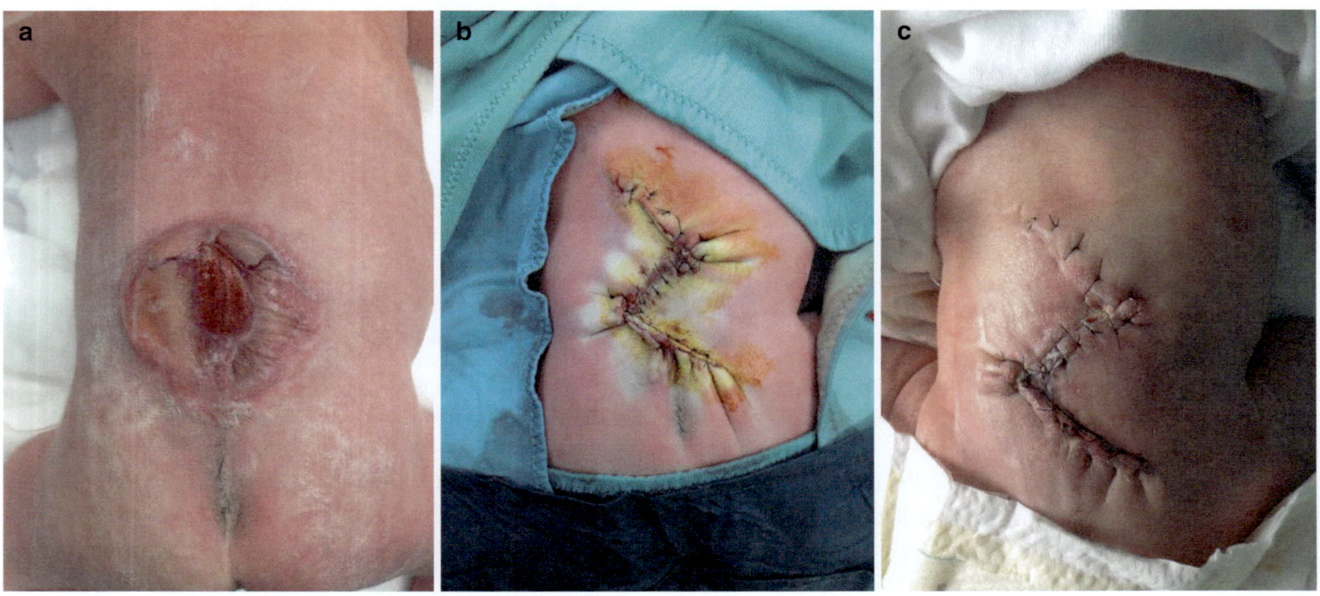

Fig. 2 (**a**) MMC with large skin defect. The skin reconstruction and closure is realized through a Z-shaped skin incision that allows us to move the edges of the surgical wound and to fill the gap (**b**). Note the discoloration immediately after the end of the procedure (**b**) and the good outcome, with regular healing and skin relaxation, after 10 days (**c**)

20–25 cm^2), reconstruction by a plastic surgeon could be necessary [9]. In our experience, this was required in less than 0.5 % of cases.

During the first 24–48 postoperative hours, the infant is recovered in the neonatal intensive care unit (ICU) to monitor the vital functions and, in particular, to detect possible signs of brainstem dysfunction. If possible, the patient is maintained prone for the first 5 postoperative days, with the lower back slightly elevated above the level of the head to reduce the risk of CSF leakage from the wound. Prophylactic intravenous antibiotics are administered for 3–5 days.

In the past few years some studies and protocols have reported that fetal repair of spina bifida is now considered a standard of care at some fetal centers. However, prenatal repair is a complex and challenging procedure, requiring the most expert, comprehensive care for both mother and fetus [36]. The surgical team's level of experience in all aspects of care surrounding the operation is of paramount importance. Therapy that is highly dependent on the provider is of limited benefit to the wider population.

In-Utero MMC Repair

In the past 20 years, in-utero treatment of MMC has been gaining greater and greater popularity in selected patients, thus representing an additional therapeutic alternative for a fetus with MMC. Studies in animal models and clinical case series laid the groundwork for a clinical trial to test the safety and efficacy of fetal MMC repair. During the late 1990s and early 2000s, results of nonrandomized clinical trials suggested that significant benefit might result from the prenatal repair of MMC [3]. At present, a prospective, randomized study (the Management of Myelomeningocele Study or MOMS trial) [2] has shown that fetal surgery for MMC before 26 weeks gestation may preserve neurological function, reverse the hindbrain herniation of the Chiari II malformation, and obviate the need for the postnatal placement of a VPS. However, this study also demonstrates that fetal surgery is associated with significant risks related to the risk of chorio-amnion separation, premature rupture of membranes, oligohydramnios, and preterm delivery, in addition to a 3 % fetal mortality rate [1].

Conclusions

Rare in developed countries, but still frequent in developing ones, MMC is still associated with significant lifelong morbidity. Adequate surgical planning is crucial to obtain an adequate repair of the defect and to minimize surgical damage of neural structures and perioperative complications. The surgical technique here summarized is the result of a literature analysis and our 30-year personal experience that could be useful to introduce young surgeons to this particular and not common procedure.

Acknowledgements Nothing to declare.

Conflict of Interest Statement We declare that we have no conflict of interest.

References

1. Adzick NS (2013) Fetal surgery for spina bifida: past, present, future. Semin Pediatr Surg 22:10–17
2. Adzick NS, Thom EA, Spong CY, Brock JW 3rd, Burrows PK, Johnson MP, Howell LJ, Farrell JA, Dabrowiak ME, Sutton LN, Gupta N, Tulipan NB, D'Alton ME, Farmer DL, Investigators M (2011) A randomized trial of prenatal versus postnatal repair of myelomeningocele. N Engl J Med 364:993–1004
3. Bruner JP, Richards WO, Tulipan NB, Arney TL (1999) Endoscopic coverage of fetal myelomeningocele in utero. Am J Obstet Gynecol 180:153–158
4. Caldarelli M, Di Rocco C, Colosimo C Jr, Fariello G, Di Gennaro M (1995) Surgical treatment of late neurological deterioration in children with myelodysplasia. Acta Neurochir 137:199–206
5. Caldarelli M, Di Rocco C, La Marca F (1996) Shunt complications in the first postoperative year in children with meningomyelocele. Childs Nerv Syst ChNS Offl J Int Soc Pediatr Neurosurg 12:748–754
6. Caldarelli M, Di Rocco C (2008) Myelomeningocele primary repair surgical technique. In: The spina bifida management and outcome. Springer, Milan/New York, pp 143–155
7. Chadduck WM, Reding DL (1988) Experience with simultaneous ventriculo-peritoneal shunt placement and myelomeningocele repair. J Pediatr Surg 23:913–916
8. Cochrane D, Aronyk K, Sawatzky B, Wilson D, Steinbok P (1991) The effects of labor and delivery on spinal cord function and ambulation in patients with meningomyelocele. Childs Nerv Syst ChNS Offl J Int Soc Pediatr Neurosurg 7:312–315
9. de Chalain TM, Cohen SR, Burstein FD, Hudgins RJ, Boydston WR, O'Brien MS (1995) Decision making in primary surgical repair of myelomeningoceles. Ann Plast Surg 35:272–278
10. Di Rocco CCM, Caldia A, Rende M (1998) Hydrocephalus and Chiari malformation. Munksgaad Copenhagen
11. Dias MS (1999) Myelomeningocele repair in utero. Pediatr Neurosurg 30:108
12. Dias MS, McLone DG (1993) Hydrocephalus in the child with dysraphism. Neurosurg Clin N Am 4:715–726
13. Fritsch MJ, Kienke S, Mehdorn HM (2004) Endoscopic aqueductoplasty: stent or not to stent? Childs Nerv Syst ChNS Offl J Int Soc Pediatr Neurosurg 20:137–142
14. Haddad FS (1997) Draping an infant for repair of meningomyelocele. Surg Neurol 48:530
15. Kumar R, Bansal KK, Chhabra DK (2002) Occurrence of split cord malformation in meningomyelocele: complex spina bifida. Pediatr Neurosurg 36:119–127
16. Lehman RA, Page RB, Saggers GC, Manders EK (1985) Technical note: the use of nitroglycerin ointment after precarious neurosurgical wound closure. Neurosurgery 16:701–702
17. Lehrman AOM (1984) Surgical repair of large meningomyeloceles. Ann Plast Surg 12:501–507

18. Luthy DA, Wardinsky T, Shurtleff DB, Hollenbach KA, Hickok DE, Nyberg DA, Benedetti TJ (1991) Cesarean section before the onset of labor and subsequent motor function in infants with meningomyelocele diagnosed antenatally. N Engl J Med 324:662–666
19. McCullough DC, Johnson DL (1994) Myelomeningocele repair: technical considerations and complications. 1988. Pediatr Neurosurg 21:83–89; discussion 90
20. McLaughlin TP, Banta JV, Gahm NH, Raycroft JF (1986) Intraspinal rhizotomy and distal cordectomy in patients with myelomeningocele. J Bone Joint Surg Am 68:88–94
21. McLone DG (1983) Results of treatment of children born with a myelomeningocele. Clin Neurosurg 30:407–412
22. McLone DG (1990) Technique for closure of myelomeningocele. Childs Nerv Syst ChNS Offl J Int Soc Pediatr Neurosurg 6:66–73
23. McLone DG (2005) Spinal dysraphism: impact of technique and technology on expectations. Clin Neurosurg 52:261–264
24. McLone DG, Dias MS (1991) Complications of myelomeningocele closure. Pediatr Neurosurg 17:267–273
25. Merrill DC, Goodwin P, Burson JM, Sato Y, Williamson R, Weiner CP (1998) The optimal route of delivery for fetal meningomyelocele. Am J Obstet Gynecol 179:235–240
26. Miller PD, Pollack IF, Pang D, Albright AL (1996) Comparison of simultaneous versus delayed ventriculoperitoneal shunt insertion in children undergoing myelomeningocele repair. J Child Neurol 11:370–372
27. Mori H, Nishiyama K, Yoshimura J, Tanaka R (2007) Current status of neuroendoscopic surgery in Japan and discussion on the training system. Childs Nerv Syst ChNS Offl J Int Soc Pediatr Neurosurg 23:673–676
28. Lanigan MW (1993) Surgical repair of myelomeningocele. Ann Plast Surg 31:514–521
29. Oi S, Di Rocco C (2006) Proposal of "evolution theory in cerebrospinal fluid dynamics" and minor pathway hydrocephalus in developing immature brain. Childs Nerv Syst ChNS Offl J Int Soc Pediatr Neurosurg 22:662–669
30. Pang D (1995) Surgical complications of open spinal dysraphism. Neurosurg Clin N Am 6:243–257
31. Parent AD, McMillan T (1995) Contemporaneous shunting with repair of myelomeningocele. Pediatr Neurosurg 22:132–135; discussion 136
32. Park T (1999) Myelomeningocele. In: Albright L, Pollack I, Adelson D (eds). Thieme, New York, pp 291–320
33. Rendeli C, Nucera E, Ausili E, Tabacco F, Roncallo C, Pollastrini E, Scorzoni M, Schiavino D, Caldarelli M, Pietrini D, Patriarca G (2006) Latex sensitisation and allergy in children with myelomeningocele. Childs Nerv Syst ChNS Offl J Int Soc Pediatr Neurosurg 22:28–32
34. Scott RM, Wolpert SM, Bartoshesky LE, Zimbler S, Klauber GT (1986) Dermoid tumors occurring at the site of previous myelomeningocele repair. J Neurosurg 65:779–783
35. Spennato P, Ruggiero C, Aliberti F, Nastro A, Mirone G, Cinalli G (2013) Third ventriculostomy in shunt malfunction. World Neurosurg 79(S22):e21–e26
36. Tarcan T, Onol FF, Ilker Y, Alpay H, Simsek F, Ozek M (2006) The timing of primary neurosurgical repair significantly affects neurogenic bladder prognosis in children with myelomeningocele. J Urol 176:1161–1165
37. Zuccaro G (2003) Split spinal cord malformation. Childs Nerv Syst ChNS Offl J Int Soc Pediatr Neurosurg 19:104–105

RETRACTED CHAPTER: Anterior Cervical Discectomy and Fusion with a Compressive C-JAWS Staple

Lei Xia, Ming-Xing Liu, Jun Zhong, Ning-Ning Dou, and Massimiliano Visocchi

Abstract

Background

Anterior cervical discectomy and fusion is currently the most commonly used technique in cervical surgery. But the implantation of a traditional plate is time-consuming and exposes the patient to additional adverse events. In this study, we analyzed results in patients who underwent anterior cervical discectomy and fusion with C-JAWS fixation. The C-JAWS device is a new cervical compressive staple developed to stabilize the spacer.

Methods

At our department, between January 2012 and December 2013, nine consecutive patients with cervical spondylopathy underwent an anterior cervical discectomy and fusion process in which we used a polyether ether ketone cervical spacer prefilled with bone substitute and secured with a cervical compressive staple. The Neck Disability Index (NDI) score and visual analog score (VAS) for neck and arm pain, as well as radiographic examinations, were adopted to assess postoperative outcome and fusion.

Results

Bony fusion was observed in all of the nine patients, and no serious surgery-related or implant-related complications were observed during the operation or in the postoperative period. The average operative time was 60.3 ± 11.6 min. The average hospital stay was 3.2 ± 0.8 days. The average skin incision length was 3.0 ± 0.3 cm. The average follow-up was 18.4 ± 4.3 months. At the last follow-up, the NDI had changed from the baseline value of 23.4 ± 10.3 to 7.1 ± 4.8, and the VAS values for neck and arm pain had changed from 6.1 ± 1.0 and 4.6 ± 1.6,respectively, to 2.3 ± 1.7 and 2.4 ± 1.1, respectively. The patients' subjective satisfaction was excellent in six and good in three.

Conclusions

Without screws, this low-profile design compressive staple, the C-JAWS, performed well in anterior cervical discectomy and fusion surgeries

Keywords C-JAWS • Anterior cervical discectomy and fusion • Cervical compression staple

Introduction

Anterior cervical discectomy and fusion (ACDF) is currently the most common process used to treat cervical disc protrusion or herniation. This procedure has produced excellent outcomes, providing relief of pain and prevention of neurologic deterioration arising from neural compression. A number of studies also suggest a statistically significant advantage of arthrodesis with anterior plating. The traditional cervical plates may indeed increase fusion rates and decrease the incidence of graft extrusion. For example, it has been proven that single-level ACDF may give rise to 90 % fusion rates [2, 9, 13, 16, 26]. However, complications such as screw pullout, plate or screw failure, plate migration, stress shielding, and injury of anterior soft-tissue structures have been reported [19, 25], and many studies have also shown an adverse effect of anterior cervical decompression and fusion on the unfused adjacent levels [14].

In this article, we introduce a new cervical compressive staple, the C-JAWS (Medicrea, Lyon, France), which, used

---

The original version of this chapter was revised. An erratum to this chapter is available at DOI 10.1007/978-3-319-39546-3_49

L. Xia, MD • M.-X. Liu, MD • J. Zhong, MD, PhD (✉)
Department of Neurosurgery, XinHua Hospital, Shanghai JiaoTong University School of Medicine, Shanghai 200092, China

Institute of Neurosurgery, Catholic University of Rome, Rome, Italy
e-mail: ZhongMDPhD@sjtu.edu.cn

N.-N. Dou, MD, MS
Department of Neurosurgery, XinHua Hospital, Shanghai JiaoTong University School of Medicine, Shanghai 200092, China

M. Visocchi, MD
Institute of Neurosurgery, Medical School, Catholic University of Rome, Rome, Italy

in internal fixation, may offer a high rate of fusion, while possibly reducing the operative time and also the postoperative complications linked to the use of plates.

Materials and Methods

Patients

Between January 2012 and December 2013, at the neurosurgical department of XinHua hospital, nine consecutive patients with cervical spondylopathy underwent the ACDF process, in which we used a polyether ether ketone cervical spacer prefilled with bone substitute and secured with a cervical compressive staple. These patients, four males and five females, with a mean age of 57.4 ± 6.9 years, had previously undergone unsuccessful conservative treatments. The diagnosis was confirmed by X-rays and magnetic resonance imaging (MRI) examination. The level of intervertebral disc protrusion or herniation was C4/5 in one patient, C5/6 in three, C3/4 and C6/7 in one, C4/5 and C5/6 in two, and C5/6 and C6/7 in two patients.

Surgical Procedures

All the operations were performed by the senior author (J. Zhong). With general anesthesia, the patient was placed supine with the neck slightly hyperextended, and the head was not turned. A transverse anterior cervical incision was made and the platysma was divided sharply. The plane between the carotid sheath laterally and the esophagus and trachea medially was dissected bluntly. With a monopolar cautery device set on low and its tip slightly bent, the longus colli muscle was gently dissected from the vertebral body approximately 2–3 mm laterally. After the correct level was identified with the help of mini-c-arm fluoroscopy (OEC Fluorostar 7900; GE, Beijing, USA), a Caspar bone spreader was positioned to increase the access to the disc space, and a spine drill burr was used to remove cartilage from end-plate vertebrae. Removal of the disc began after the annulus had been cut. The disc was removed with angled curettes and pituitary rongeurs of various sizes (Fig. 1). An IMPIX C+ polyether ether ketone cervical spacer prefilled with bone substitute (Medicrea) was placed in the intervertebral space. The C-JAWS cervical compressive staple (Medicrea) was used to secure the intersomatic spacer (Fig. 2); the compression applied thereby stabilized the fusion device. A final X-ray control confirmed the position of the spacer and the staple (Fig. 3). A drain was placed at the end of the process and removed the next day. Patients were mobilized within the first postoperative day with no external collar. A postoperative X-ray was performed before the patient was discharged.

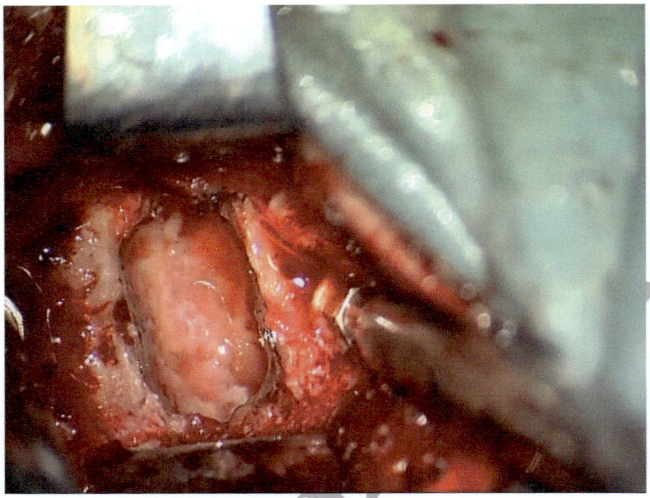

Fig. 1 An intraoperative picture showed the lesioned disc was removed

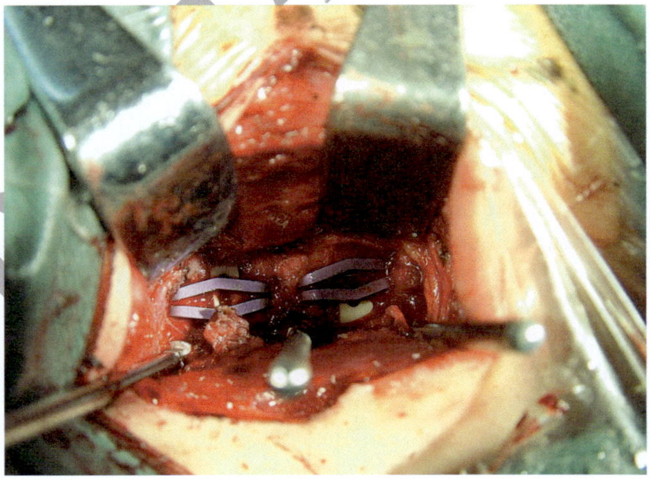

Fig. 2 Two C-JAWS staples were used to secure the intersomatic spacer

Follow-Up

The patients were followed-up every 3 months. Clinical and radiographic examinations of each patient were performed at each follow-up visit. Pain and neurological status were evaluated using the Neck Disability Index (NDI) score; neck and arm pain was assessed with a ten-point visual analog scale (VAS) with endpoint anchors of no pain and maximal pain. The patient's subjective perception

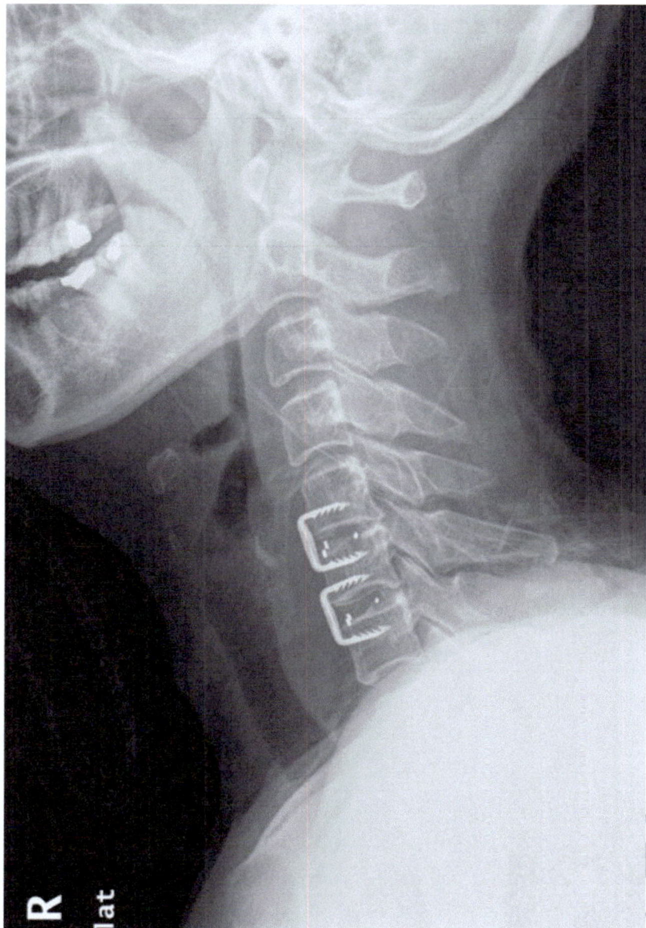

Fig. 3 Postoperative X-rays delineated two C-JAWS that were positioned properly

of overall satisfaction with the postoperative outcome was graded as excellent, good, fair, or poor. The morphologic aspects of the arthrodesis were analyzed according to the X-rays. All X-rays were reviewed by an independent observer (Ning-Ning Dou) who was not involved in the operative process.

Results

The height and arm lengths of the cervical compression staple design used in the surgery were 12.5 × 14 mm in five patients, and 14.5 × 16 mm in four. The depth, width, and height measurements of the intersomatic spacer were 12.5 × 14 × 6 mm in three patients, and 12.5 × 14 × 7 mm in six. The average operative time was 60.3 ± 11.6 min. The average hospital stay was 3.2 ± 0.8 days. The average skin incision length was 3.0 ± 0.3 cm. No serious surgery-related or implant-related complications were observed during the operation or in the postoperative period (Tables 1 and 2).

Clinical Outcomes

The average follow-up was 18.4 ± 4.3 months. At the last follow-up, the NDI had changed from the baseline of 23.4 ± 10.3 to 7.1 ± 4.8, and the VAS values for neck and arm pain had changed from 6.1 ± 1.0 and 4.6 ± 1.6, respectively, to 2.3 ± 1.7 and 2.4 ± 1.1, respectively. The patients' subjective satisfaction was excellent in six and good in three.

Radiographic Results

At the last follow-up bony fusion was observed in all of the nine patients, presenting with continuous bone growth between the vertebral endplates that was visible inside and around the spacer. Neither implant dislodgment nor device pullout was shown in any of the patients.

Discussion

Since Robinson and Smith [24] and Cloward [8] first described the anterior approaches to the cervical spine, several surgical techniques have evolved for the treatment of cervical disc disease [8]. Aiming to decompress the neural structures, to obtain a solid fusion, and to maintain or restore normal cervical alignment, the ACDF process has produced excellent outcomes, providing relief of pain and prevention of neurologic deterioration, and the process has became the gold standard in degenerative conditions of the cervical spine. According to some articles, autogenous grafts seemed to provide better fusion rates than allografts or spacer grafts [6, 12, 15, 20]. However, with ACDF there may be some complications, including donor-site pain, hematoma, and infection. Moreover, many studies have also shown the effect of ACDF on the unfused adjacent levels; the consequent symptomatic adjacent level disease and reoperation rate has been reported to be 2.9 % per year. To reduce these complications, surgeons have used cervical spacers, with or without bone substitute, to increase cervical stability. Although many authors have reported a more than 90 % of fusion rate with single-level ACDF, some articles have shown soft-tissue complications due to the migration and subsidence of the spacers, or due to pseudarthrosis [17]

To reduce the incidence of complications and pseudarthrosis rates, some surgeons have used anterior plate fixation [5, 6]. This procedure is now well established and has shown several advantages compared with a stand-alone spacer; it maximized the restoration of structural integrity and spine balance. Fusion is defined as an absence of

Table 1 Patients' data

Case no.	Sex/Age (years)	Preoperative score			Postoperative score			Surgery satisfaction	Follow-up (months)
		NDI	VAS		NDI	VAS			
			Neck	Arm		Neck	Arm		
1.	M/55	35	6.8	6.1	7	2.8	2.9	Excellent	27
2.	F/55	19	6.4	4	14	4.3	2.3	Good	23
3.	M/47	17	5.3	2.4	1	2.7	1.1	Excellent	20
4.	F/71	12	4.6	2.2	5	0.3	1.7	Excellent	18
5.	F/54	16	5.7	5.2	8	1.6	2.1	Excellent	17
6.	F/56	28	6.5	5.6	3	0.4	1.6	Excellent	17
7.	M/54	40	7.7	6.1	11	4.1	3.8	Good	16
8.	F/63	13	4.8	3.8	2	0.5	1.7	Excellent	15
9.	M/62	31	6.3	6.3	13	4.1	4.2	Good	13

NDI Neck Disability Index, *VAS* visual analog scale

Table 2 The application of C-JAWS in ACDF surgery

Case no.	Level operated	Cage size (W×D×H) (mm)	Staple Size (L×H) (mm)	Operation time (min)	Hospital stay (days)	Incision length (cm)	Radiographic outcome	Radiographic outcome
1.	C3/4, C6/7	12.5×14×7	16×14.6	52	3	3.4	Fusion	Fusion
2.	C4/5, C5/6	12.5×14×6	14×12.5	49	4	3.2	Fusion	Fusion
3.	C4/5, C5/6	12.5×14×7	16×14.6	48	2	3.2	Fusion	Fusion
4.	C5/6	12.5×14×7	14×12.5	45	3	2.7	Fusion	Fusion
5.	C5/6	12.5×14×6	14×12.5	47	4	2.8	Fusion	Fusion
6.	C5/6, C6/7	12.5×14×7	16×14.6	51	4	3.3	Fusion	Fusion
7.	C5/6, C6/7	12.5×14×7	14×12.5	57	4	3.1	Fusion	Fusion
8.	C4/5	12.5×14×7	14×12.5	52	3	3	Fusion	Fusion
9.	C5/6	12.5×14×6	16×14.6	63	2	2.7	Fusion	Fusion

ACDF anterior cervical discectomy and fusion

mobility between the instrumented vertebrae. The goal of instrumentation in a cervical arthrodesis is to stabilize the segment and to reduce the mobility to help in promoting fusion.

In the present study, the average operation time was only about an hour, while with the use of plate fixation, the time in single-level cervical spondylopathy ranged from 98.2 to 178 min according to the literature [10, 27, 28]. In our patients, the bony fusion rate was analogous to that reported with the traditional plates in some studies [1, 4, 21], while the fusion rate obtained using the C-JAWS seemed to be higher than that with the traditional plate in another study [11]. According to these results, the C-JAWS seems to be an effective device for reducing the average operation time, thus being a contributing factor for fusion. Traditional cervical plates may induce complications such as instrumentation failure, plate migration, stress shielding, compression of soft-tissue structures, esophageal perforation, and vertebral artery injury [3, 7, 22, 23]. Lim et al. stated that if we could increase the load between the graft and the endplate, we could get a better rate of fusion [18]. It is generally advised that the graft inserted in an adult cervical spine should be preloaded with a compressive force to maximize the chance of graft incorporation and to prevent its extrusion and collapse. Some authors have also noted the superiority of dynamic plating compared with static plating. Using the C-JAWS, a short operative time can be achieved and this tends to reduce the risk of postoperative complications. The thickness of the C-JAWS is only 1.5 mm; that is, it has a lower profile than most cervical plates. Due to the low profile of the C-JAWS, we only need a small skin incision, and therefore this would decrease the risk of soft-tissue injury. The most important factor is that a lower-profile plate makes it easy to visualize the space and the vertebrae during the operation and the lower profile tends to reduce the incidence of dysphagia.

Conclusions

The C-JAWS has a high rate of fusion and a low-profile design, with a shorter operative time and fewer complications than the traditional cervical plate; accordingly, this compressive staple could be recommended for the process of anterior cervical discectomy and fusion.

Conflict of Interest Statement All authors certify that they have NO affiliations with or involvement in any organization or entity with any financial interest (such as honoraria; educational; participation in speakers' bureaus; membership, employment, consultancies, stock ownership, or other equity interest; and expert testimony or patent-licensing arrangements), or non-financial interest (such as personal or professional relationships, affiliations, knowledge or beliefs) in the subject matter or materials discussed in this manuscript.

References

1. Aryan HE, Lu DC, Acosta FL Jr, Hartl R, McCormick PW, Ames CP (2007) Bioabsorbable anterior cervical plating: initial multicenter clinical and radiographic experience. Spine (Phila Pa 1976) 32:1084–1088
2. Aryan HE, Lu DC, Acosta FL Jr, Hartl R, McCormick PW, Ames CP (2007) Bioabsorbable anterior cervical plating: initial multicenter clinical and radiographic experience. Spine 32:1084–1088
3. Beutler WJ, Sweeney CA, Connolly PJ (2001) Recurrent laryngeal nerve injury with anterior cervical spine surgery: risk with laterality of surgical approach. Spine 26:1337–1342
4. Burkhardt JK, Mannion AF, Marbacher S, Kleinstuck FS, Jeszenszky D, Porchet F (2014) The influence of cervical plate fixation with either autologous bone or cage insertion on radiographic and patient-rated outcomes after two-level anterior cervical discectomy and fusion. Eur Spine J 24:113–119
5. Cabanela ME, Ebersold MJ (1988) Anterior plate stabilization for bursting teardrop fractures of the cervical spine. Spine 13:888–891
6. Caspar W, Geisler FH, Pitzen T, Johnson TA (1998) Anterior cervical plate stabilization in one-and two-level degenerative disease: overtreatment or benefit? J Spinal Disord Tech 11:1–11
7. Cho KH, Shin YS, Yoon SH, Kim SH, Ahn YH, Cho KG (2005) Poor surgical technique in cervical plating leading to vertebral artery injury and brain stem infarction—case report. Surg Neurol 64:221–225
8. Cloward RB (1958) The anterior approach for removal of ruptured cervical disks. J Neurosurg 15:602–617
9. Connolly P, Esses S, Kostuik J (1996) Anterior cervical fusion: outcome analysis of patients fused with and without anterior cervical plates. J Spinal Disord Tech 9:202–206
10. Fang Z, Tian R, Sun TW, Yadav SK, Hu W, Xie SQ (2013) Expansion open-door laminoplasty with foraminotomy versus anterior cervical discectomy and fusion for coexisting multilevel cervical myelopathy and unilateral radiculopathy. J Spinal Disord Tech 24:213–220
11. Fiere V, Greff G, Mosnier T, Faline A (2013) New cervical compressive staple: in vitro testing and early clinical results. J Spinal Disord Tech 26:385–392
12. Fraser JF, Härtl R (2007) Anterior approaches to fusion of the cervical spine: a metaanalysis of fusion rates. J Neurosurg Spine 6:298–303
13. Greg Anderson D, Albert TJ (2002) Bone grafting, implants, and plating options for anterior cervical fusions. Orthop Clin North Am 33:317–328
14. Hilibrand AS, Carlson GD, Palumbo MA, Jones PK, Bohlman HH (1999) Radiculopathy and myelopathy at segments adjacent to the site of a previous anterior cervical arthrodesis*. J Bone Joint Surg 81:519–528
15. Jacobs W, Anderson PG, Limbeek J, Willems PC, Pavlov P (2004) Single or double-level anterior interbody fusion techniques for cervical degenerative disc disease. Cochrane Database Syst Rev (4)
16. Kaiser MG, Haid RW Jr, Subach BR, Barnes B, Rodts GE Jr (2002) Anterior cervical plating enhances arthrodesis after discectomy and fusion with cortical allograft. Neurosurgery 50:229–238
17. Kim SW, Limson MA, Kim S-B, Arbatin JJF, Chang K-Y, Park M-S, Shin J-h, Ju Y-S (2009) Comparison of radiographic changes after ACDF versus Bryan disc arthroplasty in single and bi-level cases. Eur Spine J 18:218–231
18. Lim T-H, Kwon H, Jeon C-H, Kim JG, Sokolowski M, Natarajan R, An HS, Andersson GB (2001) Effect of endplate conditions and bone mineral density on the compressive strength of the graft–endplate interface in anterior cervical spine fusion. Spine 26:951–956
19. Lowery GL, McDonough RF (1998) The significance of hardware failure in anterior cervical plate fixation: patients with 2-to 7-year follow-up. Spine 23:181–186
20. Mummaneni PV, Burkus JK, Haid RW, Traynelis VC, Zdeblick TA (2007) Clinical and radiographic analysis of cervical disc arthroplasty compared with allograft fusion: a randomized controlled clinical trial. J Neurosurg Spine 6:198–209
21. Oh K, Lee CK, You NK, Kim SH, Cho KH (2013) Radiologic changes of anterior cervical discectomy and fusion using allograft and plate augmentation: comparison of using fixed and variable type screw. Korean J Spine 10:160–164
22. Pichler W, Maier A, Rappl T, Clement HG, Grechenig W (2006) Delayed hypopharyngeal and esophageal perforation after anterior spinal fusion: primary repair reinforced by pedicled pectoralis major flap. Spine 31:E268–E270
23. Pompili A, Canitano S, Caroli F, Caterino M, Crecco M, Raus L, Occhipinti E (2002) Asymptomatic esophageal perforation caused by late screw migration after anterior cervical plating: report of a case and review of relevant literature. Spine 27:E499–E502
24. Smith GW, Robinson RA (1958) The treatment of certain cervical-spine disorders by anterior removal of the intervertebral disc and interbody fusion. J Bone Joint Surg 40:607–624
25. Vaccaro AR, Balderston RA (1997) Anterior plate instrumentation for disorders of the subaxial cervical spine. Clin Orthop Relat Res 335:112–121
26. Wang JC, McDonough PW, Endow K, Kanim LE, Delamarter RB (1999) The effect of cervical plating on single-level anterior cervical discectomy and fusion. J Spinal Disord Tech 12:467–471
27. Wang ZD, Zhu RF, Yang HL, Gan MF, Zhang SK, Shen MJ, Chen C, Yuan Q (2014) The application of a zero-profile implant in anterior cervical discectomy and fusion. J Clin Neurosci 21:462–466
28. Zhu R, Yang H, Wang Z, Wang G, Shen M, Yuan Q (2014) Comparisons of three anterior cervical surgeries in treating cervical spondylotic myelopathy. BMC Musculoskelet Disord 15:233

Olfactory Groove Meningiomas: Acute Presentation and Potential

Pitfalls in Management and Functional Restoration

Alfredo Quiñones-Hinojosa, Shaan M. Raza, Christos Lazaridis, and Alessandro Olivi

Introduction

Meningiomas are typically benign, slow-growing lesions that present after an insidious onset of symptoms related to mass effect. The acute presentation of a patient who has suffered a transtentorial herniation event due to a meningioma is rare. There are only few publications describing such a presentation in the absence of hemorrhage [1]. In this case report, a patient with an olfactory groove meningioma presenting with signs and symptoms of transtentorial herniation in the absence of tumor-associated hemorrhage is discussed. This is a unique presentation of such a lesion. The patient developed Anton's syndrome—binocular visual loss with blindness denial. Management considerations for patients with meningiomas that present with acute deterioration are discussed.

A. Quiñones-Hinojosa, MD • S.M. Raza, MD
Department of Neurosurgery, Johns Hopkins School of Medicine, Baltimore, MD, USA

Department of Oncology, Johns Hopkins School of Medicine, Baltimore, MD, USA

C. Lazaridis, MD
Department of Neurosciences Critical Care, Johns Hopkins School of Medicine, Baltimore, MD, USA

A. Olivi, MD (✉)
Department of Neurosurgery, Johns Hopkins School of Medicine, Baltimore, MD, USA

Department of Oncology, Johns Hopkins School of Medicine, Baltimore, MD, USA

Department of Neurosurgery and Oncology, Phipps 1–100 – The Johns Hopkins Hospital,
600 N. Wolfe Street, Baltimore, MD 21287, USA
e-mail: aolivi@jhmi.edu

Patient History and Presentation

Preoperative Course

The patient is a 53-year-old female who was neurologically intact until 2 weeks prior to her acute decline. At this time she noticed deterioration of her visual acuity and obscured visual fields. Concurrently she developed dull headaches exacerbated by recumbency and valsalva. These symptoms brought her to an emergency room at another hospital, where imaging revealed a midline anterior cranial fossa mass with bifrontal edema. The patient was discharged home on Decadron (dexamethasone; Merck) with the intention for surgical intervention within the following 2 weeks. Four days after the initial diagnosis, she was found unresponsive in her house and was emergently transferred to our hospital.

On admission to our neurosciences critical care unit the patient was intubated for airway protection, her Glasgow Coma Scale score was 3 T. On examination, the patient was not able to open her eyes, had bilaterally dilated and fixed pupils, intact corneal reflexes, intact cough/gag reflex, and extensor posturing to stimuli. An admission head computed tomography (CT) scan demonstrated a 4 × 4 cm midline anterior cranial fossa mass with bifrontal edema, effaced lateral ventricles, and basilar cisterns. Medical interventions to lower intracranial pressure and cerebral edema were initiated, including hyperventilation, dexamethasone (20 mg q 4 h) and hyperosmolar therapy (23.4 % boluses and 3 % hypertonic saline infusion). Shortly after the initiation of medical management, including seizure medication, the patient's clinical examination improved. She now had spontaneous eye opening, dilated but minimally reactive pupils, and spontaneous, purposeful movements with the upper and lower extremities.

Once the patient was stable, gadolinium-enhanced magnetic resonance imaging (MRI) of the brain was obtained. The MRI demonstrated a homogenously enhancing mass

M. Visocchi et al. (eds.), *Trends in Reconstructive Neurosurgery*, Acta Neurochirurgica Supplement 124,
DOI 10.1007/978-3-319-39546-3_24,

most consistent with an olfactory groove meningioma (Fig. 1b). In addition, although bilateral occipital lobe infarcts were evident (Fig. 1c), MRI did not reveal evidence of brain stem injury. The presence of bilateral posterior cerebral artery infarcts was attributed to the patient's transtentorial herniation.

Operative Course

Subsequently she was taken to the operating room for an extended bifrontal craniotomy and surgical tumor resection. A gross total resection of the lesion was achieved (Simpson Grade II). Tumor-associated hemorrhage was not observed. Pathology was consistent with a meningioma.

Postoperative Course

Postoperative imaging confirmed a gross total resection with resolving bilateral frontal edema (Fig. 1d). Final pathology was a WHO grade I meningioma. At follow-up at 2 weeks and 3 months, the patient was awake and alert, cognitively intact, and with full strength in all extremities. Despite having no light perception bilaterally, she continued to deny visual loss.

Discussion

Meningiomas are typically slow-growing, extra-axial tumors that present with a gradual onset of symptoms related to mass effect. The acute presentation of a patient harboring a meningioma is rare and is almost always associated with hemorrhage [2–13] and sometimes with new onset of seizures. The hemorrhage may be intratumoral, intracerebral, subdural, or subarachnoid [2–13]. It has been proposed that extensive tumor infarction may precipitate tumor-associated hemorrhage and the acute clinical decline of such patients [2]. In this case report, we describe a patient with an olfactory groove meningioma who presented with the signs and symptoms of a herniation syndrome in the absence of tumor-associated hemorrhage. This is a unique presentation for an anterior skull base meningioma

A possible mechanism of deterioration in this patient is the acute development of central tumor necrosis. This could have precipitated an abrupt increase in cerebral edema and intracranial pressure leading to the herniation syndrome observed. On preoperative CT and MR imaging, the presence of significant bifrontal vasogenic edema was evident. Central tumor necrosis has been considered the cause of an acute decline in a previously reported patient with a meningioma who did not, though, progress to herniation [1].

Another plausible mechanism of deterioration includes a seizure event. The degree of mass effect related to cerebral edema could have placed the patient in a critical threshold, where a single seizure (with consequent hypoventilation/hypercarbia) could precipitate catastrophic herniation. This hypothesis raises the importance of prophylactic anti-epileptic medications in patients with meningiomas associated with surrounding edema.

A management nuance highlighted by this case is the timing of surgery once a patient harboring a meningioma acutely deteriorates. In this case, the patient was immediately placed on hyperosmolar therapy and steroids to mitigate the effects of increased intracranial pressure. Fortunately, she had a clinical response to this treatment and the decision was made to continue medical therapy for an additional 36 h prior to surgery. In situations where a clinical response to aggressive medical intervention does not occur and concerns for elevated intracranial pressure persist, urgent surgical decompression should be considered. For patients with meningiomas and extensive parenchymal edema, the paradigm for surgical intervention would be similar to a protocol for intractable intracranial pressure, where urgent intervention is considered when a poor clinical examination is refractory to medical therapy.

Interestingly, although the patient had no light perception on follow-up examination and had MRI evidence of bilateral occipital infarcts, she denied any visual impairment. Anton-Babinski syndrome or Anton's blindness, where the patient is blind but refuses to acknowledge the condition, is generally the result of trauma or stroke where damage has occurred to the bilateral occipital lobes, resulting in cortical blindness [14]. Although a previous report has associated an anterior skull base meningioma with the syndrome, our case is unique in regard to the acute presentation of a herniation syndrome [15].

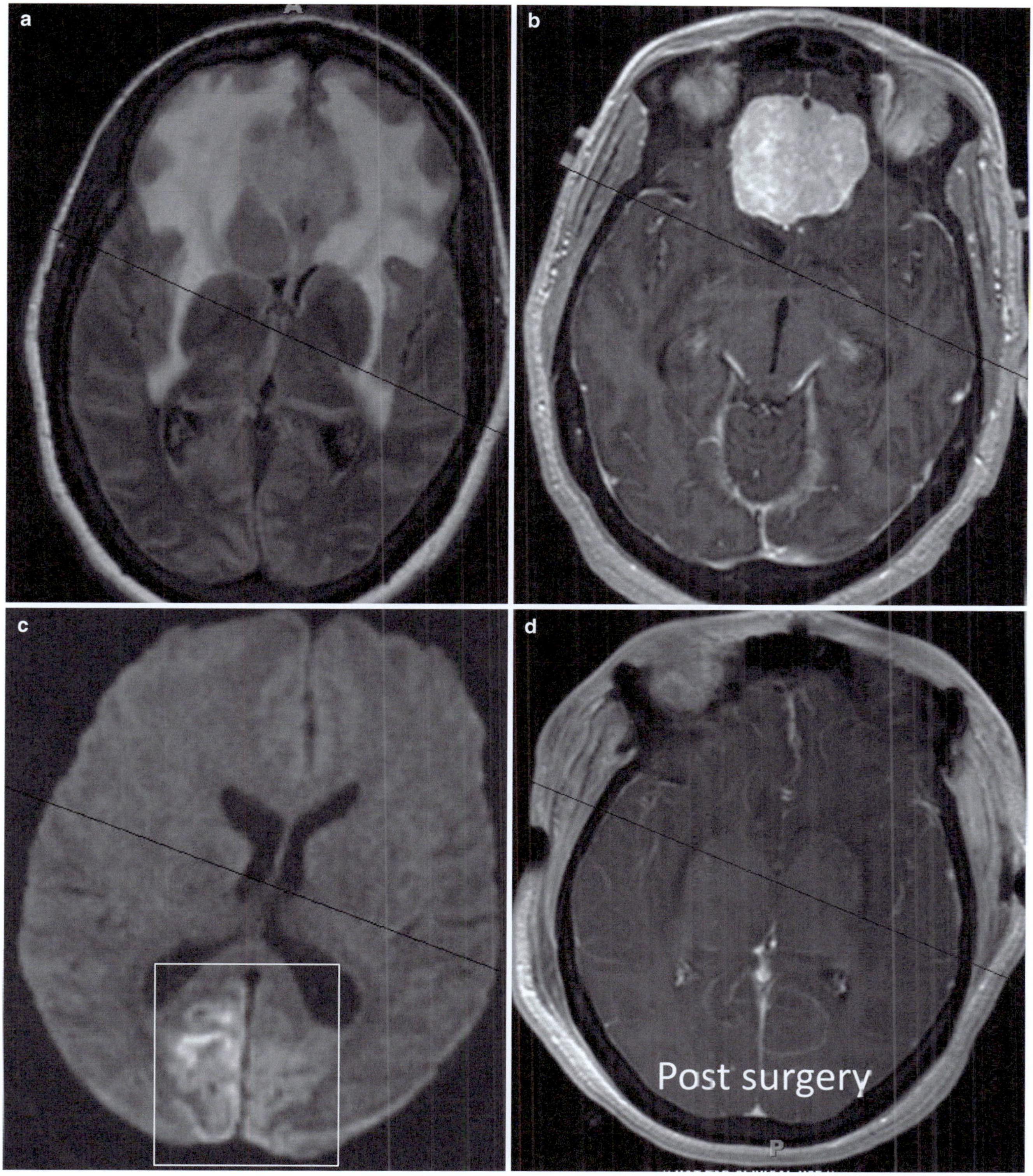

Fig. 1 (**a**) Preoperative T2-weighted magnetic resonance imaging (MRI) scan, demonstrating significant bilateral frontal lobe edema tracking posteriorly along the white fiber tracts. (**b**) Preoperative T1-weighted MRI scan with contrast, demonstrating olfactory groove meningioma. (**c**) Preoperative diffusion weighted imaging, demonstrating bilateral occipital lobe infarcts (highlighted by *box*). (**d**) Postoperative imaging demonstrating complete resection

Conclusions

To our knowledge this is the first report of a skull base meningioma presenting as an acute herniation syndrome in the absence of intratumoral hemorrhage. Acute medical and surgical management are reviewed. Central tumor necrosis exacerbating peritumoral edema is our leading hypothesis to explain herniation. Resuscitation with anti-edema therapy followed by surgical excision is the treatment of choice.

Conflict of Interest Statement We declare that we have no conflict of interest.

References

1. McDowell D, Harper CG (1991) Central tumour necrosis: a cause for the acute presentation of benign meningiomas. Aust N Z J Surg 61(7):553–556
2. Barua KK et al (2000) Spontaneous peritumoral haemorrhage associated with sinus confluence meningioma: case report. Surg Neurol 54(3):254–259
3. Hayashi K et al (2003) A case of atypical meningioma associated with acute deterioration and cerebral herniation. No Shinkei Geka 31(12):1309–1313
4. Helle TL, Conley FK (1980) Haemorrhage associated with meningioma: a case report and review of the literature. J Neurol Neurosurg Psychiatry 43(8):725–729
5. Kim DG et al (2000) Meningioma manifesting intracerebral haemorrhage: a possible mechanism of haemorrhage. Acta Neurochir (Wien) 142(2):165–168
6. Kohli CM, Crouch RL (1984) Meningioma with intracerebral hematoma. Neurosurgery 15(2):237–240
7. Lefranc F et al (2001) Intracranial meningiomas revealed by non-traumatic subdural haematomas: a series of four cases. Acta Neurochir (Wien) 143(10):977–982; discussion 982–983
8. Martinez-Lage JF et al (1991) Meningiomas with haemorrhagic onset. Acta Neurochir (Wien) 110(3–4):129–132
9. Modesti LM, Binet EF, Collins GH (1976) Meningiomas causing spontaneous intracranial hematomas. J Neurosurg 45(4):437–441
10. Niiro M et al (2003) Clinico-pathological study of meningiomas with haemorrhagic onset. Acta Neurochir (Wien) 145(9):767–772
11. Renowden SA, Hourihan MD (1992) Case report: acute subdural haematoma--an unusual presentation of a meningioma. Clin Radiol 45(5):351–352
12. Sakowitz OW et al (2005) Acute haemorrhage into a microcystic meningioma leading to cerebral herniation. Br J Neurosurg 19(3):260–264
13. Wakai S et al (1982) Spontaneous intracranial hemorrhage caused by brain tumor: its incidence and clinical significance. Neurosurgery 10(4):437–444
14. Misra M, Rath S, Mohanty AB (1989) Anton syndrome and cortical blindness due to bilateral occipital infarction. Indian J Ophthalmol 37(4):196
15. Wessling H et al (2006) Anton's syndrome due to a giant anterior fossa meningioma. The problem of routine use of advanced diagnostic imaging in psychiatric care. Acta Neurochir (Wien) 148(6):673–675; discussion 675

Middle Temporal Gyrus Versus Inferior Temporal Gyrus Transcortical Approaches to High-Grade Astrocytomas in the Mediobasal Temporal Lobe: A Comparison of Outcomes, Functional Restoration, and Surgical Considerations

Alfredo Quinones-Hinojosa, Shaan M. Raza, Ishrat Ahmed, Jordina Rincon-Torroella, Kaisorn Chaichana, and Alessandro Olivi

Abstract
Introduction
High-grade astrocytomas of the mesial temporal lobe may pose surgical challenges. Several approaches (trans-sylvian, subtemporal, and transcortical) have been designed to circumnavigate the critical neurovascular structures and white fiber tracts that surround this area. Considering the paucity of literature on the transcortical approach for these lesions, we describe our institutional experience with transcortical approaches to Grade III/IV astrocytomas in the mesial temporal lobe.

Methods
Between 1999 and 2009, 23 patients underwent surgery at the Johns Hopkins Medical Institutions for Grade III/IV astrocytomas involving the mesial temporal lobe (without involvement of the temporal neocortex). Clinical notes, operative records, and imaging were reviewed.

Results
Thirteen patients had tumors in the dominant hemisphere. All patients underwent surgery via a transcortical approach (14 via the inferior temporal gyrus and 9 via the middle temporal gyrus). Gross total resection was obtained in 92 % of the cohort. Neurological outcomes were: clinically significant stroke (2 patients), new visual deficits (2 patients), new speech deficit (1 patient); seizure control (53 %).

Conclusions
In comparison to reported results in the literature for the transylvian and subtemporal approaches, the transcortical approach may provide the access necessary for a gross total resection with minimal neurological consequences. In our series of patients, there was no statistically significant difference in outcomes between the middle temporal gyrus versus the inferior temporal gyrus trajectories.

Keywords Mesial temporal lobe • High-grade astrocytoma • Transcortical approach

Introduction

Tumors of the mediobasal temporal lobe (MTL) may represent a unique surgical challenge due to anatomic and oncologic considerations. Anatomically, the MTL is surrounded by several critical neurovascular structures, where, medially it is bounded by the carotid artery (and posterior communicating artery and anterior choroidal artery branches) and the oculomotor nerve, superiorly by the sylvian fissure and middle cerebral artery branches, and laterally by the temporal lobe neocortex, through which several crucial white fiber tracts course. As our anatomic understanding of this region has improved, surgical approaches to this region have concurrently evolved. These approaches include superior-based (i.e., trans-sylvian), lateral (i.e., transcortical), and basal (subtemporal) approaches to the MTL. Though a well-known approach, there is a paucity of literature focused on outcomes and considerations in transcortical approaches, while a great deal of literature has described trans-sylvian and subtemporal approaches to the MTL [21, 22, 26–28].

Oncologically, there is a distinct cellular architecture and unique profile of brain tumors that arise in the MTL. The greatest proportion of tumors in this area are typically low-grade brain tumors such as dysembryoplastic neuroepithelial

A. Quinones-Hinojosa, MD (✉)
Neurosurgical Oncology Outcomes Laboratory, Department of Neurosurgery, The Johns Hopkins Medical Institutions, Baltimore, MD, USA

Brain Tumor Stem Cell Laboratory, Department of Neurosurgery and Oncology, 1550 Orleans Street, Cancer Research Building II Room 247, Baltimore, MD 21231, USA
e-mail: aquinon2@jhmi.edu

S.M. Raza, MD • I. Ahmed, MS • J. Rincon-Torroella, MD
K. Chaichana, MD • A. Olivi, MD
Neurosurgical Oncology Outcomes Laboratory, Department of Neurosurgery, The Johns Hopkins Medical Institutions, Baltimore, MD, USA

M. Visocchi et al. (eds.), *Trends in Reconstructive Neurosurgery*, Acta Neurochirurgica Supplement 124,
DOI 10.1007/978-3-319-39546-3_25,

tumor (DNET) and pleomorphic xanthoastrocytoma (PXA) – over 70 % in some large published series [21]. As such, a majority of the literature focused on MTL surgery has discussed surgical approaches and complications in the context of well-demarcated lesions (i.e., low-grade gliomas or epilepsy). There are different anatomic considerations when dealing with such tumors where the tumor-brain interface is well delineated as opposed to more infiltrative lesions, such as high-grade astrocytomas.

As high-grade astrocytomas represent a small percentage of tumors affecting the MTL, there is a paucity of literature specifically focused on the surgical management of these lesions, especially in the context of oncologic outcomes (extent of resection), neurological outcomes, and functional restoration (visual deficits, seizure control). In this context, we retrospectively assessed our experience with transcortical approaches via the middle or inferior temporal gyrus for Grade III/IV astrocytomas to determine whether a selective approach could provide gross total resection while minimizing neurological morbidity.

Methods

Between 1999 and 2009, 1082 patients underwent craniotomy for the diagnosis of Grade III or Grade IV astrocytoma at The Johns Hopkins Medical Institutions. Among these patients, only those with tumors of the MTL (Schramm A – C) were identified for further study. The Schramm classification is a standardized means of anatomically describing tumors of the mesial temporal lobe [22]. Tumors involving the amygdala, uncus, and hippocampus without involvement of the neocortex were considered for this study.

Clinical data were obtained by retrospective review of patient charts, imaging, and operative notes. Operative reports were reviewed to confirm the location of corticectomy. The decision for a transcortical approach and the site of corticectomy (inferior versus middle temporal gyrus) was based primarily on surgeon preference. Postoperative complications and rehabilitation were ascertained from discharge summaries and postoperative imaging. Imaging was further reviewed for tumor location, extent of resection, and postoperative ischemia. Extent of resection was based on comparisons of pre- and postoperative contrast-enhanced T1-weighted magnetic resonance imaging (MRI) scans.

Statistical analysis to determine differences in outcomes between sites of cortical entry was performed using the Fisher exact text with a commercially available software package (GraphPad).

Results

Cohort Characteristics

Twenty-three patients during the study period met the inclusion criteria for further study (Table 1). The mean patient age was 49 years (range 25–75 years); with a male: female ratio of 15: 8. Mean Karnofsky Performance Score at the time of presentation was 80. The majority of the lesions were located in the dominant hemisphere. With regard to anatomic location, the lesions were primarily Schramm A (57 %), followed by Schramm C (35 %) and Schramm B (2 %). The mean size of the treated lesions was 2.9 × 3.5 × 2.8 cm (Table 2).

Surgical Approach and Outcomes

The inferior temporal gyrus was the most commonly employed site of cortical entry for tumors in the cohort (Table 2). Gross total resection was achieved in 92 % of the population (Table 3). With regard to complications, the most common events were: clinically significant stroke (two patients), new visual deficit (two patients), and new speech deficit (one patient). One patient without a prior history of seizures suffered a postoperative ictal event; seizures were controlled in only 53 % of those patients suffering preoperatively with seizures. The relationship between complications and site of cortical entry was further assessed (Table 3); a majority of events were encountered via the middle temporal gyrus approach. Statistical analysis revealed p values of 0.14 and 0.39 for differences in visual outcomes and speech outcomes between middle temporal gyrus and inferior temporal gyrus cortical entry.

Discussion

The MTL represents a unique structural entity within the supratentorial space with regard to the profile of brain tumors affecting this region. Due to its relatively compact structure and the fact that it is surrounded by critical vascular structures and white fiber pathways, much consideration has been paid to identifying ideal surgical approaches to the region with minimal collateral damage [1, 6, 8, 13–16, 21, 22, 26–28]. In an effort to better understand the risks of transcortical approaches through the middle/inferior temporal gyrus for high-grade gliomas, we reviewed our institutional experience with regard to the following outcomes: extent of resection, postoperative visual field deficits, and postoperative speech deficits.

Table 1 Patient demographics

Number of patients	23
Mean age	49 years (range: 25–75)
Male: female	15: 8
Presenting Karnofsky Performance Score	80
Presenting symptoms	
Speech deficit	5
Visual deficit	2
Seizure	15

Table 2 Tumor characteristics and approach

Right: left lateralization		10: 13
Mean size (CC × AP × ML)		2.9 × 3.5 × 2.8 cm
Cystic		6 (26 %)
Necrosis		12 (52 %)
Schramm distribution		
A	Tumor limited to medial temporal lobe; extends from uncus to hippocampus/parahippocampal gyrus	13 (57 %)
B	Considered to be paramediobasal tumors – they lie lateral to type A tumors and the collateral sulcus	2 (8 %)
C	These tumors occupy the same anatomic regions as type A and B tumors	8 (35 %)
Site of corticectomy		
Middle temporal gyrus		9 (39 %)
Inferior temporal gyrus		14 (61 %)
Tumor histology		
WHO grade III		14
WHO grade IV		9

CC: Carniocaudal AP: Anteroposterior ML: Mediolateral

Table 3 Surgical outcomes and complications

Outcome	**# (%)**		
Extent of resection			
Subtotal resection	2 (8)		
Gross resection	21 (92)		
Seizure control rate	8/15 (53)		
Complications by approach		**Middle temporal gyrus**	**Inferior temporal gyrus**
New postoperative stroke	2 (8)	1	1
New speech deficit	1 (4)	1	0 ($p=0.39$)
New visual deficit	2 (8)	2	0 ($p=0.14$)

White Fiber Tracts

Important white fiber tracts surround the MTL and it is important that they should be considered when selecting a surgical approach. These tracts include the visual pathways, uncinate fasciculus, superior longitudinal fasciculus, and inferior longitudinal fasciculus. While the location of Wernicke's area has been well documented and can be determined intraoperatively through awake mapping, considerable attention has been paid to elucidating the location of white fiber tracts that cannot be found via intraoperative mapping.

Considered a part of the limbic system and temporal stem, the uncinate fasciculus, which lies in close relation to the amygdala and limen insulae, is an often a neglected entity in temporal lobe surgery. This white fiber tract consists of three parts: a ventral (frontal) extension, an intermediary segment within the limen insulae, and a temporal segment [18]. This fiber tract is asymmetrical in size between both hemispheres (27 % larger in the right hemisphere), suggesting its importance even in non-dominant hemispheres [18]. Interconnecting the anterotemporal lobe with the orbitofrontal area, the uncinate fasciculus is involved in linking emotions to cognition [6, 7, 20]. It may also play a role in the retrieval of episodic memories [25], and surgical resection of lesions in this area can result in long-term difficulty with face recognition and object naming [17]. Disruption of the uncinate fasciculus with anteromedial temporal lobectomy or trans-sylvian transinsular approaches may be associated with the psychosocial and cognitive changes seen postoperatively [1, 2]. As part of the temporal stem with its origin in the hippocampal formation and the amygdala, the uncinate fasciculus may be the preferential pathway for seizure spread. This must be recognized when the goal of surgery is to control tumor-induced seizures [12].

Superior and superficial to the uncinate fasciculus, the inferior occipitofrontal fasciculus (IFOF) is believed to be involved in semantic processing [11]. The IFOF consists of two components – a dorsal subcomponent connecting the frontal, superior parietal lobe and occipital gyri, and a deeper ventral portion communicating with the visual association areas: inferior occipital gyrus and posterior temporal-basal area (fusiform gyrus, temporo-occipital sulcus, and basal inferior temporal gyrus). Intraoperative stimulation studies have shown that IFOF stimulation induces semantic paraphasias during picture-naming tasks [3, 4].

Advances in diffusion tensor imaging have further elucidated the anatomic projections of the inferior longitudinal fasciculus (ILF), which joins the posterior occipital temporal regions with the temporal lobe, where it further interacts with the uncinate fasciculus to interact with the basal frontal region. Within the dominant hemisphere, the ILF is proposed to be one of the parallel pathways of the "semantic ventral stream" that constitutes the language circuitry. Intraoperative mapping studies by Mandonnet et al. elicited semantic paraphasias when white fiber tracts underneath the superior temporal sulcus immediately above the roof of the temporal horn were stimulated [10].

Numerous recent studies utilizing Klingler dissection and diffusion tractography have helped to elucidate the course of the optic radiations as they project from the lateral geniculate body (LGB) to the calcarine cortex [6, 24]. As they leave the thalamus, the fibers can be classified into three bundles: posterior, central, and anterior – which progressively take a more curved route to the calcarine cortex. The central group of fibers follows a direct path from the LGB to the visual cortex without any anterior curve, while the central bundle takes a partial anterior curve over the superior surface of the temporal horn (just at the level of the LGB) prior to extending posteriorly. The bundle of most concern during surgical approaches to the MTL is the anterior bundle (Meyer's Loop) that passes entirely around the lateral half of the tip of the temporal horn before coursing posteriorly within the sagittal stratum to the calcarine cortex. Debate exists with regard to the precise relationship of the anterior bundle to the tip of the temporal horn [2, 5, 9, 19, 24]. Some authors have reported that Meyer's loop does not reach the anterior tip, while other studies have noted that these fibers not only reach the tip but may also extend up to 5 mm further anteriorly [5, 19, 24]. The risk of quadrantanopsia due to the location of these fibers along the lateral ependymal wall of the ventricle is one of the primary risks of transcortical approaches to the MTL.

Comparison of Different Approaches to the MTL

Selective surgical approaches to the MTL have been described in the literature, including: trans-sylvian-transcisternal/trans-insular, subtemporal, and transcortical [1, 2, 13, 14, 21, 26, 28]. Recognizing the complexity of this region, Schramm and Aliashkevich recently proposed a revised grading system based on anatomic location and tumor size in an effort to select the appropriate surgical approach [22]. In their series, a majority of Grade A patients underwent trans-sylvian resection, while Grade B tumors primarily underwent subtemporal resection, and Grade C and D tumors were resected via temporal lobectomy. Overall, a transcortical approach was used in only 6.5 % of all patients and only for Grade B and D tumors.

In his manuscript on the surgical management of limbic and paralimbic tumors, Yasargil noted that a trans-sylvian approach can be used to resect all lesions in the mesial basal temporal lobe [28]. Done via a pterional, orbitozygomatic, or supraorbital craniotomy, the trans-sylvian approach provides a superior-medial-based approach to the MTL. Both transcisternal and transinsular variants of this approach avoid manipulation of the lateral and basal neocortex by approaching the MTL structures via either the inferior limiting sulcus (transinsular approach) or through the medial surface (trancisternal approach). In their study of 235 patients with tumors in the temporal lobe and insular cortex, Schramm and Aliashkevich noted some form of visual deficit (either hemianopsia or quadrantanopsia) in approximately 32 % of patients [21]. These visual injuries are most likely a result of injury to the central visual bundle when the incision into the ventricle extends more than 2 cm posterior to the limen insulae. In addition, a 4.5 % incidence of postoperative oculomotor disturbances was noted by Schramm and Aliashkevich; this complication is typically not seen with other approaches [21]. Speech disturbances in patients with dominant hemisphere approaches were noted in 6.1 % of the cohort [21]. While providing a direct trajectory to this region, these approaches are thought to place the arterial tree and the optic radiations at risk, since they may involve transection of the superior or medial surfaces of the temporal horn of the lateral ventricle.

Primarily employed for Grade B tumors, the subtemporal route provides a basal route to the MTL. A low trajectory to the region is afforded by downward mobilization of the temporalis muscle and appropriate osteotomies. Once the temporal lobe is retracted away from the floor of the middle fossa, the temporal horn can be entered via one of several sulci or gyri (collateral, rhinal, or occipital temporal sulci; fusiform or parahippocampal gyri). While this approach may provide a direct trajectory, it is limited by the degree of retraction necessary, in addition to the risk of injury to the vein of Labbe – especially in more posteriorly extending resections. With regard to visual outcomes, quadrantanopsia is noted at a higher rate (28.6 %) than hemianopsia [21]. Likely due to venous injury, speech disturbances are reported in 7.1 % of patients postoperatively [21].

Due to the widespread adoption of neuronavigation and an improved understanding of white fiber anatomy, there have been increasing reports of transcortical approaches to the MTL for oncologic processes and epilepsy. The transcortical approaches typically proceed via the middle or inferior temporal gyri; the superior temporal gyrus is left untouched unless there is tumor involvement. Several key considerations in this approach include the posterior extent of the corticectomy and the potential entry into the ventricle. In the dominant hemisphere, a corticectomy extending more than 4 cm posteriorly can result in speech disturbances [23]. This approach provides a wide working window and opportunity for en-bloc resection of the MTL structures. New postoperative visual deficits occurred in

8 % of the present cohort, which is slightly lower than the rate reported in the literature [21]. This difference is likely due to the decreased risk of lateral ventricular surface entry and, therefore, the decreased risk of visual pathways injury – particularly to Meyer's Loop, which covers the lateral ventricular surface. Interestingly, all patients with new postoperative visual deficits had smaller Schramm A tumors, which would be less likely to cause a mass effect on surrounding structures. Tissue distortion and ventricular effacement due to tumor mass effect may create a larger corridor for resection with a reduced risk of injury to Meyer's Loop. In addition to the visual deficits, the other notable complication in the present series was acute ischemia, occurring in two patients (8 %). Both cases were a consequence of perforator infarcts occurring in tumors that had breached the medial pial surface of the MTL extending into the interpeduncular cistern and sylvian fissure. This issue highlights the fact that vascular supply to the tumor and critical neurovascular structures are encountered late in the resection.

Complications: Middle Versus Inferior Temporal Gyrus

With growing interest in transcortical approaches to MTL tumors, an important consideration is the gyrus through which surgery proceeds. In their white matter fiber dissection analysis of the optic radiations, Sincoff et al. propose that corticectomies limited to areas inferior to the inferior temporal sulcus could avoid Meyer's loop as it passes lateral to the temporal horn deep to the superior and middle temporal gyri [24]. Several groups have reported small case series in which the approach through the MTL is primarily the inferior temporal sulcus for MTL epilepsy; with well-documented Humphrey visual field examination, visual field defects were not noted postoperatively [8, 14, 15].

In our series of patients, all patients with new postoperative visual field complications underwent resection via a middle temporal gyrus approach where the ventricle was entered. While the corticectomy site (middle versus inferior temporal gyrus) was based on surgeon preference, no other variable (i.e., tumor size, Schramm classification) was found to be associated with complication occurrence. Due to the above anatomic considerations, it is probable that an approach through the middle temporal gyrus is much more likely to injure these visual fibers as they pass around the ventricle to the occipital lobe. However, statistical analysis did not reveal any significant difference in outcomes between inferior temporal gyrus and middle temporal gyrus entry in our limited series.

Conclusion

High-grade astrocytomas represent a cohort of less commonly encountered tumors in the mesial temporal lobe. From the oncologic standpoint, they represent a unique surgical challenge in comparison to the low-grade well-circumscribed lesions that typically are encountered in this region; there is a need to obtain gross total resection in order to obtain survival benefit. This need for an oncologic resection is compounded by the anatomic complexity of this region with surrounding white fiber tracts and neurovascular structures. Several approaches to a variety of lesions in the MTL have been described – including the trans-sylvian, subtemporal, and transcortical approaches, yet outcomes have not been discussed in the context of high-grade astrocytomas. Our experience indicates that the transcortical approach can be safely employed to obtain satisfactory resection of infiltrative lesions in this area; it further appears that there is no statistically significant difference in outcomes between inferior temporal gyrus as opposed to middle temporal gyrus entry. A larger study is necessary to confirm these findings.

Conflict of Interest Statement We declare that we have no conflict of interest.

References

1. Campero A, Troccoli G, Martins C, Fernandez-Miranda JC, Yasuda A, Rhoton AL Jr (2006) Microsurgical approaches to the medial temporal region: an anatomical study. Neurosurgery 59(4 Suppl 2):ONS279–ONS307 discussion ONS307-278
2. Choi C, Rubino PA, Fernandez-Miranda JC, Abe H, Rhoton AL Jr (2006) Meyer's loop and the optic radiations in the transsylvian approach to the mediobasal temporal lobe. Neurosurgery 59(4 Suppl 2):ONS228–ONS235; discussion ONS235-226
3. Duffau H (2008) The anatomo-functional connectivity of language revisited. New insights provided by electrostimulation and tractography. Neuropsychologia 46(4):927–934. doi:10.1016/j.neuropsychologia.2007.10.025
4. Duffau H, Gatignol P, Mandonnet E, Peruzzi P, Tzourio-Mazoyer N, Capelle L (2005) New insights into the anatomo-functional connectivity of the semantic system: a study using cortico-subcortical electrostimulations. Brain 128(Pt 4):797–810. doi:10.1093/brain/awh423
5. Ebeling U, Reulen HJ (1988) Neurosurgical topography of the optic radiation in the temporal lobe. Acta Neurochir (Wien) 92(1–4):29–36
6. Fernandez-Miranda JC, Rhoton AL Jr, Alvarez-Linera J, Kakizawa Y, Choi C, de Oliveira EP (2008) Three-dimensional microsurgical and tractographic anatomy of the white matter of the human brain. Neurosurgery 62(6 Suppl 3):989–1026; discussion 1026–1028
7. Gaffan D, Wilson CR (2008) Medial temporal and prefrontal function: recent behavioural disconnection studies in the macaque monkey. Cortex 44(8):928–935. doi:10.1016/j.cortex.2008.03.005

8. Ikeda K, Shoin K, Mohri M, Kijima T, Someya S, Yamashita J (2002) Surgical indications and microsurgical anatomy of the transchoroidal fissure approach for lesions in and around the ambient cistern. Neurosurgery 50(5):1114–1119; discussion 1120
9. Mahaney KB, Abdulrauf SI (2008) Anatomic relationship of the optic radiations to the atrium of the lateral ventricle: description of a novel entry point to the trigone. Neurosurgery 63(4 Suppl 2):195–202; discussion 202–293
10. Mandonnet E, Nouet A, Gatignol P, Capelle L, Duffau H (2007) Does the left inferior longitudinal fasciculus play a role in language? A brain stimulation study. Brain 130(Pt 3):623–629. doi:10.1093/brain/awl361
11. Martino J, Brogna C, Robles SG, Vergani F, Duffau H (2010) Anatomic dissection of the inferior fronto-occipital fasciculus revisited in the lights of brain stimulation data. Cortex 46(5):691–699. doi:10.1016/j.cortex.2009.07.015
12. Mayanagi Y, Watanabe E, Kaneko Y (1996) Mesial temporal lobe epilepsy: clinical features and seizure mechanism. Epilepsia 37(Suppl 3):57–60
13. Mikuni N, Miyamoto S, Ikeda A, Satow T, Taki J, Takahashi J, Ohigashi Y, Hashimoto N (2006) Subtemporal hippocampectomy preserving the basal temporal language area for intractable mesial temporal lobe epilepsy: preliminary results. Epilepsia 47(8):1347–1353
14. Miyagi Y, Shima F, Ishido K, Araki T, Kamikaseda K (2001) Inferior temporal sulcus as a site of corticotomy: magnetic resonance imaging analysis of individual sulcus patterns. Neurosurgery 49(6):1394–1397; discussion 1397–1398
15. Miyagi Y, Shima F, Ishido K, Araki T, Taniwaki Y, Okamoto I, Kamikaseda K (2003) Inferior temporal sulcus approach for amygdalohippocampectomy guided by a laser beam of stereotactic navigator. Neurosurgery 52(5):1117–1123; discussion 1123–1124
16. Miyamoto S, Kataoka H, Ikeda A, Takahashi J, Usui K, Takayama M, Satow T, Hashimoto N (2004) A combined subtemporal and transventricular/transchoroidal fissure approach to medial temporal lesions. Neurosurgery 54(5):1162–1167; discussion 1167–1169
17. Papagno C, Miracapillo C, Casarotti A, Romero Lauro LJ, Castellano A, Falini A, Casaceli G, Fava E, Bello L (2011) What is the role of the uncinate fasciculus? Surgical removal and proper name retrieval. Brain 134(Pt 2):405–414. doi:10.1093/brain/awq283
18. Peltier J, Verclytte S, Delmaire C, Pruvo JP, Godefroy O, Le Gars D (2010) Microsurgical anatomy of the temporal stem: clinical relevance and correlations with diffusion tensor imaging fiber tracking. J Neurosurg 112(5):1033–1038. doi:10.3171/2009.6.JNS08132
19. Rubino PA, Rhoton AL Jr, Tong X, Oliveira E (2005) Three-dimensional relationships of the optic radiation. Neurosurgery 57(4 Suppl):219–227; discussion 219–227
20. Schmahmann J, Pandya D (2006) Fiber pathways of the brain. Oxford University Press, New York
21. Schramm J, Aliashkevich AF (2007) Surgery for temporal mediobasal tumors: experience based on a series of 235 patients. Neurosurgery 60(2):285–294; discussion 294–295
22. Schramm J, Aliashkevich AF (2008) Temporal mediobasal tumors: a proposal for classification according to surgical anatomy. Acta Neurochir (Wien) 150(9):857–864; discussion 864
23. Schwartz TH, Devinsky O, Doyle W, Perrine K (1998) Preoperative predictors of anterior temporal language areas. J Neurosurg 89(6):962–970
24. Sincoff EH, Tan Y, Abdulrauf SI (2004) White matter fiber dissection of the optic radiations of the temporal lobe and implications for surgical approaches to the temporal horn. J Neurosurg 101(5):739–746
25. Squire LR, Zola-Morgan S (1991) The medial temporal lobe memory system. Science 253(5026):1380–1386
26. Uribe JS, Vale FL (2009) Limited access inferior temporal gyrus approach to mesial basal temporal lobe tumors. J Neurosurg 110(1):137–146
27. Wheatley BM (2008) Selective amygdalohippocampectomy: the trans-middle temporal gyrus approach. Neurosurg Focus 25(3):E4
28. Yasargil MG, Reeves JD (1992) Tumours of the limbic and paralimbic system. Acta Neurochir (Wien) 116(2–4):147–149

Tips and Tricks for Anterior Cranial Base Reconstruction

Oreste de Divitiis, Alberto Di Somma, Luigi Maria Cavallo, and Paolo Cappabianca

Abstract Reconstruction procedures come last in skull base surgery, but they are not the least important phase—rather, reconstruction is one of the most important steps required to prevent complications. In our opinion, there are *three general principles* upon which a good reconstruction of the skull base stand: (1) anatomo-surgical knowledge; (2) approach/route selection; and (3) the cooperation of the skull base surgical team. In general, three major complications may occur when a good skull base reconstruction has not been achieved, i.e., *cerebrospinal fluid (CSF) leak*, *pneumoencephalus*, and *infection*. Reconstruction of skull base defects requires a thorough knowledge of surgical anatomy, disease, and patient risk factors. Various reconstruction techniques are available, from free tissue grafting to vascularized flaps. Possible complications that can occur after these procedures need to be considered.

The reconstruction phase of the surgical procedure is a fundamental step in any surgical approach and it must not be ignored.

Keywords/Reference Phrases Reconstruction • Skull base surgery • CSF leak • Transcranial surgery • Endoscopic endonasal surgery

Introduction

Over the past 25 years skull base surgery has rapidly evolved into a highly specialized surgical discipline. As the reconstruction procedure is one of the most important steps in this kind of surgery, it has to be considered that the development of reliable reconstructive techniques has greatly facilitated the ability to access tumors that previously were not considered resectable. As a matter of fact, safe separation of the intracranial contents from the extracranial spaces is the key consideration in skull base surgery. Reconstruction comes last in skull base surgery, but it is not the least important phase of the procedure.

In any situation, failure to create adequate reconstruction may lead to significant complications, among which cerebrospinal fluid (CSF) leak, meningitis, brain herniation, and tension pneumocephalus have to be taken into serious account [7].

Skull base defects require precise and durable reconstruction in order to form a watertight dural seal, provide a barrier between the contaminated extracranial spaces and the sterile intradural compartment (thus avoiding infections), prevent airflow into the intracranial space (pneumocephalus), maintain a functional sinonasal system, and provide a good cosmetic outcome.

From the etiological point of view, it has to be kept in mind that skull base defects can arise from both traumatic and nontraumatic causes. In the traumatic group, nonsurgical trauma is the most common cause overall, and surgical (iatrogenic) damage is a minor cause. In the nontraumatic group, skull base defects can be caused by direct tumor erosions and/or as a consequence of high intracranial pressure. Less commonly, skull base defects may be caused by radiotherapy or infections. Spontaneous CSF leaks (idiopathic) can even occur [8].

As resections for skull base pathologies become more complex, the resultant defects require more difficult and extensive reconstructions. Over the past 15 years, significant advances in surgical and reconstructive techniques have evolved in the treatment of multiple extradural and intradural skull base lesions, also via expanded endoscopic endonasal approaches (EEAs), adding complexity to the reconstructive paradigm [9, 10].

O. de Divitiis (✉) • A. Di Somma • L.M. Cavallo • P. Cappabianca
Division of Neurosurgery, Department of Neurosciences, Reproductive and Odontostomatological Sciences, Università degli Studi di Napoli "Federico II", Naples, Italy
e-mail: oreste.dedivitiis@unina.it; http://www.neurochirurgia.unina.it

M. Visocchi et al. (eds.), *Trends in Reconstructive Neurosurgery*, Acta Neurochirurgica Supplement 124,
DOI 10.1007/978-3-319-39546-3_26, © Springer International Publishing Switzerland 2017

General Principles of Good Skull Base Reconstruction

We can consider *three general principles* upon which a good reconstruction of the skull base stand: (1) anatomo-surgical knowledge; (2) approach/route selection; and (3) the cooperation of the skull base surgical team.

Anatomo-Surgical Knowledge

The dissection laboratory provides a venue to improve knowledge of anatomical structures, demonstrate technique, and enhance surgical skills. Indeed, cadaveric dissection still remains the gold standard for training physicians in the field of neurosurgery [3, 4, 6].

Generally, a neurosurgeon should constantly study the microanatomy of the brain, as better knowledge of microsurgical anatomy leads to better surgery.

As a matter of fact, one of the most important steps for a good reconstruction of the skull base is to preserve tissues that are encountered during the approach. This concept is true for both the transcranial and endoscopic endonasal routes.

As a general rule, neurosurgeons should foresee what will come next. Indeed, even though the reconstruction phase comes at the end of the surgical procedure, it is of vital importance to preserve, at the beginning or during the surgical procedure itself, any autologus tissue that can be useful for repairing the skull base defect. Moreover, it is important to inform the patient about the possible additional surgical procedures that may be done to withdraw the autologous materials to be used for the reconstruction, and to obtain their signed informed consent for such procedures.

However, different heterologous materials can be useful to reinforce the reconstruction and the surgical team must ensure that such materials are present in the operating room the day before the planned surgery (Fig. 1).

Many tips and tricks may be helpful, for either the transcranial or endonasal approach, when planning a good skull base reconstruction.

Neurosurgery is no different from any sporting or art activity; only intense practice leads to good results. The skill of the individual surgeon still plays a critical role in determining outcome, and such skill must embody the concepts of tactics and technique. There is still a common misconception that simply to use a sucker, bipolar coagulation, and clips means competence in microsurgery. In fact, these modalities form only the material part of the techniques involved; the real skill has to be learned not just in the operating theater but also by meticulous laboratory training over many months. Skin incision, soft-tissue dissection, craniotomy, dural opening, intradural dissection, and respect of the sinonasal compartment are key steps that have to be taken into account during every surgical procedure.

1. Transcranial approaches
   Probably the most popular method of reconstructing the anterior skull base today is the use of the pericranial flap. After skin incision is performed, a scalp flap is elevated in the subgaleal plane to the level of the supraorbital rims [5]. At this point, the pericranium is incised and the dissection is carried forward in this plane, preserving the blood supply from the supratrochlear and supraorbital arteries (Fig. 2a). After the craniofacial resection is completed, the pericranial flap may even be sutured to the posterior remnant of the dura. It therefore may be used to repair dural defects or can be used to augment the intact or reconstructed dura by providing support, mainly because of its vascularity.

   On the other hand, if the frontal sinus is entered during a subfrontal craniotomy, the posterior wall and all sinus mucosa are carefully removed by cranialization. Subsequently, a galeal periosteal flap from the forehead, sealed with fibrin glue, can be used to cover the basal parts of the frontal sinus. The frontal sinus may also be closed with autologous fat (Fig. 2b).

   Moreover, autogenous wet bone powder collected during skull trephination may be useful for cranial reconstruction to fill in the dead space left after the craniotomy.

   In cases of extensive bone removal, a titanium mesh can be used to support the anterior skull base reconstruction.

   Coagulation of the dural edge should be avoided, as this generates its retraction and complicates a watertight closure. A clearance of several millimeters during dural opening should be allowed between the bone margin and the dural incision, to facilitate its final closure.
2. Endonasal approaches
   When performing an endonasal approach [1], the use of specific reconstructive techniques is based upon many factors apart from the corridor and target area: the size and shape of the skull base defect, the condition of the surrounding bone and remaining dura, whether or not a CSF leak was encountered intraoperatively, the nature of the CSF leak encountered (high- or low-flow), the extent of communication between the nasal and intracranial cavities, the anticipated postoperative intracranial pressure (ICP), the nature of any resected lesion, the status of the nasal septum and lateral nasal wall, a history of previous surgery or radiation therapy, and the anticipated need for adjuvant therapies (i.e., irradiation or chemotherapy). Currently, pedicled flaps offer the most reliable reconstruction of large skull base defects, ensuring the successful isolation of

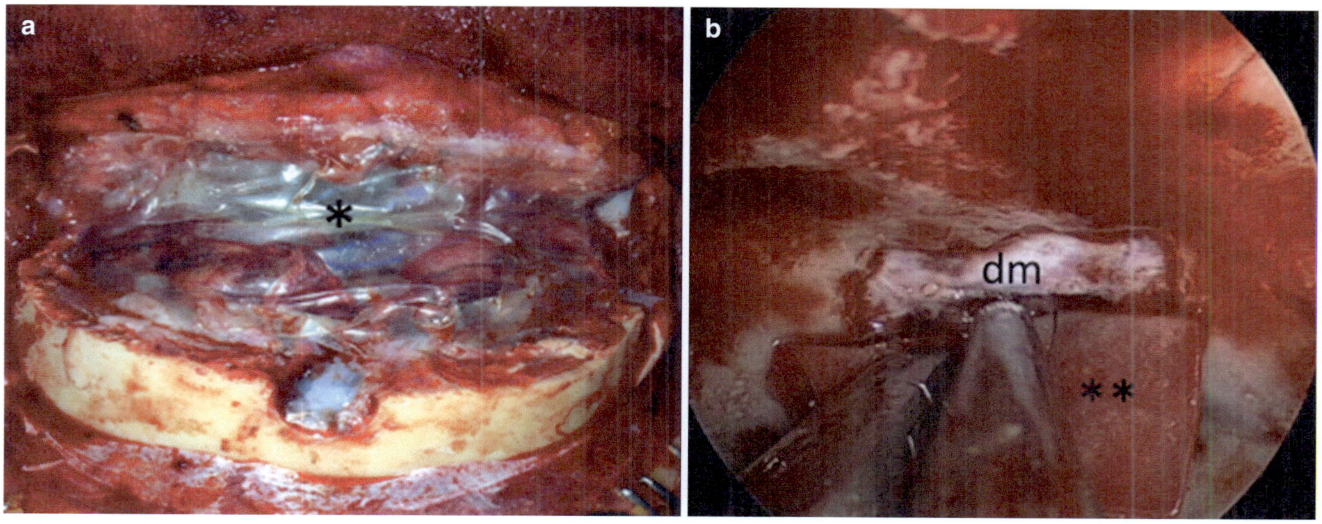

Fig. 1 (a) Subfrontal approach to the anterior cranial base; the dural substitute (*) used for dural repair has been used (Tissel®, Baxter, Deerfield, IL, USA). (b) Endoscopic endonasal approach to the planum sphenoidale; the dural substitute (**) has been placed in the epidural space (Neuro-Patch, B. Braun, Boulogne, France). dm, dura mater

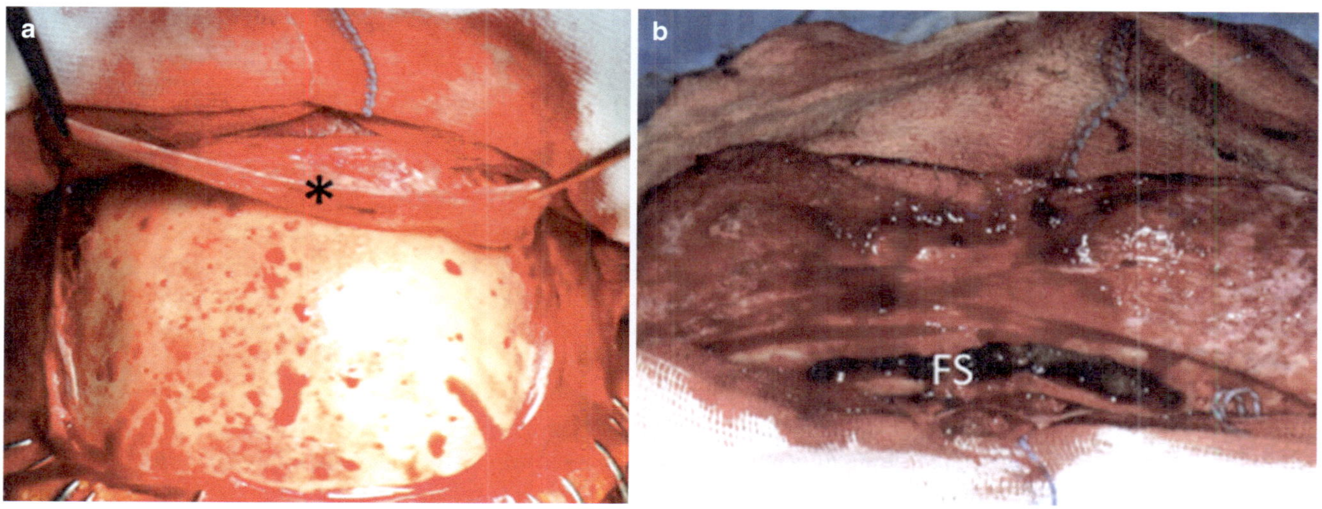

Fig. 2 The pericranial flap (*) is raised at the beginning of the surgical procedure, i.e. before the craniotomy (a). The frontal sinus is entered during a subfrontal craniotomy (b); accordingly, the posterior wall and all sinus mucosa are carefully removed and the sinus is filled with autologous and/or heterologous materials. FS, frontal sinus

the intracranial space from the sinonasal tract, to prevent complications such as meningitis, intracranial abscesses, encephaloceles, CSF leaks, and tension pneumocephalus. Pedicled flaps, such as the posterior pedicle nasoseptal flap (i.e., the Hadad-Bassagaisteguy flap) have revolutionized the endoscopic repair of the skull base, as their axial blood supply can irrorate a large surface area with a small and relatively long pedicle. In this view, it is mandatory to save the vascularization coming from the sphenopalatine artery during the nasal phase of the surgical procedure.

Furthermore, a free middle turbinate flap may be useful to reinforce the reconstruction; this has to be saved at the beginning of the procedure (Fig. 3).

Another important aspect that has to be taken into account is that, for good osteodural reconstruction, the dural and bone plane has to be preserved in order to facilitate multilayer reconstruction.

Finally, it has to be kept in mind that lumbar drainage may be helpful for treating postoperative CSF leak.

In our department, we use the "sandwich technique": in the first instance, the cistern is covered with a layer of collagen sponge coated with fibrinogen and thrombin, and the surgical cavity is filled with fat graft sutured to the inner layer of three layers of the fascia lata or dural substitute. The first layer is then positioned intradurally, the second is applied between the dura and the bone, and the

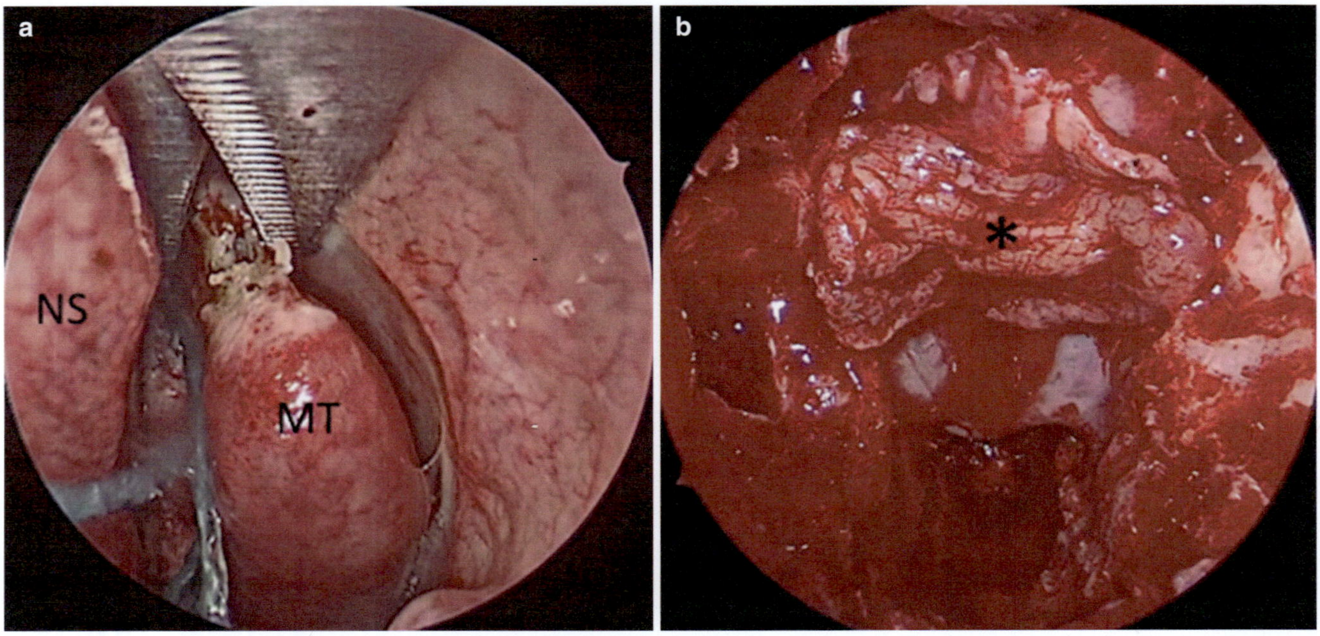

Fig. 3 Endoscopic endonasal approach. The middle turbinate is removed during the nasal step of the procedure (a) and, at the end, it is placed over the skull base defect to support the reconstruction materials (b). MT, middle turbinate; NS, nasal septum; *, middle turbinate place over the reconstruction materials.

third is applied to cover the bone. In order to support the materials used for reconstruction, the nasoseptal flap is used to cover the posterior wall of the sphenoid sinus. An inflated Foley balloon catheter, filled with 7–8 ml of saline solution, is then placed in the sphenoid sinus to support the reconstruction.

Approach Selection

The availability of a wide variety of surgical techniques poses the question whether it is better to access skull base lesions via traditional transcranial routes or via extended endoscopic endonasal approaches.

It has to be stressed, firstly, that the primary aims of skull base tumor resection are: gross total tumor resection with adequate decompression and preservation of the surrounding structures, support of the brain and orbit, complete separation of the cranial cavity from the sinonasal tract, elimination of dead space, and a watertight seal to avoid such consequences as CSF leaks and pneumocephalus. The choice of the most suitable approach for the removal of a skull base lesion should take into account not only the chance of a gross total removal but also the ability to reconstruct the pathway used to access the lesion itself.

For reconstructing skull base defects, it is important to understand the indications and limitations of each approach, together with the preoperative study of the neuroradiologic images. Preferably, the same route used for tumor removal should be used to repair the skull base defect, thus avoiding the comorbidity of a second approach.

In general, the selected surgical approach allows the most direct and maximal access to the pathologic process, with minimal morbidity to surrounding structures. Any surgical instrumentation that may be required to perform the operation should be requested and tested before starting the procedure.

The introduction of the endoscope in skull base surgery may eliminate many of the previous problems associated with microsurgical techniques. Important advantages of the extended endoscopic endonasal approach compared with classical surgical techniques and approaches are: the better access to deeply seated lesions, a more direct exposure of the midline, reduced trauma to brain parenchyma, less manipulation of the neurovascular structures, rapid decompression of the optical structures, and more efficient devascularization of neoplasms from their surroundings. It is important to acknowledge that when important neurovascular structures are above or surrounding the capsule of the tumor, the endoscopic endonasal approach is ideal; however, when a major vessel is to be encountered before reaching the surgical target, then open approaches are favored.

Given that the transcranial approach (high route) is still considered the "gold standard" in the surgical treatment of a variety of suprasellar lesions, tuberculum sellae meningiomas among them, the development of extended endonasal transsphenoidal surgery (low route) has introduced a potential alternative treatment of these tumors in certain patient populations [2].

This technique, involving special approaches to a limited array of cranial base lesions, should be recognized as an emerging reality in the neurosurgical arena. Further studies are required to ascertain the relative benefits of endosurgery over open surgery.

Comprehensive planning is an axiomatic prerequisite for any neurosurgical procedure. "Failing to prepare is preparing to fail" holds particularly true in neurosurgical cases, given the unforgiving nature of the human nervous system. Thorough preoperative consideration of the technical goals and potential pitfalls ensures the safest and most efficacious outcome for the patient. Effective planning allows the surgeon critical flexibility and latitude in managing deviations from the intended operative course. Indeed, the experience and ability to detect and handle the most adverse intraoperative events should be a goal for any surgeon. By taking the necessary steps to ensure adequate preparation for a case, the surgeon may prevent or avoid many significant neurosurgical complications.

Cooperation

The skull base is one of the most complex anatomical regions in the human body, with a variety of vital structures all within close proximity. It separates the brain from the facial skeleton and forms the floor of the cranial cavity. Skull base surgery requires a multidisciplinary team approach. Skull base surgery involves open, microscopic, and endoscopic approaches to the anterior, middle, or posterior cranial fossa. A multispecialty team approach is essential in treating patients with skull base lesions. Close cooperation between different specialists, i.e., the endocrinologist, neurosurgeon, neuroradiologist, pathologist, ophthalmologist, maxillofacial surgeon, ear nose and throat (ENT) surgeon, and others is mandatory.

Conclusions

Reconstruction of skull base defects requires a thorough knowledge of surgical anatomy, disease, and patient risk factors associated with high-flow cerebrospinal fluid leaks. Various reconstruction techniques are available, from free tissue grafting to vascularized flaps. Possible complications that can occur after these procedures need to be considered.

A better understanding of the skull base anatomy, the development of appropriate instrumentation, and the evolution of vascularized flaps have revolutionized the repair of the skull base following both open and endonasal approaches.

The purpose of any reconstruction is to achieve a watertight seal between the intracranial space and the sinonasal compartment; to supply adequate structural support for the intracranial contents—thereby preventing brain herniation into the surgical defect—and, finally but not secondarily, to achieve optimal cosmesis.

Conflict of Interest Statement The authors declare that they have no conflicts of interest.

References

1. Cavallo LM, Messina A, Esposito F et al (2007) Skull base reconstruction in the extended endoscopic transsphenoidal approach for suprasellar lesions. J Neurosurg 107:713–720
2. De Divitiis E, Esposito F, Cappabianca P et al (2008) Tuberculum sellae meningiomas: high route or low route? A series of 51 consecutive cases. Neurosurgery 62:556–563; discussion 556–563
3. De Divitiis O, Angileri FF, D'avella D et al (2002) Microsurgical anatomic features of the lamina terminalis. Neurosurgery 50:563–569; discussion 569–570
4. De Divitiis O, Conti A, Angileri FF et al (2004) Endoscopic transoral-transclival approach to the brainstem and surrounding cisternal space: anatomic study. Neurosurgery 54:125–130; discussion 130
5. De Divitiis O, Iacopino DG, Solari D et al (2010) Subfrontal approaches. In: Cappabianca P, Califano L, Iaconetta G (eds) Cranial, craniofacial and skull base surgery. Springer, Milan, pp 17–25
6. De Notaris M, Topczewski T, De Angelis M et al (2013) Anatomic skull base education using advanced neuroimaging techniques. World Neurosurg 79(S16):e19–13
7. Gil Z, Abergel A, Leider-Trejo L et al (2007) A comprehensive algorithm for anterior skull base reconstruction after oncological resections. Skull Base Off J N Am Skull Base Soc [et al.]. 17:25–37
8. Reyes C, Mason E, Solares CA (2014) Panorama of reconstruction of skull base defects: from traditional open to endonasal endoscopic approaches, from free grafts to microvascular flaps. Int Arch Otorhinolaryngol 18:S179–S186
9. Snyderman CH, Kassam AB, Carrau R et al (2007) Endoscopic reconstruction of cranial base defects following endonasal skull base surgery. Skull Base Off J N Am Skull Base Soc [et al.] 17:73–78
10. Tabaee A, Anand VK, Brown SM et al (2007) Algorithm for reconstruction after endoscopic pituitary and skull base surgery. Laryngoscope 117:1133–1137

Part V

Neurosurgical Rehabilitation and Outcome

Clinical and Neuropsychological Outcome After Microsurgical and Endovascular Treatment of Ruptured and Unruptured Anterior Communicating Artery Aneurysms: A Single-Enter Experience

Andrea Pietrantonio, Sokol Trungu, and Antonino Raco

Abstract

Background

Anterior communicating artery (ACoA) aneurysms have a high risk of rupture. Morbidity and mortality following rupture are higher than at other sites. The aim of this study was to evaluate the long-term clinical and neuropsychological outcomes of patients treated for ruptured and unruptured ACoA aneurysms: a comparison between surgical and endovascular treatment was performed.

Method

All patients surgically or endovascularly treated for ruptured and unruptured ACoA aneurysms at our institution between January 2011 and December 2013 (n=50) were retrospectively reviewed. The Glasgow outcome score and the following neuropsychological tests were used to define the clinical and neuropsychological outcomes, respectively: The **Stroop color** and word **test and the Stroop interference score** digit span forward and backward test, phonemic and semantic verbal fluency tests, Rey auditory verbal learning test, comprehensive trail making test, and the Beck Depression Inventory.

Findings

28 patients (56 %) underwent surgical treatment and 22 (44 %) endovascular coiling; there were 31 (63 %) ruptured and 19 (37 %) unruptured aneurysms. At 1 year follow-up for ruptured aneurysms, clinical outcome was better in the endovascular group; neuropsychological assessment showed a greater deterioration only in the memory domain in the patients treated surgically for ruptured aneurysms.

Conclusion

The presence of subarachnoid hemorrhage is more important than the type of treatment in determining the clinical and neuropsychological outcomes of ACoA treatment; these outcomes can be improved by adequate rehabilitation protocols.

Keywords Neuropsychological outcome • Memory • Anterior communicating artery aneurysms • Subarachnoid hemorrhage • Surgical clipping • Endovascular treatment

Introduction

Intracranial aneurysms are a common finding, with an estimated prevalence in the general population of 2.3 %; the frequency is very low in the first two decades of life and then increases steadily after the third decade [20]. The middle cerebral artery (MCA) and anterior communicating artery (ACoA) are the most common locations, while the internal carotid artery (ICA) and the vertebro-basilar system are less frequently involved, the latter accounting for 7–11 % of all intracranial aneurysms [8, 9, 20, 24]. Several studies have been performed to better define the risk of the rupture of unruptured intracranial aneurysms differently from International Study of Unruptured Intracranial Aneurysm (**ISUIA**) whose results seem to underestimate this risk, other important and well-designed studies have found that this risk ranges between 1 and 1.5 % [8, 10, 11, 20, 23, 24]. Several prognostic factors, both medical/behavioral (i.e. smoking, hypertension) and anatomical (size, shape, and location) seem to affect the likelihood of an aneurysmal rupture, allowing us to identify and preventively treat the patients with high-risk aneurysms. However, aneurysmal subarachnoid hemorrhage (aSAH) still has an incidence of approximately ten cases per 100,000 persons per year; this harmful event carries very high social costs, considering that more than one-third of the patients will die and approximately 10–15 % of all aSAH patients die before admission to hospital. During the past three decades, surgical

The original version of this chapter was revised. An erratum to this chapter is available at DOI 10.1007/978-3-319-39546-3_50

A. Pietrantonio, MD (✉) • S. Trungu, MD • A. Raco, MD
Department of Neuroscience, Mental Health and Sensory Organs – Division of Neurosurgery, S'Andrea Hospital, University of Rome "Sapienza", V. Suvereto n.240, Rome 00139, Italy
e-mail: andrea_pietrantonio@libero.it

M. Visocchi et al. (eds.), *Trends in Reconstructive Neurosurgery*, Acta Neurochirurgica Supplement 124,
DOI 10.1007/978-3-319-39546-3_27,

and medical management and rehabilitation techniques have evolved; the policy of early and ultra-early surgery, with the subsequent reduction in the incidence of preoperative rebleeding, the aggressive management of vasospasm, and the advances in surgical and endovascular treatment are the main factors involved in the continuous improvements observed in clinical outcomes [3, 12–15, 18]. Moreover, evidence from the literature shows that patients who are treated endovascularly for ruptured aneurysms show a better clinical outcome at 1 year than surgically treated patients; this advantage seems consistent only in the short- and mid-term and can be potentially and partially outweighed by the higher recanalization rate seen after coiling [2, 16, 17, 21, 22]. However, cognitive dysfunctions are frequent among survivors: approximately one-third of the patients resume their previous work entirely, while memory, executive function, and language deficits are encountered in up to 70 % of the patients.

This retrospective study evaluated and compared the 1-year clinical outcomes and quality of life in patients treated with surgical clipping or endovascular coiling for ruptured and unruptured ACoA aneurysms; this anatomical location was chosen because of its frequency, and its higher morbidity and mortality rates after rupture. This location was also chosen because, when compared with other locations, ACoA aneurysms are more prone to rupture, to rupture at a smaller size, and, following rupture, they show higher morbidity and mortality rates than those shown for other intracranial aneurysms.

Materials and Methods

All the patients surgically or endovascularly treated for ruptured and unruptured ACoA aneurysms at our institution between January 2011 and December 2013 were retrospectively reviewed. Preoperatively, all the patients underwent three-dimensional computed tomography (3D CT)-angiography (3D-CTA); 3D digital subtraction angiography (DSA) was performed in the following cases: complex and giant aneurysms, inadequate aneurysm visualization with CTA, and those with planned endovascular treatment. 3D-CTA was also used postoperatively to check aneurysm exclusion in the clipping group; in the coiling group, follow-up neuroimaging was scheduled in relation to several factors, such as occlusion grade, coil packing density, and features of the treated aneurysm. Preoperatively, the Hunt and Hess scale and Fisher grade were used in patients with ruptured aneurysms for clinical evaluation and the evaluation of subarachnoid hemorrhage, respectively; in unruptured cases, a standard neurological examination was performed. Ruptured aneurysms were treated within 48 h from the initial hemorrhage. Postoperatively, all the patients underwent neurological examination, and the Glasgow Outcome Scale (GOS) score was used to classify each patient's outcome before discharge and 3 months after treatment. GOS scores of 4 and 5, respectively, defined moderate disability and good recovery, identifying patients who achieved a good outcome; a poor outcome was defined by GOS scores of 2 and 3 (persistent vegetative state and severe disability, respectively). Neuropsychological evaluation was carried out 12 months after the initial treatment; memory, processing speed, language, working memory, attention, and mood disorders were evaluated by our neuropsychologists through the following tests:

- *The **Stroop Color** and Word **Test and the Stroop interference score***: this test evaluates cognitive flexibility and processing speed
- *Digit span forward and backward*: this test evaluates the verbal working memory, defined as the cognitive system that allows the temporary storage and manipulation of information
- *Phonemic and semantic verbal fluency tests*: these tests evaluate language in term of naming, comprehension, and production
- *(RAVLT)*: this test evaluates memory processing; namely, short-term verbal memory
- *Trail making test (TMT)*: this test consists of two parts: TMT A measures the level of attention, testing spatial organization and visual pursuit. TMT B is similar, but more difficult, and tests visual motor, spatial abilities, and mental flexibility
- *Beck Depression Inventory*: this is a 21-item self-report rating inventory that measures symptoms and attitudes of depression.

The age and education-adjusted scores obtained for each patient were expressed as a calibrated score (CS) and then converted to an equivalent score (ES): in this way, a score of 0 corresponded to compromised function and a score of 4 to normal function. For the Stroop test, the equivalent scores were expressed as percent values. For this analysis, patients were divided into four groups:

1. *Group 1*: patients treated surgically for ruptured aneurysms
2. *Group 2*: patients treated surgically for unruptured aneurysms
3. *Group 3*: patients treated endovascularly for ruptured aneurysms
4. *Group 4*: patients treated endovascularly for unruptured aneurysms.

Results

From January 2011 to December 2013, 50 patients were identified and formed the subjects of this review: 28 patients (56 %) underwent surgical treatment and 22 (44 %), endovascular coiling. The mean age of the surgical group was 56 years (range 35–67 years); there were 9 males (32 %) and

19 females (68 %). In the endovascular group, 14 patients (64 %) were females and 8 (36 %) were males, with a mean age of 59 years (range 37–71 years). There were 31 (63 %) ruptured and 19 (37 %) unruptured aneurysms; 20 of the 31 ruptured aneurysms were clipped and the remaining 11 were coiled, while 11 of the 19 unruptured aneurysms were clipped and 8 were coiled. For patients with ruptured aneurysms, the Hunt and Hess grade at presentation was 1 in 15 patients (46.9 %), 2 in 8 patients (26.5 %), 3 in 5 patients (16.7 %), and 4 in 3 patients (9.9 %); regarding Fisher grade, 3 patients (10.5 %) showed grade I, 7 (21 %) grade II, 16 (52.6 %) grade III, and 5 (15.9 %) grade IV.

The postoperative clinical outcome, assessed through the GOS, 3 months and 1 year after treatment, showed that a good recovery (GOS score 5) was achieved in all the patients with unruptured aneurysms, both in the coiling and in the clipping groups. In those with ruptured aneurysms, clinical outcomes at 3 months and 1 year are summarized in Tables 1 and 2, respectively.

Regarding the neuropsychological evaluation, the results obtained are shown in Table 3: the tests and the related cognitive domains are summarized, and scores are reported as mean values and equivalent scores.

Regarding mood disorders, 10 (37.5 %) of the 28 surgically treated patients presented mild depressive symptoms and two-thirds of these patients belonged to the group with ruptured aneurysms.

Discussion

In the neurosurgical literature, many studies regarding intracranial aneurysms can be found; the majority of these studies are focused on clinical outcomes, surgical and endovascular techniques, and/or on comparisons between surgical and endovascular treatment. However, few studies analyze the neuropsychological outcome, comparing clipping and coiling for ruptured and unruptured aneurysms at a well defined location. We have chosen to focus our attention on the ACoA, because of the high frequency and the high rupture rates of aneurysms in this location, the anatomical proximity to brain regions involved in important cognitive functions (language, memory, executive functions, etc.), and the frequent observation of neuropsychological disorders and mood disturbances during our follow-up evaluations. We also made a distinct analysis between ruptured and unruptured aneurysms and between clipping and coiling, trying to understand whether these disorders were mainly due to the aneurysm itself, to the type of treatment, or to subarachnoid hemorrhage. We have participated in a general improvement in the short-, middle-, and long-term neuropsychological outcomes of patients with ACoA aneurysms during the past 15–20 years; probably, advances in the surgical techniques, minimization of surgical trauma on the adjacent brain parenchyma, the strategy of early and ultra-early surgery, the better management of vasospasm and, most importantly, more effective and intensive neurorehabilitation strategies, are the main factors responsible for this improvement [4, 5].

In our experience, we observed no clear differences between surgically and endovascularly treated patients or between ruptured and unruptured aneurysms in regard to language; more precisely, we observed better results after coiling and in unruptured aneurysms, but these differences were not significant and all the patients showed "normal function" according to the equivalent scores. Similar results were shown for the TMT-A; this test investigates processing speed, similarly to the Stroop color test. However, in the Stroop color test, the scores of the patients in Group 1 (surgically treated for ruptured aneurysms) were worse than those of the other groups, suggesting a potentially greater negative effect of surgery than of hemorrhage. We studied executive functions through the TMT-B and the Stroop interference score: both of these tests showed worse results in patients with ruptured aneurysms than in those with unruptured aneurysms and, as a consequence, it seems that subarachnoid hemorrhage was the main factor responsible for this mild deterioration (equivalent score of 3) compared with normal subjects (equivalent score of 4). In regard to working

Table 1 Clinical outcome at 3 months in patients with ruptured aneurysms

		Glasgow Outcome Scale (3 months)				
		1	2	3	4	5
Ruptured aneurysms	Surgery (20 patients)	–	2 (11 %)	6 (30 %)	8 (39 %)	4 (20 %)
	Endovascular (11 patients)	–	1 (9 %)	3 (26 %)	3 (26 %)	4 (39 %)

Table 2 Clinical outcome at 1 year in patients with ruptured aneurysms

		Glasgow Outcome Scale (1 year)				
		1	2	3	4	5
Ruptured aneurysms	Surgery (20 patients)	–	1 (5 %)	5 (25 %)	9 (45 %)	5 (25 %)
	Endovascular (11 patients)	–	1 (9 %)	2 (18 %)	2 (18 %)	6 (55 %)

Table 3 Neuropsychological tests and cognitive domains explored

Test	Group 1			Group 2			Group 3			Group 4			
	Mean value	ES	SD	Mean value	ES	SD	Mean value	ES	SD	Mean value	ES	SD	
Stroop Test	73 s	90 %	19.45	63 s	100 %	17.8	67 s	100 %	2.1	60 s	100 %	5.4	*Processing speed*
TMT A	37 s	4	17.2	26	4	10.8	35.3	4	10.5	16	4	4.4	
PF	32	4	9.54	47	4	22.9	33	4	6.5	37	4	8.9	*Language*
SF	41.67	4	8.14	50.2	4	6.6	42	4	5.9	57	4	7.6	
RAVLT	34.3	2	6.9	43.6	4	10	44.3	4	7.1	54.3	4	9.7	*Short-term verbal memory*
DSF	5.18	3	1.38	5.31	4	5.3	5.28	4	1.9	5.75	4	4.1	
DSB	4.80	3	0.74	5.30	4	0.43	5.32	4	0.6	5.70	4	1.2	*Working memory*
Stroop interference	149 s	90 %	44.04	122 s	100 %	17.4	138 s	90 %	37.3	118 s	100 %	24.6	*Executive functions*
TMT B	107	3	32	43	4	33.4	83.5	3	28.4	17	4	29.6	

Group 1 patients treated surgically for ruptured aneurysms; *Group 2* patients treated surgically for unruptured aneurysms; *Group 3* patients treated endovascularly for ruptured aneurysms; *Group 4* patients treated endovascularly for unruptured aneurysms. *RAVLT* Rey auditory verbal learning test *ES* equivalent score, *SD* standard deviation, *PF* phonemic fluency, *SF* semantic fluency, *DSF* digit span forward, *DSB* digit span backward, *TMT* trail making test

memory, the digit span backward test showed worse scores in Group 1; however, this function was globally preserved even though surgery seemed to be a little more traumatic than endovascular treatment in this domain.

Short-term verbal memory proved to be the most deteriorated domain in this series. This observation was confirmed by the results in our patients during the follow-up examinations: memory deficits were considered the most limiting factors in daily activities and were reported with higher frequencies. Both surgery and subarachnoid hemorrhage seem to negatively affect this function, although the role of subarachnoid hemorrhage seems to be less important: indeed, endovascularly treated patients showed better results than surgically treated patients when considering the ruptured aneurysm group as a whole.

In the literature, several studies report better results after coiling than after clipping, in terms of neuropsychological functions: Chan et al. compared nine surgically treated patients with nine patients treated endovascularly for ACoA aneurysms, the latter group had better outcomes for verbal memory and executive functions [4]; Fontanella et al. had results similar to those of Chan et al. regarding memory and executive functions (20 patients in the surgical group and 17 in the endovascular group) and Bellebaum et al. found a slightly worse outcome after surgery [1, 6]. Our results are similar to the findings of Proust et al.: they performed neuropsychological evaluation and an MRI study 14 months after surgical (36 patients) or endovascular (14 patients) treatment for ACoA aneurysm rupture [19]. Language, visual memory, and executive functions were equally preserved in both groups, while verbal memory was more compromised in the surgical group [19]. Frazer et al. reached the same conclusion, but added an important consideration: the differences in cognitive functions observed between the two groups, with a better short-term outcome after endovascular treatment, were no longer evident in the middle- and long-term periods [7]. In other words, surgically treated patients may experience an immediate neuropsychological deterioration that frequently improves with time and with adequate rehabilitation therapies [7]. Moreover, although short-term outcome is strongly associated with the Hunt and Hess grade and Fisher grade at presentation, this association becomes less strong in the long term.

Our results are quite good and show slight differences in the neuropsychological outcomes between patients with ruptured and unruptured aneurysms 12 months after treatment; among all the functions explored, memory was the domain where the biggest differences were seen, and the patients who were surgically treated for ruptured aneurysms had the worst results. We conclude that the influence of surgery and endovascular treatment on the neuropsychological outcome is less relevant than the presence of subarachnoid hemorrhage, which seems to affect the outcome more strongly.

This single-center study is limited by the small number of patients and by its retrospective nature; the results should be considered preliminary and further improvements in clinical and neuropsychological outcome cannot be excluded with longer follow-up. Nonetheless, its results are interesting, providing a different perspective in the evaluation of patients treated for intracranial aneurysms; however, the better clinical outcome described in the literature after endovascular treatment, and confirmed by our results, does not correspond to a better neuropsychological outcome in the same group of patients.

Conflict of Interest Statement The authors declare that they have no conflicts of interest

References

1. Bellebaum C, Schäfers L, Schoch B, Wanke I, Stolke D, Forsting M, Daum I (2004) Clipping versus coiling: neuropsychological follow up after aneurysmal subarachnoid haemorrhage (SAH). J Clin Exp Neuropsychol 26(8):1081–1092
2. Campi A, Ramzi N, Molyneux AJ, Summers PE, Kerr RS, Sneade M, Yarnold JA, Rischmiller J, Byrne JV (2007) Retreatment of ruptured cerebral aneurysms in patients randomized by coiling or clipping in the International Subarachnoid Aneurysm Trial (ISAT). Stroke 38(5):1538–1544
3. Cantore G, Raco A (1998) Early versus delayed surgery for treatment of aneurysmal subarachnoid hemorrhage. Minerva Anestesiol 64(4):155–158
4. Chan A, Ho S, Poon WS (2002) Neuropsychological sequelae of patients treated with microsurgical clipping or endovascular embolization for anterior communicating artery aneurysm. Eur Neurol 47(1):37–44
5. Diamond BJ, DeLuca J, Kelley SM (1997) Memory and executive functions in amnesic and non-amnesic patients with aneurysms of the anterior communicating artery. Brain 120:1015–1025
6. Fontanella M, Perozzo P, Ursone R, Garbossa D, Bergui M (2003) Neuropsychological assessment after microsurgical clipping or endovascular treatment for anterior communicating artery aneurysm. Acta Neurochir (Wien) 145(10):867–872
7. Frazer D, Ahuja A, Watkins L, Cipolotti L (2007) Coiling versus clipping for the treatment of aneurysmal subarachnoid hemorrhage: a longitudinal investigation into cognitive outcome. Neurosurgery 60(3):434–441
8. International Study of Unruptured Intracranial Aneurysms Investigators (1998) Unruptured intracranial aneurysms – risk of rupture and risks of surgical intervention. N Engl J Med 339(24): 1725–1733
9. Iwamoto H, Kiyohara Y, Fujishima M, Kato I, Nakayama K, Sueishi K, Tsuneyoshi M (1999) Prevalence of intracranial saccular aneurysms in a Japanese community based on a consecutive autopsy series during a 30-year observation period. The Hisayama study. Stroke 30(7):1390–1395
10. Juvela S, Porras M, Heiskanen O (1993) Natural history of unruptured intracranial aneurysms: a long-term follow-up study. J Neurosurg 79(2):174–182
11. Juvela S, Poussa K, Lehto H, Porras M (2013) Natural history of unruptured intracranial aneurysms: a long-term follow-up study. Stroke 44(9):2414–2421
12. Kassell NF, Torner JC, Haley EC Jr, Jane JA, Adams HP, Kongable GL (1990) The International Cooperative Study on the Timing of Aneurysm Surgery. Part 1: Overall Management Results. J Neurosurg 73(1):18–36
13. Kassell NF, Torner JC, Jane JA, Haley EC Jr, Adams HP (1990) The International Cooperative Study on the Timing of Aneurysm Surgery. Part 2: Surgical results. J Neurosurg 73(1):37–47
14. Laidlaw JD, Siu KH (2002) Ultra-early surgery for aneurysmal subarachnoid hemorrhage: outcomes for a consecutive series of 391 patients not selected by grade or age. J Neurosurg 97(2):250–258
15. Laidlaw JD, Siu KH (2003) Poor-grade aneurysmal subarachnoid hemorrhage: outcome after treatment with urgent surgery. Neurosurgery 53(6) 1275–1280
16. McDougall CG, Spetzler RF, Zabramski JM, Partovi S, Hills NK, Nakaji P, Albuquerque FC (2012) The barrow ruptured aneurysm trial. J Neurosurg 116(1):135–144
17. Molyneux A, Kerr R, International Subarachnoid Aneurysm Trial (ISAT) Collaborative Group, Stratton I, Sandercock P, Clarke M, Shrimpton J, Holman R (2002) International Subarachnoid Aneurysm Trial (ISAT) of neurosurgical clipping versus endovascular coiling in 2143 patients with ruptured intracranial aneurysms: a randomized trial. J Stroke Cerebrovasc Dis 11(6):304–314
18. Park J, Woo H, Kang DH, Kim YS, Kim MY, Shin IH, Kwak SG (2015) Formal protocol for emergency treatment of ruptured intracranial aneurysms to reduce in-hospital rebleeding and improve clinical outcomes. J Neurosurg 122(2):383–391
19. Proust F, Martinaud O, Gérardin E, Derrey S, Levèque S, Bioux S, Tollard E, Clavier E, Langlois O, Godefroy O, Hannequin D, Fréger P (2009) Quality of life and brain damage after microsurgical clip occlusion or endovascular coil embolization for ruptured anterior communicating artery aneurysms: neuropsychological assessment. J Neurosurg 110(1):19–29
20. Rinkel GJ, Djibuti M, Algra A, van Gijn J (1998) Prevalence and risk of rupture of intracranial aneurysms: a systematic review. Stroke 29(1):251–256
21. Spetzler RF, McDougall CG, Zabramski JM, Albuquerque FC, Hills NK, Russin JJ, Partovi S, Nakaji P, Wallace RC (2015) The barrow ruptured aneurysm trial: 6-year results. J Neurosurg 123(3):609–617
22. Spetzler RF, McDougall CG, Albuquerque FC, Zabramski JM, Hills NK, Partovi S, Nakaji P, Wallace RC (2013) The barrow ruptured aneurysm trial: 3-year results. J Neurosurg 119(1):146–157
23. UCAS Japan Investigators, Morita A, Kirino T, Hashi K, Aoki N, Fukuhara S, Hashimoto N, Nakayama T, Sakai M, Teramoto A, Tominari S, Yoshimoto T (2012) The natural course of unruptured cerebral aneurysms in a Japanese cohort. N Engl J Med 366(26):2474–2482
24. Wiebers DO, Whisnant JP, Huston J 3rd, Meissner I, Brown RD Jr, Piepgras DG, Forbes GS, Thielen K, Nichols D, O'Fallon WM, Peacock J, Jaeger L, Kassell NF, Kongable-Beckman GL, Torner JC, International Study of Unruptured Intracranial Aneurysms Investigators (2003) Unruptured intracranial aneurysms: natural history, clinical outcome, and risks of surgical and endovascular treatment. Lancet 362(9378):103–110

Ventricular Central Neurocytoma: Rate of Shunting and Outcome 2 Years After Total and Subtotal Excision

Wessam Samir Soliman

Abstract

Background

Central neurocytoma is an intraventricular tumor that affects young adults. It has a favorable prognosis after adequate surgical intervention; however, an aggressive course may take place in some cases.

Objective

The objective of the study was to evaluate the rate of shunting and the outcome of control measures in patients with central neurocytoma submitted to total and subtotal excision.

Methods

Twelve patients were included in this study, with a follow-up of 24 months. Data collected included: age, sex, clinical presentation, early morbidity and mortality, radiological findings (tumor location, features, residual, recurrence, and hydrocephalus). All patients underwent surgery for total or subtotal excision through a transcortical approach. External Ventricular Drain (EVD) was inserted then removed or replaced by a shunt. Histopathology and the MIB index were used to confirm diagnosis and guide the follow-up; adjuvant radiotherapy or Gamma Knife radiosurgery were used for residual tumor or recurrence.

Results

The ages of the patients ranged from 14 to 48 years. Two patients died early, after total and subtotal excision, from sepsis and thalamic infarction, respectively. Six patients (60 %) had a total excision; two of them had a high MIB index and showed small recurrence at 12 months and 18 months, respectively, and received Gamma Knife radiosurgery. One of the six patients with total excision needed a shunt, and no shunt was needed in the four otherpatients; a subtotal excision was done for four patients (40 %) An early shunt was inserted for two of these patients. radiosurgery-controlled for one patient, while radiotherapy was used for control in the other three patients; radiotherapy control failed in one patient, who underwent a second surgery at 13 months

Conclusion

Central neurocytoma may have a favorable prognosis, with a lower incidence of shunt insertion throughout its course than that for other intraventricular tumors, if total removal is achieved.

Keywords Central Neurocytoma • Gamma Knife • Shunt

Introduction

Neurocytomas are rare World Health Organization (WHO) grade II neuronal tumors, which were first identified as histologically distinct entities in 1982 [7, 8]. They likely have an incidence of less than 1 %, given that a range of 0.1–0.5 % has been reported [3, 5, 20, 21]. These tumors occur most commonly within the ventricles, with a generally favorable prognosis as a result of indolent growth [4, 7].

Immunohistochemistry is frequently used to help distinguish this tumor from other central nervous system (CNS) neoplasms; the MIB-1 proliferation index is commonly used in an attempt to predict biologic behavior and may help direct adjuvant therapy. An MIB-1 labeling index (LI) of >2 % often heralds poor prognosis and tumor recurrence [3, 11].

Owing to its rare incidence, the diagnosis and management of this neoplasm remain controversial [17].

Little is known about the management of patients with this tumor, because most reports are from the pathologic literature and contain sparse information regarding clinical management [17, 24]. Safe maximal resection is considered the ideal therapeutic option at present, with the best long-term prognosis in terms of local control and survival. And an

W.S. Soliman, MD
Faculty of Medicine, Cairo University, Cairo, Egypt
e-mail: wessamsoliman75@yahoo.com

M. Visocchi et al. (eds.), *Trends in Reconstructive Neurosurgery*, Acta Neurochirurgica Supplement 124,
DOI 10.1007/978-3-319-39546-3_28, © Springer International Publishing Switzerland 2017

excellent prognosis can be expected if a gross total resection is achieved. Postoperative radiation therapy (RT) may be considered after subtotal resection. Otherwise, RT is an option for medically inoperable or recurrent disease [16].

Aim

We aimed to evaluate the rate of shunt insertion and the outcome of control measures after 2-year follow-up in patients with central neurocytoma who received total or subtotal excision (75 % to 80 % removal).

Patients and Methods

This study was conducted on 12 patients who presented with intraventricular central neurocytoma between January 2010 and January 2013, who were managed surgically and followed-up for 24 months after pathological and radiological confirmation of central neurocytoma.

Data included age, sex, symptoms at presentation, and location of the tumor within the ventricles. All patients were managed surgically through a transcortical or transcallosal approach if needed; an external ventricular drain was inserted and then removed or replaced by a shunt.

MIB histopathology and proliferation indexes were used to confirm the diagnosis and guide further management. Computed tomography (CT) and magnetic resonance imaging (MRI) were used to detect the extent of tumor removal (total or subtotal) after surgery and in the follow-up period for determining recurrence, progression, and/or residual tumor.

Adjuvant radiotherapy was given for all patients with residual tumors and for some patients showing progression; Gamma-Knife radiosurgery was an option for some patients with recurrent or residual tumors, and ventriculoperitoneal (V/P) shunt insertion and reoperation were done if indicated.

Data abstracted from medical records included: sex; symptoms at presentation; Location of tumor within ventricles; extent of tumor resection—i.e., gross total resection (GTR) or subtotal resection (STR); pathological details, including features associated with atypical neurocytoma; date of tumor progression; and treatment at progression.

Radiotherapy was considered as adjuvant therapy if it was given immediately postoperatively and it was considered as salvage therapy if the patient was initially observed and treated for clinical or radiographic progression post-surgery. Progression was detected on CT or MRI scans.

Local control was defined as the absence of any tumor regrowth or progression on imaging, and excluded the patient who died in the perioperative period. Survival was calculated from data at diagnosis, and local control was calculated from the date of first surgery.

Results

A total of 12 patients (7 males and 5 females) were included in this study; age ranged from 18 to 42 years. The most common presentation was headache (83.3 %), with blurring of vision (50 %), unsteadiness (16.6 %), and seizures (8.3 %).

Tumor location was confined to the body and/or showed extension to the frontal or occipital horns or third ventricle. Tumor locations are shown in Table 1 and Fig. 1.

The transcortical approach was used for ten patients (83.3 %) and the transcallosal for two patients (16.6 %); two patients died early after total and subtotal removal, from sepsis and thalamic infarction, respectively, and they were excluded from the follow-up study. Six patients had a total removal (60 %), while subtotal resection was done for four patients (40 %).

Pathological confirmation aided by immunohistochemistry was diagnostic for central neurocytoma in all patients (100 %). An MIB LI of >2 % was detected in three patients (30 %; two total resections and one subtotal).

In the subtotal group adjuvant radiotherapy was given for three patients (30 %) early postoperatively, while one patient with small residual tumor received stereotactic radiosurgery (SRS) using a Gamma Knife (GK).

A V/P shunt was inserted for three patients (30 %; two early postoperatively, with subtotal removal, and one with recurrence after total removal and SRS).

A small recurrence was detected in two patients in the total resection group, at 12 and 18 months, respectively, in whom a GK was used to control the disease. One patient (10 %) in the subtotal group showed progression that required reoperation at 18 months after progression of the residual tumor and failure of RT.

Table 1 Tumor location

Site	No. of patients	%
Right	2	16.6 %
Left	5	41.6 %
Biventricular	4	33.3 %
Extending to third ventricle	1	8.3 %

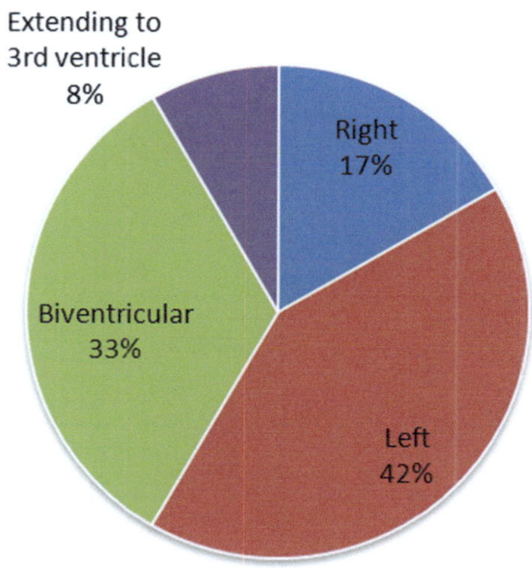

Fig. 1 Tumor location

At the time of the last follow-up, 24 months after the initial surgery, one patient had died (10 %) from hemorrhage during reoperation, while nine patients (90 %) had a favorable course. Local control was achieved in four patients (40 %; no evidence of disease), while five patients (50 %, with stable disease) were controlled with GK and RT, and three patients (30 %) had a V/P shunt. The rate of local control with total removal was (40 %), while the GK and RT control rate was 50 %; (Table 2).

Case Illustration

Case (1) (Fig. 2)

Case (2) (Fig. 3)

Case (3) (Fig. 4)

Discussion

There is a lack of studies reporting on the outcome of treatment control of central neurocytomas. The management of neurocytomas has been guided by retrospective case reports, institutional case series, and meta-analysis of institutional experiences [6, 14, 25]. Surgery is the primary modality of initial intervention, with watchful waiting (surveillance) not documented as a common primary option [1, 19]. Despite the indolent nature of these tumors, most patients are symptomatic at presentation, with increased intracranial pressure due to mass effect or hydrocephalus, and thus they require intervention rather than surveillance. Clinically, the tumor causes signs of increased intracranial pressure, visual and mental disturbances and occasionally, pyramidal or endocrine signs and symptoms. Rarely, neurocytomas may be discovered incidentally with no clinical symptoms.

Table 2 Summary of results and outcome

		Total	(75 % to 80 % removal)
No. of patients		6 (60 %)	4 (40 %)
MIB labeling index >2		2 (20 %)	1 (10 %)
Residual		0	4 (40 %)
Recurrence		2 (20 %) (12 and 18 months)	1 (10 %)
Radiotherapy (RT)		0	3 (30 %)
Gamma Knife (GK)		0	1 (10 %)
Local control (24 months)	Free	4 (40 %)	0
	GK	2 (100 %)	1 (100 %)
	RT	0	2 (65 %)
Reoperation	0		1 (10 %)
Failure of control	0		1 (10 %)
Ventriculoperitoneal (VP) shunt	1 (10 %)		2 (20 %)

In our study, the tumor was located predominantly in the left ventricle, in 41.6 % of cases, while it was biventricular in 33.3 % of cases. All the tumors were in the body of the ventricles attached to the septum, while Shin et al. [22] reported that 50 % of central neurocytomas had a typical location in the lateral ventricle around the foramen of Monro and 15 % had biventricular location.

The majority of neurocytomas are benign. In our study 30 % of cases showed an MIB LI of >2 % and were considered to be of an aggressive nature.

Brat et al. [2] and Soylemezoglu et al. [23] found that approximately 25 % of these rare CNS tumors were more aggressive, with an MIB-1 LI of >2 % or atypical histological features.

In a study of 14 cases by Mackenzie et al. [13], clinical follow-up was available for all 14 patients. The proliferation potential of central neurocytoma was found to be a useful predictor of clinical outcome, whereas histological atypia alone was not prognostically significant.

Therapeutic options in the treatment of central neurocytomas are complete resection (CR), complete resection plus RT, incomplete resection (IR), and IR plus RT and SRS (with a GK). Several studies suggest that CR leads to significantly better local control and survival than IR. After IR, patients benefit from postoperative RT and chemotherapy.

Rades et al. [15] reported that 25 % of these CNS tumors were more aggressive, with an MIB LI of >2 % or atypical histological features, with their local control rate being 70 % at 3 years after CR.

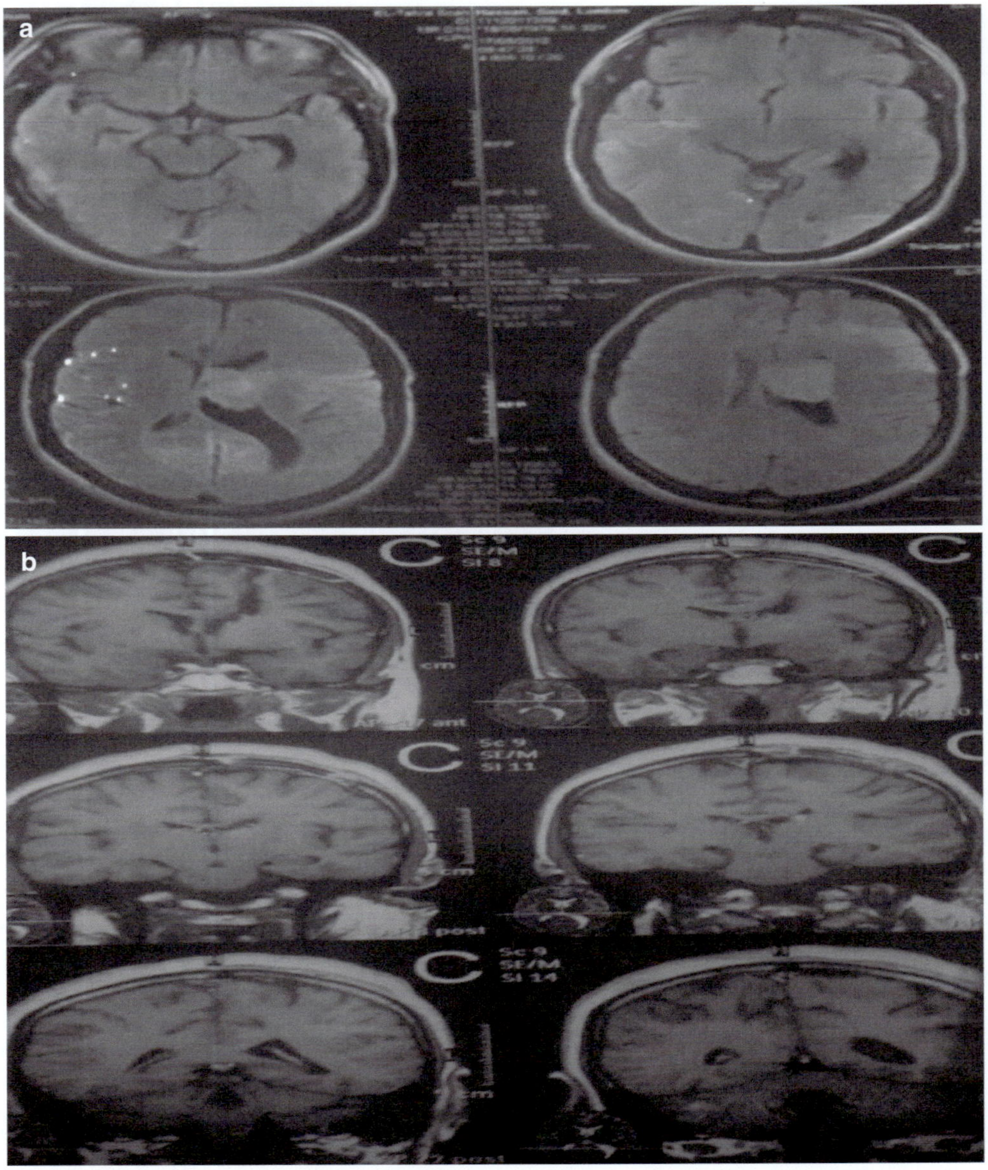

Fig. 2 (**a**) Magnetic resonance imaging (MRI) showing left ventricular central neurocytoma (CN). (**b**) MRI showing total resection of the tumor during the follow-up period

In our study, total resection was achieved in six patients (60 %) and in four patients (40 %) subtotal resection was performed. After 2 years of follow-up, four of the six patients with total resection (66.6 %) were locally controlled without adjuvant therapy, while two patients (33.3 %) with MIB LI >2 % showed control of recurrence after GK treatment.

In our study, it was also found that, for patients with incomplete resection (subtotal), after 2-year follow-up, RT was effective in two cases (66 %) to control progression, while RT failed in one patient and GK treatment was effective in one patient (100 %); this result may agree with and be comparable with results in several studies.

Rades and colleagues [16, 18] found that local control was significantly better after CR, CR + RT, and IR + RT than after IR and that overall survival rates were 99.2 % and 86.1 % after total and subtotal removal, respectively, after 12-month follow-up.

We also found that RT was sufficient after incomplete resection (IR), with local control being 83 %.

The largest single institutional experience, from the Mayo Clinic, as reported by Leenstra et al. [13], showed that 35 % of their patients received adjuvant radiation as part of their initial management, but that one-third of their patients recurred. For consideration of adjuvant therapy,

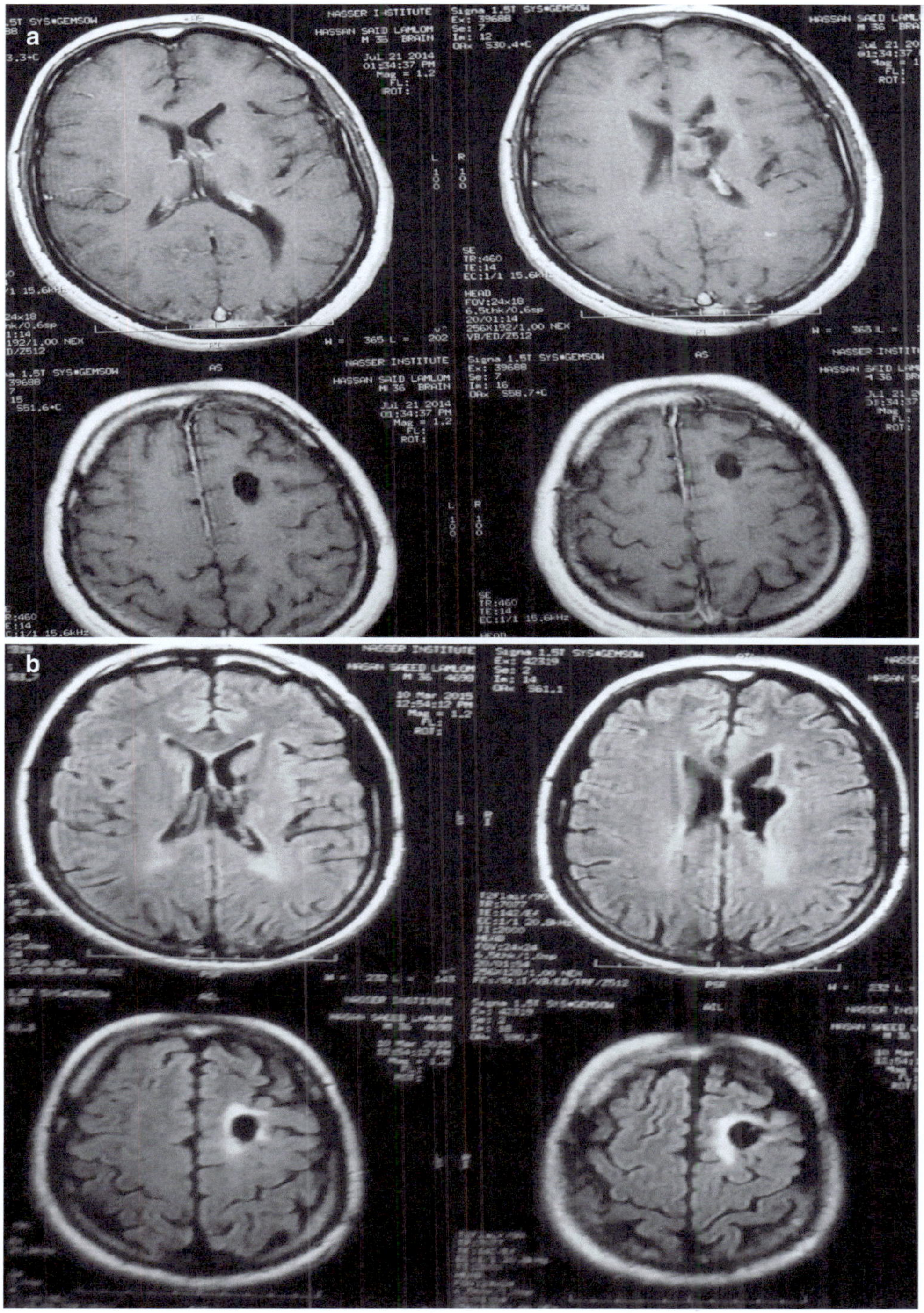

Fig. 3 (**a**) Recurrence at 18 months after total resection; (**b**) local control with Gamma Knife radiosurgery and its effect after 6 months

they proposed patient selection on the basis of atypical neurocytomas. Our results support the concept that atypical neurocytomas are at higher risk of recurrence.

In the present study, SRS was very effective, controlling 100 % of incompletely resected cases and 100 % of recurrent cases after CR.

Rades et al. [18] commented on the value of postoperative SRS and RT for IR of typical neurocytomas and found that local control was significantly better with SRS after IR+RT (87 %) and after IR+SRS (100 %) after 5-year follow-up].

Rades et al. [18] found that in very indolent tumors, SRS may produce significant local control even though the tumor

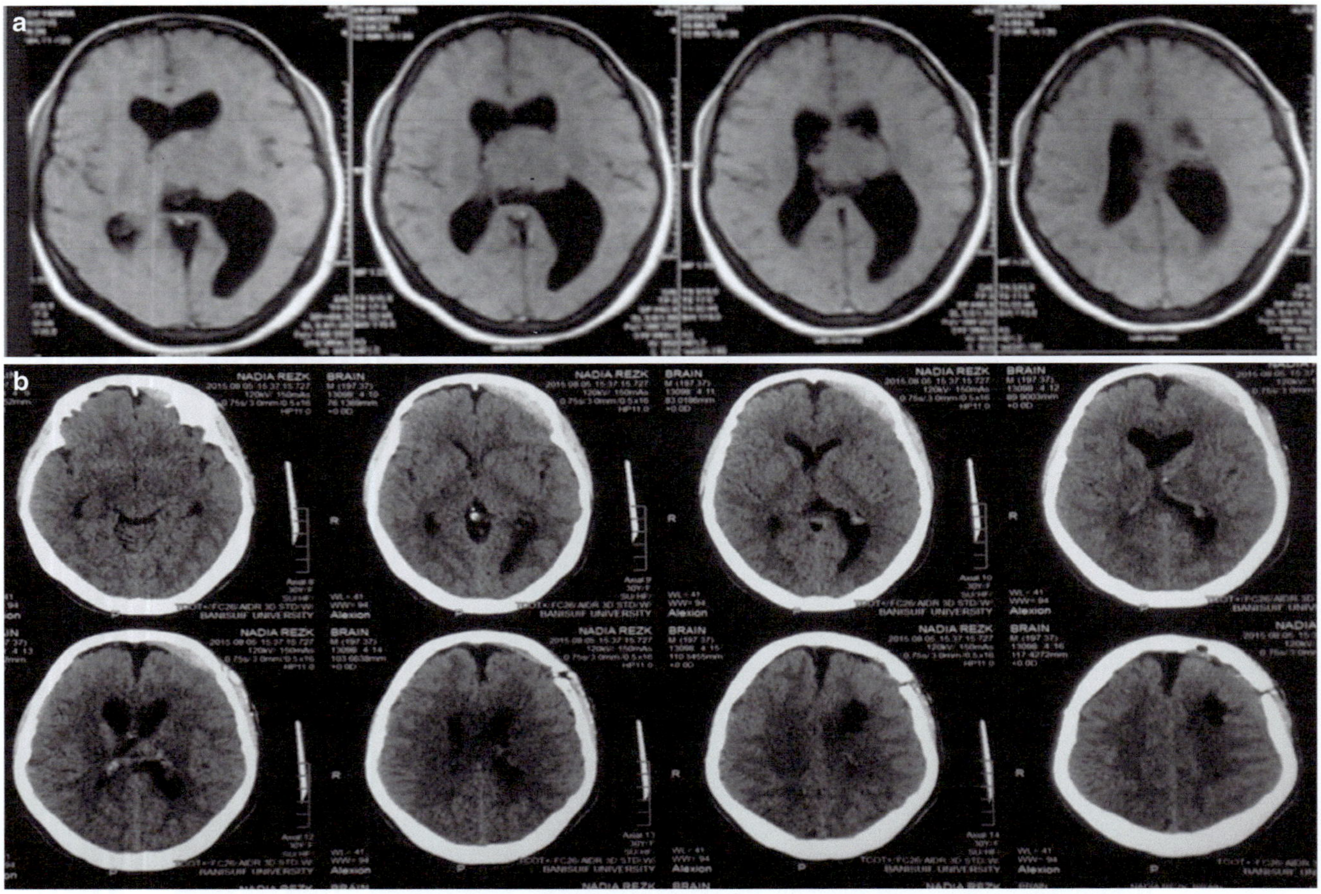

Fig. 4 (**a**) MRI showing huge CN and hydrocephalus, (**b**) Computed tomography (CT) postoperatively, showing total removal of the tumor

may eventually progress. With regard to the adjuvant treatment debate, a 2002 analysis of published institutional reports of 504 patients from 91 centers concluded that adjuvant radiation was beneficial after IR. The debate arises given that the extent of resection influences local control only, but not survival, and the impact of symptomatic recurrences is not well established in the literature.

Kim and colleagues [9, 10] support the concept of GK SRS being useful as primary or secondary postoperative therapy for the treatment of central neurocytomas.

In the present study, three patients (30 %) required shunting; two patients (20 %) required V/P shunt in the early postoperative period after the EVD was removed, while in the third patient a shunt was inserted, due to a small recurrence near the foramen, obstructing the cerebrospinal fluid (CSF), prior to GK treatment. We do not yet have much data about the rate of shunting in central neurocytoma, but the pathological features of this tumor, because it is located in the body of the ventricle, mean the foramen has a plane of cleavage from the ependyma, while the tumor is soft, extractable, moderately vascular, and attached to the septum; these features allow septostomy during surgery. All the above factors may be involved in the low incidence of hydrocephalus developing postoperatively, and may be associated with the ease of draining the ventricles at the level of the foramen, with little blood loss.

In our study, the survival rate after 2 years was 90 %,with an overall favorable prognosis, and local control rates were comparable to those of Leenstra et al. [13] and Rades et al. [15–18].

Conclusion

Complete resection is much more effective for the treatment of central neurocytomas than incomplete resection; after incomplete resection postoperative adjuvant RT and radiosurgery significantly improve local control.

Conflict of Interest Statement We declare that we have no conflict of interest.

References

1. Bertalanffy A, Roessler K, Dietrich W et al (2001) Gamma knife radiosurgery of recurrent central neurocytomas: a preliminary report. J Neurol Neurosurg Psychiatry 70:489–493
2. Brat DJ, Scheithauer BW, Eberhart CG, Burger PC (2001) Extraventricular neurocytomas: pathologic features and clinical outcome. Am J Surg Pathol 25:1252–1260. [PubMed]
3. Brown DM, Kariovits S, Lee LH, Kim K, Rothfus WE, Brown HG (2001) Management of neurocytomas: case report and review of the literature. Am J Clin Oncol 24:272–278
4. Chen CM, Chen KH, Jung SM et al (2008) Central neurocytoma: 9 case series and review. Surg Neurol 70:204–209
5. Shravan Kumar C, Sharma DN, Sharma K, Haresh KP, Rath GK (2010) Youngest case of third ventricular anaplastic neurocytoma. Indian J Med Paediatr Oncol 31(2):69–71
6. Dirk R, Fabian F, Steven E (2004) Schild, original article treatment of atypical neurocytomas. Cancer 100:814–817
7. Hassoun J, Gambarelli D, Grisoli F et al (1982) Central neurocytoma. An electron-microscopic study of two cases. Acta Neuropathol 56:151–156
8. Hassoun J, Soylemezoglu F, Gambarelli D et al (1993) Central neurocytoma: a synopsis of clinical and histological features. Brain Pathol 3:297–306
9. Kim CY (2007) Gamma-knife radiosurgery for central neurocytoma: primary and secondary treatment. Cancer 110(10):2276–2284
10. Kim DG, Chi JG, Park SH et al (1992) Intraventricular neurocytoma: clinicopathological analysis of seven cases. J Neurosurg 76:759–765
11. Kim DG, Kim JS, Chi JG et al (1996) Central neurocytoma: proliferative potential and biological behavior. J Neurosurg 84:742–747
12. Leenstra JL, Rodriguez FJ, Frechette CM et al (2007) Central neurocytoma: management recommendations based on a 35-year experience. Int J Radiat Oncol Biol Phys 67:1145–1154
13. Mackenzie IR (1999) Central neurocytoma: histologic atypia, proliferation potential, and clinical outcome. Cancer 85:1606–1610
14. Maiuri F, Spaziante R, De Caro ML et al (1995) Central neurocytoma: clinico-pathological study of 5 cases and review of the literature. Clin Neurol Neurosurg 97:219–228
15. Rades D, Fehlauer F, Schild S, Lamszus K, Alberti W (2003) Treatment for central neurocytoma: a meta-analysis based on the data of 358 patients. Strahlenther Onkol 179(4):213–218
16. Rades D, Fehlauer F, Schild SE (2004) Treatment of atypical neurocytomas. Cancer 100:814–817
17. Rades D, Schild SE, Ikezaki K et al (2003) Defining the optimal dose of radiation after incomplete resection of central neurocytomas. Int J Radiat Oncol Biol Phys 55:373–377
18. Rades D, Schild SE (2006) Treatment recommendations for the various subgroups of neurocytomas. J Neurooncol 77(3):305–309. Epub 2006 Mar 31
19. Schild SE, Scheithauer BW, Haddock MG et al (1997) Central neurocytomas. Cancer 79:790–795
20. Sharma MC, Deb P, Sharma S, Sarkar C (2006) Neurocytoma: a comprehensive review. Neurosurg Rev 29(4):270–285; discussion 285. Epub 2006 Aug 29
21. Sharma MC, Rathore A, Karak AK et al (1998) A study of proliferative markers in central neurocytoma. Pathology 30:355–359
22. Shin JH, Lee HK, Khang SK et al (2002) Neuronal tumors of the central nervous system: radiologic findings and pathologic correlation. Radiographics 22(6):1473–1505
23. Söylemezoglu F, Scheithauer BW, Esteve J, Kleihues P (1997) Atypical central neurocytoma. J Neuropathol Exp Neurol 56:551–556. [PubMed]
24. Yang GF, Wu SY, Zhang LJ et al (2009) Imaging findings of extraventricular neurocytoma: report of 3 cases and review of the literature. AJNR: Am J Neuroradiol 30:581–585. see comment
25. Yasargil MG, von Ammon K, von Deimling A et al (1992) Central neurocytoma: histopathological variants and therapeutic approaches. J Neurosurg 76:32–37

Complications in Craniovertebral Junction Instrumentation: Hardware Removal Can Be Associated with Long-Lasting Stability. Personal Experience

Massimiliano Visocchi, Pier Paolo Mattogno, Francesco Signorelli, Jun Zhong, Gerardo Iacopino, and Giuseppe Barbagallo

Abstract

Background

The causes of craniovertebral junction (CVJ) instabilities include trauma, rheumatological diseases, tumors, infections, congenital malformations, and degenerative disease processes; these complex pathologies often require CVJ instrumentation. Hardware complications were analyzed in a personal series of 48 treated patients. In light of the analysis of very unusual radiological and clinical findings, the authors tried to better investigate the related mechanisms and to reach possible useful conclusions.

Methods

In a series of 48 patients who underwent CVJ instrumentation and fusion procedures in our Institution, we describe three cases of hardware failure, due to: (1) infection; (2) radio- and chemotherapy; and (3) incorrect surgical procedure.

Results

1. A stable bone CVJ fusion can occur after instrumentation removal for infection, since this removal can enhance bone fusion mechanisms;
2. Radio- and chemotherapy can cause hardware failure due to interference with local bone metabolism;
3. Although old-fashioned, wiring techniques still deserve consideration, mostly in CVJ re-do surgery after screwing technique failures; nevertheless, although the procedure is simple, safe, and effective, care must be taken in the preparation of the cranial holes in order to avoid sliding complications of the U-shaped rods.

Conclusions

CVJ instrumentations provide reasonably good mechanical stabilization with a high rate of bony fusion. Complications, such as dislocation or rupture of the fixation system, screw loosening, dural fistula, neural or vascular damage, and wound infection, are relatively infrequent. Knowledge and prevention of these complications is fundamental to improve surgical results and outcomes.

Keywords Craniovertebral junction • Occipito-cervical fusion • Wiring • Screwing • Bone infections

M. Visocchi • P.P. Mattogno (✉) • F. Signorelli
Institute of Neurosurgery, Medical School, Catholic University of Rome, Rome, Italy
e-mail: pierpaolo.mattogno@gmail.com

J. Zhong
Department of Neurosurgery, XinHua Hospital, Shanghai JiaoTong University School of Medicine, Shanghai 200092, China

G. Iacopino
Department of Experimental Biomedicine and Clinical Neurosciences, School of Medicine, Neurosurgical Clinic, University of Palermo, Palermo, Italy

G. Barbagallo
Division of Neurosurgery, Department of Neurosciences, Policlinico "G. Rodolico" University Hospital, Catania, Italy

Introduction

Over the past 40 years, various methods and instrumentation types have been developed for craniovertebral junction (CVJ) fixation in the management of several diseases. The indications for CVJ fixation are inflammatory diseases (of which rheumatoid arthritis is the most common etiology), congenital abnormalities, tumors, trauma, and other nonspecific indications. Complications directly related to surgical procedures are mainly neural injuries, vascular injuries, and hardware malposition; on the other hand, the main complications indirectly related to the surgical procedures are infective [13, 28]. In a series of 48 patients who underwent CVJ instrumentation and fusion procedures in our Institution, we report three cases

M. Visocchi et al. (eds.), *Trends in Reconstructive Neurosurgery*, Acta Neurochirurgica Supplement 124,
DOI 10.1007/978-3-319-39546-3_29, © Springer International Publishing Switzerland 2017

of hardware failure due to different causes. In light of this analysis, we try to highlight the related mechanisms and to show some unusual radiological and clinical findings.

Materials and Methods

All 48 patients had been referred to the senior author (M.V.) between 2000 and 2013 and operated for different CVJ diseases. Three of the patients underwent subsequent evaluation and surgical management of their clinical and radiological complications. Patient records were reviewed for demographics, clinical presentation, neuroimaging studies, and operative case notes from the primary surgery to the final clinical outcome. Informed patient consent was obtained for each patient.

Case Presentation 1

An 18-year-old male patient with a medical history of neck pain and rhinolalia, since 2009, underwent a magnetic resonance (MR) examination that demonstrated platybasia, cerebellar tonsil herniation into the foramen magnum, and upper displacement of the odontoid process, along with compression of the medulla oblongata (Fig. 1a, b). In May 2011, the patient underwent surgical decompression by the transoral approach, along with external CVJ stabilization with the Halo-Vest System, since he refused a one-stage anterior and posterior approach (Fig. 1c). In July 2011, after removal of the Halo Vest System (De Puy Acromed, Leeds, England) and under intraoperative Mayfield fixation, C0-C2-C3 mass screw (Vertex System; Medtronic, Minneapolis, USA) instrumentation, along with synthetic bone graft substitute fusion (beta-tricalcium phosphate [ß-TCP]; VITOSS® Synthetic Cancellous Bone Void Filler, Stryker Kalamazoo, MI USA), was performed. Two weeks later, due to infective dehiscence of the surgical wound, the hardware was revised, and the synthetic bone graft substitute was removed, along with local surgical debridement and short-lasting (72 h) external draining. The patient was discharged with broad-spectrum polyantibiotic therapy (daptomycin 350 mg/day IV, rifampicin 600 mg/day per os) to be continued for 2 months (no bacterial identification). Polyantibiotic therapy was started again in January 2013 due to evidence of recurrent wound infection by *Staphylococcus aureus*. In May 2013, soon after the appearance of a painful subcutaneous tumescence at the upper cranial surgical wound, the patient

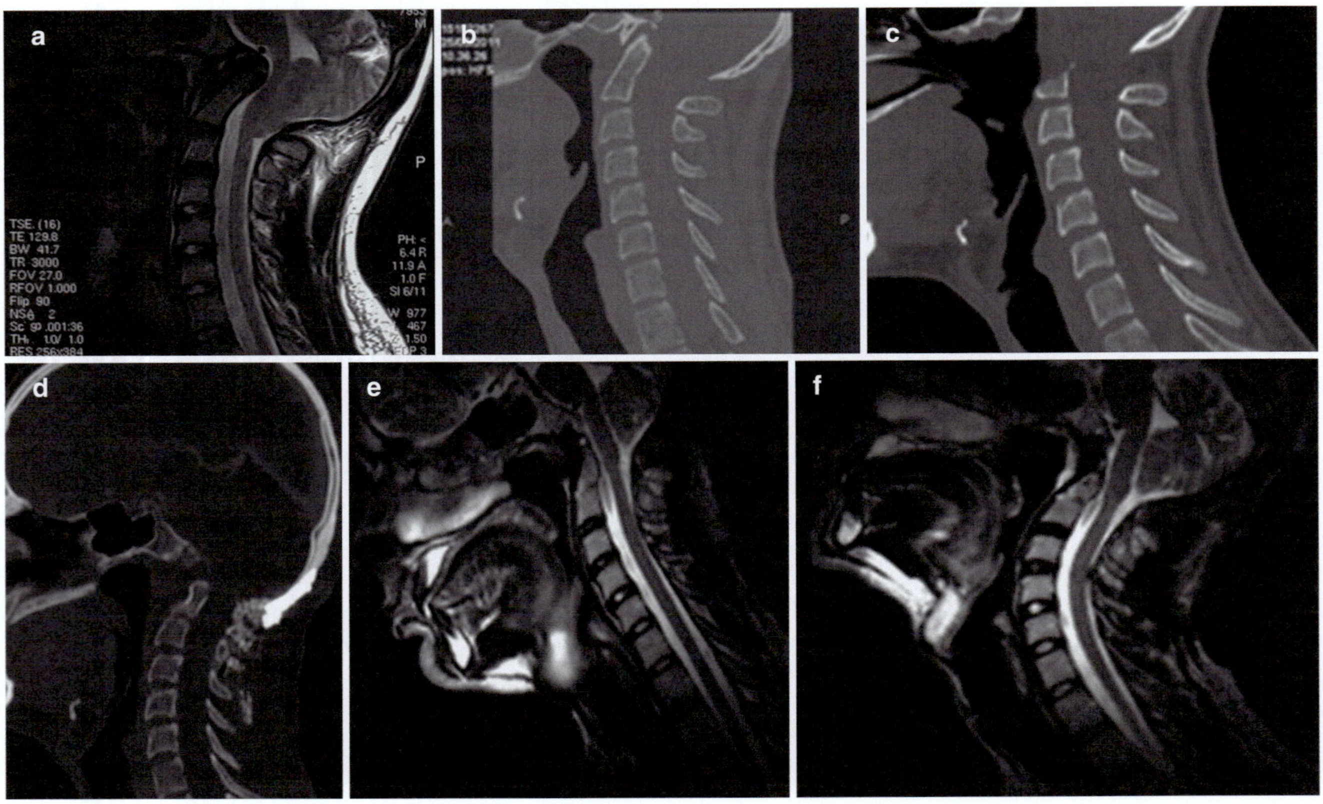

Fig. 1 Sagittal T2-weighted magnetic resonance (MR) image (**a**) and computed tomography (CT) scan sagittal reconstruction; (**b**) demonstrates platybasia, cerebellar tonsil herniation into the foramen magnum, and upper displacement of the odontoid process, along with compression of the medulla oblongata. Postoperative CT scan sagittal reconstruction (**c**) shows craniovertebral junction (CVJ) decompression by means of odontoidectomy performed with the transoral approach. CT scan sagittal reconstruction (**d**) shows complete bone fusion and secondary stabilization of C0-C1-C2-C3. *Dynamic sagittal T2-weighted MR images* (**e–f**) *confirm the CVJ stability*

underwent new CT and MR scans of the CVJ, which showed the presence of a subcutaneous fluid collection spanning from the skull base down to the C5 body. Under polyantibiotic therapy (teicoplanin 400 mg/day, linezolide 1200 mg/day) the patient underwent a second surgical evacuation. In March 2014, a painful subcutaneous tumescence at the surgical wound appeared again. Since a CT of the CVJ demonstrated a complete bone fusion and a secondary stabilization of C0-C1-C2-C3 (Fig. 1d), the removal of the instrumentation system was indicated, and this was performed without complications 33 months after the onset of the infection. The patient was supported with a rigid cervical collar for 3 months. Antibiotic therapy (teicoplanin 400 mg/day, linezolide 1200 mg/day, ciprofloxacin 1000 mg/day) was then tailored, according to new evidence of *Enterobacter cloacae* isolated in the latter surgical specimens, and discontinued 2 months later. One month after the collar removal, a dynamic cervical MR examination confirmed CVJ stability (Fig. 1e–f).

Case Presentation 2

A 52-year-old man, with a few months' history of cervical pain, underwent CT and MR scans of the CVJ with contrast medium, showing an osteolytic process involving the axis, with complete infiltration and ballooning of C2 extending towards the anterior arch of C1 and the inferior edge of the clivus (Fig. 2a–b). Possible multiple localizations of the neoplastic disease were ruled out by a total-body CT scan and Technetium-99 m sestamibi bone scintigraphy. Neither monoclonal gammopathy nor Bence-Jones proteinuria was found. Radiological findings were suggestive of a high risk of axis fracture with potential CVJ instability. Biopsy of the lesion was performed and extemporary histopathological examination was consistent with plasmacytoma. A CVJ instrumentation and fusion procedure was performed via C0-C3-C4-C5 with lateral mass screws (SUMMIT™ SI OCT System; Codman Johnson & Johnson, Leeds, England) and rods; also in this patient a synthetic bone graft substitute (ß-TCP; VITOSS® Synthetic Cancellous Bone Void Filler) was locally applied to achieve fusion. A post-operative CT scan documented the correct placement of the device, in the absence of CVJ dislocation (Fig. 2c). After 1 month, the patient underwent radiotherapy (RT) consisting of the administration of 200 centigray (cGy) for 20 consecutive days (6 MV-photons) to C1–C2–C3 vertebrae (total dose on the tumor core 4,000 cGy). Prednisone administration was started (25 mg daily for the first week; 12.5 mg daily for the second week; 12.5 mg on alternate days for 3 weeks). Six months after the RT, the patient complained of left cervical pain and local paresthesia. Cervical CT scan showed partial bone resorption with the dislocation of screws and rods, along with atlantoaxial rotatory subluxation (Fig. 2d). The dislocated screws and rods were removed and Songer's titanium sublaminar wires (SUMMIT™ SI OCT System; Codman Johnson & Johnson USA) were placed at the C3–C4–C5 levels and connected with an occipital plate with rods. Two cross-link bars were also placed, at the C0 and C5 levels, in order to reduce the risk of rotatory subluxation. Autologous bone was harvested from the posterior left iliac crest, cut in a double-wing shape, and fixed to the CVJ using a silk suture, synthetic bone dust, and human fibrin glue (Tissucol; Baxter, West Lake Village, CA, USA). A Halo-Vest System was then applied. No signs of local infection were found. The patient regained walking ability quickly. Postoperative CT scan confirmed the correct placement of the device and the restoration of vertebral alignment. Oral bisphosphonate (BP) therapy with zoledronic acid was instituted (Zometa®; Novartis Pharmaceuticals) at a dose of 4 mg IV every 4 weeks. The Halo-Vest was maintained for 3 months. A 30-month follow-up CT scan and dynamic X-ray study documented correct vertebral alignment and fusion due to successful instrumented surgery, with no signs of CVJ instability. A cervical MR scan showed tumor remission, and a new bone marrow needle biopsy excluded metastatic diffusion.

Case Presentation 3

A 54-year-old patient presented with a medical history of neck pain with lower limb weakness, ataxia, and paresthesia in four limbs, of several months' duration. A CVJ MRI examination demonstrated os odontoideum instability with compression of the medulla oblongata. In May 2000, CVJ instrumentation with C0 suboccipital and C2-C3 sublaminar wires (Songer cables) and U-shaped rods was performed, along with synthetic bone graft substitute fusion (ß-TCP; VITOSS® Synthetic Cancellous Bone Void Filler). The patient was advised to wear a hard collar. Some days after the surgery patient experienced acute neck pain associated with the fixed neck position in hyperextension. Neck X-ray examination documented a displacement of the occipital wires, sliding caudally on the U-shaped rods, which had resulted in deformation—hyperlordosis (Fig. 3a). The patient underwent a second surgical procedure to correct the dislocation; this was done by performing a new operation with a different U-shaped construct (Fig. 3b–d).

Discussion

The causes of CVJ instability include CVJ trauma, rheumatological diseases, tumors, infections, congenital malformations, and degenerative disease processes. Considering

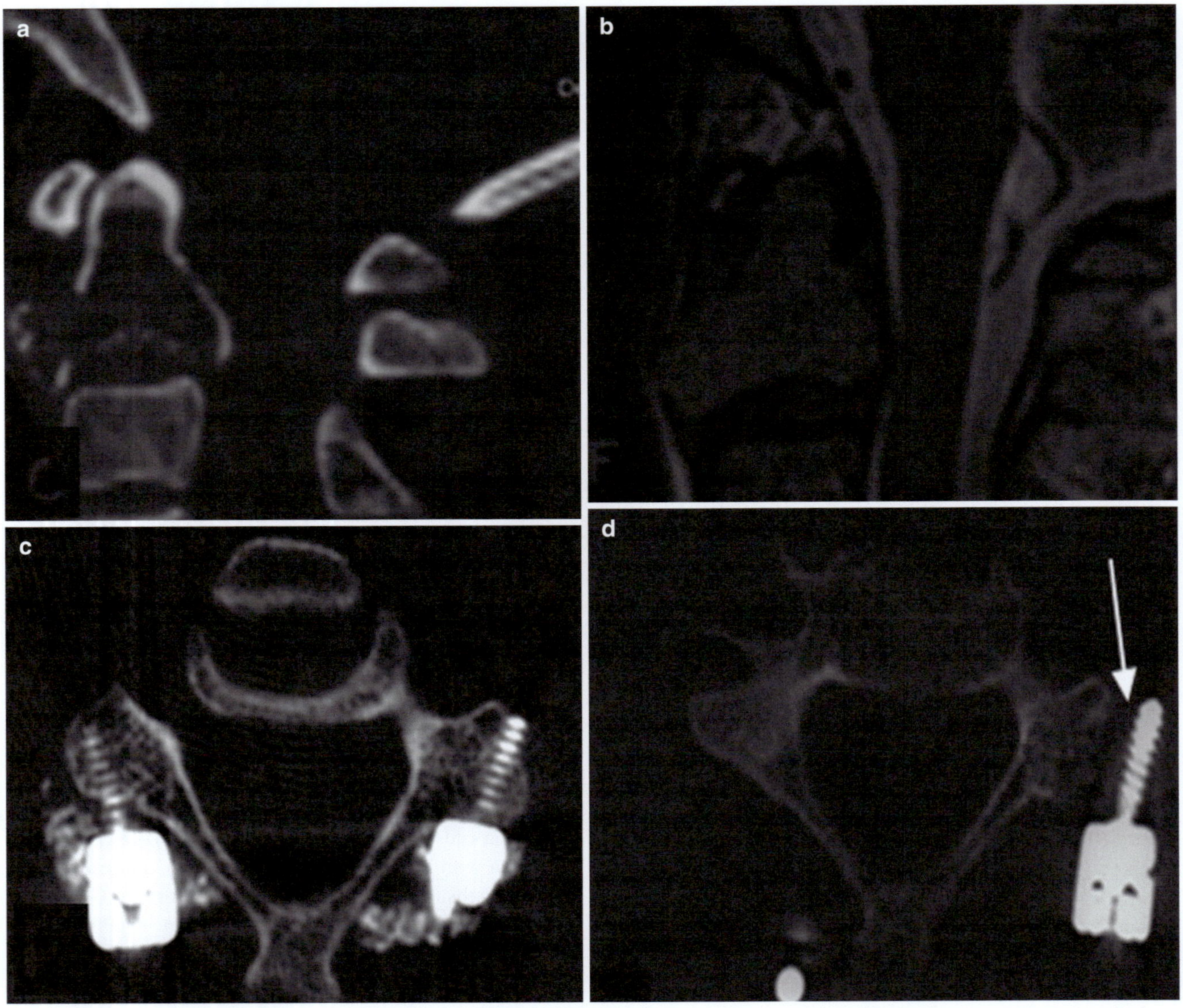

Fig. 2 Sagittal CT scan (**a**) and MR studies (**b**) show the advanced osteolytic process exclusively involving the axis at the level of the soma and the dens, especially the trabecular spongiosa region, evidencing the distorted profile of the axis without spinal cord compression. Specific axial CT scan study of the C5 vertebra after the first intervention demonstrates the right position of the screws in the lateral masses; absence of osteolytic areas around the intraosseous screw course is also evident (**c**). Differently, axial CT scan after radiotherapy (RT) shows signs of bone resorption and evidence of left screw dislocation on the C5 vertebra (*white arrow*) (**d**)

dysmorphic pathologies, basilar invagination is the most common bony abnormality (38–74 %) of the CVJ, followed by platybasia, often associated with other findings such as Chiari malformation, syringomyelia, and os odontoideum [23, 30]. Instability of the CVJ may lead to significant pathological problems, including cervical pain at first, followed by all the consequences of cervical cord compression: respiratory distress, cranial nerve dysfunction, paresis, and plegia, or even sudden death. Fusion procedures at the CVJ must be capable of withstanding the forces of compression, axial loading, flexion, extension, lateral rotation, and lateral bending. A variety of techniques are used for CVJ instrumentation and fusion: rigid rod-screw fixation, rod-wire systems, occipital hooks, and cervical claws. These techniques have shown comparable effectiveness (fusion rates from 89 to 100 %). Despite possible alternative less invasive surgical procedures sparing occipital bone fusion, CVJ instrumentation and fusion remains the most appropriate treatment when dealing with instability or in cases of widespread bone destruction, fractures, or progressive inflammatory or metabolic diseases [6, 8, 17, 20, 21]. The choice of surgical procedure is mainly based on radiological parameters, such as occiput bone density or the firmness of posterior cervical elements, but also considering the surgeon's experience. Biomechanical in-vitro experiments have demonstrated for many years that screws provide better immediate

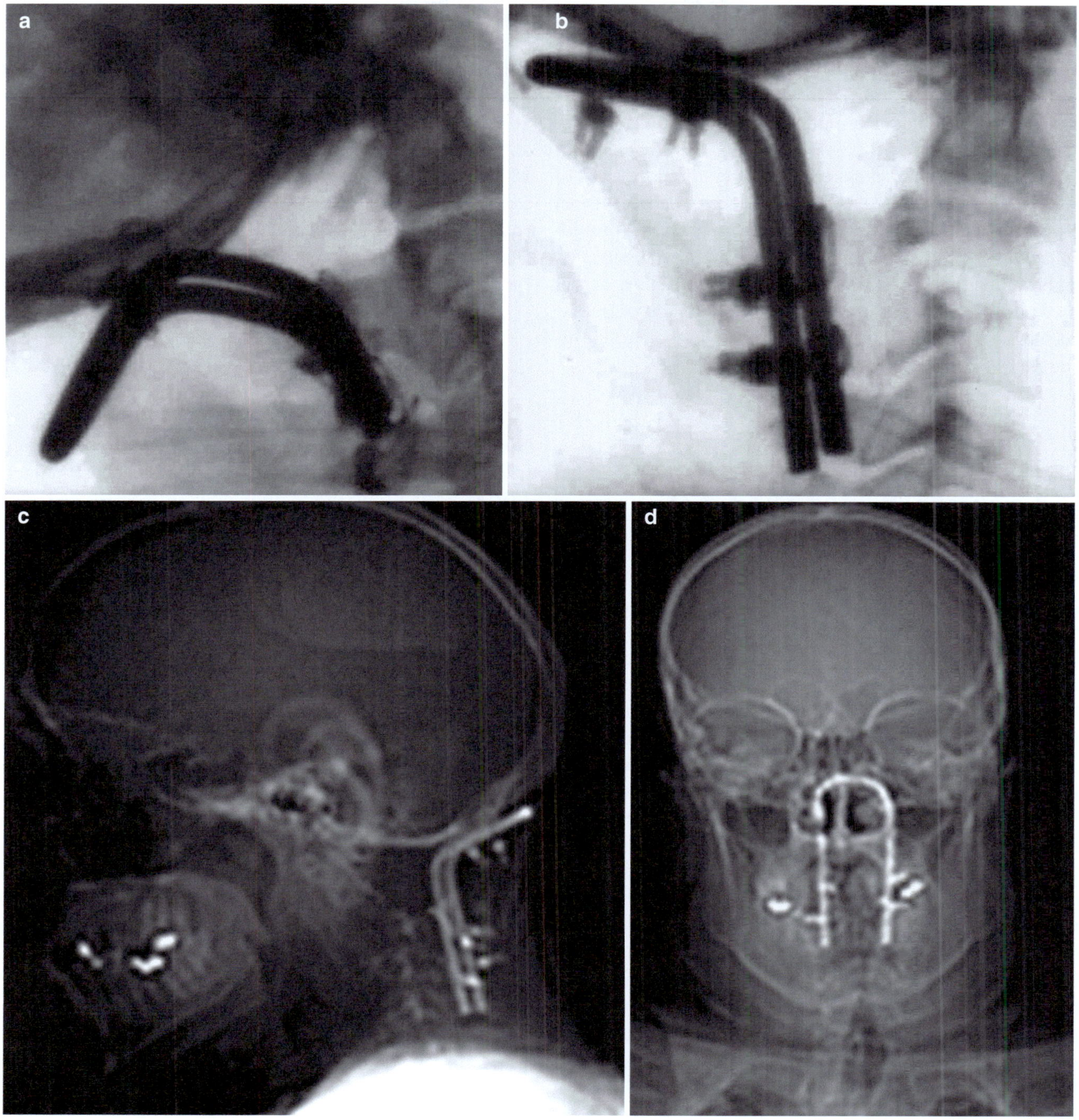

Fig. 3 Early postoperative lateral X-ray demonstrates the sliding of the titanium construct associated with the pulling out of the hardware (**a**). Postoperative X-rays (CT scout view) (**b–d**) show control after reoperation

stability than wires, but it is still debated whether this results in a true higher rate of fusion. Reports of a fusion rate of 80 % after wiring and of 94 % after screwing do not seem to influence markedly the final clinical results [9, 18, 19, 24]. Wiring constructs remain an excellent method for stabilizing the CVJ and upper cervical spine. When wired to the spine and skull, bone struts or metal implants provide reasonably good mechanical stabilization properties with a low incidence of complications [2]. Ductility, resistance to stress, and the possibility of using MR postoperatively convey advantages in using titanium, compared with other metals, in osteosynthesis. The most frequent complications related to CVJ instrumentation and fusion include dislocation and rupture of the fixation system, screw loosening, dural fistula, neural or vascular damage, wound infection, and the persistence of neurological pain [1, 13]. In more detail, the most commonly encountered perioperative complications have been related to instrumentation failure after nonunion, with rates as high as 7 % during CVJ instrumentation and fusion and 6.7 % during atlantoaxial fusion (quite the same!). Other commonly encountered complications have included injury to the vertebral artery (1.3–4.1 % during the placement of C1-C2 transarticular screws, most commonly in the case of high-riding vertebral artery [13]. In our experience, we can affirm that CVJ instrumentation procedures can be considered efficient and substantially safe; in the 48 patients who underwent CVJ posterior instrumented procedures at our Institution, the three cases of instrumentation failure, are discussed and compared with related results in the current literature as outlined below.

Case Considerations

Case 1

Postoperative infection of a stabilization system is considered a serious complication, with an incidence rate of 0.1–3 %, and it mostly occurs 3–7 days after surgery, with a higher incidence via the transoral approach [10]. Pediatric patients with postoperative wound infections requiring surgical debridement have higher surgical failure rates after CVJ instrumentation and fusion. Those with skeletal dysplasia and congenital spinal anomalies are more likely to require reoperation for hardware failure, as confirmed in our case [15]. Moreover, in a series of pediatric patients, the efficacy of surgical wash-out associated with antibiotics was demonstrated; this treatment was successful in curing wound infections in many patients without hardware removal [11, 13]. Bathia et al., in a total of 100 patients, reported three cases of infections treated with wound washout, and in one case only, hardware removal was performed (3 %), without any subsequent radiological confirmation of long-lasting stability. Choi et al. also reported one case of hardware removal with referred CVJ bone fusion, but no radiological evidence of the fusion was shown in their report [5, 7]. Similar conclusions were reached by Ahmed et al., who presented the case of a 20-year-old patient who developed postoperative infection after CVJ instrumentation and fusion in Chiari I malformation. After removal of the instrumentation, solid bony fusion was evidenced, but it was not confirmed by later dynamic cervical radiographs [1]. As far as we know, our case is the *first* documented condition of hardware removal followed by stability due to bone fusion *confirmed* by postoperative dynamic neuroradiological investigations. These data support our finding: we can hypothesize that the infection, "long-lasting" as in our patient, had a role in the ossification process involved in the CVJ fusion, since the fusion occurred 33 months after onset of the infection. Finally, and very surprisingly, in our case, the post-infective bone fusion not only produced a good fixation but also resulted in *a sort of odontoid regeneration*, never reported before. Although we recently reported a "true" odontoid process regeneration (along with clival regeneration and Chiari malformation recurrence) after transoral decompression, in the present case we observed the union of the remaining C2 with C3 bodies, strongly mimicking a concomitant quite complete axis and clival regeneration [26].

Case 2

Spinal plasmocytoma is a rare neoplasm (representing <5 % of primary tumors of the spine) that is characterized by high response to radiation therapy: effective local control of the disease, achieving remission in more than 86 % of cases after 5 months, along with regression of neurological impairment and pain, is described in some non-comparative retrospective studies [12]. Local recurrence is reported in 3–26 % of patients within 5 years; no recurrence has been documented in small series of patients treated with more than 3,500–4,000 cGy [12]. In our case, considering the massive involvement of the C2 vertebra, the anterior arch of C1, and the inferior edge of the clivus, fixation and fusion surgery was indicated to prevent potential CVJ instability due to the evolving nature of the invasive plasmocytoma. CVJ instrumentation via C3–C4–C5 lateral mass screws connected to an occipital plate by rods with cement was initially considered the best treatment in respect of semirigid fusions ("rods and wires", "rib grafts"), with evidence at the first radiological follow-up of good surgical results. The influence of RT on bony metabolism is still a matter of debate. Some authors have hypothesized that RT could reduce the risk of pathologic fractures, while others have observed that long-term remodelling of bone after RT does not re-establish the original bony architecture [14, 16]. It is known that skeletal RT may alter and damage osteoblastic cells, causing uncontrolled resorption of the bone matrix; the bone alterations produced by radiation have been defined as "radiation osteitis", "radiation osteonecrosis", or "osteoradionecrosis" [27].

Radiation seems to influence the growth pattern in children, and is involved in chondrogenesis failure in adults. Moreover, RT may decrease deposition of the bony matrix by damaging osteoblastic cells [4]. Quantifying the biological influence of RT in relation to the failure of instrumented fusion is difficult, as there is a lack of prospective comparative studies. The time of trabecular remodelling in humans is about 150 days; it is probable that RT may alter this biological process [29].

Case 3

The os odontoideum, or mobile odontoid apophysis, is a malformation of the CVJ caused by missing unity in the ossification center of the dens on the body of the axis. This malformation induces atlantoaxial instability and exposes the subject to the risk of bulbar-medullary compression. The true incidence of this condition is difficult to determine because many cases are asymptomatic. Congenital and traumatic theories have been advanced, although the pathogenesis is still being discussed. Consequently, the surgical management of os odontoideum should aim at achieving both neural decompression and stabilization of the CVJ [23]. Our patient with os odontoideum refused the anterior approach, so posterior instrumentation only was performed [3, 22, 25]. According to the literature, dislocation of the hardware is another important cause of the failure of CVJ stabilization; in our patient, failure was due to the sliding of the U-shaped rod.

In order to understand and analyze this event, it is important to consider that the two burr holes made into the occipital bone were placed 0.5 cm cranially to the rim of the foramen magnum, on the same axis of wire and hyperlordotic rod passage as the C2 and C3 laminar arches, thus not securing the hardware or preventing the caudo-cranial sliding of the hardware and the pulling out of the construct. In other words, the distance between the occipital burr holes was too great. The patient needed further surgery to improve occipital fixation, with further suboccipital wiring of the U-shaped rod performed by creating two new burr holes that were closer than the original ones.

Conclusions

On the basis of our experience we can conclude that:

1. A stable bone CVJ fusion can occur after instrumentation removal for infection;
2. Radio- and chemotherapy can cause hardware failure due to interference with local bone metabolism;
3. Although old-fashioned, wiring techniques still deserve consideration, mainly in re-do CVJ surgery after screwing technique failures. Nevertheless, although wiring is a simple, safe, and effective procedure, care must be taken in the preparation of the cranial holes in order to avoid sliding complications with the U-shaped rods.

Acknowledgments Nothing to declare.

Conflict of Interest Statement The authors declare that they have no conflicts of interest.

References

1. Ahmed R, Menezes AH (2013) Management of operative complications related to occipitocervical instrumentation. Neurosurgery 72(2 Suppl Operative):214–228; discussion ons228
2. Apostolides PJ, Sonntag WHK, Dickman CA (1998) Occipitocervical wiring techniques. In: Dickman CA et al (eds) Surgery of the craniovertebral junction. Thieme, New York, pp 795–808
3. Barbagallo GM, Certo F, Visocchi M, Palmucci S, Sciacca G, Albanese V (2013) Disappearance of degenerative, non-inflammatory, retro-odontoid pseudotumor following posterior C1-C2 fixation: case series and review of the literature. Eur Spine J 22(Suppl 6):S879–S888
4. Benvenuti S, Brandi ML (2000) Corticosteroid-induced osteoporosis: pathogenesis and prevention. Clin Exp Rheumatol 18(4 Suppl 20):S64–S66
5. Bhatia R, Desouza RM, Bull J, Casey AT (2013) Rigid occipitocervical fixation: indications, outcomes, and complications in the modern era. J Neurosurg Spine 18(4):333–339
6. Brooks AL, Jenkins EB (1978) Atlanto-axial arthrodesis by the wedge compression method. J Bone Joint Surg Am 60(3):279–284
7. Choi SH, Lee SG, Park CW, Kim WK, Yoo CJ, Son S (2013) Surgical outcomes and complications after occipito-cervical fusion using the screw-rod system in craniocervical instability. J Korean Neurosurg Soc 53(4):223–227
8. Gallie WE (1939) Fractures and dislocation of the cervical spine. Am J Surg 46:495–499
9. Grob D, Dvorak J, Panjabi MM, Antinnes JA (1994) The role of plate and screw fixation in occipitocervical fusion in rheumatoid arthritis. Spine 15:2545–2551
10. He B, Yan L, Xu Z, Chang Z, Hao D (2014) The causes and treatment strategies for the postoperative complications of occipitocervical fusion: a 315 cases retrospective analysis. Eur Spine J 23(8):1720–1724
11. Hwang SW, Gressot LV, Chern JJ, Relyea K, Jea A (2012) Complications of occipital screw placement for occipitocervical fusion in children. J Neurosurg Pediatr 9(6):586–593
12. Jyothirmayi R, Gangadharan VP, Nair MK, Rajan B (1997) Radiotherapy in the treatment of solitary plasmacytoma. Br J Radiol 70(833):511–516
13. Lall R, Patel NJ, Resnick DK (2010) A review of complications associated with craniocervical fusion surgery. Neurosurgery 67(5):1396–1402; discussion 1402–1403
14. Lecouvet F, Richard F, Vande Berg B, Malghem J, Maldague B, Jamart J, Ferrant A, Michaux JL (1997) Long-term effects of localized spinal radiation therapy on vertebral fractures and focal lesions appearance in patients with multiple myeloma. Br J Haematol 96(4):743–745
15. Mazur MD, Sivakumar W, Riva-Cambrin J, Jones J, Brockmeyer DL (2014) Avoiding early complications and reoperation during occipitocervical fusion in pediatric patients. J Neurosurg Pediatr 14(5):465–475
16. Mendoza S, Urrutia J, Fuentes D (2004) Surgical treatment of solitary plasmocytoma of the spine: case series. Iowa Orthop J 24:86–94
17. Menezes AH (1996) Rheumatological disorders. In: Menezes AH, Sonntag WHK (eds) Principles of spinal surgery. Mc Graw Hill, New York, pp 705–722
18. Oda I, Abumi K, Sell LC, Haggerty CJ, Cunningham BW, Mc Afee PC (1999) Biomechanical evaluation of five different occipito-atlanto-axial fixation technique. Spine 24:2377–2382
19. Ryken TC, Menezes AH (1996) Abnormailities of the craniovertebral junction in Down's syndrome. In: Menezes AH, Sonntag WHK (eds) Principles of spinal surgery. Mc Graw Hill, New York, pp 395–40
20. Sonntag WHK, Dickman CA (1996) Posterior occipital C1–C2 instrumentation. In: Menezes AH, Sonntag WHK (eds) Principles of spinal surgery. Mc Graw Hill, New York, pp 1067–1079
21. Vale FL, Oliver M, Cahill DW (1999) Rigid occipitocervical fusion. J Neurosurg 91(2 Suppl):144–150
22. Visocchi M, Pietrini D, Tufo T, Fernandez E, Di Rocco C (2009) Preoperative irreducible C1–C2 dislocations: intraoperative

reduction and posterior fixation. The always posterior strategy. Acta Neurochir 151(5):551–559
23. Visocchi M, Di Rocco C (2014) Os odontoideum syndrome: pathogenesis, clinical patterns and indication for surgical strategies in childhood. Adv Tech Stand Neurosurg 40:273–293
24. Visocchi M, Di Rocco F, Meglio M (2003) Craniocervical junction instability: instrumentation and fusion with titanium rods and sub-laminar wires. Effectiveness and failures in personal experience. Acta Neurochir (Wien) 145(4):265–272; discussion 272
25. Visocchi M, Doglietto F, Della Pepa GM, Esposito G, La Rocca G, Di Rocco C, Maira G, Fernandez E (2011) Endoscope-assisted microsurgical transoral approach to the anterior craniovertebral junction compressive pathologies. Eur Spine J 20:1518–1525
26. Visocchi M, Trevisi G, Iacopino DG, Tamburrini G, Caldarelli M, Barbagallo GM (2015) Odontoid process and clival regeneration with Chiari malformation worsening after transoral decompression: an unexpected and previously unreported cause of "accordion phenomenon". Eur Spine J 24(Suppl 4):S564–S568
27. Williams HJ, Davies AM (2006) The effect of X-rays on bone: a pictorial review. Eur Radiol 16:619–633
28. Winegar CD, Lawrence JP, Friel BC, Fernandez C, Hong J, Maltenfort M, Anderson PA, Vaccaro AR (2010) A systematic review of occipital cervical fusion: techniques and outcomes. J Neurosurg Spine 13(1):5–16
29. Xue Q, Li H, Zou X, Dalstra M, Lind M, Christensen FB, Bünger C (2010) Alendronate treatment improves bone-pedicle screw interface Wxation in posterior lateral spine fusion: an experimental study in a porcine model. Int Orthop 34(3):447–451
30. Zileli M, Cagli S (2002) Combined anterior and posterior approach for managing basilar invagination associated with type I Chiari malformation. J Spinal Disord Tech 15(4):284–289

Management of Cerebral Radiation Necrosis: A Retrospective Study of 12 Patients

Chenlong Liao, Massimiliano Visocchi, Wenchuan Zhang, Min Yang, Wenxiang Zhong, and Pengfei Liu

Abstract

Background

Cerebral radiation necrosis (RN) is a severe complication of radiotherapy for cerebral pathologies. This study discusses the radiographic and pathological features of 12 patients with RN and investigates the management strategy.

Methods

Eleven patients with brain tumors, and one with cerebral cavernous angioma, treated by surgical resection or Gamma Knife alone before radiotherapy developed RN during follow-up. Surgical resection for the cerebral RN was performed in nine patients, and the other three patients received medical treatment. The clinical features, magnetic resonance imaging (MRI), surgical findings, and pathological sections are reviewed.

Results

The diagnosis of RN was confirmed by histological study in all the patients; those with surgical and medical treatment recovered.

Conclusion

As a major complication of radiotherapy, from the clinical and neuroradiological points of view, RN may simulate tumor recurrence. Due to the increasing number of patients with RN who will need to be treated in future years, the definite diagnosis and appropriate treatment of RN remain critical.

C. Liao • W. Zhang (✉) • M. Yang
W. Zhong • P. Liu
Department of Neurosurgery, XinHua Hospital, affiliated to Shanghai JiaoTong University School of Medicine, NO.1665 Shanghai Kongjiang Road, Yangpu, Shanghai, People's Republic of China
e-mail: zhangwench88@hotmail.com

M. Visocchi
Institute of Neurosurgery, Medical School, Catholic University of Rome, Rome, Italy

Keywords Brain tumor • Cavernous angioma • Cerebral radiation necrosis • Radiotherapy

Introduction

Radiotherapy is an important modality option for treating brain tumors, arteriovenous malformations, and head and neck cancers. Clinically, there are three forms of radiation injuries acute, early delayed, and late delayed reactions [10, 14]. The main manifestation of a late delayed reaction is radiation necrosis (RN), which was first described in 1930 [8]. Since then, numerous reports have documented RN as a major complication of radiotherapy to the brain. The exact incidence of RN after radiotherapy for brain tumors and arteriovenous malformations remains undetermined. Risk factors associated with the development of RN include total radiation dose, fraction size, treatment time, and radiation field and volume. Although late cerebral RN has been seen at a radiotherapy dose of less than 50 Gy, generally, it more often develops with a higher irradiation dose, larger fraction size, and when radiation is combined with chemotherapy [35].

The diagnosis of RN has been challenging since the necrosis and tumor recurrence share similar manifestations, both in clinical presentation and in imaging appearance. Therefore, biopsy and pathological study have been recommended as the diagnostic gold standard. Making a definite diagnosis, which is directly related to the choice of treatment, is of great importance.

Twelve patients, previously treated for brain tumor and arteriovenous malformation, with a neuropathological diagnosis of cerebral RN, were included in this study. The aim of this study was to compare the radiographic and pathological features of RN and report our experience in the management of late cerebral RN by focusing on the therapeutic options of medical and surgical therapy.

M. Visocchi et al. (eds.), *Trends in Reconstructive Neurosurgery*, Acta Neurochirurgica Supplement 124,
DOI 10.1007/978-3-319-39546-3_30, © Springer International Publishing Switzerland 2017

Material and Methods

This study was approved by the XinHua Hospital Medical Ethics Committee.

Twelve patients, 7 males and 5 females, ranging in age from 41 to 73 years (mean age 55.8 years) were retrospectively reviewed. Between January 2005 and December 2011, 11 of the 12 patients were surgically treated in our Department. Pathological examinations were performed by two certified neuropathologists and the diagnosis was glioblastoma multiforme (GBM) in 3 cases (27 %) and glioma in 8 (73 %).

One patient, who had a cavernous angioma, found by MRI, preferred stereotactic Gamma Knife radiosurgery to surgical resection. The lesion sites were as follows: 3 temporal; 2 frontal; 3 parietal; 2 frontotemporal; 1 occipital, and 1 parieto-occipital All the patients underwent external-beam fractionated radiation therapy; 1.8–2.0 Gy per day was administered with a 6-MV linear accelerator for 5 consecutive days, for a total dose of 45–60 Gy. Seven patients also received temozolomide as chemotherapy (Table 1).

Magnetic resonance imaging (MRI) was performed at 3- to 6-month intervals after completion of the radiation therapy. When clinical deterioration occurred, MRI was performed. All the patients, including the one with cavernous angioma, developed a newly formed lesion mimicking tumor recurrence (Fig. 1). The mean time between the administration of radiation therapy and the appearance of the newly formed lesion was 18 months (range 8–33 months). Nine patients underwent surgical resection of the lesion to alleviate the severe symptoms, while three patients received only medical treatment after stereotactic biopsy, as they had relatively mild symptoms. Pathological study of the specimens showed RN (Fig. 2). MRI or computed tomography (CT) scans of the brain were then performed at 3- to6 -month intervals during the follow-up. The mean duration of follow-up was 16 months (range 4–36 months) (Table 1).

Results

In our series the median preoperative Karnofsky performance score (KPS) was 80. Apparent total surgical removal of the lesion was performed in nine patients, all of whom had a significant reduction in intracranial pressure within a few days postoperatively. No major complications occurred.

Two patients presented with a severe motor deficit of the left arm and two had postoperative seizures. Transient dysphasia was observed in two more patients. After surgery, brain edema progressively resolved in all the patients within 3 weeks, allowing a reduction or suspension of corticosteroid therapy by that time.

Three patients developed mild neurological symptoms a few weeks postoperatively. MRI showed a worsening of cerebral edema, which recovered after corticosteroid administration.

Discussion

The treatment of brain tumors remains challenging, although neurosurgery, radiotherapy, and chemotherapy are the current options, and they can be integrated. However, prolongation of survival can be accompanied by the appearance of new features, such as RN, which has increased in incidence since radiotherapy started to be considered an outstanding treatment opportunity for brain tumors, arteriovenous malformations, and some head and neck cancers [7].

The primary goal of brain radiotherapy is to deliver a therapeutic dose of radiation, sparing the surrounding normal brain tissue; in fact, irradiation occasionally affects the normal tissue, damaging normal brain tissue near the tumor site [13, 36]. The tolerance of normal tissue has been a limiting factor in the radiation therapy of cerebral pathologies. Patients vary in their individual responses to radiotherapy: some may develop severe adverse reactions, while others receiving comparable radiation doses for similar pathologies in similar locations do not. The exact reason for this variability in response remains unclear, although several researchers have tried to address the issues of intrinsic tissue sensitivity over the past two decades [1, 9, 20, 28, 33], and a median dose of 20 Gy in a single fraction has been advocated to obtain an optimal balance between therapeutic efficacy and the risk of complications [17, 23].

Currently, the mechanisms of RN are still an open question. Theories of vascular injury; glial injury; autoimmune reactions; and oxyradical damage of cell membrane lipids have been advanced so far (15, 21). These mechanisms may generally coexist. The main target of RN is neuroglial cells, especially oligodendroglial and endothelial cells, rather than neurons [31]. Endothelial cell damage caused by abnormal microvascular circulation, "nutritional" insufficiency, and disruption of the blood-brain barrier, promoted by the activated immunological system, contribute to the development of gliosis, vascular injury, and progressive necrosis of the surrounding brain parenchyma [4, 6, 19, 22, 26, 29, 34].

As previously stated, efforts have long been made by a great many investigators to apply radiographic imaging studies to the differential diagnosis of recurrent tumor and RN [5, 24, 32], which can be a radiologic dilemma, since they share the following features at CT or MRI: the original tumor site, mass effect, and contrast enhancing with surrounding edema,

Table 1 Characteristics and management of the 12 patients

Patient number	Sex/Age (years)	Pathological diagnosis	Side/Location	Primary treatment	RT dose/fraction number	Total RT time (days)	Chemotherapy	Time to progression (months)	Preoperative KPS	Treatment for RN	Follow-up (months)
1	M/41	glioma	L/temporal	Surgery	50 Gy/25 fr	38	/	24	80	Surgery	36
2	F/49	Cavernous angioma	L/frontal	Gamma Knife	60 Gy/30 fr	46	/	18	90	Surgery	36
3	M/64	Glioma	L/temporal	Surgery	50 Gy/25 fr	37	TMZ	8	70	Surgery	19
4	M/59	Glioma	L/temporal	Surgery	60 Gy/30 fr	44	TMZ	13	70	Surgery	14
5	F/67	GBM	R/frontal	Surgery	60 Gy/30 fr	48	TMZ	15	90	Medical	9
6	F/73	GBM	R/parietal	Surgery	60 Gy/30 fr	50	TMZ	14	90	Medical	15
7	M/57	GBM	L/parietal	Surgery	60 Gy/30 fr	40	TMZ	28	80	Surgery	14
8	M/57	Glioma	L/fronto-temporal	Surgery	60 Gy/30 fr	50	/	26	70	Surgery	12
9	M/48	Glioma	L/occipital	Surgery	50 Gy/25 fr	38	/	14	80	Surgery	4
10	M/55	Glioma	R/fronto-temporal	Surgery	60 Gy/30 fr	48	TMZ	14	80	Surgery	22
11	F/53	Glioma	R/parietal	Surgery	50 Gy/25 fr	39	TMZ	12	90	Medical	18
12	F/47	Glioma	R/parietal-occipital	Surgery	45 Gy/20 fr	34	/	33	70	Surgery	6

F female, *GBM* glioblastoma multiforme, *KPS* Karnofsky performance score, *L* left, *M* male, *R* right, *RN* radionecrosis, *RT* radiotherapy, *TMZ* temozolomide

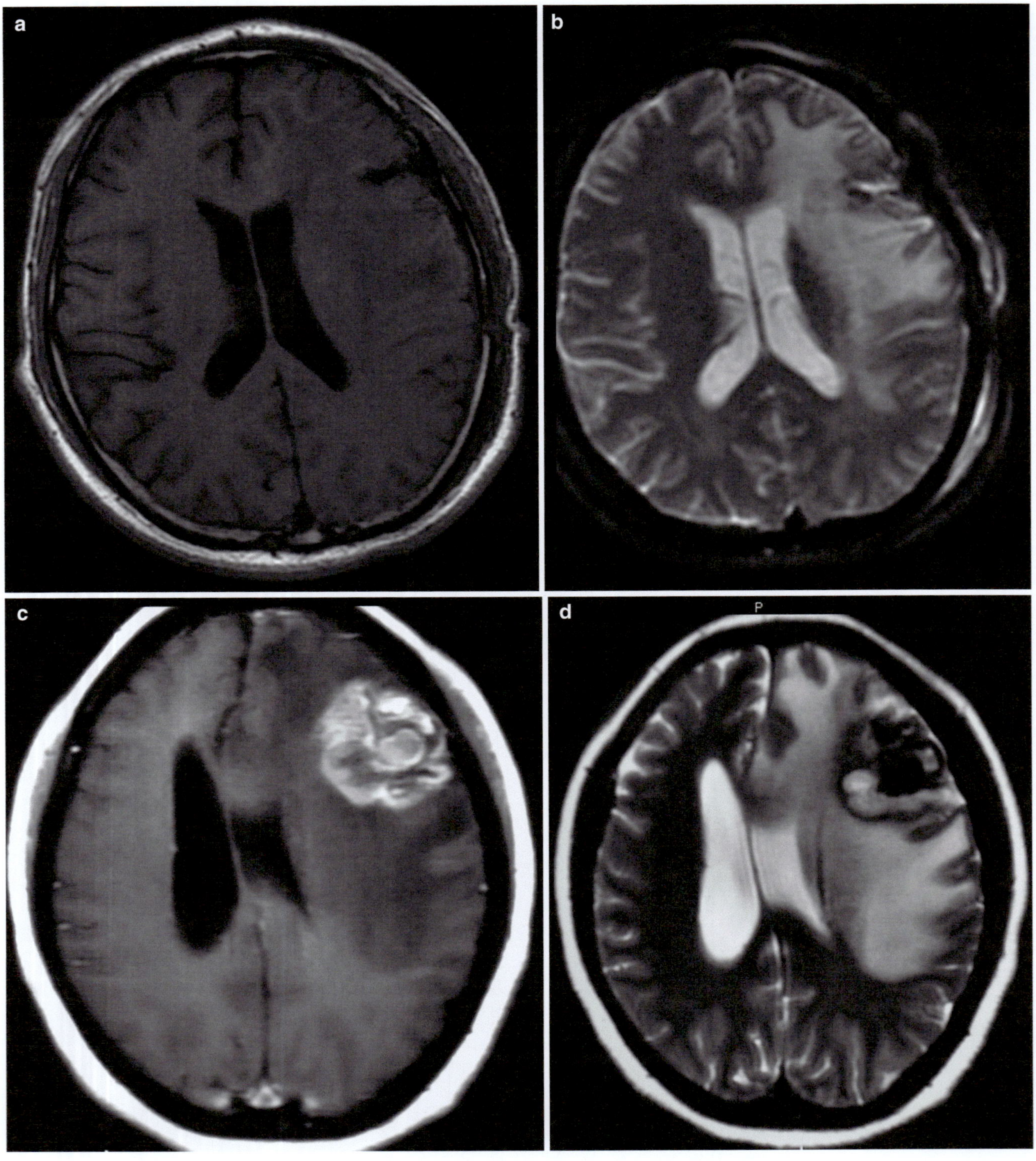

Fig. 1 (**a**, **b**) Axial magnetic resonance (MR) images of patient with primary glioma. The solid portion of radionecrosis and the perilesional edema in the left temporal lobe have low signal intensity on axial T1-MRI and high signal intensity on axial T2-MRI (**c**, **d**) axial MR images of patient with primary cavernous angioma. The solid portion of radionecrosis and the perilesional edema in the frontal lobe show iso-to-hypointense signal intensity on T1 Fluid attenuated inversion recovery (FLAIR) image; the lesion was irregularly enhancement. Hyperintense signal with patches of a hypointense signal area are shown on T2-MR image

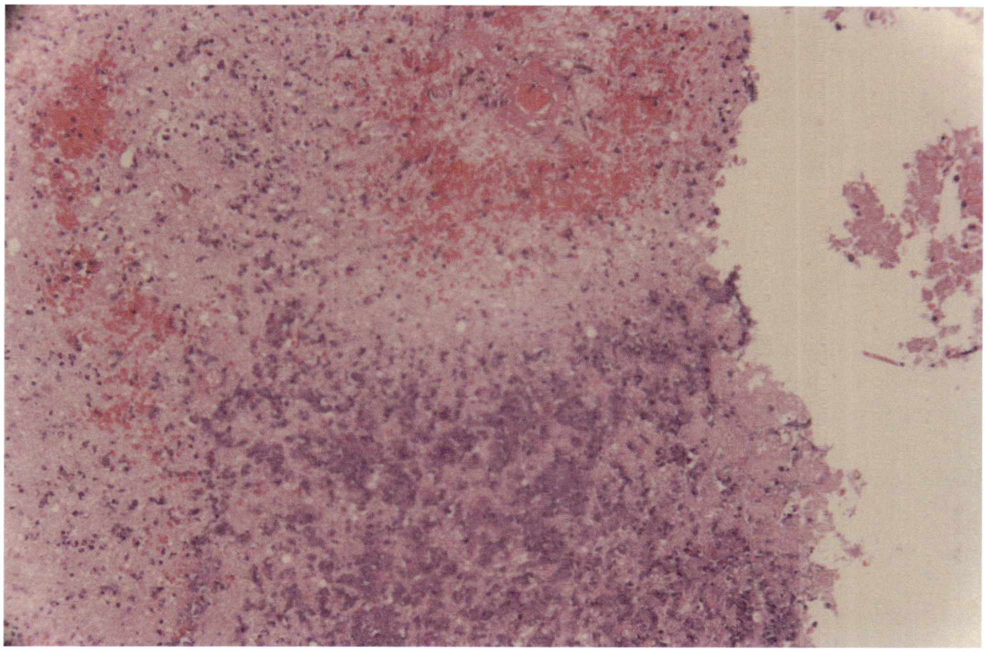

Fig. 2 Pathological section of radionecrosis shows proliferation of surrounding gliocytes, coagulative necrosis of large areas, formation of a glial scar, and infiltration of inflammatory cells around the blood vessels

and may increase in size over time. Other techniques, such as spectroscopy and perfusion MR, perfusion CT, positron emission tomography (PET) and single-photon emission CT (SPECT) have also been widely used. Nevertheless, so far no evidence has been provided that any of these investigations is apparently superior to any other modalities in terms of diagnostic sensitivity or specificity [2, 5, 6, 21, 25]. With no best option recommended, the decision to use one or more imaging techniques always depends on a series of factors, such as the availability of various imaging modalities at an institution, the location and size of the tumor, the neurological findings, and the cost.

A differential diagnosis is very important to illuminate the appropriate management: recurrent tumor might be treated with surgical resection, radiation, or chemotherapy, while RN may benefit from corticosteroids, other medical therapies, or surgery [5, 30]. The definitive diagnosis of RN requires pathological studies, despite the existence of sampling error from a stereotactic biopsy, due to the frequent mixed area of tumoral cells and necrosis. However, in patients with previously irradiated tumor or other pathologies in whom RN or tumor recurrence was clinically or radiographically suspected, results from stereotactic biopsy or surgical biopsy could be used to differentiate tumor recurrence, RN, a mixture of both lesions, and radiation-induced tumor [10].

Spontaneous resolution of cerebral RN may happen, but in most patients symptoms would develop and can be progressive, calling for treatment to provide symptomatic relief [11]. It is reported that resolution or improvement may be obtained following medical treatment with high-dose corticosteroids [16]. However, Gutin and colleagues reported that the effect of steroids for acute episodes in patients with lateral cerebral RN proved to be doubtful. The steroid level at radiation is reported to have an adverse effect on the outcome, owing to severe systemic complications, and as a consequence, increased susceptibility to RN [12, 18]. According to our personal experience, although no definitive conclusion can currently be drawn, we suggest that the management of patients with RN is predominantly surgical in those with elevated intracranial pressure, or if symptoms require prompt control or they progress with conservative treatment. Moreover, surgical intervention may also provide a biopsy specimen and confirm the diagnosis. This statement seems to be in accord with the literature [15, 16, 22, 27, 34].

Regarding the patient with cavernous angioma in our series, we strongly supported surgery to remove both the RN and the remaining lesion, since cavernous angioma can rebleed and support epileptic seizures; in our opinion such a surgery it is not an especially demanding operation, particularly when the nidus is located on the convexity. Similarly to treatment in those patients with long-term recurrent epilepsy unresponsive to antiepileptic drugs, surgery should be resolutely carried out in order to remove the nidus, and to prevent massive hemorrhage and expansion of the epileptic focus. Moreover, surgery should also be considered in the following conditions: (1) acute or progressive functional nervous damage; (2) a single nidus in a non-eloquent area or in the eloquent area in cases of hemorrhage; and (3) a serious focal symptom arising out of multiple encephalic pathological changes [3].

In our study, surgical treatment appeared to be a favorable strategy in patients with good KPS and accessible location of the necrotic mass. We believe that the number of patients requiring treatment of RN will rise in the coming years with the increasing population receiving radiotherapy. Accordingly, prospective randomized, multicenter studies and even complete guidelines are necessary for the management of RN.

Conclusion

As a major complication of radiotherapy, RN may simulate tumor recurrence in both its clinical features and on MRI. The definite diagnosis and appropriate treatment of RN are critical, and will be even more important, given that there will be increasing numbers of patients with RN to be treated in the coming years.

Conflict of Interest Statement The authors declare that they have no conflicts of interest.

References

1. Alter BP (2002) Radiosensitivity in Fanconi's anemia patients. Radiother Oncol : J Eur Soc Therape Radiol Oncol 62:345–347
2. Barajas RF, Chang JS, Sneed PK, Segal MR, McDermott MW, Cha S (2009) Distinguishing recurrent intra-axial metastatic tumor from radiation necrosis following gamma knife radiosurgery using dynamic susceptibility-weighted contrast-enhanced perfusion MR imaging. AJNR Am J Neuroradiol 30:367–372
3. Bertalanffy H, Benes L, Miyazawa T, Alberti O, Siegel AM, Sure U (2002) Cerebral cavernomas in the adult. Review of the literature and analysis of 72 surgically treated patients. Neurosurg Rev 25:1–53; discussion 54–55
4. Coderre JA, Morris GM, Micca PL, Hopewell JW, Verhagen I, Kleiboer BJ, van der Kogel AJ (2006) Late effects of radiation on the central nervous system: role of vascular endothelial damage and glial stem cell survival. Radiat Res 166:495–503
5. Dequesada IM, Quisling RG, Yachnis A, Friedman WA (2008) Can standard magnetic resonance imaging reliably distinguish recurrent tumor from radiation necrosis after radiosurgery for brain metastases? A radiographic-pathological study. Neurosurgery 63:898–903; discussion 904
6. Dropcho EJ (2010) Neurotoxicity of radiation therapy. Neurol Clin 28:217–234
7. Eisele SC, Dietrich J (2015) Cerebral radiation necrosis: diagnostic challenge and clinical management. Rev Neurol 61:225–232
8. Eyster EF, Nielsen SL, Sheline GE, Wilson CB (1974) Cerebral radiation necrosis simulating a brain tumor. Case report. J Neurosurg 40:267–271
9. Gatti RA (2001) The inherited basis of human radiosensitivity. Acta Oncol 40:702–711
10. Giglio P, Gilbert MR (2003) Cerebral radiation necrosis. Neurologist 9:180–188
11. Glass JP, Hwang TL, Leavens ME, Libshitz HI (1984) Cerebral radiation necrosis following treatment of extracranial malignancies. Cancer 54:1966–1972
12. Gutin PH, McDermott MW, Ross G, Chan PH, Chen SF, Levin KJ, Babuna O, Marton LJ (1990) Polyamine accumulation and vasogenic oedema in the genesis of late delayed radiation injury of the central nervous system (CNS). Acta Neurochir Suppl 51:372–374
13. Hopewell JW, Millar WT, Ang KK (2007) Toward improving the therapeutic ratio in stereotactic radiosurgery: selective modulation of the radiation responses of both normal tissues and tumor. J Neurosurg 107:84–93
14. Iwai Y, Yamanaka K, Oda J, Tsuyuguchi N, Ochi H (2001) Tracer accumulation in radiation necrosis of the brain after thallium-201 SPECT and [11C]methionine PET – case report. Neurol Med Chir 41:415–418
15. Kano H, Kondziolka D, Zorro O, Lobato-Polo J, Flickinger JC, Lunsford LD (2009) The results of resection after stereotactic radiosurgery for brain metastases. J Neurosurg 111:825–831
16. Lorenzo ND, Nolletti A, Palma L (1978) Late cerebral radionecrosis. Surg Neurol 10:281–290
17. Maldaun MV, Aguiar PH, Lang F, Suki D, Wildrick D, Sawaya R (2008) Radiosurgery in the treatment of brain metastases: critical review regarding complications. Neurosurg Rev 31:1–8; discussion 8–9
18. Morris JG, Grattan-Smith P, Panegyres PK, O'Neill P, Soo YS, Langlands AO (1994) Delayed cerebral radiation necrosis. Q J Med 87:119–129
19. Pena LA, Fuks Z, Kolesnick RN (2000) Radiation-induced apoptosis of endothelial cells in the murine central nervous system: protection by fibroblast growth factor and sphingomyelinase deficiency. Cancer Res 60:321–327
20. Roberts SA, Spreadborough AR, Bulman B, Barber JB, Evans DG, Scott D (1999) Heritability of cellular radiosensitivity: a marker of low-penetrance predisposition genes in breast cancer? Am J Hum Genet 65:784–794
21. Ross DA, Sandler HM, Balter JM, Hayman JA, Archer PG, Auer DL (2002) Imaging changes after stereotactic radiosurgery of primary and secondary malignant brain tumors. J Neurooncol 56:175–181
22. Ruben JD, Dally M, Bailey M, Smith R, McLean CA, Fedele P (2006) Cerebral radiation necrosis: incidence, outcomes, and risk factors with emphasis on radiation parameters and chemotherapy. Int J Radiat Oncol Biol Phys 65:499–508
23. Shehata MK, Young B, Reid B, Patchell RA, St Clair W, Sims J, Sanders M, Meigooni A, Mohiuddin M, Regine WF (2004) Stereotatic radiosurgery of 468 brain metastases < or =2 cm: implications for SRS dose and whole brain radiation therapy. Int J Radiat Oncol Biol Phys 59:87–93
24. Soffietti R, Sciolla R, Giordana MT, Vasario E, Schiffer D (1985) Delayed adverse effects after irradiation of gliomas: clinicopathological analysis. J Neurooncol 3:187–192
25. Sundgren PC (2009) MR spectroscopy in radiation injury. AJNR Am J Neuroradiol 30:1469–1476
26. Szeifert GT, Massager N, DeVriendt D, David P, De Smedt F, Rorive S, Salmon I, Brotchi J, Levivier M (2002) Observations of intracranial neoplasms treated with gamma knife radiosurgery. J Neurosurg 97:623–626
27. Takeuchi J, Hanakita J, Abe M, Handa H (1976) Brain necrosis after repeated radiotherapy. Surg Neurol 5:89–93
28. Taylor AM, Harnden DG, Arlett CF, Harcourt SA, Lehmann AR, Stevens S, Bridges BA (1975) Ataxia telangiectasia: a human mutation with abnormal radiation sensitivity. Nature 258:427–429
29. Torcuator R, Zuniga R, Mohan YS, Rock J, Doyle T, Anderson J, Gutierrez J, Ryu S, Jain R, Rosenblum M, Mikkelsen T (2009) Initial experience with bevacizumab treatment for biopsy confirmed cerebral radiation necrosis. J Neurooncol 94:63–68
30. Truong MT, St Clair EG, Donahue BR, Rush SC, Miller DC, Formenti SC, Knopp EA, Han K, Golfinos JG (2006) Results of surgical resection for progression of brain metastases previously treated by gamma knife radiosurgery. Neurosurgery 59:86–97; discussion 86–97

31. Tsui EY, Chan JH, Leung TW, Yuen MK, Cheung YK, Luk SH, Tung SY (2000) Radionecrosis of the temporal lobe: dynamic susceptibility contrast MRI. Neuroradiology 42:149–152
32. van Dellen JR, Danziger A (1978) Failure of computerized tomography to differentiate between radiation necrosis and cerebral tumour. S Afr Med J 53:171–172
33. West CM, Elliott RM, Burnet NG (2007) The genomics revolution and radiotherapy. Clin Oncol 19:470–480
34. Wong ST, Loo KT, Yam KY, Hung WM, Fok KF, Yuen SC, Fong D (2010) Results of excision of cerebral radionecrosis: experience in patients treated with radiation therapy for nasopharyngeal carcinoma. J Neurosurg 113:293–300
35. Yoshii Y (2008) Pathological review of late cerebral radionecrosis. Brain Tumor Pathol 25:51–58
36. Yu JB, Schulder M, Knisely J (2012) Radiosurgical dose selection for brain metastasis. Prog Neurol Surg 25:139–147

The Rehabilitation of Spinal Cord Injury Patients in Europe

Giorgio Scivoletto, Massimo Miscusi, Stefano Forcato, Luca Ricciardi, Mariano Serrao, Roberto Bellitti, and Antonino Raco

Abstract In Western European countries there is an incidence of traumatic spinal cord injury (SCI) of 16 to 19.4 new cases per million inhabitants per year. Since World War II, European physicians have been fundamental in the development of SCI medicine, starting from Sir Ludwig Guttman, who developed the idea of the integrated treatment of these patients. More recently, scientists from Germany and Switzerland have developed a new rehabilitative approach, Body Weight Support Treadmill Training, based on the concept of activity-based therapy and aimed at restoring walking in SCI patients. This review highlights issues concerning different organizational systems and health policies within and outside Europe.

Keywords Spinal Cord Injury • Rehabilitation • Historical perspective

G. Scivoletto, MD, PhD (✉)
Spinal Cord Unit, IRCCS S. Lucia Foundation, Rome, Italy
e-mail: g.scivoletto@hsantalucia.it

M. Miscusi, MD, PhD
Department of Medico-Surgical Sciences and Biotechnologies, Sapienza University of Rome, Rome, Italy

S. Forcato, MD
Department of Neursciences NESMOS, Sapienza University of Rome, Rome, Italy

L. Ricciardi, MD
Institute of Neurosurgery, Catholic University of Rome, Rome, Italy

M. Serrao, MD, PhD
Department of Medico-Surgical Sciences and Biotechnologies, Sapienza University of Rome, Rome, Italy

R. Bellitti, MD
Department of Radiology, Cerignola Civic Hospital, Cerignola(FG), Italy

A. Raco, MD
Departement of Neurosciences NESMOS, Sapienza University of Rome, Rome, Italy

Introduction and Epidemiology

Spinal cord injury (SCI) has a worldwide incidence of between 10.4 and 83 cases per million per year [39], with a great impact on health care systems; although recently a number of epidemiological studies of SCI have become available, most of these studies have only considered traumatic SCI. However, a substantial part (although not perfectly known) of the population with SCI consists of patients with non-traumatic SCI; furthermore, non-traumatic spinal cord lesions will probably increase in the near future as the European population is getting older. This group of patients may have divergent rehabilitation needs due to their age and due to the higher frequency of comorbidities than in the traumatic population [30]. Consequently, there is a need for more insight into the characteristics of the populations affected with traumatic and non-traumatic SCI in different settings and countries, as well as the need for a standardized method of reporting personal and injury characteristics in clinical studies [4, 30, 37, 39].

In Europe, the mean incidence of SCI is between 16 and 19.4 new cases per million inhabitants per year, and the incidence is similar in various European countries (Table 1). In most countries, SCI incidence is approximately 20 new cases per million inhabitants per year, with the remarkable exceptions of the Netherlands, Spain, and Denmark, where the incidence is lower, and Greece and Russia, where the incidence is approximately 30 new cases per year. European countries also demonstrate substantial similarity regarding other features of the SCI population (male/female ratio, age, and neurological presentation), particularly regarding etiology, with motor vehicle accidents and falls being the primary causes of injury.

There are fewer prevalence studies than incidence studies, and they show that prevalence (average number per million inhabitants) varies from 280 in Finland [9] to 355 in Norway [21]. The estimated mean European prevalence is 250 cases per million inhabitants [39].

M. Visocchi et al. (eds.), *Trends in Reconstructive Neurosurgery*, Acta Neurochirurgica Supplement 124,
DOI 10.1007/978-3-319-39546-3_31,

Table 1 Mean incidence of new cases of spinal cord injury (SCI) per million inhabitants per year in various European countries

	Europe	Europe	Denmark	Holland	France	Finland	Germany	Greece	Ireland	Italy	Norway	Portugal	Romania	Russia	Sweden	Spain
Incidence (no./million)	16	19.4	9.2	7.5	19.4	13.8		33.6	13.1	19	21.2	25.4		29.7	19.5	8.1
Sex																
Male (%)			73	74			72	88	87	80.1	75.6	77	77	78	77	72
Female (%)			27	26			28	12	13	19.9	24.4	23	23	22	23	28
Age (years)				40.8		42.4		43.4	37	38.5	42.9	50		33.5	47.4	41.8
Level of Injury																
Paraplegia (%)			49	59		50.6	62	52	50	56.6	47.6	51	40	51	55	62
Tetraplegia (%)			51	41		49.4	38	48	50	39.9	52.4	49	60	49	45	38
Neurological classification																
Complete (%)			48	68					39	50.7	41.4	56	40	44		
Incomplete (%)			52	32					61	49.3	58.6	44	60	56		
Etiology																
MVA	46		47	42		39.5	35	51	50	58	34.2	57	13	25	23	52
Work-related	21			11			14	20	15	15				11	11	
Fall	37		26	23		41		37	2	5	45.5	33	59	19	47	27
Sport accident	9		11	15		6.6	4	4	9	8	8.6		7	33	17	3
Violence	8		2	4		2.7	1	2		3				2	2	3
Other			14	5		10	5	2		10	11.6	5.3	21	10	3	14

MVA motor vehicle accident

The incidence of new SCI cases, and therefore the prevalence of SCI, is lower in Europe than in other areas of the world: North America has an incidence (approximately 39 new cases per million) more than twice that of Western Europe [8]. Fall-related SCIs are more frequent in Europe, while violence-related SCIs are much more frequent in other areas of the world, including North and South America, the Middle East, North Africa, and southern Africa [8]. The features of the SCI population are changing, as the age of SCI patients, the number of fall-related SCIs, the number of female cases, and the proportion of incomplete cervical lesions all appear to be increasing [1, 10, 22].

With regard to non-traumatic SCI, the data are much more confused, with an incidence rate ranging from 6 new cases per million population per year (Europe) to 76 new cases (North America) [28]. However, the same author, comparing nine SCI units around the world (including four from Europe) reported that about 40 % of the admitted patients had a non-traumatic lesion [29]: therefore, the real incidence of non-traumatic SCIs is probably higher than reported and deserves further studies.

European Contribution to the Rehabilitation of SCI Patients

In the twentieth century, it was proposed and understood that complete treatment of patients with SCI should include acute hospital care, as well as both acute and ongoing rehabilitation. In general, we can state that, in the past, because of the lack of effective medical and surgical protocols, SCI was considered an untreatable disease; however, medical efforts have been directed toward its cure, by improving the management of both the acute and long-term phases [15]. Indeed, we have passed from an era of therapeutic nihilism to a new concept of SCI, and are able to treat the needs of individuals with SCI by competently treating the direct injury as well as all affected organ systems, including the psychological aspects of SCI.

Although the concept of integrated treatment was also known in the United States after World War II, European figures such as Sir Ludwig Guttman pioneered and made tremendous contributions to the multispecialist treatment of SCI patients. Guttmann (born in Silesia, Germany, in 1899) in 1944 was placed in charge of the SCI Unit at the Stoke Mandeville Hospital, in the United Kingdom. Guttmann introduced postural reduction, divided his patients' neurological lesions into "complete" and "incomplete", and then subdivided them into improved, unchanged, and deteriorated. This organization called attention to two basic concepts: the difference between a complete and an incomplete spinal cord lesion, and the prognostic implications of this difference on recovery. Later, this concept was expanded by Dr. Frankel, a pupil and co-worker of Guttmann's, who reported clinical data from a Stoke Mandeville Hospital series, introducing the scale of neurological involvement that is now universally known as the Frankel Scale [17].

Furthermore, Guttman proposed a model of integrated care of SCI patients from the very acute phase to discharge. He created the very first SCI unit in the world at Stoke Mandeville, and this model of care was accepted and developed only later in the United States and Canada [17]. Guttmann realized that, contrarily to the usual division of specialties, doctors treating this illness had to be rehabilitationists with a commitment to all the needs of the patient, not only those within the scope of one's specialty. Already in 1944 Guttmann had established "the fundamental rules for the care of SCI patients:

(a) Management of a unit by an experienced physician who is prepared to give up part, or all, of their own specialty.
(b) Sufficient numbers of allied health professionals, e.g., nurses and therapists, to cope with details of care.
(c) Technical facilities to establish workshops and vocational outlets.
(d) Attention to social, domestic, and industrial resettlement.
(e) Regular aftercare, or extended care, over the lifetime of each individual."

Guttmann was also a great believer in wheelchair sports, and is remembered for founding the Paralympics [20, 34]. Although only a small number of SCI subjects take part in elite and Paralympic sport, recreational sport plays an important part in the rehabilitation and life of paralyzed persons. Furthermore, elite sports and the Paralympics are fundamental in the reintegration of paralyzed persons and in changing the perception of wheelchair users.

Although little of Guttmann's work was published in the medical literature [20], he was a great teacher and was experienced in what is now known as public relations. Many doctors visited Stoke Mandeville and trained in SCI medicine under the supervision of Guttmann; among these were most of the physicians who then developed SCI units in European countries, as well as in the rest of the world.

At present, after 70 years, SCI rehabilitation is strongly influenced by what we learnt from the Stoke Mandeville experience. Rehabilitation has traditionally focused on maximizing the person's recovery from the injury and returning them home as independently as possible. However, once the milestone of returning home with a new SCI is crossed, the individual must conquer the barriers of societal reintegration with their altered body habitus. Traditionally, management of the person with SCI involves the determination of the person's level of injury and functional capacity and then the generation of a problem list and prescription for therapies, with the overall goal of having the patient achieve their maximum functional potential. Therefore, determining the potential neurological and functional outcome of an individual post-SCI is the basis of the rehabilitation process [5]. Functional outcomes are traditionally determined based on the level and severity of the SCI [7] in conjunction with other factors, such as age [33]. The determination of the severity of the lesion is based on the International Standards for Neurological Classification of Spinal Cord Injury, which are a development of the Frankel Classification created in Stoke Mandeville in 1969 [17].

Always, according to Guttmann's principles, SCI-specific rehabilitation should be initiated as soon as possible (ideally beginning in the intensive care setting) and in a specialized unit capable of providing comprehensive management of a person with SCI with an interdisciplinary approach of the rehabilitation team, which includes the patient and their family [35]. In fact it has been demonstrated that a delay in starting these interventions may negatively influence a patient's ultimate functional capability and increase their length of rehabilitation stay [32]. Furthermore, early admission to an SCI unit may help in preventing early medical complications, thus facilitating the course of inpatient rehabilitation and reducing the total cost of care. The rehabilitation of SCI patients is a multifaceted problem and needs a multispecialty approach. SCI can disrupt upper and/or lower motor and sensory pathways, and can result in either a complete or an incomplete lesion. Although recent advances in primary damage healing, rehabilitation, and the prevention of complications have improved the prognosis of SCI [2], its consequences are still traumatic and disabling. Paralysis and loss of sensation are only two of several physical consequences of SCI. Therefore, the care of persons with SCI is not limited to the treatment of the neurologic injury. The medical consequences resulting from altered neurologic input, such as the development of respiratory failure, or neurogenic bladder and its resultant complications (pneumonia and urinary tract infections), affect most organ systems. These consequences not only represent an important health issue, as they could be major causes of morbidity and mortality (e.g., respiratory and urinary problems), but they also represent an obstacle to the social and vocational reintegration of patients (e.g., because of urinary incontinence). In addition, the psychological impact of SCI is vast, and complications such as depression and anxiety must be addressed.

The influence of the Stoke Mandeville experience is particularly evident in the care of two of the major causes of morbidity and mortality in SCI patients: pressure sores and urological complications.

With regard to pressure sores, Guttmann was the first to demonstrate that this complication may not only be healed, but may also be prevented, with a careful program of bed-turning (every 2 h) of the patient [16]. Very recently, a group of SCI experts agreed that the best practice to prevent pressure sores is still the bed-turning program established by Frankel [6].

With regard to urological complications, most of the patients admitted to Stoke Mandeville in the 1940s were managed with a suprapubic catheter that was changed only when closed. Guttmann had the intuition of removing the indwelling catheter and beginning to catheterize SCI patients every 4–6 h [16]. Therefore, he started the practice of clean (sterile) intermittent catheterization that later on was demonstrated to be the best practice to prevent renal failure in SCI patients [11].

In the past 30 years the rehabilitation of SCI patients has involved a second aim, which will probably increase in importance as incomplete lesions seem to be increasing: the restoration of impaired functions, in particular of walking. Recently some rehabilitative approaches have received particular attention, most of all in the field of activity-based therapy. After years of evidence from the basic science data, the old rehabilitative principles of compensation and adaptation are changing. Strategies for stimulating the nervous system to optimize functional recovery and elicit lost abilities are becoming more and more important [31]. Activity-based therapy refers to "interventions that target activation of the neuromuscular system below the level of the lesion, with the goal of retraining the nervous system to recover a specific motor task". Intense physical activity has been shown to improve physiological function and health outcomes in individuals with chronic (> 1 year post-injury) SCI [3].

In this field of rehabilitation, European scientists have made important contributions. Following the concept of a central pattern generator (CPG) introduced by Grilner [19] and according to the idea of "spinal locomotion" [38], scientists from Germany and Switzerland introduced body weight supported treadmill (BWST) training [12, 38]. BWST training holds promise for walking recovery. In motor-complete SCI patients, unsupported walking seldom, if ever, recovers. However, this therapy showed the ability to induce a locomotor pattern even in patients with complete paraplegia, in conjunction with increases in leg extensor electromyographic responses [38]. In motor-incomplete SCI patients, daily locomotor training with BWST training often results in significant improvements in locomotor function [12]. More recently, robotic orthoses have been used for training SCI patients on the treadmill [25]. These orthoses have been designed to address some of the problems of BWST training (in particular the personnel costs) and to ensure that patients have a constant, safe, and regular repetition of the exercise. Although the cost-effectiveness and benefits of these devices compared with conventional therapies targeting the improvement of gait post-SCI are still to be demonstrated [26, 36], the robotic rehabilitation of walking is experiencing increasing success and is, at present, an integral part of the treatment in several SCI units.

State of the Art of Rehabilitation Systems in Europe for Patients with Spinal Cord Injury

A comparison of the different systems of care is particularly difficult and therefore data on this issue are scarce. However, recently, the European Spinal Cord Injury Federation (ESCIF) produced an acute care and rehabilitation report to gather and collate information on the acute care and primary rehabilitation services offered to people with SCI [24]. They sent questionnaires to 18 European countries and, although the response rate was good, most respondents submitted estimates of the figures requested. The difficulty in collecting data from local or national registries was often due to a lack of organization and coordination in primary and rehabilitative care. The ESCIF collected data about the destinations of new SCI patients (Table 2) and the primary rehabilitation organizations (Table 3) in European countries. And New [29] offered a comparison of SCI units around the world, including four from Europe. These studies clearly show similarities as well as differences in the standards of care in the various European countries. Notably, Europe does not have a common conduit for the management of SCI patients from acute care through rehabilitation; individual countries take different organizational approaches according to their own health systems and traditions. However, rehabilitation approaches to SCI patients are substantially similar in western European countries, which seem to have accepted and applied the principles of care dictated by Guttmann. Most patients (varying from 60 to 100 % depending on the country) are cared for in specialized SCI hospitals/units/centers [24]. According to New [29] these units show substantial similarities with regard to referral processes and criteria at admission (in particular with regard to some categories of patients: those with malignancies and spina bifida), as well as showing similarities in key personnel and services available.

The main differences concern the hours of treatment per week and the length of stay (LOS) of the patients. The report by the ESCIF [24] clearly shows differences in both these parameters between the European countries (Table 4). An important question is: do these differences have an impact on the outcomes of SCI patients?

With regard to the effect of the intensity of treatment, the available data are really scarce and are limited to an article by Heinemann [23]. This author calculated the therapeutic intensity provided for each patient and examined whether the amount of therapy was associated with the outcome in SCI patients, but failed to demonstrate a correlation in these patients.

With regard to LOS, it seems that this parameter may have an effect on the outcome of SCI patients, as well as on the behavior of professionals. Although comparison studies between the various countries are quite limited, some data can be retrieved from the available bibliography (Table 5). The study by Fromovich-Amit et al. examined the characteristics of spinal rehabilitation units in four countries, and

Table 2 Acute care – destinations of new SCI patients, with percentages where given

Country	Specialized hospital	SCI unit or ward	Neurosurgical ward	Trauma ward	Non-specialized ward
Austria			✓	✓	
Belgium		60		20	20
Croatia	✓		✓	✓	
Denmark		✓	✓		
Finland			50	50	
Germany		80	5	15	
Ireland	100				
Italy		50	25	25	
Netherlands				✓	
Portugal		✓			
Scotland	✓		✓	✓	
Slovenia	✓	✓		✓	
Spain	40	15	10	10	25
Sweden		✓			
Switzerland	50	25	25	✓	✓
England and Wales	✓	✓	✓	✓	

Reproduced with permission from Horsewell [24]

Table 3 Primary rehabilitation of SCI patients, with percentages and numbers of beds available where given

Country	Specialized SCI hospital	SCI unit or ward in general hospital	SCI rehab center	Generalist rehab unit	Beds avail in Specialised SCI hospital and SCI rehab centre
Austria			85 %	15 %	180 %
Belgium		60 %		40 %	126
Croatia			90 %	10 %	35*
Denmark			90 %	10 %	67
Finland			60 %	40 %	43
Germany	✓	✓	✓	✓	?
Ireland	100 %				50
Italy		50 %	25 %	25 %	500
Netherlands			✓	✓	?
Portugal	✓		✓	✓	?
Scotland	✓			✓	48
Slovenia			99 %	1 %	70
Spain	✓	✓	✓	✓	650
Sweden		✓	✓	✓	?
Switzerland	60 %	35 %	5 %		235
England and Wales	75 %	✓	✓	✓	400

*Croatian Institute for Health Insurance. In fact, 50 beds are available.
Reproduced with permission from Horsewell [24]

Table 4 Individual therapy, sports activities, and access to individual training facilities

	One-on-one physiotherapy (average hours per week)		One-on-one occupational therapy (average hours per week)		Sports	Individual exercise	
Country	Quad	Para	Quad	Para		Yes	no
Austria	5	3	5	2	x	x	
Belgium	15	10	15	10	x	x	
Croatia	5	5	5	5	75/25	x	
Denmark	5	3	5	1–2	x	50/25	
Finland	7	5	3	0	x	75/25	
Germany	5	5	2	2	75/25	75/25	
Ireland	10	10	10	10	x	50/50	
Italy	12	8	10	6	75/25	x	
Netherlands	5	4	6	3	x	50/50	
Portugal	?	?	?	?	50		x
Scotland	3	3	2	2	75	x	
Slovenia	30	20	?	?	X	X	
Spain	14	20	20	14	X	X	
Sweden	5	5	4	4	75/25	75/25	
Switzerland	7	5	5	2	x	X	
UK	?	?	?	?	x	X	

Reproduced with permission from Horsewell J. Acute care and primary rehabilitation in ESCIF member countries 2007. ESCIF: European Spinal Cord Injury Federation Web site. 2007. http://www.escif.org/files/documents/members_downloads/infoproject.pdf. [24]
x means "*always*", 75/25 means "*often*" and 50/50 means "*sometimes*"

Table 5 Mean length of stay in rehabilitation

Country	Days (mean)	Notes
Netherlands	188	
Denmark	176	
Russia	47	
Lithuania	66,5	
Israel	132	
Canada	55	
Australia	54	
Germany	120	Paraplegic patients without any further lesion
	165	Paraplegic patients with polytrauma

clearly showed that, in Russia and Lithuania, where rehabilitation time is predetermined and limited, rehabilitation objectives are determined by the time allowed for rehabilitation [18]. In Denmark and Israel, which each allow a longer LOS, the rehabilitation team tends to achieve the maximal possible independence allowed by the lesion. Differences in LOS are even more striking when comparing European and United States practices. Although direct comparisons between European and United States systems of care are not available, some data can be derived from the literature. In the article by Ditunno and colleagues, who compared the preferences of rehabilitation professionals for alternative functional goals during SCI rehabilitation [13], Italian and Canadian rehabilitation professionals demonstrated a preference for walking over wheelchair mobility at lower stages of assumed recovery, while professionals in the United States set wheelchair independence at a higher priority than walking. In the United States, LOS and, consequently, rehabilitation goals, are mostly determined by third-party payer requirements for discharge. Patients must be discharged as soon as they achieve a minimally acceptable level of mobility. For example, with SCI patients for whom walking is feasible, independent wheelchair mobility can typically be achieved before independence in walking; therefore, the patient is discharged at the level of independent wheelchair mobility. In Italy and Canada, LOS in rehabilitation is primarily based on when patients achieve the highest level of independence feasible, or maximal mobility independence, such as walking.

In another study, Ditunno and colleagues demonstrated a difference in the use of braces and devices between United States and European practices, and postulated that this difference was due to differences in the health care systems, and particularly due to LOS [14]. The LOS is shorter in the United States (typically 4–6 weeks for paraplegics and 8–10 weeks for quadriplegics) [27] than in Europe (typically 4 months for paraplegics and 6 months for quadriplegics) [24]. Therefore, in the United States, therapists are required to progress patients more aggressively.

Another important question that will deserve further studies is: do these differences in LOS depend on health policies (for example, payment modality)? While it is quite easy to answer this question when examining the United States reality, where payment is primarily based on insurances, the answer is not so easy in Europe, where payment is mostly based on the national health systems. Although one hopes that this drive toward health economy is not affecting patient outcomes, this is obviously an area of great concern that requires conscientious and rigorous investigation. Furthermore, these nonclinical variables must be taken into account when projecting and evaluating international clinical trials, as they might influence the outcomes of SCI patients [29].

Conflict of Interest Statement We declare that we have no conflict of interest.

References

1. Ahoniemi E, Alaranta H, Hokkinen E-M, Valtonen K, Kautiainen H (2008) Incidence of traumatic spinal cord injuries in Finland over a 30-year period. Spinal Cord 46(12):781–784
2. Anderson DK, Beattie M, Blesch A, Bresnahan J, Bunge M, Dietrich D (2005) Recommended guidelines for studies of human subjects with spinal cord injury. Spinal Cord 43(8):453–458
3. Behrman AL, Harkema SJ (2007) Physical rehabilitation as an agent for recovery after spinal cord injury. Phys Med Rehabil Clin N Am 18(2):183–202
4. Biering-Sørensen F, Charlifue S, DeVivo M, Noonan V, Post M, Stripling T, Wing P (2006) International spinal cord injury data sets. Spinal Cord 44(9):530–534
5. Chan SC, Chan AP (2005) Rehabilitation outcomes following traumatic spinal cord injury in a tertiary spinal cord injury centre: a comparison with an international standard. Spinal Cord 43:489–498
6. Chhabra HS, Harvey LA, Muldoon S, Chaudhary S, Arora M, Brown DJ (2013) www.elearnSCI.org: a global educational initiative of ISCoS. Spinal Cord 51:176–182
7. Consortium for Spinal Cord Medicine. Outcomes following traumatic spinal cord injury: clinical practice guidelines for health-care professionals. J Spinal Cord Med. 2000 Winter;23(4):289–316
8. Cripps RA, Lee BB, Wing P, Weerts E, Mackay J, Brown D (2011) A global map for traumatic spinal cord injury epidemiology: towards a living data repository for injury prevention. Spinal Cord 49(4):493–501
9. Dahlberg A, Kotila M, Leppanen P, Kautiainen H, Alaranta H (2005) Prevalence of spinal cord injury in Helsinki. Spinal Cord 43(1):47–50
10. de Groot S, Dallmeijer AJ, Post MWM, van Asbeck FWA, Nene AV, Angenot ELD, van der Woude LHV (2006) Demographics of

the Dutch multicenter prospective cohort study 'Restoration of mobility in spinal cord injury rehabilitation'. Spinal Cord 44(11):668–675
11. Di Benedetto P (2011) Clean intermittent self-catheterization in neuro-urology. Eur J Phys Rehabil Med 47(4):651–659
12. Dietz V, Colombo G, Jensen L, Baumgartner L (1995) Locomotor capacity of spinal cord in paraplegic patients. Ann Neurol 37:574–582
13. Ditunno JF, Scivoletto G, Patrick M, Biering-Sorensen F, Abel R, Marino R (2008) Validation of the walking index for spinal cord injury in a US and European clinical population. Spinal Cord 46(3):181–188
14. Ditunno PL, Patrick M, Stineman M, Morganti B, Townson AF, Ditunno JF (2006) Cross-cultural differences in preference for recovery of mobility among spinal cord injury rehabilitation professionals. Spinal Cord 44(9):567–575
15. Donovan WH (2007) Donald Munro Lecture Spinal cord injury--past, present, and future. J Spinal Cord Med 30(2):85–100
16. Frankel HL (2012) The Sir Ludwig Guttmann lecture 2012: the contribution of Stoke Mandeville Hospital to spinal cord injuries. Spinal Cord 50(11):790–796
17. Frankel HL, Hancock DO, Hyslop G, Melzak J, Michaelis LS, Ungar GH, Vernon JD, Walsh JJ (1969) The value of postural reduction in the initial management of closed injuries of the spine with paraplegia and tetraplegia. I. Paraplegia 7(3):179–192
18. Fromovich-Amit Y, Biering-Sørensen F, Baskov V, Juocevicius A, Hansen HV, Gelernter I, Hart J, Baskov A, Dreval O, Terese P, Catz A (2009) Properties and outcomes of spinal rehabilitation units in four countries. Spinal Cord 47(8):597–603
19. Grillner S (1975) Locomotion in vertebrates: central mechanisms and reflex interaction. Physiol Rev 55(2):247–304
20. Guttmann L (1973) Spinal cord injuries, comprehensive management and research. Blackwell Scientific Publications, Oxford Press, London
21. Hagen EM, Aarli JA, Grønning M (2001) Patients with traumatic spinal cord injuries at the department of neurology, Haukeland Hospital 1952-99. Tidsskr Nor Laegeforen 121(28):3273–3275
22. Hagen EM, Eide GE, Rekand T, Gilhus NE, Gronning M (2010) A 50-year follow-up of the incidence of traumatic spinal cord injuries in Western Norway. Spinal Cord 48(4):313–318
23. Heinemann AW, Hamilton B, Linacre JM, Wright BD, Granger C (1995) Functional status and therapeutic intensity during inpatient rehabilitation. Am J Phys Med Rehabil 74:315–326
24. Horsewell J (2007) Acute care and primary rehabilitation in ESCIF member countries 2007. ESCIF: Eur Spinal Cord Injury Federation Web site. Available at: http://www.escif.org/files/documents/members_downloads/infoproject.pdf
25. Jezernik S, Colombo G, Keller T, Frueh H, Morari M (2003) Robotic orthosis lokomat: a rehabilitation and research tool. Neuromodulation 6:108–115
26. Mehrholz J, Kugler J, Pohl M (2008) Locomotor training for walking after spinal cord injury. Cochrane Database Syst Rev 16(2):CD006676
27. National Spinal Cord Injury Statistical Center (2010). Annual report for the spinal cord injury model systems. Birmingham. Available at: goo.gl/rQEmNp
28. New PW, Cripps RA, Bonne Lee B (2014) Global maps of non-traumatic spinal cord injury epidemiology: towards a living data repository. Spinal Cord 52(2):97–109. Review. Erratum in: Spinal Cord 52(5):417
29. New PW, Townson A, Scivoletto G, Post MWM, Eriks-Hoogland I, Gupta A (2013) International comparison of the organisation of rehabilitation services and systems of care for patients with non-traumatic spinal cord injury. Spinal Cord 51(1):33–39
30. O'Connor RJ, Murray PC (2006) Review of spinal cord injuries in Ireland. Spinal Cord 44(7):445–448
31. Sadowsky CL, McDonald JW (2006) Activity-based restorative therapies: concepts and applications in spinal cord injury-related neurorehabilitation. Dev Disabil Res Rev 15:112–116
32. Scivoletto G, Morganti B, Molinari M (2005) Early versus delayed inpatient spinal cord injury rehabilitation: an Italian study. Arch Phys Med Rehabil 86(3):512–516
33. Scivoletto G, Morganti B, Ditunno P, Ditunno JF, Molinari M (2003) Effects of age on spinal cord lesion patients' rehabilitation. Spinal Cord 41:457–464
34. Silver JR (2005) History of the treatment of spinal injuries. Postgrad Med J 81(952):108–114
35. Smith M (2002) Efficacy of specialist versus non-specialist management of spinal cord injury within the UK. Spinal Cord 40(1):11–16
36. Swinnen E, Duerinck S, Baeyens JP, Meeusen R, Kerckhofs E (2010) Effectiveness of robot-assisted gait training in persons with spinal cord injury: a systematic review. J Rehabil Med 42(6):520–526
37. Van Asbeck FWA, Post MWM, Pangalila RF (2000) An epidemiological description of spinal cord injuries in The Netherlands in 1994. Spinal Cord 38(7):420–424
38. Wernig A, Muller S, Nanassy A, Cagol E (1995) Laufband therapy based on `rules of spinal locomotion' is effective in spinal cord injured persons. Eur J Neurosci 7:823–829
39. Wyndaele M, Wyndaele J-J (2006) Incidence, prevalence and epidemiology of spinal cord injury: what learns a worldwide literature survey? Spinal Cord 44(9):523–529

ReAbility: Complex External Prosthesis Systems to Rehabilitate Movement

Bice Previtera, Carlo Jovine, and Massimiliano Visocchi

Abstract The *ReAbility* Project is a solution that provides an alternative to wheelchair mobility for people with serious disabilities of the lower limbs, such as paraplegics, allowing them to stand and walk once again. This solution is achieved by utilizing an instrument called an "exoskeleton".

The *ReAbility* device is a robotic-assisted system designed to improve the quality of life of people with a movement disability and/or reduced movement; it is meant both for daily domestic use and for rehabilitation therapy in hospitals and rehabilitation centers.

The principal characteristics of the device are its modularity, manageability, and wearability for the patient, who is actively involved and has full control of all the movement functions. Its light weight (16 kg) renders it easy to wear for the patient and competitive on the market. From an esthetic perspective it can be worn under clothes, with obvious and important psychological and social advantages. Its cost is also well contained.

With the use of this device, there is also a very real positive effect on healthcare costs.

Keywords Neurorehabilitation • Exoskeleton • Paraplegy • Spinal cord injury • Movement • *ReAbility* • Healthcare costs

B. Previtera (✉)
Azienda Sanitaria Unica Regionale delle Marche, Area Vasta n. 2, Ancona, Italy
e-mail: biceprevitera@gmail.com

C. Jovine
Department of Neurology, Ospedale S.Giovanni Battista del Sovrano Militare Ordine di Malta, Rome, Italy

M. Visocchi
Institute of Neurosurgery, Medical School, Catholic University of Rome, Rome, Italy

Introduction

New knowledge has allowed old barriers to be broken. A spinal cord injury has extremely serious consequences for those afflicted, as well as for society in general. People with these injuries often need the help of others in order to conduct their lives, and they face innumerable obstacles along the course of their lives, be it in study, their professions, or other areas. With the right sociopolitical responses, it is, however, possible to enable people with a spinal cord injury to overcome these obstacles in every part of the world and to actively participate in society. Research commitments to paraplegics, in collaboration with the World Health Organization (WHO), are therefore an essential contribution, in their favor, to facilitate self-determination in their daily lives. While it is true that a spinal cord injury changes the life of a person, it is equally true that, notwithstanding this radical change, it is possible to maintain an elevated standard of life and an active daily routine. The WHO report "International Perspectives on Spinal Cord Injury" makes an important contribution in this sense, emphasizing the essential role of science in this pathology. The report presents a summary of the most important scientific factors and the most up-to-date knowledge relevant to the topic of spinal cord injury, offering, in particular, a vision of the fields of epidemiology, sanitary assistance, and medical interventions and an outline of the relevant policies for these conditions. Furthermore, actual experiences in the lives of para- and tetraplegics from across the world are reported. On the basis of this knowledge, concrete recommendations have been made in accordance with the United Nations Declaration on the Rights of Disabled Persons [1].

After the acute phase of a spinal cord injury, it is possible to work on stabilized medullar injuries. In that phase, the *ReAbility* Project has a place.

M. Visocchi et al. (eds.), *Trends in Reconstructive Neurosurgery*, Acta Neurochirurgica Supplement 124,
DOI 10.1007/978-3-319-39546-3_32,

Materials and Methods

The *ReAbility* device is a solution that provides an alternative to wheelchair mobility for people with serious disabilities of the lower limbs, such as paraplegics, allowing them to stand and walk once again. This solution is reached by utilizing an instrument called an "exoskeleton".

The instrument (Fig. 1) is a robotic-assisted system designed to improve the quality of life of people with a movement disability and/or reduced movement, both for daily domestic use and for rehabilitation therapy in hospitals and rehabilitation centers.

The design of the device was made possible by scientific collaboration among international university schools of science and technology, bringing together the world's leading experience on robotics applied to bioengineering.

The *ReAbility* system consists of a wearable exoskeleton integrated with a motor activated in correspondence with the wearer's joints, a series of movement sensors, and an IT system based on sophisticated controls and safe and secure algorithms, as well as a rechargeable battery.

The patient is actively involved and has full control of all the movement functions, through unique control processes. Walking around is controlled by variations in the center of gravity, and stability is guaranteed by the use of crutches.

In detail, the device is a motorized exoskeleton that is worn on the lower limbs either over or under clothes. The electric motors commanding the knee and hip joints, and powered by a battery carried in a backpack on the shoulders, are controlled by a computerized system that is also carried in the backpack. The exoskeleton, designed to be used with two Canadian crutches to guarantee stability, both when walking and when stationary in the erect position, is activated by sensors positioned on the front of the upper body, and controlled by the patient through slight changes in their center of gravity, or, in other words, through inclination of the upper body.

Fig. 1 The instrument, a robotic assisted system as a wearable exoskeleton integrated

The device is available in two versions, one for rehabilitation centers, to be used during the rehabilitation treatment, and the other personalized for the patient at the end of the rehabilitation period.

Results

The device allows for multiple results.

From a bioethical and psychological perspective the person is reborn, able to determine their daily life for themselves and dedicate themselves to their preferred activities, and those who are sporty excel, with surprising results.

From a social perspective, a return to socializing and integration between equals is evident.

From a strictly medical point of view, the advantages include regaining the capacity to walk around and the prevention of complications connected to paraplegia (in particular, rehospitalization, bedsores, osteoporosis and consequent pathological fractures and muscular atrophy, spasms, and chronic pain), not to mention a net reduction in the percentage of patients abandoning their classic braces (currently between 15 and 71 %).

From an economic aspect, there are also benefits due to the great reduction in healthcare costs (it is enough to think of the costs of the multidisciplinary treatment needed for difficult wounds such as bedsores, femoral fractures, and daily physiotherapy to reduce spasticity and muscular atrophy). In the United States, the cost of medical and intensive rehabilitative treatment for those with spinal cord injuries is estimated to be ten times that necessary for tumor treatment, six times that of a heart attack and three times that of a stroke. The American Food and Drug Administration (FDA) has, accordingly, approved the principles of the *ReAbility* device technology.

To highlight the enormous potential that this innovation holds, we note that a young paraplegic boy wearing an exoskeleton kicked the first ball (Fig. 2) [2] at the World Football Cup on 12 June 2014 in Sao Paolo, Brazil, thereby symbolically kicking off the great sporting event.

Discussion

Advantages of the Device

We can offer paraplegic patients and those with serious paraparesis the opportunity to discover the sensation of standing and walking once more. In fact the exoskeleton enables peo-

Fig. 2 The first ball that a young paraplegic boy wearing an exoskeleton kicked on 12 June 2014 at the World Football Cup in Sao Paolo, Brazil

ple with disabilities of the lower limbs to reinstate the lost functions, and thereby improve their physical health and quality of life.

The principal characteristics of the device are modularity, manageability, and wearability for the patient. Its light weight (16 kg) renders it easy to wear for the patient without help and makes it competitive on the market. From an esthetic perspective it can be worn under clothes, with obvious and important psychological and social advantages. The instrument is thus innovative, modular, light, flexible, and easy to use. The cost is also well contained.

There is also a very real positive effect on healthcare costs: in fact, on the one hand the device reduces the need for physiotherapy and rehospitalization caused by immobility, which many patients are constricted by, and on the other hand, by maintaining the patient in a vertical position daily, it alleviates many of the problems caused by long-term wheelchair use.

Within the realm of international collaboration, the research team has made their field of competence the modelling of walking and the development of dedicated information communication technology (ICT) applications available to facilitate specialization of the device, which is currently generic, for specific pathologies such as ictus.

Target of the Initiative

The *ReAbility* device is intended for people with a disability of the lower limbs, paraplegia, or serious paraparesis arising from multiple causes; in particular, from spinal cord injuries and spina bifida.

War veterans deserve particular mention. In the film *The Men*, Fred Zinnemann illustrates how these men are left half men: war veterans, paraplegics condemned to a wheelchair. A stylized and relentless opening scene depicts the action in the battle in which Bud Wilczek suffers a spinal injury and moves his legs for the last time. We see Bud, played by Marlon Brando in his tremendous cinematographic debut, with his sweating face and grimace of pain and dismay, together with the desperate voice describing the scene from beyond, twisting in a rise and fall of irony and torment. Immediately after, seen contorted in a hospital bed, Bud wonders: "I survived. Should I feel lucky?". At that time, disabled war veterans had few prospects in life. It has been reported that the Italian Paralympic Committee has signed an agreement with the Ministry of Defence to introduce sport for disabled military personnel. These soldiers, who have, to date, been confined to military hospitals, have been left alone in their search for a sports club.

The *ReAbility* device could similarly also be used for the aged. It could, in fact, be a resource for patients with serious walking difficulties, which are frequently seen in the aged. Considering the epidemiology of the aged population, the potential in this sector could be of extreme interest.

Estimate of Potential Users

Knowledge of the incidence and prevalence of spinal cord injuries makes it possible to organize prevention programs to sensitize the populations most at risk; for example, in schools, work environments, and domestic environments. Moreover, this information is fundamental for the national and regional planning of the Unipolar Spinal Unit in Italy, an organization that is totally dedicated to spinal cord injury treatment and scientific research in this realm [3].

Traumatic medullary lesions are a condition affecting almost one in every thousand people annually (0.721–0.906 per thousand in the United States; thus, about 40 new cases per million inhabitants per year). In Italy the incidence of spinal cord injury is about 18-20 new cases per year per million inhabitants. In a recent Italian epidemiology study [3] (conducted by GISEM: Gruppo Italiano Studio Epidemiologico Mielolesioni), involving the 37 principal centres dealing with spinal cord injuries, 1014 new cases of spinal cord injury were reported in two years.

The average age of the people affected by medullary lesions varies; 30 % are in the age group from 10 to 40 years, with young adults (between the ages of 16 and 30) being the most affected; however, the average age at trauma has increased since the 1970s, from 28.7 to 39.5 years. About 80 % are men, and the male/female ratio is 4: 1).

The majority of cases are the result of trauma (67.5 %), but there is a continual increase in the number of cases that are non-trauma-related (32.5 %). Among the causes, road accidents are the leading problem (42 %), followed by falls (27.1 %) and violence (15.3 %), with a large proportion of cases caused by violence being the result of firearms.

In the United States, non-traumatic medullary lesions account for 39 % of all spinal cord injuries. They have an incidence equal to 5–10 new cases for every million inhabitants and are principally the result of primary neoplasms, vascular pathologies, inflammations, infections, and degenerative diseases. The etiology puts degenerative pathologies at 53 %, neoplasms at 25 %, infections at 12 %, and vascular pathologies at 3 % of the cases. The average age at the time of injury is higher than in the case of traumatic injury (around 60 years old), while there are no gender differences and frequently the paralysis is incomplete.(34.1 %), with complete paraplegia at 23 %, incomplete paraplegia at 18.5 %, and complete tetraplegia at 18.3 %.

In Europe, spinal cord injuries are estimated to be principally of a traumatic origin. In northern and central Europe the incidence varies between 9.2 and 20 cases per million inhabitants/year, depending on the country and the study methodologies undertaken. The incidence of traumatic spinal cord injury is diminishing in some countries, such as Sweden, where major investment has been made in preventing road accidents, improving the infrastructure, and reinforcing driver education.

The incidence of traumatic cases in southern European countries is between 8 and 12 cases per million inhabitants/year in Spain and 58 cases per million/year in Portugal, while the total number of spinal cord injuries—traumatic and non-traumatic—is between 12 and 20 cases per million inhabitants/year.

In the Mediterranean countries the prevalence of all spinal cord injuries, independent of their cause, is around 350 cases per million inhabitants, with this trend rising as a result of the increased life expectancy of people with spinal cord injuries; in the Nordic countries the estimate of only traumatic injuries is 280 cases per million inhabitants.

In Italy it is estimated that every year there are around 1,800 new cases of paraplegia (mostly the consequence of trauma resulting from road accidents or accidents at work) and there is a paraplegic community of around 80,000 people.

Similar estimates can be made for war veterans and for the aged population.

Estimate of the Direct and Indirect Public

Direct Public

The direct public is represented by paraplegic patients and those with serious paraparesis or serious walking difficulties.

Indirect Public

The indirect public is composed of the whole scientific-medical community, with particular reference to neurologists, neurosurgeons, orthopedic specialists, physiatrists, and rehabilitation therapists, as well as the social, sporting, tourist, and economic-financial sectors.

The Objectives

The objectives of the ***ReAbility*** Project are to provide those who are confined to a wheelchair with restored psycho-physical integrity and freedom.

The Project also aims to promote the adoption of a framework for paraplegia that can equilibrate and harmonize the healthcare, social, organizational, economic, bioethical, psychological, and legal implications of this condition.

The Future

The Medical Role

The doctor is not only a professional provider of technical services, nor a mere unit of production, but rather a protagonist in the process of change in the healthcare system, and the evolution of this figure in society takes on a strategic role in the service of the said society.

The Project Aims

The capacity to walk again, provided by *ReAbility*, is revolutionary in that it has an exceptionally positive impact on the patient and their family, as well as having such an impact on society.

The aim of the Project is to relegate the wheelchair to the attic, entrusting in restored mobility for those who cannot walk. The technology that led to the exoskeleton represents the results of an enormous challenge, together with the great impulse of innovation. This sector is in such expansion that, according to the data in *Nova*, *Sole 24 Ore*, a growth rate of 68 % per year is expected in exoskeleton diffusion by the year 2020. This will have a phenomenal impact on the quality of life of disabled people, as well as on healthcare costs.

Fig. 3 Sunrise

Future Developments

The future developments are to be found in the concepts of "extended medicine": the passage from clinical medicine to molecular medicine, the microelectronic evolution, the info-telematic evolution, the "extended hospital", and the "extended doctor".

For sure the exoskeleton is an innovative device letting us go to the future.

"There is no night so long as to prevent the sun to rise again" (Fig. 3).

Conflict of Interest Statement The authors declare that they have no conflict of interest.

References

1. International Perspectives on Spinal Cord Injury WHO Report (2013) http://www.paraplegiker-stiftung.ch/files/pdf4/SPF_WHO-Bericht-Querschnittlahmung_I.pdf. Accessed 3 Dec 2013
2. Corriere della sera (2014) http://www.corriere.it/salute/disabilita/14_giugno_03/mondiali-ragazzo-tetraplegico-7eabea18-eb11-11e3-9008-a2f40d753542.shtml. Accessed 3 June 2014
3. Scivoletto G, Di Lucente L, Fuoco U, Di Donna V, Laurenza L, Macellari V, Giacomozzi C, Molinari M (2008) Riabilitazione e valutazione dei pazienti mielolesi: l'esperienza della Fondazione S. Lucia di Roma. Istituto Superiore di Sanità, Roma, Rapporti ISTISAN 08/39

Part VI

Neurosurgical Rehabilitation and Outcome: Special Issues

Focus on Functional Delayed Central Sleep Apnea Following Cervical Laminectomy. An Example of Respiratory Dysfunction in Restorative Neurosurgical Procedures

Massimiliano Visocchi, Gerardo Iacopino, Antonino Germanò, and Giuseppe Barbagallo

Abstract In sleep-related breathing disorders, sleep apnea is a clinical symptom that can be categorized as obstructive sleep apnea (OSA) or mixed apnea by analysis using polysomnography. The occurrence of delayed central sleep apnea (CSA) is an extremely rare complication of cervical laminectomy for spondylotic myelopathy. So far only three studies concerning such an event have been reported in the literature. Naim-ur-Rahman, in 1994, reported a case of postoperative CSA following C3-C6 laminectomy, and Visocchi and colleagues, in 2014, in two studies, stressed the lack of association with any other neurological sign of spinal cord damage. No definitive mechanism has been recognized so far for delayed CSA after cervical laminectomy. A transient dysfunction of the reticulo-spinal fibers directed to the nucleus of the phrenic nerve can be speculated, although neither emidiaphragm paralysis, nor any prominent nocturnal sleep-related disorders are associated with this delayed CSA.

Keywords Cervical laminectomy • Central sleep apnea • Obstructive sleep apnea

M. Visocchi, MD (✉)
Institute of Neurosurgery, Medical School, Catholic University of Rome, Rome, Italy

Division of Neurosurgery, Department of Neurosciences, Policlinico "G. Rodolico" University Hospital, Catania, Italy
e-mail: mvisocchi@hotmail.com

G. Iacopino, MD
Department of Experimental Biomedicine and Clinical Neurosciences, School of Medicine, Neurosurgical Clinic, University of Palermo, Palermo, Italy

A. Germanò, MD
Department of Neuroscience, Neurosurgical Clinic, University of Messina, Messina, Italy

G. Barbagallo, MD
Division of Neurosurgery, Department of Neurosciences, Policlinico "G. Rodolico" University Hospital, Catania, Italy

Introduction

In sleep-related breathing disorders, sleep apnea is a clinical symptom that can be categorized as obstructive sleep apnea (OSA) or mixed apnea by analysis using polysomnography. OSA is defined as a cessation of airflow for at least 10 s during sleeping. The event is obstructive if, during apnea, there is effort to breathe. On the other hand, in central sleep apnea (CSA), which is a less common clinical problem, there is no effort to breathe during sleep.

CSA has been reported, although rarely, in patients with upper cervical lesions caused by rheumatoid arthritis, Arnold-Chiari type 1 malformation, anterior C1-2 osteochondroma, and os odontoideum, but this complication has rarely been described following cervical laminectomy.

The occurrence of delayed sleep apnea is an extremely rare complication of cervical laminectomy for spondylotic myelopathy. Ventilatory insufficiency has been, of course, described as a complication following cervical vertebral surgery and spinal cord surgery. However, the event of breathing disorders occurring after an operation such as cervical laminectomy has been more rarely described, and additionally, the event of a *delayed* CSA syndrome occurring *after* cervical laminectomy is to be considered exceptional.

So far only three studies concerning such an event have been reported in the literature. Naim-ur-Rahman, in 1994, reported a case of postoperative CSA following C3-C6 laminectomy, occurring right after surgery and associated with spyncterial incontinence, that spontaneously recovered 3 weeks after onset [1]. My group (Visocchi et al.) reported in two studies, the occurrence of delayed onset (nearly 3 weeks after surgery) CSA not associated with any other neurological sign of spinal cord damage (postoperative neurophysiological tests showed, instead, an improvement compared with preoperative tests) [2, 3].

The mechanisms underlying such an event are difficult to interpret.

M. Visocchi et al. (eds.), *Trends in Reconstructive Neurosurgery*, Acta Neurochirurgica Supplement 124,
DOI 10.1007/978-3-319-39546-3_33,

Breathing Anatomy and Physiology

Breathing is a rhythmic motor behavior generated and controlled by hindbrain neuronal networks.

Neural circuits controlling breathing in mammals are organized within serially arrayed and functionally interacting brainstem compartments extending from the pons to the lower medulla. The core circuit components that constitute the neural machinery for generating respiratory rhythm and shaping inspiratory and expiratory motor patterns are distributed among three adjacent structural compartments in the ventrolateral medulla: the Bötzinger complex (BötC), pre-Bötzinger complex (pre-BötC), and the rostral ventral respiratory group (rVRG). The respiratory rhythm and inspiratory-expiratory patterns emerge from dynamic interactions between: (i) excitatory neuron populations in the pre-BötC and rVRG active during inspiration that form the inspiratory motor output; (ii) inhibitory neuron populations in the pre-BötC that provide inspiratory inhibition within the network; and (iii) inhibitory populations in the BötC active during expiration that generate expiratory inhibition.

More recent models describe interacting populations of respiratory neurons spatially distributed within the BötC and pre-BötC and rostral ventrolateral medulla that contain core circuits of the respiratory central pattern generator (CPG). Network interactions within these circuits, along with the intrinsic rhythmogenic properties of neurons, form a hierarchy of multiple rhythm generation mechanisms. The functional expression of these mechanisms is controlled by input drives from other brainstem components, including the retrotrapezoid nucleus and pons, which regulate the dynamic behavior of the core circuitry. The emerging view is that the brainstem respiratory network has rhythmogenic capabilities at multiple levels of circuit organization. This allows the flexible, state-dependent expression of different neural pattern-generation mechanisms under various physiological conditions, enabling a wide repertoire of respiratory behaviors. Some models consider control of the respiratory CPG is maintained by pulmonary feedback and network reconfiguration during defensive behaviors such as cough. The location and fiber arrangement of the descending respiratory pathways (involuntary respiratory pathway) in the ventral reticulospinal tract is close to the descending micturition pathways within the upper cervical cord [3].

Interestingly, the CSA syndrome in pathologies involving the craniovertebral junction (CVJ), such as axis rheumatoid arthritis, Arnold-Chiari type 1 malformation, anterior C1-2 osteochondroma, os odontoideum, and occipital encephalocele, can be related to a respiratory center dysfunction. More precisely, it might be postulated that a dysfunction in or adjacent to the pre-BötC in the lower medulla oblongata determines a loss of normal autonomic response to chemical changes in the blood.

On the other hand, spondylotic compression of the spinal cord is definitely anatomically far from the lower medulla (C3-C6), and also surgery at this level is too distant to hypothesize a direct compressive/traumatic mechanism determining a disturbance to the pre-Bötzinger area respiratory centers. If we critically analyze postoperative magnetic resonance imaging (MRI) after posterior decompression, we do not find that the spinal cord is displaced posteriorly, thus determining an angular deformation of the lower medulla oblongata that could justify such a mechanism.

In conclusion, no definitive mechanism has been recognized so far for delayed CSA after cervical laminectomy. A transient dysfunction of the reticulo-spinal fibers directed to the nucleus of the phrenic nerve can be speculated, although neither emi-diaphragm paralysis, nor any prominent nocturnal sleep-related disorders are associated with this delayed CSA.

Conflict of Interest Statement We declare that we have no conflict of interest.

References

1. Naim-Ur R (1994) Post-operative central sleep apnoea complicating cervical laminectomy: case report. Br J Neurosurg 8(5):621–625
2. Visocchi M, Della Pepa GM, Barbagallo G, Spallone A (2014) Reversible and delayed isolated central sleep apnea after cervical laminectomy: report of the first case. Acta Neurochir (Wien) 156(2):267–268
3. Della Pepa GM, La Rocca G, Barbagallo G, Spallone A, Visocchi M (2014) Transient breathing disorders after posterior cervical surgery for degenerative diseases: pathophysiological interpretation. Eur Rev Med Pharmacol Sci 18(1 Suppl):89–92

Brainstem and Autonomic Nervous System Dysfunction: A Neurosurgical Point of View

A. Martín-Gallego, L. González-García, A. Carrasco-Brenes, M. Segura-Fernández-Nogueras, A. Delgado-Babiano, A. Ros-Sanjuán, L. Romero-Moreno, M. Domínguez-Páez, M.S. Dawid-Milner, and M.A. Arráez-Sánchez

Abstract Central autonomic control nuclei and pathways are mainly integrated within the brainstem, especially in the medulla oblongata. Lesions within these structures can lead to central dysautonomia.

Central autonomic control structures can be damaged by tumors, during surgery, or by other neurosurgical pathologies. These may elicit clinical or subclinical autonomic complications that can constitute a serious clinical problem.

The authors present a broad review of the central autonomic nervous system, its possible dysfunctions, and the relation between neurosurgery and this "not-well-known system". Preliminary results of an autonomic study of brainstem lesions that is currently being carried out by the authors are also shown.

Keywords Autonomic nervous system • Brainstem • Neurosurgery • Dysautonomia • Tumors

Introduction

The major concern of a neurosurgeon during surgery that is inside or close to the brainstem is damage to vital motor and sensory nuclei and pathways; however, autonomic pathways and centers are less well known and considered. Central autonomic control structures are mostly located at different levels within the brainstem and can also be damaged during surgery, thus eliciting clinical or subclinical autonomic complications that can constitute a serious clinical problem.

Despite the important advances in neuroimaging and autonomic laboratory tests developed in the past two decades, it remains very difficult to make a diagnosis of a specific focal brainstem lesion and the subsequent autonomic dysfunction. The complexity of the brainstem anatomy and its autonomic network and pathways, added to the well-known neuroplasticity after chronic distortion of neural structures, may explain this notable difficulty.

Autonomic complications due to brainstem lesions or those occurring after tumor surgery are rarely reported. The literature is limited to very few case reports and short case series [3, 19, 22, 39, 50], and most of them did not confirm the autonomic dysfunction with appropriate tests. As brainstem tumors and surgery are quite common in our practice, and the generation of secondary autonomic dysfunction may be possible, it has become necessary to study and review the clinical manifestations of this type of lesion to gain an appropriate view of this clinical problem.

Central Autonomic Network (CAN) and the Brainstem

The central autonomic network (CAN), as defined by Benarroch [5], has been thoroughly studied during recent decades. The CAN consists of a group of interconnected areas, located between the telencephalon, diencephalon, and brainstem, that control and integrate sympathetic and parasympathetic visceral tone, input, and outflow.

The CAN includes several supratentorial areas: the prefrontal and insular cortex, the amygdala and stria terminalis, and the hypothalamus. The infratentorial areas include the periaqueductal gray matter (PGM) in the midbrain, the parabrachial Kölliker-Fuse nucleus region (PB) in the pons, and

A. Martín-Gallego (✉) • L. González-García • A. Carrasco-Brenes
M. Segura-Fernández-Nogueras • A. Delgado-Babiano
A. Ros-Sanjuán • L. Romero-Moreno • M. Domínguez-Páez
M.A. Arráez-Sánchez
Department of Neurosurgery, Regional University Hospital of Málaga, Avenida Carlos Haya s/n, 29010 Málaga, Spain
e-mail: alvaro1710@hotmail.com

M.S. Dawid-Milner
Department of Autonomic Nervous System, CIMES, University of Málaga Foundation (FGUMA), Malaga, Spain

M. Visocchi et al. (eds.), *Trends in Reconstructive Neurosurgery*, Acta Neurochirurgica Supplement 124,
DOI 10.1007/978-3-319-39546-3_34,

several important nuclei in the medulla oblongata. Considered the most important region of the CAN, the medulla oblongata includes relevant nuclei that control cardiovascular and respiratory responses. These nuclei are: the nucleus of the solitary tract (NST), the rostral and caudal ventrolateral medulla (RVLM and CVLM) with sympathetic function, the nucleus ambiguus, and the dorsal motor nucleus of the vagus nerve (DMNV) with parasympathetic function (Fig. 1).

The NST is the first central relay for afferent inputs controlling cardiovascular functions, including the baroreflex and cardiac reflexes, respiratory and laryngeal reflexes [11], and gastrointestinal motility [9]. Due to its anatomical location, close to the floor of the fourth ventricle, the NST can be easily damaged during surgery over the floor of the fourth ventricle. Its bilateral lesion produces acute hypertension in rats [14], while in bigger animals this lesion can lead to systemic blood pressure (BP) lability, chronic hypertension, and exaggerated BP responses after environmental stress [35, 44].

The RVLM is considered the key area for BP regulation and it has been established as the major sympathetic pressor center, providing sympathetic tonic outflow activation to peripheral vessels of the skeletal muscles and visceral organs, as well as to the coronary vessels and cardiac muscle [6]. This nucleus receives, among others, inhibitory signals from the NST and CVLM and it is located in the ventrolateral medulla, at the entry zone of the IX and X cranial nerves [45]. RVLM lesions can lead to BP collapse and orthostatic hypotension [20]. RVLM chronic compression can result in chronic overstimulation of the sympathetic nervous system, leading to chronic arterial hypertension [25, 32].

The CVLM maintains tonic inhibitory control of the RVLM and mediates the sympathoinhibitory component of the arterial baroreflex. This nucleus is within the ventrolateral bulbo-medullary region and its lesion is related to hypertensive states, similar to those observed in NST lesions [10].

The bulbar region also contains two groups of parasympathetic preganglionic neurons: (1) the DMNV, located in the dorsomedial medulla, close to the NST, at the floor of the fourth ventricle; and (2) the nucleus ambiguus, located in the bulbo-medullary region with a ventrolateral location. The DMNV innervates the local ganglia in the respiratory tract, enteric nervous system, liver, and pancreas, and its lesion can lead to gastrointestinal dysmotility and other visceral disorders [37]. The nucleus ambiguus is the main origin of cardiomotor inhibitory output [27] and it also controls respiratory function. Unilateral lesion of this nucleus produces central hypoventilation, laryngeal stridor, impaired baroreflex cardioinhibition, and respiratory sinus arrhythmia [9].

The PB in the pons is a coordinating center. It receives visceral, thermoreceptive, and nociceptive inputs from the spinal cord and sends this information to the hypothalamus, amygdala, and thalamus [43]. The PB participates in the control of respiratory, cardiovascular, gastrointestinal, and micturition reflexes [13]. The PGM, in the midbrain, has a role in the integration of autonomic and somatic responses to stress, pain modulation, and other adaptive functions. It participates in cardiovascular responses associated with pain modulation, thermoregulation, coordination of micturition reflexes, and respiratory rhythm [9].

Autonomic Tests

The autonomic nervous system regulates vital homeostatic functions such as BP, heart rate, respiration, thermoregulation, gastrointestinal activity, and urinary bladder and sexual functions.

Recent advances have resulted in the development of new quantitative, non-invasive, and reproducible autonomic tests. The main goals of these tests are to easily evaluate the severity and the distribution of autonomic neuropathies, and to monitor their course and response to treatment.

There are tests that evaluate cardiovascular function and tests that quantify sudomotor function [31]. Cardiovascular tests can be divided into cardiovagal and adrenergic tests. The most commonly used cardiovagal tests are: quantification of the heart rate response to deep breathing and to the Valsalva maneuver. Adrenergic function can be measured by the following methods: (1) BP and heart rate responses to the Valsalva maneuver, (2) head-up tilt test, (3) plasma norepinephrine, and (4) quantitave cardiac uptake of MIBG (iodine-123 meta-iodobenzylguanidine).

Tests that analyze sudomotor function are: the quantitative sudomotor axon reflex test (QSART) and the thermoregulatory sweat test (TST) [31]. The QSART evaluates the integrity of the postganglionic sympathetic sudomotor axon and can define the distribution of sweat loss in the forearm, proximal leg, distal leg, and foot. The TST evaluates pre- and postganglionic thermoregulatory sympathetic pathways from the hypothalamus to the eccrine sweat glands. This test can detect patterns of anhydrosis.

Autonomic Dysfunction

Autonomic dysfunction can be classified, according to its extent, as generalized autonomic failure, selective or partial autonomic failure, distal neuropathies, and orthostatic intolerance syndromes [6].

Symptoms that suggest the existence of autonomic failure include: orthostatic hypotension (OH); digestive discomfort; gastrointestinal, sexual, and bladder dysfunction; increase or

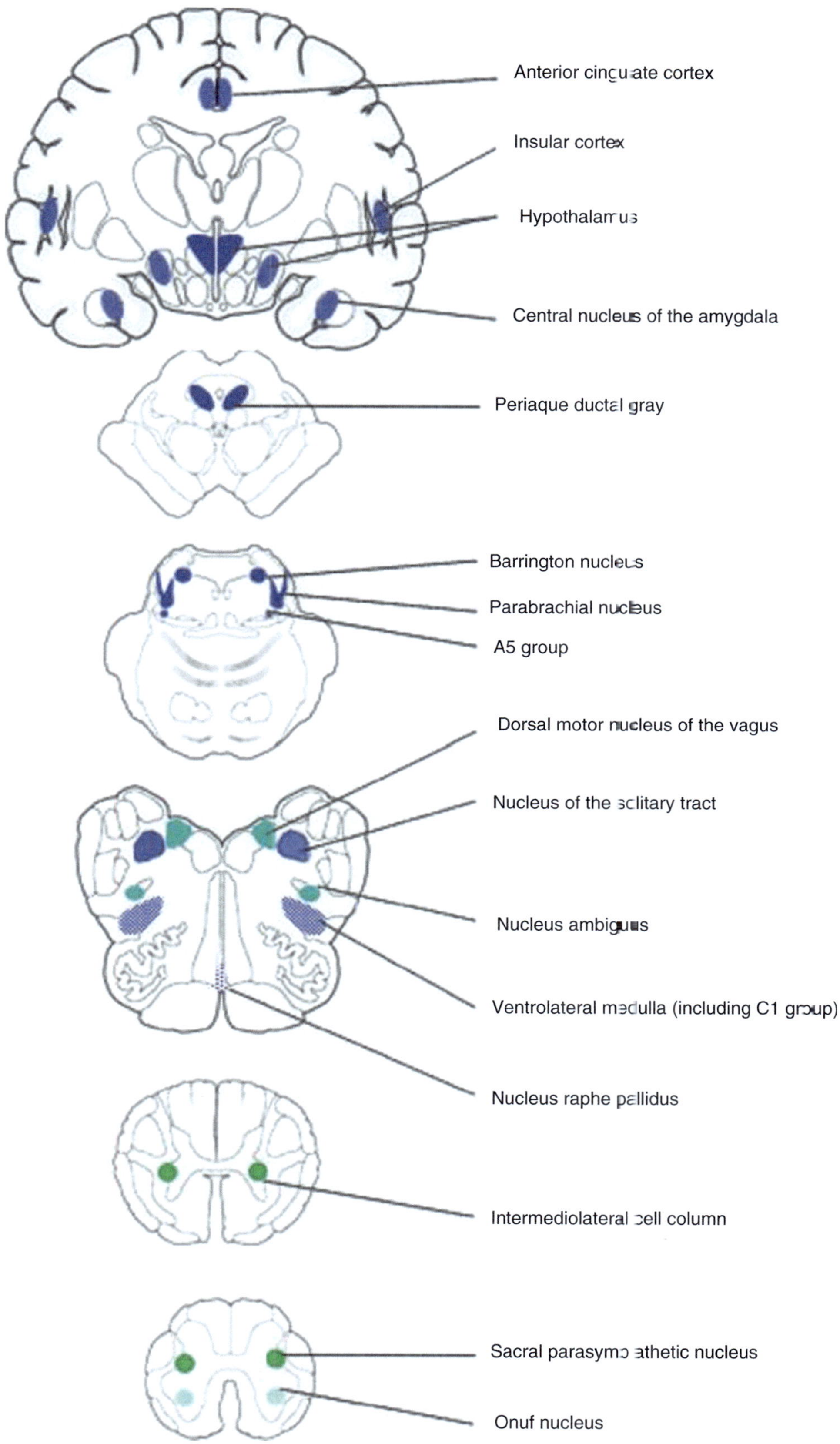

Fig. 1 Central autonomic control areas (From Cersosimo et al. [9])

decrease of sweating; dry mucous membranes; and changes in skin temperature and color [36].

OH can explain some of the symptoms that neurosurgical patients develop before and after brainstem surgery. Lightheadedness, blurred vision, cognitive difficulties, headaches, tremulousness, and palpitations can be related to OH.

As previously mentioned, autonomic nuclei can be damaged during brainstem surgery. Surgery over the floor of the fourth ventricle can easily injure the NST, which may elicit baroreflex failure. Baroreflex failure can lead to acute severe arterial hypertension, bradycardia, and cardiorespiratory arrest, as well as chronic cardiovascular instability and OH. If the injury is produced within the region of the ventrolateral surface of the pontomedullary junction, as in acoustic schwannoma surgery, the RVLM may be damaged and patients may develop severe postoperative OH. On the other hand, if the RVLM is stimulated by a pulsatile arterial loop or a neural structure, sympathetic hyperstimulation and subsequent chronic arterial hypertension may be elicited [26].

Which Neurosurgical Pathologies Can Affect CAN Structures?

Brainstem lesions associated with dysautonomia include tumors, multiple sclerosis [42], stroke [1, 7], syringobulbia [21], Chiari malformation [47], and vascular compression [15].

Tumors

The brainstem can be affected by intrinsic tumors (including cavernous malformations) and extrinsic compressing posterior fossa tumors.

The authors are currently carrying out a study of brainstem tumors (intrinsic and extrinsic); 43 patients have been studied. Intrinsic and extrinsic gliomas, ependymomas, medulloblastomas, cavernous malformations, epidermoid cysts, cerebellopontine angle schwannomas, choroid plexus papillomas, chondrosarcomas, pituitary adenomas, lymphomas, chordomas, arachnoid cysts, and meningiomas have been collected (Fig. 2).

In Fig. 1 we show six cases of patients from our study with brainstem compression caused by tumors and presenting autonomic dysfunction. Relevant data were collected after our preliminary analysis: 92.9% of the patients tested showed some grade of autonomic dysfunction. The most frequent diagnoses were: autonomic instability (57.1%), sympathetic hyperactivity (35.7%), and OH (31%).

The first case of a brainstem tumor causing autonomic dysfunction was reported by Wagner et al., who described a tumor located in the floor of the fourth ventricle [22]. The first authors to describe the relation between the removal of tumors from the floor of the fourth ventricle and physiological abnormalities were Baker et al., in 1964 [2], who stated: "the floor of the fourth ventricle is considered no man's land, and from an anatomical and physiological standpoint it is a very important region". Baker and colleagues reported their experience with 11 subependymomas, which they classified into three types: (1) those that can be removed completely without any physiological changes, (2) others that cannot be approached because respiratory paralysis or vasomotor instability develops with manipulation, and (3) those that can be completely removed but with a residue of severe long-lasting physiological abnormalities. The autonomic abnormalities collected were: respiratory paralysis, OH, postural tachycardia, and digestive disorders.

Cameron and Doig described two patients with systemic arterial hypertension that improved after posterior fossa tumor surgical resection [8]. They included the posterior fossa tumors in the differential diagnosis of pheochromocytoma. Evans et al. published the first pediatric case of astrocytoma near the floor of the fourth ventricle, mimicking the features of pheochromocytoma. These clinical findings disappeared after tumor removal [16]. Cushing syndrome has also been included in the differential diagnosis of posterior fossa tumors [51].

Riedel et al. were the first to describe OH after the surgery of posterior fossa tumors: a foramen magnum meningioma and an ependymoma of the floor of the fourth ventricle [40]. Wörner et al. also showed another foramen magnum meningioma, with compression of the ventrolateral medulla oblongata, in which resection solved the chronic arterial hypertension [50].

A decade later, Hsu et al. [22], in 1984, published the first autonomic study in patients with a brainstem tumor. They described three patients with posterior fossa tumors and brainstem compression. They carried out an autonomic evaluation in one of the patients with a diagnosis of dorsal medulla oblongata astrocytoma with pontine and cervical medulla invasion. The study revealed OH caused by baroreflex dysfunction. Yamashita et al. described similar findings in autonomic tests in a patient with a diagnosis of bulbar lymphoma with OH [52]. These authors postulated damage of the NST or the DMNV as the origin of the dysfunction.

Rouchoux et al. reported, for the first time, the possible relation between hydrocephalus and brainstem lesions [41]. Subsequently, in 2010, Gomez-Esteban published a case of an epidermoid cyst associated with hydrocephalus that

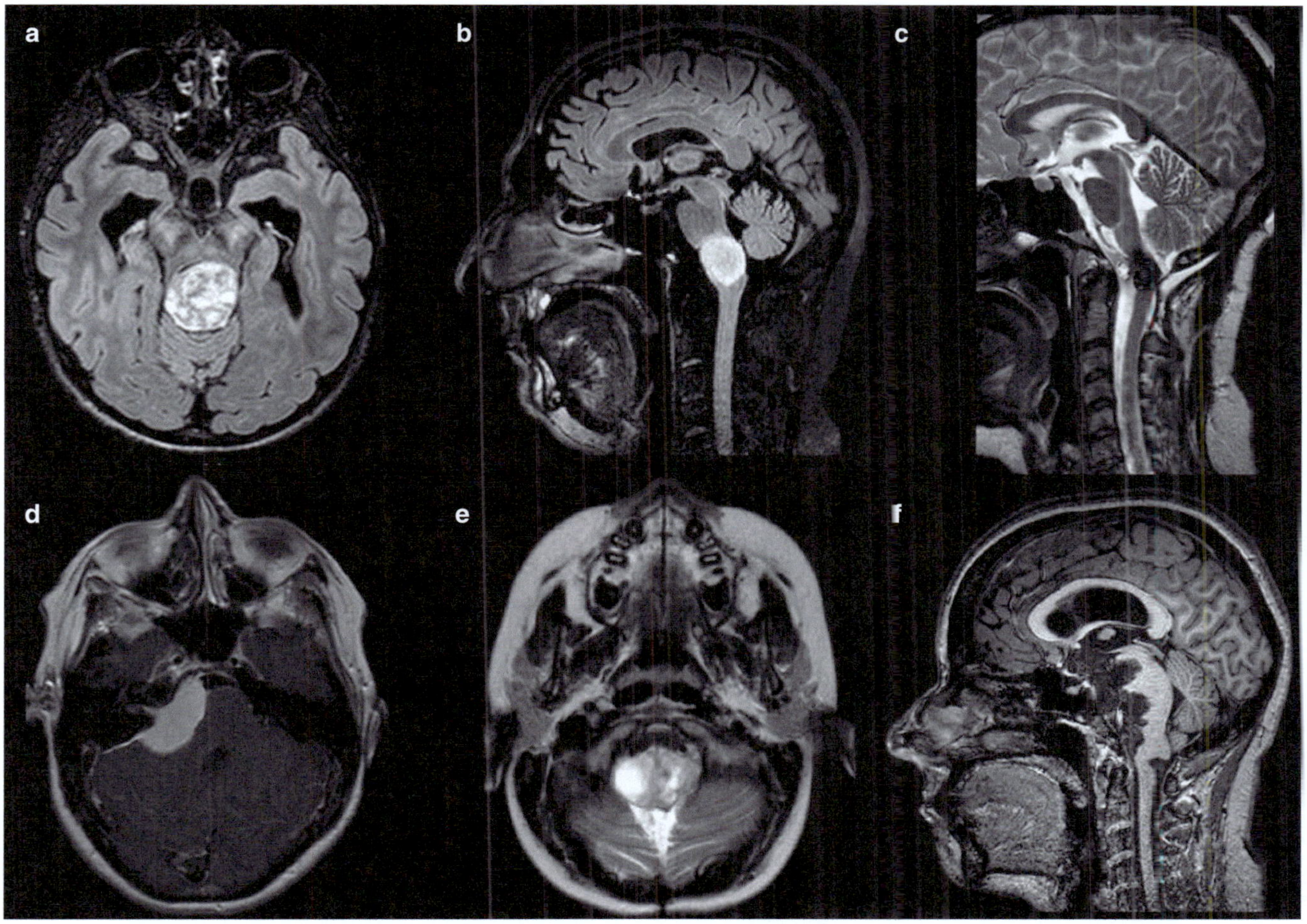

Fig. 2 (**a**) Patient with a low-grade glioma compressing the dorsal surface of the mesencephalon, presenting orthostatic hypotension and postural orthostatic tachycardia; (**b**) patient with a bulbar intrinsic lesion presenting sympathetic hyperactivity, autonomic instability, orthostatic hypotension, and postural orthostatic tachycardia; (**c**) patient operated for a bulbo-medullary cavernous malformation, diagnosed with autonomic instability and orthostatic hypotension; (**d**) patient with a cerebello-pontine meningioma with ventrolateral pontine compression, presenting sympathetic instability and hyperactivity; (**e**) pediatric patient operated for a pilocytic astrocytoma with bulbar distortion, presenting arterial hypertension and orthostatic hypotension; (**f**) patient with an epidermoid cyst causing compression of all brainstem levels, presenting autonomic instability, sympathetic hyperactivity, and orthostatic hypotension

caused OH. The autonomic dysfunction was resolved after the placement of a ventriculoperitoneal shunt [19].

Monge Argilés et al. published a case-control study of 14 patients with brainstem lesions (ischemic strokes, vascular malformations, and neoplastic and post-traumatic lesions) and 25 control patients. The authors demonstrated a decrease in heart rate variability in the patients with brainstem lesions [32].

More recently, Idiaquez et al. described a woman with syncopal episodes with a diagnosis of a cavernous malformation of the dorsal and posterior medulla oblongata. After a partial resection of the tumor, the patient experienced a worsening of the OH. The authors proposed a lesion of the catecholaminergic transtegmental tract as the possible origin of the disease [24]. Ideguchi et al. reported a case of sympathetic hyperactivity with severe arterial hypertension and tachycardia that developed after surgery for a hemangioblastoma with bulbar and pontine compresssion. The authors postulated bilateral NST damage as the origin of the autonomic dysfunction [23].

From our preliminary study, we can report the case of a 9-year-old male patient with a recurrent fourth ventricle anaplastic ependymoma who developed severe arterial hypertension and BP lability during and after surgery. A punctual bilateral lesion located within the mid-dorsal medulla oblongata, caused by both infiltration and surgical resection, was observed on postoperative magnetic resonance imaging (MRI) (Fig. 3). On autonomic evaluation, we observed an increase in sympathetic outflow with tachycardia, together with OH caused by baroreceptor reflex dysfunction. We postulate that a bilateral injury to both nuclei of the solitary tract may have caused central dysautonomia.

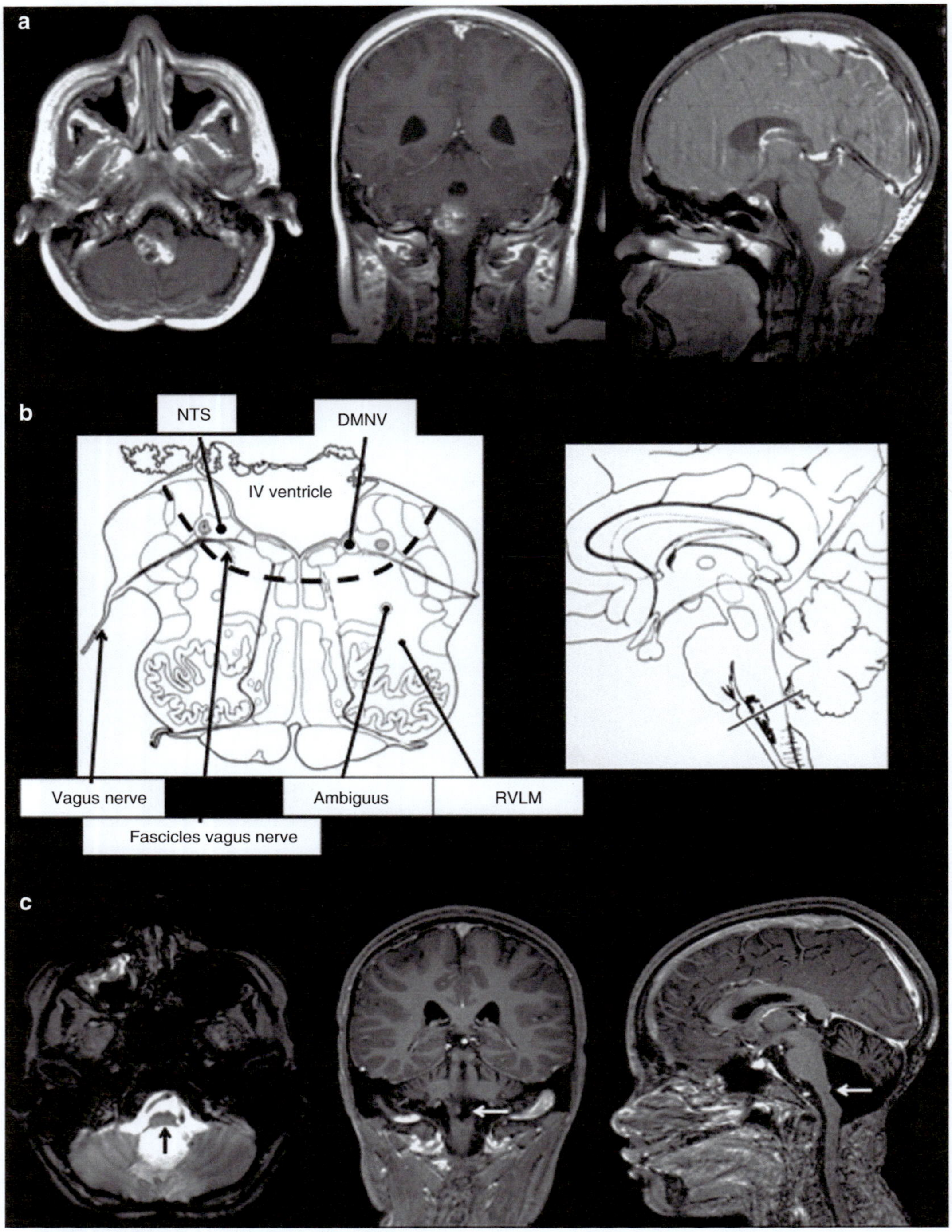

Fig. 3 (**a**) Axial, sagittal, and coronal preoperative magnetic resonance imaging (MRI) views, showing new recurrence of anaplastic ependymoma with infiltration of the floor of the fourth ventricle. (**b**) Axial view of the upper medulla oblongata on anatomical atlas [12]. *Black circle* shows the possible location of the injury. (**c**) MRI showing post-operative distortion/injury of the medulla oblongata. *Black and white arrows* show the area of injury. This MRI was performed on the same day as the autonomic tests

Vascular Compression

In 1979, Jannetta et al. observed that brainstem compression caused by arterial loops could be responsible for some cases of essential neurogenic arterial hypertension [25]. The authors found that compression of the left ventrolateral surface of the medulla oblongata, specifically over the retro-olivary sulcus, close to the exit of the IX and X cranial nerves, may elicit activation of the RVLM, evoking a subsequent increase in sympathetic tone and BP elevation. The authors showed the results in five patients, with a significant improvement of the hypertension after vascular microdecompression surgery.

In the following years, several studies in animals concluded that chronic compression of RVLM elicits arterial hypertension [26, 33, 45]. The proposed theories for the arterial pulsatile beat effect on sympathetic pathways have focused on: (1) direct hyperstimulation of the RVLM and, (2) mechanical damage of the afferent pathways of the IX and X nerves that may cause deafferentation of the baroreceptor reflex arch with subsequent sympathetic hyperactivity due to RVLM liberation [34, 48].

Neuroimaging studies have revealed controversial results. Naraghi et al., in their study of 24 patients with a diagnosis of arterial hypertension, found 20 patients with vascular compression of the left RVLM [34]. Morimoto et al. observed that 78% of their hypertensive patients showed compression, compared with 34.5% of normal patients [33]. On the other hand, Watters did not find any statistically significant difference between cases and controls [49]. Finally, Levy et al. (2001), in their review, concluded: "current MRI technology is not adequate to image the microvasculature of the brainstem" [30].

Regarding the surgical findings and outcomes, Jannetta et al. reported their results in 53 hypertensive patients treated with microvascular decompression for a primary diagnosis of trigeminal neuralgia or hemifacial spasm; 36 of 42 surgically treated patients had normal postoperative BP (follow-up 2–9 years) [26]. Levy et al. reported that 8 eight of 12 patients with refractory arterial hypertension achieved an improvement of at least 20 mmHg in their systolic BP after decompression. Six of them presented sustained normotension [29]. Geiger et al. published, in *The Lancet* in 1998, their results in eight patients with essential arterial hypertension. Postoperative BP was lowered in seven patients, but with a very short-lasting follow-up [18]. Three years later, these authors reported, in *Stroke*, their results with a longer follow-up (>4 years): three patients remained normotensive (one patient died with a follow-up of 2 years), and five patients remained hypertensive [17]. More recently, Barley et al. published a review of eight studies that included 107 patients with refractory arterial hypertension of neurogenic origin and concluded that the observed positive short-term outcomes indicated potential benefits from surgical intervention in a small subset of patients with refractory or resistant hypertension [4]. Sindou et al. have published the latest report of microvascular decompression surgery [46]. These authors reported 48 patients with hemifacial spasm and severe arterial hypertension who underwent microvascular decompression 28 patients had returned to normal BP, and no differences between left and right sides were found, at a mean follow-up of 7 years.

Regarding the conclusions of the experts about microvascular decompression, Levy and Jannetta reported that this technique has a therapeutic role in the treatment of refractory hypertension, and its success lies in the adequate selection of patients [30]. Frank et al. concluded that the technique is a successful alternative therapy in certain subgroups of patients with intractable hypertension [17]. Kaplan made the following comment on this theme: "no patient should be subjected to this procedure until a properly controlled trial is conducted" [28]. Pickering et al. concluded: "neurovascular compression as a surgically treatable cause of hypertension is an intriguing possibility, but is not ready for prime time" [38]. Zaidi et al., from the group of Dr. Spetzler, commented that larger series are needed to determine the viability of the surgical approach compared with medical treatment [53]. Sindou et al. suggested the current indications of surgery: "essential hypertension, likely to be neurogenic, in patients in whom high-resolution MRI shows clear-cut images of neurovascular compression at the CN IX-X REZ (root entry/exit zone) and adjacent ventrolateral medulla and in whom BP cannot be controlled by medical treatment" [46]. Levy and Jannetta reported, in 2001, that a multi-institutional, randomized, prospective trial was in progress [30]. But these results have not been published yet.

Other Pathologies

Chiari malformation [47] and syringobulbia [21] have also been described as possible causes of autonomic dysfunction.

Conclusion

The central nuclei or pathways of the autonomic nervous system can be impaired during brainstem surgery, leading to severe autonomic complications. We suggest that patients can benefit from pre- and postoperative autonomic studies, in order to diagnose and treat these clinical and subclinical complications as early as possible. The authors are currently carrying out a clinical study in order to analyze the possible autonomic alterations caused by brainstem tumors or surgery. The preliminary results show relevant findings.

Conflict of Interest Statement The authors declare that they have no conflict of interest.

References

1. Altavilla R, Landi D, Altamura C, Bussone G, Maggio P, Corbetto M et al (2012) Dorsolateral medullary ischemic infarction causing autonomic dysfunction and headache: a case report. J Headache Pain 13(4):327–330. doi:10.1007/s10194-012-0427-8
2. Baker GS (1965) Physiologic abnormalities encountered after removal of brain tumors from the floor of the fourth ventricle. J Neurosurg 23(3):338–343
3. Barbizet J, Degos JD, Poirier J, Marsault C, Meyrignac C (1981) Dysautonomia and behavioural disorders from periventricular metastasis. Rev Neurol (Paris) 137(5):319–331
4. Barley J, Ellis C (2013) Microvascular decompression: a surgical option for refractory hypertension of neurogenic etiology. Expert Rev Cardiovasc Ther 11(5):629–634. doi:10.1586/erc.13.30
5. Benarroch EE (1993) The central autonomic network: functional organization, dysfunction, and perspective. Mayo Clin Proc 68(10):988–1001
6. Benarroch EE (2014) The clinical approach to autonomic failure in neurological disorders. Nat Rev Neurol 10(7):396–407
7. Brozman B, Romano JG, Tusa RJ, Forteza AM (2002) Postural vertigo and impaired vasoreflexes caused by a posterior inferior cerebellar artery infarct. Neurology 59(9):1469–1470
8. Cameron SJ, Doig A (1970) Cerebellar tumours presenting with clinical features of phaeochromocytoma. Lancet 1(7645):492–494
9. Cersosimo MG, Benarroch EE (2013) Central control of autonomic function and involvement in neurodegenerative disorders. Handb Clin Neurol 117:45–57. doi:10.1016/B978-0-444-53491-0.00005-5
10. Cravo SL, Morrison SF, Reis DJ (1991) Differentiation of two cardiovascular regions within caudal ventrolateral medulla. Am J Physiol 261:R985–R994
11. Dawid-Milner MS, Silva-Carvalho L, Goldsmith GE, Spyer KM (1995) Hypothalamic modulation of laryngeal reflexes in the anaesthetized cat: role of the nucleus tractus solitarii. J Physiol 487(Pt 3):739–749
12. DeArmond SJ, Fusco MM, Dewey MM (1989) Structure of the human brain. A photographic atlas, 3rd edn. Oxford university Press, Inc, New York
13. Díaz-Casares A, López-González MV, Peinado-Aragonés CA, González-Barón S, Dawid-Milner MS (2012) Parabrachial complex glutamate receptors modulate the cardiorespiratory response evoked from hypothalamic defense area. Auton Neurosci 169(2):124–134. doi:10.1016/j.autneu.2012.06.001, Epub 2012 Jun 28
14. Doba N, Reis DJ (1973) Acute fulminating neurogenic hypertension produced by brainstem lesions in the rat. Circ Res 32:584–593
15. Emanuele MA, Dorsch TR, Scarff TB, Lawrence AM (1981) Basilar artery aneurysm simulating pheochromocytoma. Neurology 31(12):1560–1561
16. Evans CH, Westfall V, Atuk NO (1972) Astrocytoma mimicking the features of pheochromocytoma. N Engl J Med 286(26):1397–1399
17. Frank H, Schobel HP, Heusser K, Geiger H, Fahlbusch R, Naraghi R (2009) Long-term results after microvascular decompression in essential hypertension. Stroke 40(1):47–51
18. Geiger H, Naraghi R, Schobel HP, Frank H, Sterzel RB, Fahlbusch R (1998) Decrease of blood pressure by ventrolateral medullary decompression in essential hypertension. Lancet 352(9126):446–449, Stroke. 2001 Dec 1;32(12):2950-5
19. Gómez-Esteban JC, Berganzo K, Tijero B, Barcena J, Zarranz JJ (2009) Orthostatic hypotension associated with an epidermoid tumor of the IV ventricle. J Neurol 256(8):1357–1359
20. Guyenet PG (2006) The sympathetic control of blood pressure. Nat Rev Neurosci 7(5):335–346
21. Heidel KM, Benarroch EE, Gené R, Klein F, Meli F, Saadia D et al (2002) Cardiovascular and respiratory consequences of bilateral involvement of the medullary intermediate reticular formation in syringobulbia. Clin Auton Res 12(6):450–456
22. Hsu CY, Hogan EL, Wingfield W, Webb JG, Perot PL Jr, Privitera PJ et al (1984) Orthostatic hypotension with brainstem tumors. Neurology 34:1137–1143
23. Ideguchi M, Kajiwara K, Yoshikawa K, Kato S, Ishihara H, Fujii M, Fujisawa H, Suzuki M (2010) Continuous hypertension and tachycardia after resection of a hemangioblastoma behind the dorsal medulla oblongata: relationship to sympathetic overactivity at the neurogenic vasomotor center. J Neurosurg 113(2):369–373. doi:10.3171/2009.10.JNS09504
24. Idiaquez J, Araya P, Benarroch E (2009) Orthostatic hypotension associated with dorsal medullary cavernous angioma. Acta Neurol Scand 119(1):45–48
25. Jannetta PJ, Gendell HM (1979) Clinical observations on etiology of essential hypertension. Surg Forum 30:431–432
26. Jannetta PJ, Segal R, Wolfson SK (1985) Neurogenic hypertension: etiology and surgical treatment; observa- tions in 53 patients. Ann Surg 201:391–398
27. Jordan D (2005) Vagal control of the heart: central serotonergic (5-HT) mechanisms. Exp Physiol 90(2):175–181
28. Kaplan NM (2001) Microvascular decompression: hype or hoax? Editorial comment. Stroke 32:2955–2956
29. Levy EI, Clyde B, McLaughlin MR, Jannetta PJ (1998) Microvascular decompression of the left lateral medulla oblongata for severe refractory neurogenic hypertension. Neurosurgery 43(1):1–6; discussion 6–9
30. Levy EI, Scarrow AM, Jannetta PJ (2001) Microvascular decompression in the treatment of hypertension: review and update. Surg Neurol 55(1):2–10; discussion 10–1
31. Low PA (1993) Autonomic nervous system function. J Clin Neurophysiol 10(1):14–27
32. Monge Argilés JA, Palacios Ortega F, Vila Sobrino JA, Bautista Prados J, Pérez Vicente JA, Morales Ortiz A, Palao SA (2000) Brainstem lesions decrease heart rate variability. Neurologia 15(4):158–163
33. Morimoto S, Sasaki S, Miki S, Kawa T, Iotoh H, Nakata T et al (1997) Pulsatile compression of the rostral ventrolateral medulla in hypertension. Hypertension 29(part 2):514–518
34. Naraghi R, Fahlbusch R (1999) Microvascular decompression of the left lateral medulla oblongata for severe refractory neurogenic hypertension. Neurosurgery 45(4):955
35. Nathan MA, Reis DJ (1977) Chronic labile hypertension produced by lesions of the nucleus tractus solitarii in the cat. Circ Res 40:72–81
36. Navarro X (2016) Functional evaluation of peripheral nerve regeneration and target reinnervation in animal models: a critical overview. Eur J Neurosci 43(3):271–286. doi:10.1111/ejn.13033. Epub 2015 Aug 26. Review
37. Pfeiffer RF (2003) Gastrointestinal dysfunction in parkinson's disease. Lancet Neurol 2(2):107–116
38. Pickering TG (2007) Neurovascular compression of the medulla: can it cause neurogenic hypertension? J Clin Hypertens (Greenwich) 9(1):63–66
39. Rhodes RH, Wightman HR (2000) Nucleus of the tractus solitarius metastasis: relationship to respiratory arrest? Can J Neurol Sci 27(4):328–332

40. Riedel G, Frewin DB, Gladstone L, Downey JA (1974) Orthostatic hypotension following surgery on brain-stem neoplasms: report of two cases. Arch Phys Med Rehabil 55(10):471–473
41. Ruchoux MM, Gray F, Gherardi R, Schaeffer A, Comoy J, Poirier J (1986) Orthostatic hypotension from a cerebellar gangliocytoma (Lhermitte-Duclos disease). Case report. J Neurosurg 65(2):245–248
42. Sakakibara R, Mori M, Fukutake T, Kita K, Hattori T (1997) Orthostatic hypotension in a case with multiple sclerosis. Clin Auton Res 7:163–165
43. Saper CB (2002) The central autonomic nervous system: conscious visceral perception and autonomic pattern generation. Annu Rev Neurosci 25:433–469, Epub 2002 Mar 25
44. Schmitt H, Laubie M (1979) Destruction of the nucleus tractus solitarii in dogs: acute effects on blood pressure and haemodynamics, chronic effects on blood pressure: importance of the nucleus for the effects of drugs. In: Meyer P, Schmitt H (eds) Nervous System and Hypertension. Willey-Flammarion, New York, pp 173–201
45. Segal R, Gendell HM, Canfield D, Dujovny M, Jannetta PJ (1982) Hemodynamic changes induced by pulsatile compression of the ventrolateral medulla. Angiology 33(3):161–172
46. Sindou M, Mahmoudi M, Brînzeu A (2015) Hypertension of neurogenic origin: effect of microvascular decompression of the CNIX-X root entry/exit zone and ventrolateral medulla on blood pressure in a prospective series of 48 patients with hemifacial spasm associated with essential hypertension. J Neurosurg 31:1–9
47. Straus D, Foster K, Zimmerman F, Frim D (2009) Chiari drop attacks: surgical decompression and the role of tilt table testing. Pediatr Neurosurg 45(5):384–389. doi:10.1159/000260909, Epub 2009 Nov 26
48. Thorén P (1979) Role of cardiac vagal C-fibers in cardiovascular control. Rev Physiol Biochem Pharmacol 86:1–94
49. Watters MR, Burton BS, Turner GE, Cannard KR (1996) MR screening for brain stem compression in hypertension. AJNR Am J Neuroradiol 17(2) 217–221
50. Wörner BA, Rahim T, Lange M, Fink U, Oeckler R (2002) Long-lasting improvement of arterial hypertension after surgical treatment of a foramen magnum meningioma: case report. Surg Neurol 58(3-4):189–192; discussion 193
51. Yagil Y, Futterweit W, Krakoff LR, Rubin H, Weinrauch H (1989) Cerebellar tumor causing hypertensive crisis and simulating pheochromocytoma and Cushing's syndrome. Mt Sinai J Med 56(1):56–58
52. Yamashita A, Hirayama M, Koike Y, Nukata M, Hashizume M, Takahashi A (1996) Orthostatic hypotension caused by a localised dorsal medullary tumour. J Neurol Neurosurg Psychiatry 61(1):118–119
53. Zaidi HA, Awad AW, Chowdhry SA, Fusco D, Nakaji P, Spetzler RF (2015) Microvascular decompression for hemifacial spasm secondary to vertebrobasilar dolichoectasia: surgical strategies, technical nuances and clinical outcomes. J Clin Neurosci 22(1):62–68. doi:10.1016/j.jocn.2014.09.008, Epub 2014 Dec 12

Correlation Between Timing of Surgery and Outcome in ThoracoLumbar Fractures: Does Early Surgery Influence Neurological Recovery and Functional Restoration? A Multivariate Analysis of Results in Our Experience

Alessandro Landi, Nicola Marotta, Angela Ambrosone, Emiliano Prizio, Cristina Mancarella, Fabrizio Gregori, Giuseppe La Torre, Antonio Santoro, and Roberto Delfini

Abstract

Background

Treatment for spinal trauma is affected by both nonmodifiable and modifiable variables. The aim of this study was to compare early surgery with intermediate and late surgery to determine the benefits of spinal reconstruction in neurological recovery and functional restoration in patients with thoracolumbar fractures.

Methods

In order to identify correlations between treatment timing, fracture site, neurological recovery, American Spinal Injury Association (ASIA) score restoration, and rehabilitation prognosis in patients with thoracic and lumbar fractures, we conducted a multivariate analysis of the results of surgery, at our institution, in 166 consecutive patients with unstable thoracolumbar fractures with or without neurological impairment. We conducted a literature review (1988–2012) and compared our results with those already published.

Results

Regardless of the location and type of fracture, early surgery resulted in a reduction of median hospital and intensive care unit (ICU) length of stay, as well as a reduction of nosocomial complications. Regardless of the type of fracture and preoperative ASIA score, thoracic fractures had the worst outcome. Early treatment seemed to have better results, depending on the preoperative ASIA score.Conclusion

Early surgery in patients with thoracolumbar fractures with incomplete neurological damage could positively affect neurological recovery, functional restoration, length of hospital and ICU stay, and associated comorbidity. Thoracic fractures had the worst outcome. Early surgery seemed to have better results if the initial ASIA score was good. The better the ASIA score on admission, the better was the outcome. Surgical timing did not affect the outcome when the ASIA score was A or E.

Keywords Thoracolumbar fracture early surgery • Spinal trauma • Neurological recovery • Spine surgery • Vertebral fractures

Introduction

Spinal trauma is a very common disease, associated with spinal cord injury in 15–30 % of cases [9]. The treatment of this disease is affected both by nonmodifiable variables (fracture morphology and trauma biomechanics, fracture site, neurological status, and associated primary or secondary comorbidities) and modifiable variables (first-aid and transport, hospital, supportive therapy, surgical timing, and other variables directly dependent to the operator). The role of decompressive and reconstructive surgery, and its timing, after acute spinal cord injury in relation to neurological restoration and rehabilitative prognosis is controversial and is still being debated . Surgical treatment depends both on the general health condition of the patient and on the extent of the neurological deficit. In the light of this we can identify three surgical timing groups: early surgery, which is performed within the first 48 h after injury; intermediate surgery, which is performed between 48 h and 7 days after the trauma; and late surgery, which is performed more than 7 days after the injury. When the vertebral fracture is unstable, according to the international classification (AO Spine), surgery is indicated with the aim of creating a good reconstruction and adequate stabilization, so as to prevent pain and secondary

A. Landi, MD, PhD (✉) • N. Marotta, MD • A. Ambrosone, MD • E. Prizio, MD • C. Mancarella, MD • F. Gregori, MD • A. Santoro, MD • R. Delfini, MD
Department of Neurology and Psychiatry, Division of Neurosurgery, "Sapienza" University of Rome, Viale del Policlinico 155, 00161 Rome, Italy
e-mail: dott.alessandro.landi@gmail.com

G. La Torre, MD
Department of Public Health and Infectious Disease, "Sapienza" University of Rome, Rome, Italy

M. Visocchi et al. (eds.), *Trends in Reconstructive Neurosurgery*, Acta Neurochirurgica Supplement 124,
DOI 10.1007/978-3-319-39546-3_35, © Springer International Publishing Switzerland 2017

deformity [12]. The presence of incomplete or progressive neurological deficits (American Spinal Injury Association [ASIA] score B, C, or D) is, for some authors [2, 7, 9], the only indication for an early surgery, performed by an instrumented reconstruction after careful decompression of the neural elements. When neurological deficits are present, early surgery guarantees a reduction of comorbidities (respiratory failure and sepsis), shorter stay in intensive care, and less hospitalization time [4, 5, 10, 22]. When neurological deficits are absent (ASIA score E) or there is complete neurological involvement (ASIA score A), the surgical indication is extremely controversial, especially in regard to the decompression phase. The objective of the present study was to compare early surgery with intermediate and late surgery by performing a multivariate statistical analysis of our results, trying to identify correlations between treatment timing, fracture site , neurological recovery, ASIA score restoration, and rehabilitation prognosis, in patients with thoracic and lumbar fractures.

Methods

We conducted a descriptive longitudinal retrospective study in a cohort of consecutive patients with unstable thoracolumbar fractures enrolled between January 2002 and January 2010 in the Neurosurgery Department of "Sapienza" University of Rome. All enrolled patients underwent surgery. The patients were periodically studied in follow-up periods ranging between 3 years and 11 years (mean 70 years SD+−12.7 months). To perform our statistical multivariate study, for each patient we analyzed the following variables: age, sex, AO spine classification of fracture, type of surgical treatment, timing of surgery after the trauma (<48 h [early], between 48 h and 7 days [intermediate], and>7 days [late]), and pre- and postoperative neurological status, expressed as the ASIA score [1]. Patients with osteoporotic or pathological fractures and patients treated in a conservative way were excluded. We conducted three multivariate analyses. In the first multivariate analysis, we examined the timing of surgery in relation to preoperative neurological status and patient outcome. In the second multivariate analysis, we examined the timing of surgery in relation to the preoperative ASIA score, the location of the fracture, and the patient outcome, expressed as days of hospitalization and presence of comorbidity. In the third multivariate analysis, we examined the fracture site in relation to patient outcome, in terms of change in the ASIA score.

We also conducted an extensive literature review, using PubMed, related to our study parameters, with the aim of validating our results in the light of the evidence-based literature.

Results

The study included a cohort of 166 consecutive patients (45 female [27 %] and 121 male [73 %]) with unstable thoracolumbar fractures, surgically treated, with a mean age of 44.5 years (SD+−18.3). According to the AO spine classification, 89.8 % of patients had a type A lesion, 9.6 % had a type B, and 0.6 % of patients had a type C. In 84 cases (50.6 %) the lesion was lumbar, in 46 patients (27.7 %) it was thoracic, and in 36 patients (21.7 %) it was at the thoracolumbar junction. From a preoperative neurological point of view, 61.4 % of patients were ASIA E, 15.7 % ASIA D, 4.2 % ASIA C, 9.6 % ASIA B, and 9.1 % ASIA A. According to these data we conducted three different multivariate analyses.

First Multivariate Analysis: Correlation Between Timing of Surgery, ASIA Scores, and Patient Outcome

For this first statistical analysis (Table 1), the timing of surgery was examined in relation to ASIA scores and patient outcome, expressed as days of hospitalization and presence of comorbidity. According to the timing of surgery, we were able to divide the results into three groups:

Early Surgery In 42 patients (25.3 %), 12 females (28.6 %) and 30 males (71.4 %), with a mean age of 42.3 years (SD+−20.1) surgery was performed within 48 h of the injury. In 9 cases (21.4 %) the fracture was thoracic, in 26 cases (61.9 %) it was lumbar, and in 7 cases (16.7 %) it was located at the thoracolumbar junction. In 35 cases (83.5 %), the fracture was type A according to the AO classification and in 7 cases (16.7 %) it was type B. The preoperative neurological status was classified as ASIA E in 17 patients (40.5 %), ASIA D in 7 patients (16.7 %), ASIA B in 9 patients (21.4 %), and ASIA A in 9 patients (21.4 %). The median length of hospitalization was 23 days, with a median length of ICU stay of 17 days; 11 % of patients developed nosocomial complications.

Intermediate surgery In 47 patients (28.3 %), 8 females (17 %) and 39 males (83 %), with a mean age of 44.9 years (SD +−16.1); surgery was performed from 2–7 days after injury. In 11 patients (23.4 %) the fracture was thoracic, in 24 patients (51.1 %) it was lumbar, and in 12 patients (25.5 %) it was located at the thoracolumbar junction. Forty-four patients (93.6 %) had a type A fracture and three patients (6.4 %) a type B. The preoperative neurological status in 32 patients (68.1 %) was classified as ASIA E, with ASIA D in 5 patients (10.6 %), ASIA C in 3 patients (6.4 %), ASIA B in 4 patients (8.5 %) , and ASIA A in 3 patients (6.4 %). The median length of hospitalization was 29 days, with a median

Table 1 Multivariate analysis: time of surgery in relation to American Spinal Injury Association (ASIA) score, days of hospitalization, and comorbidity

Timing	Site	ASIA score preop	AO spine	Sex F=45 (27.5%) M=121 (72.5%)	Median age (years)+SD	Recovery
Early <48h 42 PTS (25.3%)	T(Thoracic)=9 (21.4%) L(Lumbar)=26 (61.9%) TL(Thoracolumbar)=7 (16.7%)	E=17 (40.5%) D=7 (16.7%) C=0 B=9 (21.4%) A=9 (21.4%)	A=35 (83.3%) B=7 (16.7%) C=0	F=12 (28.6%) M=30 (71.4%)	42.3 20.1±SD	Median HospitAl stay=23 Median ICU stay=17 Complications=11%
Intermediate 2–7 days 47 PTS (28.3%)	T(Thoracic)=11 (23.4%) L(Lumbar)=24 (51.1%) TL(Thoracolumbar)=12 (25.5%)	E=32 (68.1%) D=5 (10.6%) C=3 (6.4%) B=4 (8.5%) A=3 (6.4%)	A=44 (93.6%) B=3 (6.4%) C=0	F=8 (17%) M=39 (83%)	44.9 ±16.1 SD	Median hospital stay.=29 Median ICU stay=18 Complications=21%
Late >7 days 46.4% PTS	T(Thoracic)=26 (33.8%) L(Lumbar)=34 (44.1%) TL(Thoracolumbar)=17 (22.1%)	E=53 (68.8%) D=14 (18.2%) C=4 (5.2%) B=3 (3.9%) A=3 (3.9%)	A=70 (90.9%) B=6 (7.8%) C=1 (1.3%)	F=25 (32.5%) M=52 (67.5%)	47.6 ±16.9 SD	Median hospitAl stay=37 Median ICU stay=26 Complications=33%

length of ICU stay of 18 days; 21% of the patients developed nosocomial complications.

Late surgery In 77 patients (46.4%), 25 females (32.5%) and 52 males (67.5%), with a median age of 47.6 years (SD +−16.9), surgery was performed more than 7 days after the injury. In 26 patients (33.8%) the fracture was thoracic, in 34 patients (44.1%) it was lumbar, and in 17 patients (22.1%) it was at the thoracolumbar junction. In 70 patients (90.9%) the fracture was type A, in 6 patients (7.8%) it was type B, and in 1 patient (1.3%) it was type C. The preoperative neurological status in 53 patients (68.8%) was classified as ASIA E, with ASIA D in 14 patients (18.2%), ASIA C in 4 patients (5.2%), ASIA B in 3 patients (3.9%), and ASIA A in 3 patients (3.9%). The median length of hospitalization was 37 days, with a median ICU length of stay of 26 days; 33% of the patients developed nosocomial diseases.

This statistical analysis shows that, concerning the relation between fracture location and type, early surgical treatment resulted in a reduction of the median lengths of hospital and ICU stay, as well as in a reduction in cases of nosocomial complications(p value 0.001, $\chi^2 = 32.10$) showing statistical significance and good predictive value.

Second Multivariate Analysis: Correlation Between Timing of Surgery and Neurological Changes

For the second multivariate analysis, we determined whether timing of surgery was related to any change in the neurological status, expressed as the ASIA score (Table 2).

ASIA E: One hundred and two patients had an ASIA E score: in 32 patients (31.4%) the fracture was thoracic, in 50 patients (49%) it was lumbar, and in 20 patients (19.6%), it was located at the thoracolumbar junction. Seventeen patients underwent early surgery; 32 patients underwent surgery with an intermediate timing, and 53 patients underwent late surgery. The ASIA score was unchanged in relation to the timing of surgery.

ASIA D: Twenty-six patients had an ASIA D score: in 5 patients (19.2%) the fracture was thoracic, in 15 patients (57.7%) it was lumbar, and in 6 patients (23.1%) it was located at the thoracolumbar junction. Seven patients underwent early surgery, with an improvement of one grade in the ASIA score in 100% of these patients. Five patients underwent surgery with an intermediate timing: in 1 patient (20%) the ASIA score was unchanged, and in 4 patients (80%) the score was improved by one grade. Fourteen patients underwent late surgery: in three 3 patients (21.4%) the ASIA score was unchanged and in 11 patients (78.6%) it was improved by one grade.

ASIA C: Seven patients had an ASIA C score, with a thoracic fracture in two patients (28.6%), lumbar fracture in four patients (57.1%), and fracture at the thoracolumbar junction in one patient (14.3%). None of these patients underwent early surgery. Three patients underwent surgery with an intermediate timing: in one patient (33.3%) the ASIA score was unchanged, in one patient (33.3%) the ASIA score was improved by one grade, and in one patient (33.3%) the ASIA score was improved by two grades. Four patients underwent late surgery: in two patients (50%) the ASIA score was unchanged, in one patient (25%) it was improved by one grade, and in one patient (25%) it was improved by two grades.

ASIA B: Sixteen patients had an ASIA B score; in four patients (25%) the fracture was thoracic, in six patients (37.5%)

Table 2 Multivariate analysis: change in neurological status (ASIA score) was related to surgical timing

ASIA score	Site	Timing: early <48h	Intermediate 2–7 days	Late >7 days
E=102 PTS	T(Thoracic)=32 (31.45 %) L(Lumbar)=50 (49 %) TL(Thoracolumbar)=20 (19.6 %)	17 PTS Unchanged = 17 (100 %)	32 PTS Unchanged = 100 %	53 PTS Unchanged 100 %
D=26 PTS	T=5 (19.2 %) L=15 (57.7 %) TL=6 (23.1 %)	7 PTS Unchanged=0 Improved 1grade=7 (100 %)	5 PTS Unchanged=1 (20 %) Improved 1 grade=4 (80 %)	14 PTS Unchanged=3 (21.4 %) Improved 1 grade=11 (78.6 %)
C=7 PTS	T=2 (28.6 %) L=4 (57.1 %) TL=1 (14.3 %)	0 PTS	3 PTS Unchanged 1 (33.3 %) Improved 1 grade=1 (33.3 %) Improved 2 grade=1 (33.3 %)	4 PTS Unchanged=2 (50 %) Improved 1 grade=1 (25 %) Improved 2 grade=1 (25 %)
B=16 PTS	T=4 (25 %) L=6 (37.5 %) TL=6 (37.5 %)	9 PTS Unchanged=1 (11.10 %) Improved 1 grade=3 (33.3 %) Improved 2grade=5 (55.6 %)	4PTS Unchanged=1 (25 %) Improved 1 grade=2 (50 %) Improved 2 grade=1 (25 %)	3 PTS Unchanged=2 (66.7 %), Improved 1 grade=1 (33.3 %) Improved 2 grade=0
A=15 PTS	T=3 (20 %) L=9 (60 %) TL=3 (20 %)	9 PTS Unchanged=9 (100 %)	3 PTS Unchanged 3 (100 %)	3 PTS Unchanged = 3 (100 %)

it was lumbar, and in six patients (37.5 %) it was at the thoracolumbar junction. Nine patients underwent early surgery: in one patient (11.10 %) the ASIA score was unchanged, in three patients (33.3 %) it was improved by one grade, and in five patients (55.6 %) it was improved by two grades. Four patients were operated with an intermediate timing: in one patient (25 %) the ASIA score was unchanged, in two patients (50 %) it improved by one grade, and in one patient (25 %) it improved by two grades. Three patients underwent late surgery: in two patients (66,7 %) the ASIA score was unchanged, in one patient (33,3 %) it improved by one grade.

ASIA A: Fifteen patients had an ASIA A score, with thoracic fracture in three patients (20 %), lumbar fracture in nine patients (60 %), and fracture at the thoracolumbar junction in three patients (20 %). The ASIA score remained unchanged regardless of the surgical timing.

This analysis suggests that early surgical treatment appears to provide a better neurological recovery with statistical significance and good predictive value. The better the ASIA score on admission, the better the outcome (p value=0.001009, $\chi^2 = 26.10$).

Third Multivariate Analysis: Correlation Between Site of Fracture and Patient Outcome

In the third statistical analysis we examined the relation between site of fracture and patient outcome.

Thoracic Site In 46 patients (27.7 %), 10 women (21.7 %) and 36 men (78.3 %), with a median age of 41.1 years (SD + −16.8), the fracture was localized in the thoracic spine: 43 patients (93.5 %) had a type A fracture and and 3 patients (6.5 %) had a type B; 32 patients (69.6 %) had an ASIA E score, 5 patients (10.9 %) an ASIA D, 2 patients (4.3 %) an ASIA C, 4 patients an ASIA B (8.7 %), and 3 patients an ASIA A (6.5 %). In 72.70 % of the patients the ASIA score appeared to be unchanged, while in 27.30 % of the patients it was improved by one grade.

Lumbar Site In 84 patients (50.6 %), 24 women (28.6 %) and 60 men (71.4 %), with median age of 44.5 years (SD+ 17.4), the fracture was localized in the lumbar spine: in 77 patients (91.7 %) there was a type A fracture, in 6 patients (7.1 %) a type B, and in 1 patient (1.2 %) a type C. Fifty patients (60.4 %) had an ASIA E, 15 (17 %) an ASIA D, 4 (4.8 %) an ASIA C, 6 (7.1 %) an ASIA B, and 9 (10.7 %) an ASIA A. In 4 % of the patients the ASIA score appeared to be unchanged; in 72 % of the patients there was an improvement of one grade and in 24 % an improvement of two grades.

ThoracoLumbar Junction In 36 patients (21.7 %), 11 women (30.6 %) and 25 men (69.4 %), with a median age of 48.5 years (SD +−18.9) the fracture was located at the thoracolumbar junction: in 29 patients (80.6 %) there was a type A fracture, and in 7 patients the fracture was type B (19.4 %). Twenty patients (55.5 %) had an ASIA E score, 6 (16.7 %) had an ASIA D, 1 (2.8 %) an ASIA C, 6 (16.7 %) an ASIA B, and 3 (8.3 %) an ASIA A. In 15.40 % of the patients the ASIA score appeared to be unchanged; in 76.90 % of the patients the score improved by one grade, and in 7.70 % of the patients the score improved by two grades.

This analysis shows that regardless of the type of fracture and preoperative ASIA score (p value=0.357033, $\chi^2 = 4.38$), fractures located in the thoracic spine had the worst outcome,

Table 3 Third multivariate analysis: change in neurological status, expressed by the ASIA score, was related to the site of fracture

Site	ASIA score	AO spine	Sex	Median age (years)	Outcome
Thoracic 46 pts 27.7 %	E=32 69.7 % D=5 10.9 % C=2 4.3 % B=4 8.7 % A=3 6.5 %	A=43 93.5 % B=3 6.5 %	10 F 21.7 % 36 M 78.3 %	41.1 (SD±16.8)	Unchanged = 72.7 % Improved 1 grade = 27.3 %
Lumbar 84 pts 50.6 %	E=50 60.4 % D=15 17 % C=4 4.8 % B=6 7.1 % A=9 10.7 %	A=77 91.7 % B=6 7.1 % C=1 1.2 %	24 F 28.6 % 60 M 71.4 %	44.5 (SD±17.4)	Unchanged=4 % Improved 1 grade=72 % Improved 2 grade=24 %
Thoracolumbar 36 pts 21.7 %	E=20 55.5 % D=6 16.7 % C=1 2.8 % B=6 16.7 A=3 8.3	A=29 80.6 % B=7 19.4 %	11 F 30.6 % 25 M 69.4 %	48.5 (SD±18.9)	Unchanged=15.4 % Improved 1 grade=76.9 % Improved 2 grades = 7.7 %

with results showing good predictive value and statistical significance (Table 3; Fig. 1).

Discussion

Spinal cord injury is one of the primary causes of neurological damage. Currently in the scientific community, there is no clear opinion about the optimal timing of surgical decompression and reconstruction in patients with thoracic and lumbar fractures. Surgery is crucial in order to reduce secondary damage and to improve the patient's outcome in terms of neurological recovery and functional restoration [24]. Most of the concerns about the timing of treatment are related to insufficient information about the pathophysiology of the neurological damage and the beneficial effects of the surgery. Trauma of the spinal cord induces a combination of signs and symptoms associated with earlier and later damage [6, 8, 16]. The forces of traction and compression cause a primitive injury to the central gray matter with extensive metabolic damage. Damage to the gray matter is irreversible from the first hours after the trauma, while damage to the white matter becomes irreversible 72 h after the trauma [8, 26]. Experimental studies show that the time period of spinal cord compression is related to the severity of the pathological damage and, consequently the degree of neurological recovery and functional restoration: the greater the period of compression, the smaller is the possibility of clinical recovery [3, 11, 18, 22]. Surgical decompression potentially allows a reduction of the intradural pressure and an increase in blood flow through the spinal cord, thus reducing the risk of ischemia and preventing oxidative secondary damage [14–17]. However, the best timing of surgical decompression is unclear. In the literature, there are many reports about the validity and effectiveness of the implementation of early surgery (within 8 h of injury). Early surgery may be associated with increased intraoperative blood loss (compared with later procedures) and the onset of hypotension with the possibility of spinal cord injury and then the onset of neurological deficits. On the other hand, early surgery guarantees a better outcome for the patient, thanks to early mobilization, shorter hospitalization, and a lower risk of thromboembolism and pulmonary complications [16–18].

Some studies have shown that early treatment (within 8 h of injury) led to a sudden improvement in neurological status and a better outcome in the subsequent follow-up [13, 15, 21, 23]. Chenzing et al. [19] have shown that in early surgery the neurological outcome is closely tied to the preoperative neurological status, assuming that there is a greater chance of neurological recovery in patients with ASIA scores of C and D. In their retrospective study, Boakye et al. [27] postulated that independent variables, such as age, comorbidities, and original ASIA score, also influenced the outcome [16]. La Rosa et al. [16] conducted a systematic review and concluded that decompressive surgery, performed up to 24 h after injury, resulted in a better outcome.

Some studies show that in pateints with complete or incomplete neurological deficits, the improvement of the deficits is independent of the timing of surgery. In a prospective study of 106 patients performed by Pointillart et al. [20], approximately half of the patients underwent early surgery (<8 h), with no improvement in neurological status. McKinley et al. [25] concluded that there were no differences in neurological status between patients with early surgery (<72 h after the trauma) and those with late surgery [12, 23, 25–27]. Bliemel et al. [6] believe that the outcome of the patient depends on the fracture site: lumbar fractures have a better outcome.

Our statistical analysis, performed with a large cohort of patients, has three important aspects:

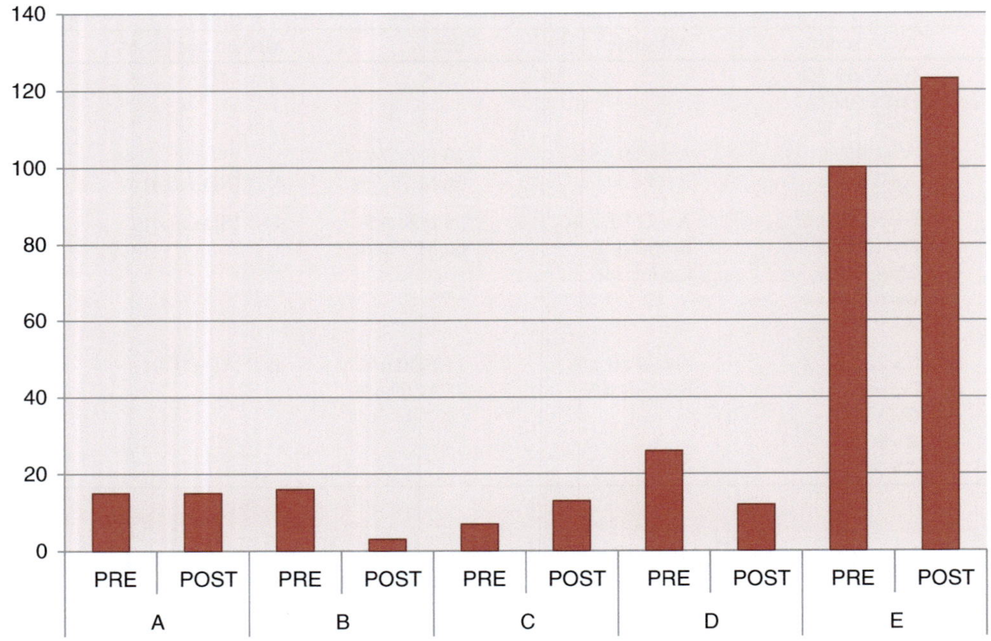

Fig. 1 Total American Spinal Injury Association (ASIA) score improvement in the entire cxohort of 164 patients

1. Concerning the relation between timing of surgery, ASIA score, and functional outcome, early surgical treatment seems to result in a greater reduction in the median hospital and ICU length of stay, as well as in a reduction of cases of nosocomial complications (p value 0.001, $\chi^2 = 32.10$) with statistical significance and good predictive value. So, in the light of these findings, early surgery seems to be effective in reducing hospital stay and nosocomial complications. It is obvious that this result may influence neurological recovery and functional restoration, with the patient being able to begin rehabilitation programs early.
2. Concerning the relation between timing of surgery and ASIA score improvement, our analysis suggests that early surgical treatment seems to provide better neurological recovery in terms of ASIA score improvement, with statistical significance and good predictive value. We conclude that the better the ASIA score on admission, the better the outcome (p value=0.001009, $\chi^2 = 26.10$).
3. Concerning the relation between the site of fracture and ASIA score improvement, our analysis suggests that fractures located in the thoracic spine have the worst outcome, in terms of ASIA score improvement, compared with lumbar and thoracolumbar fractures, with good predictive value and statistical significance (p value=0.357033, $\chi^2 = 4.38$). Probably this result is related to the vascularization of the thoracic tract of the spinal cord. In terms of neurological recovery and functional restoration, it can be concluded that a patient with a thoracolumbar or lumbar fracture with neurological involvment is more suitable for undertaking a prolonged rehabilitation program than a patient with a thoracic fracture, in terms of the possibility of functional restoration.

Our results, in accordance with the scientific literature, seem to suggest that the timing of surgery is the most important factor that can influence patient outcome, in terms of ASIA score improvement, when there is incomplete neurological damage preoperative. In patients with complete neurological damage or total absence of damage, the timing of surgery does not affect the outcome.

Conclusion

Our multivariate analyses suggest that early decompressive surgery in the thoracic and lumbar spine with incomplete neurological damage could positively affect the patient's outcome in terms of neurological recovery, functional restoration, length of hospital and ICU stay, and associated comorbidity. Thoracic fracture location has the worst outcome compared with other fracture locations. Moreover, independently of the location of the fracture, early surgery seems to have better results when the initial ASIA score is good; the better the ASIA score on admission, the better the outcome. Surgical timing does not affect the outcome when the ASIA score is A or E. We note that in the multivariate analyses in this study we did not consider many other variables related to thoracolumbar trauma.

Conflicts of Interest Statement

Ethical Standards All human and animal studies have been approved by the appropriate ethics committee and have therefore been performed in accordance with the ethical standards laid down in the 1964 Declaration of Helsinki and its later amendments.

All persons gave their informed consent prior to their inclusion in the study. Any details that might disclose the identity of the subjects under study have been omitted.

The first author is
Alessandro Landi

Disclaimer Landi We certify that this manuscript is a unique submission and is not being considered for publication with any other source in any medium.

We certify that the work has not been previously published or submitted elsewhere for review.

We certify that we do not have any financial relationships with any manufacturers of equipment used during the study' and there is no conflict of interests.

The first author
Alessandro Landi

References

1. American Spinal Injury Association (ed) (2000) International standards for neurological classification of spinal cord injury. American Spinal Association, Chicago
2. Aysegul B, Kemal Ilik AE, Alper Basefer AE (2008) A real prospective (rct) randomized controlled study. Arch Orthop Trauma Surg 128:959–966. doi:10.1007/s00402-007-0518-1
3. Bracken MB, Holford TR (1993) Effects of timing of methylprednisolone or naloxone administration on recovery of segmental and long-tract neurological function in NASCIS 2. J Neurosurgery 79:500–507
4. Cengiz SL, Kalkan E, Bayir A et al (2008) Timing of thoracolombar spine stabilization in trauma patients; impact on neurological outcome and clinical course. A real prospective (RCT) randomized controlled study. Arch Orthop Trauma Surg 128(9): 959–966
5. Chipman JG, Deuser WE, Beilman GJ (2004) Early surgery for thoracolumbar spine injuries decreases complications. J Trauma 56:52–57
6. Christopher B, Rolf L, Benjamin B, Michael F, Johannes S, Antonio K, Steffen R, Frangen TM (2014) Early or delayed stabilization in severely injured patients with spinal fractures? Current surgical objectivity according to the trauma registry of DGU: treatment of spine injuries in polytrauma patients. J Trauma Acute Care Surg 76(2):366–373
7. Croce MA, Bee TK, Pritchard E, Miller PR, Fabian TC (2001) Does optimal timing for spine fracture fixation exist? Ann Surg 233:851–858
8. Xing D, Chen Y, Ma JX, Song DH, Wang J, Yang Y, Feng R, Lu J, Ma XL (2013) A methodological systematic review of early versus late stabilization of thoracolumbar spine fractures. Eur Spine J 22(10):2157–2166 Published online 2012 Dec 22. doi: 10.1007/s00586-012-2624-1PMCID: PMC3804693
9. Decq P, Kéravel Y (1995) Neurochirurgie: traumatismes vertébro-médullaires. Universités francophones ed, Paris
10. Dimar JR 2nd, Glassman SD, Raque GH, Zhang YP, Shields CB (1999) The influence of spinal canal narrowing and timing of decompression on neurologic recovery after spinal cord contusion in a rat model. Spine 16:1623–1633
11. Harrop JS, Sharan A, Ratliff J (2006) Central cord injury: pathophysiology, management, and outcomes. Spine J 6:198–206
12. Heidari P, Zarei MR, Rasouli MR, Vaccaro AR, Rahimi-Movaghar V (2010) Spinal fractures resulting from traumatic injuries. Chin J Traumatol 13(1):3–9
13. Holmes JF, Miller PQ, Panacek EA et al (2001) Epidemiology of thoracolumbar spine injury in blunt trauma. Acad Emerg Med 8:866–872
14. Krengel WF, Anderson PA, Henley MB (1993) Early stabilization and decompression for incomplete paraplegia due to a thoracic-level spinal cord injury. Spine (Phila Pa 1976) 18:2080–2087
15. Launay O, Charles YP, Steib JP (2012) Should post-traumatic thoracolumbar Frankel a paraplegia be operated as an emergency? Report of three cases and review of the literature. Orthop Traumatol Surg Res 98(3):352–358. doi:10.1016/j.otsr.2011.08.019, Epub 2012 Mar 21
16. La Rosa G, Conti A, Cardali S, Cacciola F, Tomasello F (2004) Does early decompression improve neurological outcome of spinal cord injured patients? appraisal of the literature using a meta-analytical approach Spinal Cord 42(9):503–512
17. Maynard FM Jr, Bracken MB, Creasey G, Ditunno JF Jr, Donovan WH, Ducker TB et al (1997) International standards for neurological and functional classification of spinal cord injury. Spinal Cord 35:266–274
18. McLain RF, Sparling E, Benson DR (1993) Early failure of short-segment pedicle instrumentation for thoracolumbar fractures. A preliminary report. J Bone Joint Surg Am 75(2):162–167
19. Nystrom B, Berglund JE (1988) Spinal cord restitution following compression injuries in rats. Acta Neurol Scand 78: 467–472
20. Pointillart V, Petitjean ME, Wiart L, Vital JM, Lassié P, Thicoipé M, Dabadie P (2000) Pharmacological therapy of spinal cord injury during the acute phase. Spinal Cord 38(2):71–76
21. Cengiz SL, Kalkan E, Bayir A, Ilik K, Basefer A (2008) Timing of thoracolomber spine stabilization in trauma patients; impact on neurological outcome and clinical course. A real prospective (rct) randomized controlled study. Arch Orthop Trauma Surg 128(9):959–966
22. Schinkel C, Frangen TM, Kmetic A et al (2006) Timing of thoracic spine stabilization in trauma patients: impact on clinical course and outcome. J Trauma 61 156–60

23. Stefan Arthur R, John Festo K, Thomas K, Ulrich N, Hans-Peter R, Gregor A (2005) Neurological recovery and its influencing factors in thoracic and lumbar spine fractures after surgical decompression and stabilization. Neurosurg Rev 28:44–52. doi:10.1007/s10143-004-0356-3
24. Tator CH (2006) Review of treatment trials in human spinal cord injury: issues, difficulties, and recommendations. Neurosurgery 59:957–82
25. William MK, Meade MA, Steven K, Barbara B (2004) Outcomes of early surgical management versus late or no surgical intervention after acute spinal cord injury. Arch Phys Med Rehabil 85:1818–25
26. Vaccaro AR, Daugherty RJ, Sheehan TP, Dante SJ, Cotler JM, Balderston RA et al (1997) Neurologic outcome of early versus late surgery for cervical spinal cord injury. Spine 15:2609–2613, Arch Orthop Trauma Surg 2008;128:959–66
27. Boakye M, Arrigo R, Gephart MG, Zygourakis C, Lad SP (2012) Retrospective, propensity score-matched cohort study examining timing of fracture fixation for traumatic thoracolumbar fractures. J Neurotrauma 29(12):2220–5. doi: 10.1089/neu.2012.2364. Epub 2012 Jul 24

Part VII

Diagnostic in Surgery

Functional Magnetic Resonance Imaging (fMRI), Pre-intraoperative Tractography in Neurosurgery: The Experience of Sant'Andrea Rome University Hospital

Giancarlo D'Andrea, Giuseppe Trillo', Veronica Picotti, and Antonino Raco

Abstract

Background

The goal of neurosurgery for cerebral intraparenchymal neoplasms of the eloquent areas is maximal resection with the preservation of normal functions, and minimizing operative risk and postoperative morbidity.

Currently, modern technological advances in neuroradiological tools, neuronavigation, and intraoperative magnetic resonance imaging (MRI) have produced great improvements in postoperative morbidity after the surgery of cerebral eloquent areas.

The integration of preoperative functional MRI (fMRI), intraoperative MRI (volumetric and diffusion tensor imaging [DTI]), and neuronavigation, defined as "functional neuronavigation" has improved the intraoperative detection of the eloquent areas.

Methods

We reviewed 142 patients operated between 2004 and 2010 for intraparenchymal neoplasms involving or close to one or more major white matter tracts (corticospinal tract [CST], arcuate fasciculus [AF], optic radiation).

All the patients underwent neurosurgery in a BrainSUITE equipped with a 1.5 T MR scanner and were preoperatively studied with fMRI and DTI for tractography for surgical planning.

The patients underwent MRI and DTI during surgery after dural opening, after the gross total resection close to the white matter tracts, and at the end of the procedure.

We evaluated the impact of fMRI on surgical planning and on the selection of the entry point on the cortical surface.

We also evaluated the impact of preoperative and intraoperative DTI, in order to modify the surgical approach, to define the borders of resection, and to correlate this modality with subcortical neurophysiological monitoring.

We evaluated the impact of the preoperative fMRI by intraoperative neurophysiological monitoring, performing "neuronavigational" brain mapping, following its data to localize the previously elicited areas after brain shift correction by intraoperative MRI.

Results

The mean age of the 142 patients (89 M/53 F) was 59.1 years and the lesion involved the CST in 66 patients (57 %), the language pathways in 24 (21 %), and the optic radiations in 25 (22 %).

The integration of tractographic data into the volumetric dataset for neuronavigation was technically possible in all cases.

In all patients intraoperative DTI demonstrated a shift of the bundle position caused by the surgical procedure; its dislocation was both outward and inward in the range of +6 mm and −2 mm.

Conclusion

We found a high concordance between fMRI/DTI and intraoperative brain mapping; their combination improves the sensitivity of each technique, reducing pitfalls and so defining "functional neuronavigation", increasing the definition of eloquent areas and also reducing the time of surgery.

Keywords Intraoperative MRI • DTI • fMRI

Introduction

The goal of neurosurgery for cerebral intraparenchymal neoplasms of the eloquent areas is maximal resection with the preservation of normal functions, minimizing the operative risk and postoperative morbidity [1, 4, 6, 32].

Neurophysiological monitoring has been traditionally used to define the exact localization of perirolandic gyri

G. D'Andrea, MD (✉) • G. Trillo' • V. Picotti, MD • A. Raco, MD
Institute of Neurosurgery, S Andrea Hospital, University of Rome "La Sapienza", V. L. Mantegazza 8, 00152 Rome, Italy
e-mail: gdandrea2002@yahoo.it

and the corticospinal tract (CST), while awake surgery allows the detection of the phasic cortical areas and the arcuate fasciculus (AF) [6].

Currently, modern technological advances in neuroradiological tools, neuronavigation, and intraoperative magnetic resonance imaging (MRI) have produced great improvements in postoperative morbidity after the surgery of cerebral eloquent areas.

Nevertheless, preoperative functional MRI (fMRI) and diffusion tensor imaging (DTI) are insufficient for accurate neuronavigation because of the brain shift during surgery; thus, their best results are achieved when integrated with intraoperative MRI.

The integration of preoperative fMRI, intraoperative MRI (volumetric and DTI), and neuronavigation, defined as "functional neuronavigation", improves the intraoperative detection of the eloquent areas by comparison of the previously detected functional areas and neurophysiological brain mapping [6].

We analyzed three points: the benefit of preoperative fMRI during surgery with asleep patients; the role of intraoperative DTI in relation to neuronavigation and brain shift; and the relation between intraoperative DTI and neurophysiological recordings.

Methods and Materials

We reviewed 142 patients operated between 2004 and 2010 for intraparenchymal neoplasms involving or close to one or more major white matter tracts (CST, arcuate fasciculus [AF], optic radiation).

This series included cases affecting the sensorimotor area, the optic radiations, and the AF.

All the patients underwent neurosurgery in the BrainSUITE, BrainLAb Germany, equipped with a 1.5 T MR scanner and were preoperatively studied with fMRI and DTI for tractography for surgical planning.

Neurological examination was performed preoperatively, immediately after surgery, at discharge, and 1 month after surgery.

The patients underwent MRI and DTI during surgery after dural opening, after the gross total resection close to the white matter tracts, and at the end of the procedure.

We evaluated the impact of fMRI on surgical planning and on the selection of the entry point on the cortical surface.

We also evaluated the impact of preoperative and intraoperative DTI in order to modify the surgical approach, to define the borders of resection, and to correlate this modality with subcortical neurophysiological monitoring.

We evaluated the impact of the preoperative fMRI by intraoperative neurophysiological monitoring, performing neuronavigational brain mapping, following its data to localize the previously elicited areas after brain shift correction by intraoperative MRI.

We track the mono- or bipolar probe for neuronavigation through a reference star that is recognized and "neuronavigated"; therefore using it for the dual role of pointer and electrical probe (Fig. 1). Motor-evoked potentials were recorded with subdermal platinum/iridium needle electrodes [5] (MRI-compatible), positioned on muscles of the limbs and face (Fig. 2).

Our neurophysiologist prepared the patients some days before the functional MRI to correctly perform the requested movements; the selected muscles were the same as those elicited during preoperative functional MRI (fMRI).

In all the patients, evaluation of the motor area was confirmed by the study of phase reversal and with "neuronavigational" brain mapping.

Preoperative fMRI was performed to localize the Wernicke cortical area in all the neoplasms affecting the cortical and subcortical language pathways.

Before the study was initiated, the hospital ethics committee was consulted and all patients gave their informed written consent to participate.

Preoperative MR Protocol

The MR examinations were performed by using a 1.5-T magnet (Sonata; Siemens, Erlangen, Germany), identical to the one in the operating room. The following sequences were acquired: T2, FLAIR, and isotropic volumetric T1-weighted magnetization-prepared rapid acquisition gradient echo (MPRAGE) before and after the intravenous administration of paramagnetic contrast material and DT sequences. The DTI study was performed with 12 non-collinear directions (*b* value = 0 and 1,000 s/mm^2) and echo planar sequences (TE 86 ms,TR 9,200 ms, matrix 128 × 128, FOV 240 mm, slice thickness 1.9 mm, bandwidth 1,502 Hz/Px, 60 slices, no gap, acquisition time 5 min 31 s).

Tractography post-processing was performed with a method similar to those presented by Basser et al., Mori et al., and Stieltjes et al. [2, 20, 33], using the planning software iPlan 2.6 (BrainLAB AG, Feldkirchen, Germany).

Color maps were used to define an appropriate region of interest (ROI) for the subsequent tractography procedure. The fiber-tracking technique contemplates the three-dimensional (3D) reconstruction of white matter trajectories of the CST by using a fractional anisotropy threshold of 0.17

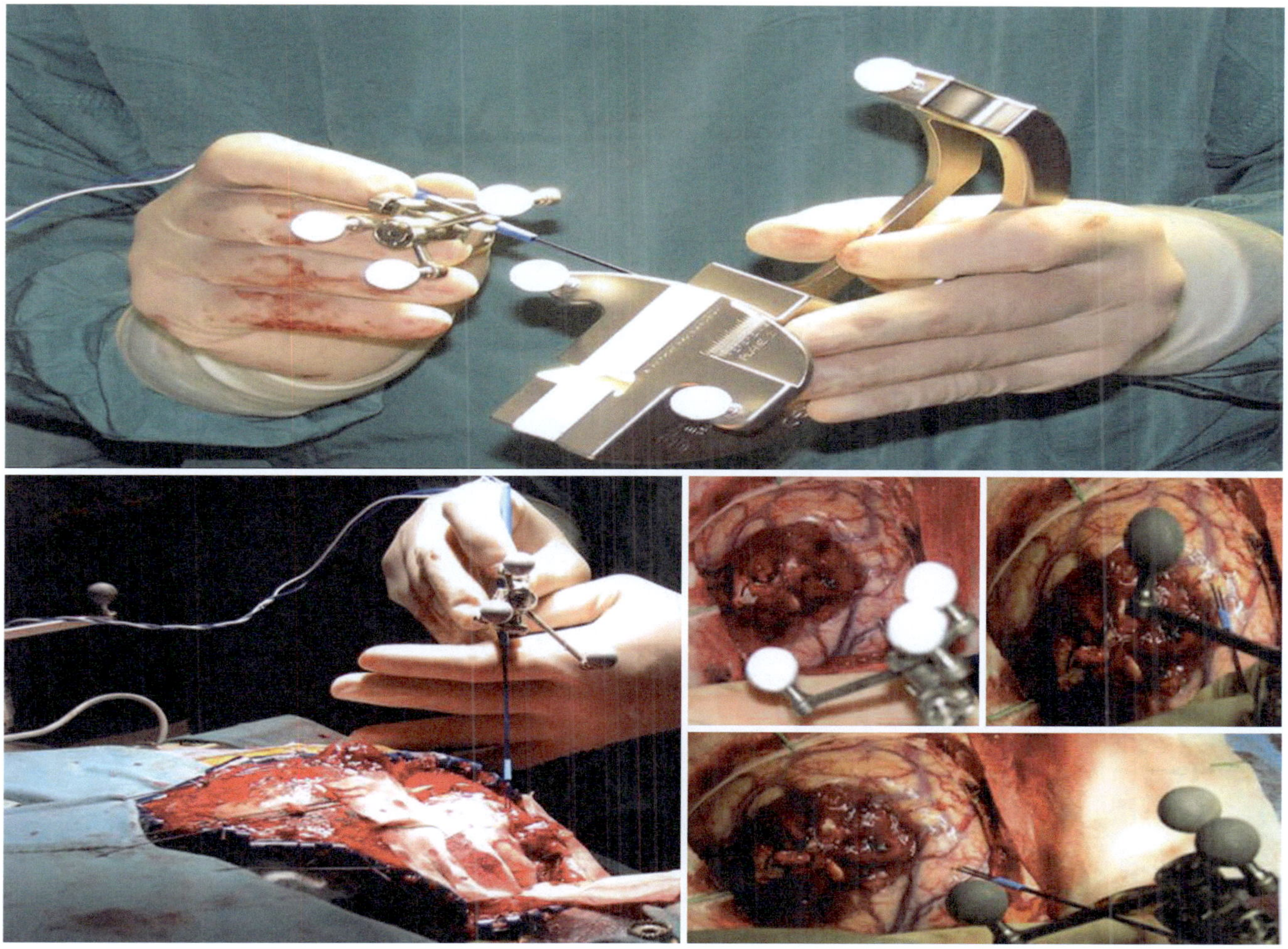

Fig. 1 We track the mono- or bipolar probe for neuronavigation through a reference star that is recognized and "neuronavigated'; therefore using it for the dual roles of pointer and electrical probe

and a processing angle above 55°. The positioning of the ROI for the fiber tracking changed according to the trajectories of the fibers to be reconstructed (posterior arm of the internal capsule for the pyramidal tract, geniculate ganglion for the optic radiation and, only in the right-handed patients with lesions on the left side, the ROI encompassed the horizontal fibers lateral to the corona radiata and medial to the cortex of the posterior part of the ventrolateral frontal lobe). Tracking was initiated in both the retrograde and orthograde directions according to the direction of the principal eigenvector in each voxel of the ROI. The reconstructed trajectories were transformed into 3D objects.

Mean data processing time of the CST was 2–3 min, while the mean data processing time of the AF was 8-10 min. Tractography results were saved in a file containing the x/y/z coordinates for each fiber. These data were imported together with the b=0 diffusion images into the navigation software (import module for iPlan 1.0 programmed by U. Mezger [Brain Lab, Heimstetten, Germany]). After rigid registration of the b=0 images with the anatomical volumetric package, and after having verified that there were no discrepancies between data (differences greater than 3 mm) in the region of the tumor the white matter tracts could be displayed in standard anatomical images. Fiber margins were then segmented to allow them to be defined as objects in the navigation system and to be depicted intraoperatively. These objects were automatically enlarged by the software 2 mm in every direction. The trajectories were considered suitable for surgical planning if there were no interruptions on any of the layers at the level of the lesion. The boundaries of the tumors were defined, considering the outer rim of enhancement for grade III and IV gliomas and the T2 signal for grade II gliomas. The distance between the CST and tumor margins was calculated. The mean overall time for this data processing, which was generally performed the day prior to surgery, was 30 min. During surgery the position of the CST was used as a reference for tumor resection.

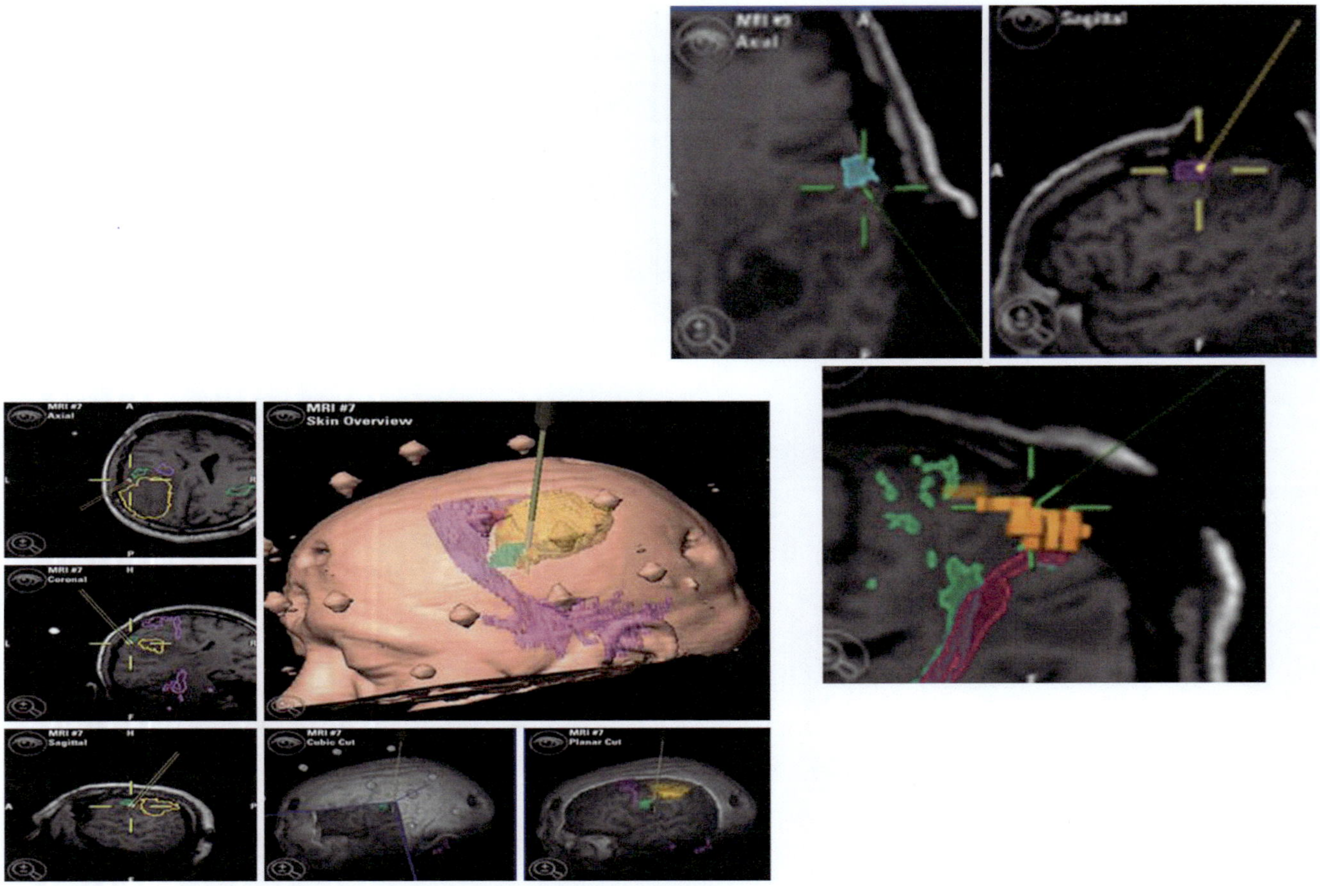

Fig. 2 Motor evoked potentials were recorded with subdermal platinum/iridium needle electrodes positioned on muscles of the limbs and face

Intraoperative MR Protocol (BrainSUITE)

Currently we do not perform intraoperative presurgical MRI because of infrared matching of head's patients with previously acquired volumetric exams, but for prone position.

Total acquisition time and time for sending preoperative images to the neuronavigation system for surgical planning was limited and was less than 2 min.

We perform the first intraoperative MRI after the dural opening, acquiring an intraoperative volumetric MRI for navigation and an intraoperative DTI for tractography. This step is fundamental for correcting possible brain shift.

Average total acquisition and processing time was about 15 min.

The same neuroradiologist reconstructed the white matter tracts and uploaded the new tractographic data in the neuronavigation system; these data were subsequently used for further surgery.

When the neurosurgeon was confident that most of the lesion was removed, intraoperative MRI and DTI were again performed and the same neuroradiologist reconstructed the tractographic data with the previously mentioned technique.

The new tractographic data were again uploaded into the neuronavigation system and subsequently used for further surgery and eventual intraoperative subcortical stimulation.

At the end of the procedure, we performed the last MRI to check the amount of resection and the sparing of the white matter tracts.

Brain Shift Evaluation

Pre- and intraoperative DTI were registered with automatic image fusion software (iPlan 2.6; BrainLAB AG), which was used to perform semiautomatic rigid registration. After registration, the images could be displayed side by side or in an overlay mode. The extent of shifting was considered as the maximum distance between the preoperative and intraoperative contours of the trajectories of the evaluated white matter tracts on identical/registered axial slices.

According to the direction of the shift we assigned positive or negative values in relation to the craniotomy opening. A positive value was assigned if movement was outward (toward the surface), and a negative value was assigned if movement was inward.

We measured peritumoral edema and tumor volume using the planning software noted above, and, as well, we calculated the craniotomy size and the distance of the tumor from the cortical surface.

Results

The mean age of the 142 patients (89 M/53 F) was 59.1 years and the lesion involved the CST in 66 patients (57 %), the language pathways in 24 (21 %), and the optic radiations in 25 (22 %).

All the patients were affected by intraparenchymal neoplasms and histological examination revealed low-grade gliomas, oligodendroglioma, metastases, and high-grade gliomas.

Cortical motor-evoked potentials (MEPs) followed preoperative fMRI and were performed after brain shift correction to choose the optimal site for corticectomy and brain retraction.

Preoperative fMRI permitted us to localize the Wernicke cortical area in all the patients with neoplasms affecting the language areas, except for three who had severe preoperative dysphasic disturbances; AF reconstruction was achieved in all the patients.

Less severe dysphasia did not prejudice the possibility of identifying the Wernicke area and, moreover, did not play any role in DTI.

The integration of tractographic data into the volumetric dataset for neuronavigation was technically possible in all cases (Fig. 3).

In all patients intraoperative DTI demonstrated the shift of the bundle position caused by the surgical procedure (Figs. 4 and 5) and its dislocation was both outward and inward, in the range of +6 mm and −2 mm.

We considered that the neuronavigation system had an average error of 0.79 ± 0.25 mm and a maximum error of 2.0 mm..

In 40 % of the cases an outward shift was observed during surgery; an inward shift was observed in 50 %, while in 10 % no intraoperative displacement was detected.

The maximum intraoperative shifting of the CST ranged from an inward value of 9.7 mm to an outward value of 13.8 mm. Only peritumoral edema showed a statistically significant correlation with the amount of shift ($P = 0.001$), indicating a more pronounced outward shift correlated with larger edema. In a comparison of patients showing inward and outward shifting, statistically significant differences were evident, and peritumoral edema was more pronounced in patients with outward shifting ($P = 0.001$), as was the craniotomy size ($P = 0.038$). No statistically significant differences were evident when comparing the tumor sizes.

A direct correlation was evident between craniotomy size and shifting after dura mater opening ($P = 0.05$).

Evaluation of the visualized trajectories related to the lesion produced an a-priori modification of the surgical approach to corticectomy in 21 % of our cases, and in 64 % had an important impact on the definition of the resection margins during surgery. The overall percentage impact on the surgical procedure was 82 %.

In all patients, pre- and intraoperative tractography demonstrated the white matter bundle containing the pyramidal tract, and the motor function was preserved in all but three patients, who showed a transient weakness of the contralateral side, which dramatically improved between 1 month and 3 months after the procedure. In all three cases the lesions were considered to be in contact with the motor cortex and the CST.

During surgery, close to the pyramidal tract, several electrical "neuronavigational" stimulations were repeatedly performed following the neuronavigation after intraoperative volumetric MRI and DTI, demonstrating a deep residual tumor volume and correcting the eventual brain shift. On preoperative tractography, the average overall distance between the lesion and the CST was 2.26 mm and the average distance between the CST and the site of subcortical MEP was 3.5 mm.

We performed subcortical MEPs in 61 % of the cases between 0 and 2 mm from the CST, in 33 % between 4 and 8 mm, and in 6 % between 12 and 15 mm, detecting the CST in all cases by intraoperative neurophysiology, except for two cases, in which the CST was far from the lesion.

Intraoperative postsurgical MRI demonstrated complete tumor removal in all the patients in this series and the postoperative outcome was excellent in all the patients.

The immediate postoperative overall outcome was excellent in 89 % of the patients and we registered a transient worsened motor deficit in only 11 % of the patients.

In 77.7 % of the patients with neoplasms of the language areas a complete resection was performed, while in the remaining patients we performed a subtotal resection.

A subtotal resection was defined in relation to a safe distance of 0–2 mm from the AF or to its neoplastic involvement.

At immediate postoperative examination and at discharge the neurological status was normal in 81.5 % of the patients and preoperative symptoms were improved in 74 % ; the asymptomatic patients maintained their negative neurological status and we did not register any case of deteriorated clinical status, while, in particular, dysphasic disturbances recovered in 81.2 % of the patients.

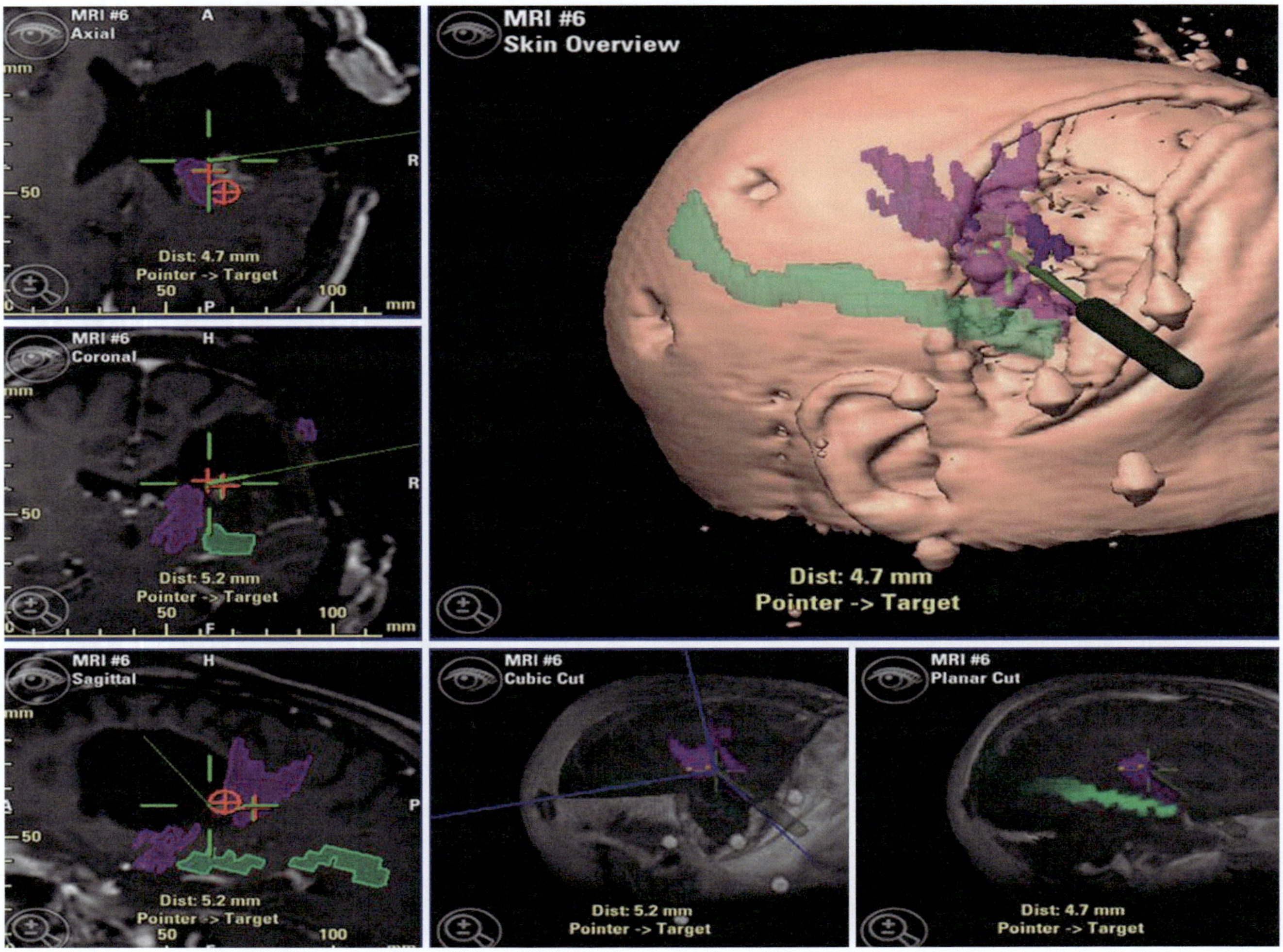

Fig. 3 Integration between intraoperative diffusion tensor imaging (DTI) and neuronavigation

The outcome was excellent, without deficits affecting a normal quality of life, in 85.1 % of the patients, and good (moderate deficit affecting the normal quality of life) in 14.8 %.

In particular, the patients who underwent complete tumor removal (95.2 %) obtained an excellent outcome despite the aggressive resection (χ^2 $p<0.05$).

Discussion

The quality of life in patients affected by malignant neoplasms in eloquent areas must be a priority parameter [8] for choosing surgery rather than biopsy because of their expected short survival.

This goal can be achieved through the exact localization and consequent preservation of the cortical and subcortical functional areas.

It was recently reported that MR tractography may have an impact on surgical planning, leading to changes in the surgical approach and in the limits of resection [31], and in our experience, tractography modified the surgical approach to corticotomy (in 21 % of cases), permitted definition of the resection margins (in 64 %), and resulted in an overall modification of the procedure in 82 % of cases.

The intraoperative DTI for reconstruction of white matter tracts after dural opening and during surgery with subcortical "neuronavigational" neurophysiological monitoring allowed a better resection of these lesions.

We strongly agree with Bello et al. [3], who suggested a strong connection between DTI and subcortical mapping.

Many authors consider resection close to the white matter tracts and, moreover, close to the CST, to be risky and they prefer to maintain a "safe distance" to reduce transient/permanent postoperative morbidity; however, our data disagree with their findings, probably because we had the opportunity of updating DTI data when we were quite close to the tract [6].

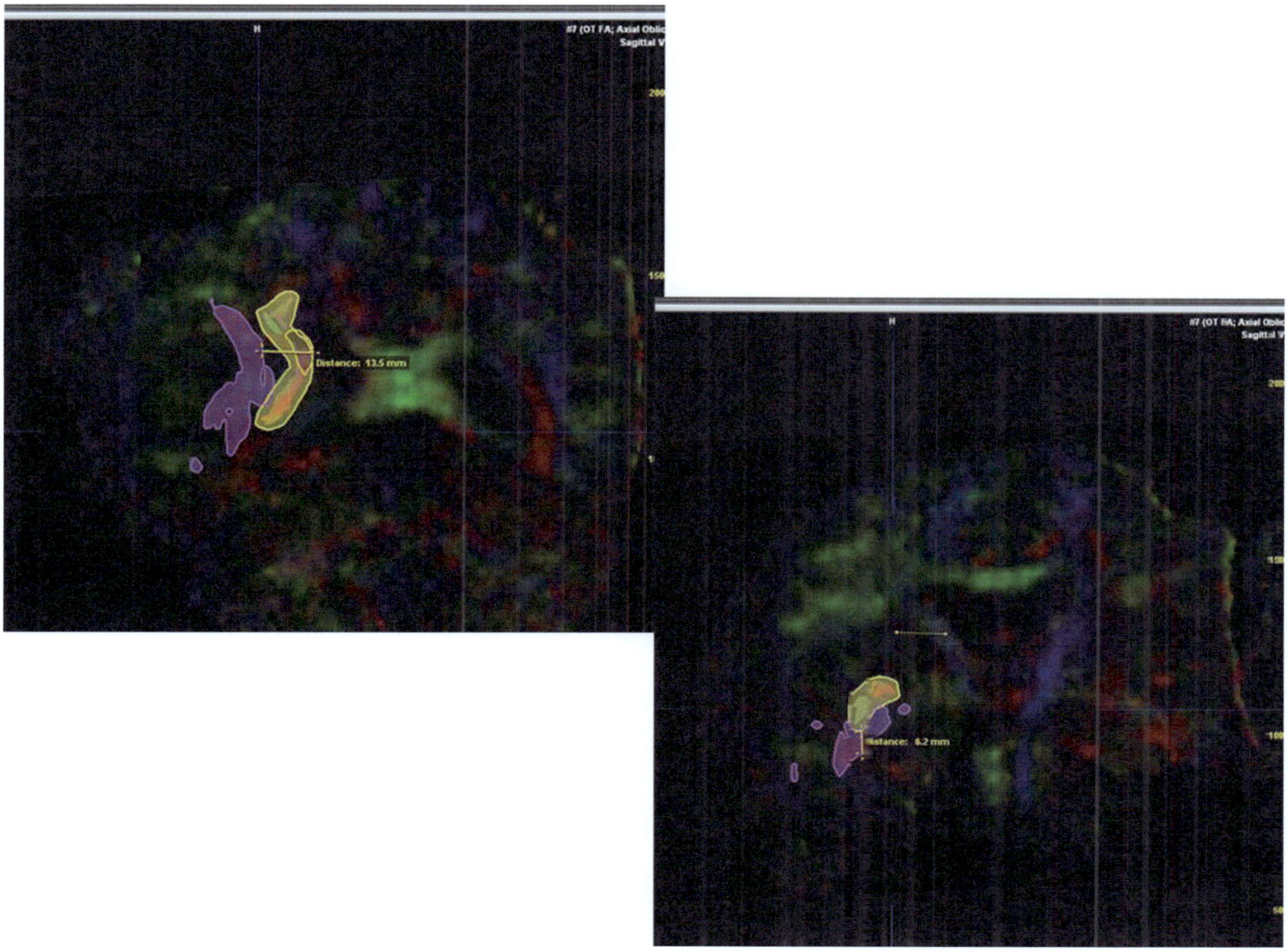

Fig. 4 Intraoperative DTI demonstrated the shifting of the bundle position caused by the surgical procedure

DTI and tractography may affect the surgical management of patients with brain tumors [30] and match with electrical mapping in the range of 0–2 cm [1].

Li et al. [18] demonstrated a good correlation between the CST estimated with DTI and the subcortical electrical stimulation; Mikuni et al. [19] also reported this.

Subcortical MEPs were consistently produced at distances of less than 7 mm and were absent at distances of more than 13 mm from the fiber tracking of the pyramidal tracts in the present study.

Gambini et al. [10] reported correspondences between intraoperative subcortical neurostimulation sites and tractography of 84% and 79 % for the motor tract and the speech circuit, respectively.

Intraoperative diffusion tractography solves the problem of ? so well, correlating with subcortical stimulation, as recently demonstrated by Ozawa et al. [28], because intraoperative DTI corrected the brain shift, thus allowing image validation and reporting positive MEPs between 0 and 4.7 mm from the stimulation site to the depicted bundle.

We always updated intraoperative tractography, because, in our experience, brain shift can cause dislocation of the pyramidal tract, in the range of 8 mm; shifting of white matter tracts during neurosurgical procedures has been demonstrated since 2005 by Nimsky et al. [22, 23], ranging from 8 mm inward to 15 mm outward.

In their studies, shifting was observed in 89 % of cases, and shifting was confirmed in our study [1] during tumor removal.

Other authors [16] believe that fiber tracking cannot accurately estimate the size of the white matter tracts using navigation based on preoperative DTI, because it is affected by brain shift both in the deep white matter and the cortex.

Shifting of the deep tumor portions during resection and, consequently, shifting of the white matter tracts, is currently accepted, so that preoperative MRI and DTI should be considered inaccurate [27] because of brain shift.

Standard neuronavigation, without intraoperative upgrade, is inaccurate because of brain shift and because preoperative MRI is a good but insufficient tool for intraoperative "functional" neuronavigation [6].

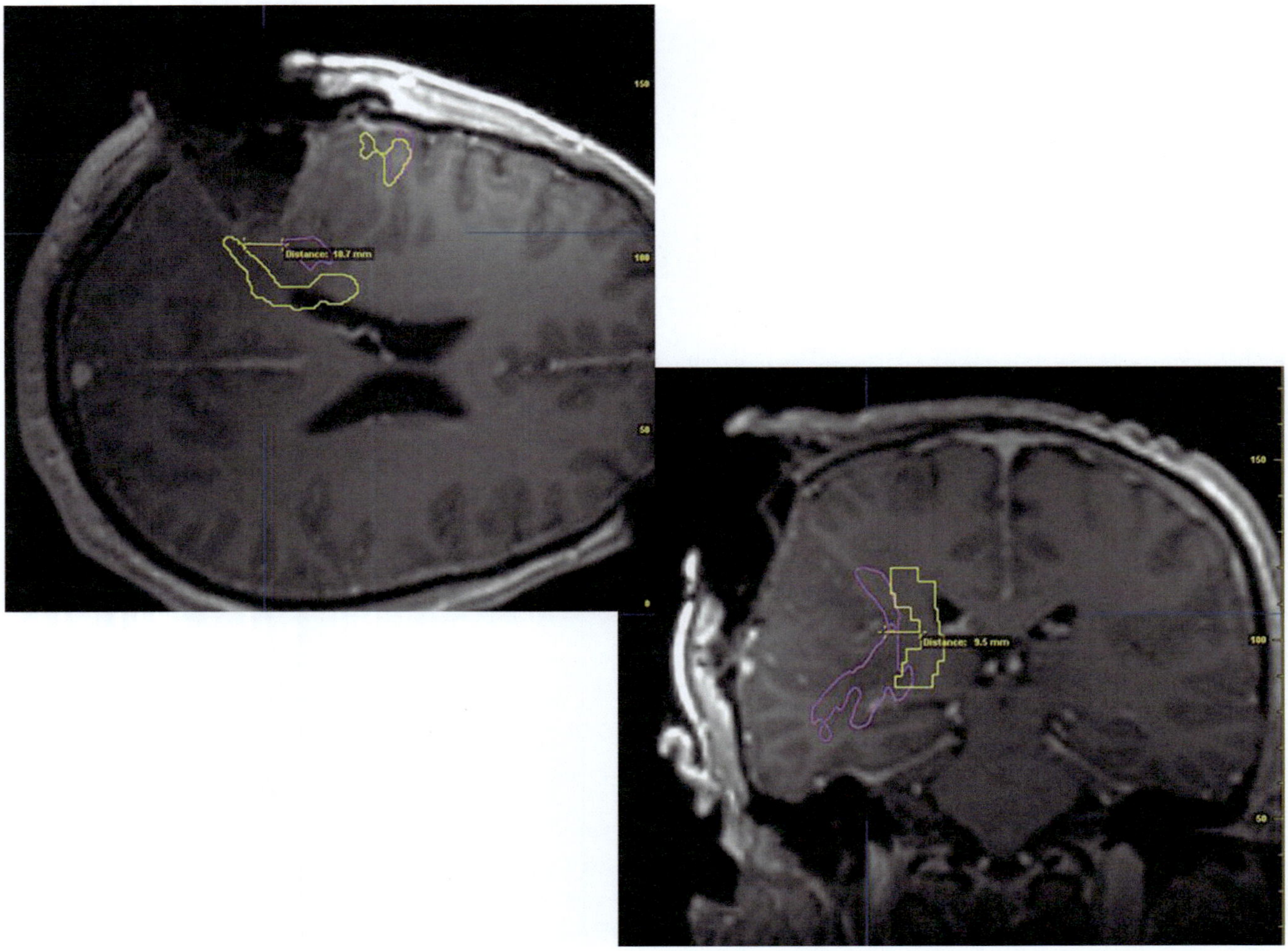

Fig. 5 Intraoperative integration between DTI and volumetric magnetic resonance imaging (MRI), demonstrating the shifting of the bundle position caused by the surgical procedure

Brain shift is a serious problem and, even after a simple craniotomy, shifting of up to 0.5-1 cm has been described, and shifting has also been described after the dura was opened [7, 12, 14].

We registered a shift of the white matter bundles of up to 29.7 mm simply after dural opening and Nimsky and colleagues reported a shift of up to 20-24 mm [13, 22, 24].

During surgery the shifting of deep structures, so-called subsurface shifting, seems to be much more relevant [25, 29] than the shifting of cortical structures, which is clearly visible during surgery.

We documented that shift involving the CST can be observed in 90 % of cases during surgery, similarly to other studies [7, 22, 28].

Nimsky et al. [22] described a shift ranging from 28 to +15 mm; for Ozawa et al. [28] the shift ranged from 28 to +8.7 mm, and in our study it ranged from 29.7 to +11 mm.

However, intraoperative tractography has some pitfalls [6], so that we advocate neurophysiological "neuronavigational" monitoring to compensate for these drawbacks.

Duffau et al. [9] stated that direct fiber stimulation was safe, accurate, and reliable, while Kamada et al. [15] observed some technical difficulties, such as the selection of the optimal stimulus point, visually indistinguishable subcortical pathways, continuously interrupted surgery, and long wasted time.

Again, we note that the integration of the neurophysiological techniques with the neuronavigation system, based on the intraoperative DTI, improves the precision of stimulation and reduces the time of the surgical procedure.

These techniques allowed us to reach a verified average distance between the CST and the site of subcortical MEP of 3.93 mm, with an excellent postoperative course.

Similarly to our experience, Nimsky et al. [26] reported mild postoperative deficits, which completely recovered during the postoperative course, and they had only one case of motor deficit, after 3 months.

We believe that the higher incidence of postoperative deficits that other studies report, although this could be due to surgical traction, heat from bipolar coagulation, cytotoxic edema, and/or microvascular reorganization [11], is caused by non-updated and unreliable information regarding the spatial (DTI tractography) and functional (subcortical electric stimulation) anatomy of the CST.

In particular, the intraoperative updating of anatomical and functional data allows us to reverse the surgical strategy, while the use of a stimulator probe "neuronavigated" onto updated tractographic objects guarantees control of the integrity of the tract before removal of the residual lesion volume, in contrast to verifying the integrity of the tract after a potential injury.

The goal is not to remove and to check, but to verify after an intraoperatively guided removal.

In regard to the CST, we recognized parameters—such as the amount of peritumoral edema and the craniotomy size—that predict its shift independent of direction [30], but shifting is not necessarily unidirectional, because the subsurface motion during resection is driven not by external pressure, but by the relief of weight and intraparenchymal pressure [7, 21, 29].

The direction of white matter tract shifting in the outward or inward direction seems to be unpredictable [30].

In our study, an outward shifting of the CST was mostly related to a large amount of edema and to a large craniotomy size, but in patients showing inward shifting, tumor size played the most important role, although we did not find any statistical significance for this parameter [30].

We strongly agree with Nabavi et al. [21], who suggest even more frequent or, if possible, continuous imaging for brain shift tracking, because only serial imaging with high spatial resolution allows the elucidation of deformation patterns with differing reactions to surgical manipulations.

We aim to achieve these results with neuronavigated brain mapping with fMRI and direct cortical stimulation for the surgery of lesions involving the motor and the language cortex.

The identification of the motor cortex is the first step in this surgery, and even if perirolandic gyri have been traditionally identified by phase reversal and by direct motor cortex stimulation, many studies report the limitations and failure of these techniques, especially for perirolandic mass lesions, in which localization was questionable or impossible in 10-18 %, because these tumors produce sensory evoked potential (SEP) latencies and amplitudes of high variability [6].

The neuronavigation with preoperative fMRI helps to better identify the stimulation sites and the precentral gyrus [6, 34].

Electrophysiological difficulties can also be related to neoplastic desynchronization of the afferent electrical impulses and to the mass effect distortion on the cortical electrical dipoles on the brain surface, and these difficulties may eventually lead to the surgeon choosing an inappropriate recording site.

Brain mapping alone allows identification of the primary motor cortex in only 60 % of cases [17], while the combination of neurophysiological monitoring and fMRI improves its accuracy [6], with a high concordance between the procedures.

We absolutely agree with these findings, and we also found a high concordance between fMRI and intraoperative brain mapping; their combination improves the sensitivity of each technique, reducing pitfalls (such as the Blood-oxygen-level dependent (BOLD) effect; motion-related artifacts caused by heartbeat, breathing, or head motion; and a too sensitive signal to large draining veins with poor spatial resolution on fMRI and electrical artifacts on direct cortical stimulation).

This combination (i.e., fMRI and intraoperative brain mapping), which is also defined as "functional neuronavigation", increases the definition of eloquent areas by comparison of the previously detected functional areas (fMRI) and the brain mapping recordings, and so reduces the time of surgery.

Conflict of Interest Statement We have no conflict of interest.

References

1. Bozzao A, Romano A, Angelini A, D'Andrea G, Calabria LF, Coppola V, Mastronardi L, Fantozzi LM, Ferrante L (2010) Identification of the pyramidal Tract by neuronavigation based on intraoperative magnetic resonance tractography: correlation with subcortical stimulation. Eur Radiol 20:2475–2481
2. Basser PJ, Pajevic S, Pierpaoli C, Duda J, Aldroubi A (2000) In vivo fiber tractography using DT-MRI data. Magn Reson Med 44(4):625–632
3. Bello L, Castellano A, Fava E, Casaceli G, Riva M, Scotti G et al (2010) Intraoperative use of diffusion tensor imaging fiber tractography and subcortical mapping for resection of gliomas: technical considerations. Neurosurg Focus 28(2):E6, 1–14
4. Berger MS, Rostomily RC (1997) Low grade gliomas: functional mapping resection strategies, extent of resection and outcome. J Neurooncol 34:85–101
5. D'Andrea G, Angelini A, Foresti C, Familiari P, Caroli E, Frati A (2014) Platinum-iridium subdermal magnetic resonance imaging-compatible needle electrodes are suitable for intraoperative neurophysiological monitoring during image-guided surgery with high-field intraoperative magnetic resonance imaging: an experimental study. Neurosurgery 10(3):387–392
6. D'Andrea G, Angelini A, Romano A, Di Lauro A, Sessa G, Bozzao A, Ferrante L (2012) Intraoperative DTI and brain mapping for surgery of neoplasm of the motor cortex and the corticospinal tract: our protocol and series in BrainSUITE. Neurosurg Rev 35:401–412
7. Dorward NL, Alberti O, Velani B, Gerritsen FA, Harkness WF, Kitchen ND et al (1998) Postimaging brain distortion: magnitude, correlates, and impact on neuronavigation. J Neurosurg 88(4):656–662
8. Duffau H (2009) Surgery of low-grade gliomas: towards a 'functional neurooncology'. Curr Opin Oncol 21(6):543–549
9. Duffau H, Capelle L, Sichez N, Denvil D, Lopes M, Sichez JP et al (2002) Intraoperative mapping of the subcortical language path-

ways using direct stimulations. An anatomo-functional study. Brain 125(Pt 1):199–214
10. Gambini A, Bello L, Falini A (2006) Corrispondenza tra fiber tracking e siti di neurostimolazione sottocorticale intraoperatoria dei circuiti motori e del linguaggio in pazienti affetti da tumori cerebrali. Neuroradiol J 19(Suppl 1):70–71
11. González-Darder JM, González-López P, Talamantes F, Quilis V, Cortés V, García-March G et al (2010) Multimodal navigation in the functional microsurgical resection of intrinsic brain tumors located in eloquent motor areas: role of tractography. Neurosurg Focus 28(2):E5, 1–10
12. Hall WA, Liu H, Truwit CL (2005) Functional magnetic resonance imaging-guided resection of low-grade gliomas. Surg Neurol 64(1):20–27
13. Hastreiter P, Rezk-Salama C, Soza G, Bauer M, Greiner G, Fahlbusch R et al (2004) Strategies for brain shift evaluation. Med Image Anal 8(4):447–464
14. Hill DL, Maurer CR Jr, Maciunas RJ, Barwise JA, Fitzpatrick JM, Wang MY (1998) Measurement of intraoperative brain surface deformation under a craniotomy. Neurosurgery 43(3):514–528
15. Kamada K, Todo T, Masutani Y, Aoki S, Ino K, Takano T et al (2005) Combined use of tractography integrated functional neuronavigation and direct fiber stimulation. J Neurosurg 102(4):664–672
16. Kinoshita M, Yamada K, Hashimoto N, Kato A, Izumoto S, Baba T, Maruno M, Nishimura T, Yoshimine T (2005) Fiber-tracking does not accurately estimate size of fiber bundle in pathological condition: initial neurosurgical experience using neuronavigation and subcortical white matter stimulation. Neuroimage 25(2):424–429
17. Kombos T, Suess O, Funk T, Kern BC, Brock M (2000) Intraoperative mapping of the motor cortex during surgery in and around the motor cortex. Acta Neurochir (Wien) 142(3):263–268
18. Li ZX, Jp D, Jiang T et al (2006) Function magnetic resonance imaging and diffusion tensor tractography in patients with brain gliomas involving motor areas: clinical application and outcome. Zhonghua Wai Ke Za Zhi 44(18):1275–1279
19. Mikuni N, Okada D, Enatsu R et al (2007) Clinical impact of integrated functional neuronavigation and subcortical electrical stimulation to preserve motor function during resection of brain tumours. J Neurosurg 106:593–598
20. Mori S, VanZijl PC (2002) Fiber tracking: principles and strategies- a technical review. NMR Biomed 15:468–480
21. Nabavi A, Black PM, Gering DT et al (2001) Serial intraoperative magnetic resonance imaging of brain shift. Neurosurgery 48(4):787–798
22. Nimsky C, Ganslandt O, Hastreiter P, Wang R, Benner T, Sorensen AG, Fahlbusch R (2005) Intraoperative diffusion-tensor MR imaging: shifting of white matter tracts during neurosurgical procedures – initial experience. Radiology 234(1):218–225
23. Nimsky C, Ganslandt O, Merhof D, Sorensen AG, Fahlbusch R (2006) Intraoperative visualization of the pyramidal tract by diffusion-tensor-imaging-based fiber tracking. Neuroimage 30(4):1219–1229
24. Nimsky C, Ganslandt O, Hastreiter P, Wang R, Benner T, Sorensen AG et al (2005) Preoperative and intraoperative diffusion tensor imaging-based fiber tracking in glioma surgery. Neurosurgery 56(1):130–138
25. Nimsky C, Ganslandt O, Cerny S, Hastreiter P, Greiner G, Fahlbusch R (2000) Quantification of, visualization of, and compensation for brain using intraoperative magnetic resonance imaging. Neurosurgery 47(5):1070–1080
26. Nimsky C, Ganslandt O, Fahlbusch R (2006) Implementation of fiber tract navigation. Neurosurgery 58(ONS Suppl 2):292–304
27. O'Shea JP, Whalen S, Branco DM, Petrovich NM, Knierim KE, Golby AJ (2006) Integrated image- and function-guided surgery in eloquent cortex: a technique report. Int J Med Robot 2(1):75–83
28. Ozawa N, Muragaki Y, Nakamura R et al (2009) Identification of he pyramidal yract by neuronavigation based on intraoperative diffusion weighted imaging combined with subcortical stimulation. Stereotact Funct Neurosurg 87:18–24
29. Paulsen KD, Miga MI, Kennedy FE, Hoopes PJ, Hartov A, Roberts DW (1999) A computational model for tracking subsurface tissue deformation during stereotactic neurosurgery. IEEE Trans Biomed Eng 46(2):213–225
30. Romano A, Ferrante M, Cipriani V, Fasoli F, Ferrante L, Fantozzi LM, D' Andrea G, Bozzao A (2007) Role of magnetic resonance tractography in the preoperative planning and intraoperative assessment of patients with intra-axial brain tumours. Radiol Med 112(6):906–920
31. Romano A, D'Andrea G, Minniti G et al (2009) Pre-surgical planning and MR-tractography utility in brain tumour resection. Eur Radiol 19(12):2798–2808
32. Sanai N, Berger MS (2008) Glioma extent of resection and its impact on patient outcome. Neurosurgery 62:753–766
33. Stieltjes B, Kaufmann WE, van Zijl PC, Fredericksen K, Pearlson GD, Solaiyappan M et al (2001) Diffusion tensor imaging and axonal tracking in the human brainstem. Neuroimage 14(3):723–735
34. Suess O, Suess S, Brock M, Kombos T (2006) Intraoperative electrocortical stimulation of brodman area 4: a 10-year analysis of 255 cases. Head Face Med 2(20):1–13

Functional Reconstruction of Motor and Language Pathways Based on Navigated Transcranial Magnetic Stimulation and DTI Fiber Tracking for the Preoperative Planning of Low Grade Glioma Surgery: A New Tool for Preservation and Restoration of Eloquent Networks

Giovanni Raffa, Alfredo Conti, Antonino Scibilia, Carmela Sindorio, Maria Catena Quattropani, Massimiliano Visocchi, Antonino Germanò, and Francesco Tomasello

Abstract

Background

Surgery of low-grade gliomas (LGGs) in eloquent areas still presents a challenge. New technologies have been introduced to enable the performance of "functional", customized preoperative planning aimed at maximal resection, while reducing the risk of postoperative deficits. We describe our experience in the surgery of LGGs in eloquent areas using preoperative planning based on navigated transcranial magnetic stimulation (nTMS) and diffusion tensor imaging (DTI) tractography.

Methods

Sixteen patients underwent preoperative planning, using nTMS and nTMS-based DTI tractography. Motor and language functions were mapped. Preoperative data allowed for tailoring of the surgical strategy. The impact of these modalities on surgical planning was evaluated. Influence on functional outcome was analyzed in comparison with results in a historical control group.

Results

In 12 patients (75 %), nTMS added useful information on functional anatomy and surgical risks. Surgical strategy was modified in 9 of 16 cases (56 %). The nTMS "functional approach" provided a good outcome at discharge, with a decrease in postoperative motor and/or language deficits, as compared with controls (6 vs. 44 %; $p=0.03$).

Conclusions

The functional preoperative mapping of speech and motor pathways based on nTMS and DTI tractography provided useful information, allowing us to plan the best surgical strategy for radical resection; this resulted in improved postoperative neurological results.

Keywords Arcuate fasciculus • Brain tumors • Corticospinal tract • DTI tractography • Low-grade gliomas • Navigated transcranial magnetic stimulation

G. Raffa, MD (✉)
Department of Neuroscience, Neurosurgical Clinic, University of Messina, Messina, Italy

Department of Clinical and Experimental Medicine, University of Messina, Messina, Italy

A.O.U. Policlinico "G. Martino",
Via Consolare Valeria, 1, Messina 98125, Italy
e-mail: mark83g@alice.it

A. Conti, MD, PhD • A. Scibilia, MD • A. Germanò, MD
F. Tomasello, MD
Department of Neuroscience, Neurosurgical Clinic, University of Messina, Messina, Italy

C. Sindorio, PsyD
Department of Neuroscience, Neurosurgical Clinic, University of Messina, Messina, Italy

Department of Clinical and Experimental Medicine, University of Messina, Messina, Italy

M.C. Quattropani
Department of Cognitive Sciences, Psychology, Pedagogy and Cultural Studies, University of Messina, Messina, Italy

M. Visocchi, MD
Institute of Neurosurgery, Medical School,
Catholic University of Rome, Rome, Italy

Introduction

Low-grade gliomas (LGGs) in eloquent areas still present a surgical challenge, with a marked risk of postoperative deficits [9]. However, the use of modern intraoperative techniques for monitoring brain function, such as direct cortical/subcortical stimulation (DCS/DSS) and awake surgery, has actually made LGG resection more successful [11, 12]. In the past few years several new techniques have been developed to perform "customized" noninvasive preoperative mapping for the surgery of LGGs in eloquent areas [8, 13]. These techniques allow for an improved understanding of the spatial relationship existing between lesions and functional structures, so helping the surgeon to select the best surgical strategy. Navigated transcranial magnetic stimulation (nTMS) provides reliable preoperative cortical mapping of motor and speech areas and has turned out to be as accurate as DCS [6, 7]. Moreover, it can be used as a seeding region for "functional" diffusion tensor imaging (DTI) tractography of the corticospinal tract (CST) [2] and the arcuate fasciculus (AF), providing a preoperative reconstruction of motor and language pathways that can also be easily used intraoperatively as a guide for resection. The aim of the present study was to describe a new technique for a "functional" reconstruction of motor and language cortico-subcortical networks, based on nTMS and DTI tractography for the preoperative planning of surgery for LGGs located in eloquent areas, and to analyze the impact of the technique on surgical strategy and functional outcome.

Materials and Methods

Patients and Mapping Protocol

We prospectively analyzed the clinical data of adult patients who underwent surgical resection of LGGs, located in the proximity of the motor pathway and/or language pathways, at the Neurosurgical Clinic of the University of Messina, Italy, between 2012 and 2014. Data were compared with data in a control group consisting of a consecutive series of 16 patients operated for LGGs in eloquent areas at our Institution before nTMS was available. Patients signed an informed consent for the scientific use of their data according to the requirements of the local Institutional Review Board (IRB). All patients underwent preoperative mapping of the eloquent cortex (motor and/or language areas) using our nTMS protocol [2]. Then the nTMS mapping was used as a seeding region for DTI tractography of the main eloquent white matter tracts (CST and AF).

nTMS Mapping of Motor and Speech Areas

All patients underwent a preoperative brain magnetic resonance (MR) scan using a 3 T scanner (Achieva 3 T; Philips Medical Systems, Amsterdam The Netherlands). For each patient we obtained T1-weighted, gadolinium-enhanced multiplanar reconstruction (MPR); 3D Fluid Attenuation Inversion Recovery (FLAIR) acquisition; and diffusion weighted sequences for DTI computation. The MPR sequence was imported into the nTMS machine for the mapping of the eloquent cortex. The nTMS mapping was performed using the NBS system 4.3 (NexstimOy; Elimäenkatu 9 B, Helsinki, Finland) for both motor and language areas, with the protocol already adopted at our institution [2]. Briefly, for motor mapping, we used a single pulse stimulation at 120 % of the resting motor threshold (RMT), using a navigated coil that is able to elicit a motor response recorded as a motor evoked potential (MEP) with standard electromyography electrodes. We recorded motor response from at least three channels, choosing at least one muscle for each body segment (face, arm, leg) in order to obtain the somatotopic organization of the motor cortex. Muscles usually used were the first dorsal interosseous (FDI), tibialis anterior (TA), and mentalis (MEN). With reference to language mapping, we used a repetitive stimulation task able to interfere with language comprehension and production, causing speech errors. Stimulation was performed during a simple object-naming task. Parameters of repetitive stimulation were: nTMS trains composed of five pulses at 5 Hz with an intensity of 100 % of the RMT. The display time was 4 s, the inter-picture interval 4 s, and the stimulation was set to start simultaneously with the picture presentation. During mapping, all sites were stimulated three times. We considered a site as eloquent when at least two of three stimulations caused an error response. The presence of speech errors was considered as a *positive mapping*, whereas the absence of errors was considered as a *negative mapping*. At the end of the procedure, the final nTMS map of motor and language areas was exported in a Digital Imaging and COmmunication in Medicine (DICOM) format to the neuronavigation system for DTI tractography computation.

nTMS-Based DTI Tractography of the CST and AF

The nTMS maps were used as seeding regions for computing the DTI tractography of the CST and AF. In cases of LGGs located near the motor pathway, we performed somatotopic reconstruction of the CST as previously described [2]. Briefly, using this technique, we were able to visualize the functionally different fibers composing the CST, distinguishing among face, arm, and leg fiber bundles. This allowed us to obtain a functional reconstruction of the CST and to better analyze its spatial relationship with the tumor. In tumors located near language pathways, the nTMS map of language areas was used as a seeding region for the reconstruction of the AF. In cases of negative speech mapping, standard DTI tractography of the AF was performed. The workflow for the tractography was performed by using StealthViz software (Medtronic Navigation; Coal Creek Circle, Louisville, CO, USA). A deterministic approach based on the *fiber assignment by continuous tracking* (FACT) algorithm was used, with a fractional anisotropy (FA) threshold of = 0.20; vector step length = 1 mm; minimum fiber length = 30 mm; and seed density = 1.0.

Impact on Surgical Strategy

The functional preoperative mapping was presented to the surgeon after the general principles of the surgical strategy had already been planned using only the standard MR scans. The surgeon was then asked to state the impact of nTMS mapping on surgical strategy, using a three-category scale: (1) useful information that increased knowledge of the functional anatomy and surgical risks, modifying the previously designed surgical plan; (2) helpful information that increased awareness of the functional anatomy and surgical risks without modifying the previous surgical plan; and (3) information that only confirmed the expected functional anatomy and surgical risks without modifying the surgical plan.

Surgery

Surgery was performed under general anesthesia. DCS and DSS were performed, in cases of tumors located in motor areas, according to Berger and Ojemann [1]. During surgery, the functional reconstruction of motor and language pathways was displayed in the neuronavigation system and used as an additional support, guiding tumor resection. DCS was performed by bipolar stimulation delivering square-wave pulses in 4-s trains at 60 Hz, with a duration of 1 ms. Stimulation intensity was progressively increased in 1-mA steps, from 0 up to 20 mA, until a motor response was obtained. DSS consisted of monopolar stimulation (trains of 3-4 stimuli for cranial surgery; pulse duration, 50-500 μs; inter-stimuli interval [ISI], 250-500 Hz [4-2 ms]) with increasing intensity until a motor response was obtained, with an upper limit of 20 mA. The resection was terminated when a MEP was elicited by using an intensity of 5 mA. In the study group, DCS and DSS were guided by the preoperative functional nTMS-based motor pathway mapping, as displayed in the neuronavigation system.

Impact on Intraoperative Neurophysiological Monitoring of the Motor Pathway

We evaluated the impact of the nTMS-based preoperative mapping of motor pathways on the intraoperative neurophysiological monitoring findings by comparing the mean DSS intensity at which lesion resection was stopped in both groups. Moreover, in order to verify the accuracy of the functional CST mapping, we analyzed the concordance between the type of MEP recorded during DSS (arm, face, or leg response) and the somatotopic organization of the CST provided by the nTMS-based DTI fiber tracking as displayed in the neuronavigation system.

Impact on Functional Outcome

We assessed and compared motor and/or language performances, on the first postoperative day and at discharge, with data from the control group. Motor performance was evaluated by using the Medical Research Council (MRC) scale. Language performance was evaluated by distinguishing different categories: i.e., no deficits, mild/moderate/severe motor deficits, and sensory or global aphasia, as rated by the participating neuropsychologist.

Statistical Analysis

Analysis of the homogeneity between the study group and controls was performed using the unpaired Student *t*-test for continuous data, and the Fisher exact test and the Chi-square test for categorical data. Analysis of differences in DSS findings was performed by using the unpaired Student *t*-test. Comparison of functional outcome was performed by using the Fisher exact test. Statistical significance was defined as a p value of <0.05. Analysis was realized using GraphPad Prism version 6.00 for Windows (GraphPad Software, La Jolla, California, USA, www.graphpad.com).

Results

Sixteen patients (10 male; 6 female; mean age 50.2 ± 3.59 years) were included in the study. Eight patients were affected by diffuse astrocytomas (50 %), 4 by oligoastrocytomas (25 %), and 4 by oligodendrogliomas (25 %). Table 1 summarizes the clinical and demographic characteristics of the patients. Mapping of the motor cortex was performed in all cases, obtaining its somatotopic organization (representation of face, arm, and leg muscles). Language areas were involved by tumors in 7 cases. Positive language mapping was achieved only in 3 patients (43 %). The most common type of error consisted of performance errors (54 %), followed by circumlocutions (21 %), and semantic paraphasia (14 %). A speech arrest was achieved three times (11 %). Language errors were obtained primarily when stimulating the fronto-opercular region (65 %). The other spots were localized in the angular gyrus and temporal lobe (36 %). In the remaining cases of tumors suspected to involve language cortex (4 patients; 57%) we observed a negative

Table 1 Salient clinical and demographic data of the study group and controls

Patients	Age (years)	Sex	Dominant hemisphere	Tumor location	Motor and/or language deficits on admission	Motor and language deficits on first postop day	Motor and language deficits at discharge	Type of nTMS-based mapping	MEP intensity and type at which resection was stopped (mA)	Histological diagnosis
nTMS + DTI										
#1	50	F	Left	Right, fronto-temporo-insular	No deficits	No deficits	No deficits	MC + CST	5 mA	Diffuse astrocytoma
#2	71	F	Left	Right, fronto-opercular	No deficits	No deficits	No deficits	MC + CST	6 mA	Oligodendroglioma
#3	44	F	Left	Left, fronto-insular	No deficits	No deficits	No deficits	MC + LC + AF	–	Diffuse astrocytoma
#4	35	M	Left	Right, fronto-temporo-insular	No deficits	Left f-b-c hemiparesis (4/5)*	No deficits	MC + CST	5 mA	Diffuse astrocytoma
#5	46	M	Right	Right, fronto-opercular	No deficits	No deficits	No deficits	MC + LC + CST + AF	6 mA	Diffuse astrocytoma
#6	49	M	Left	Left, fronto-opercular	No deficits	No deficits	No deficits	MC + LC + CST + AF	5 mA	Oligoastrocytoma
#7	41	M	Left	Right, temporal	No deficits	No deficits	No deficits	MC + CST	7 mA	Diffuse astrocytoma
#8	28	M	Left	Right, temporo-insular	No deficits	No deficits	No deficits	MC + CST	5 mA	Diffuse astrocytoma
#9	39	F	Left	Right, fronto-insular	No deficits	No deficits	No deficits	MC + CST	7 mA	Oligoastrocytoma
#10	53	F	Left	Right, frontal	No deficits	No deficits	No deficits	MC + CST	5 mA	Oligodendroglioma
#11	70	M	Left	Left, temporal	Mild sensory aphasia	Mild sensory aphasia	Mild sensory aphasia	MC + LC + AF	/	Diffuse astrocytoma
#12	76	M	Left	Left, fronto-opercular	Mild motor aphasia	Moderate motor aphasia	Mild motor aphasia	MC + LC + CST + AF	5 mA (face)	Oligodendroglioma
#13	58	F	Left	Right, frontal with invasion of corpus callosum	Left b-c hemiparesis (3/5)	Left b-c hemiparesis (3/5)	Left b-c hemiparesis (3/5)	MC + CST	4 mA	Oligoastrocytoma
#14	38	M	Left	Left, frontal	No deficits	Severe motor aphasia	Mild motor aphasia	MC + Lang + AF	–	Oligoastrocytoma
#15	67	M	Left	Right, parietal	No deficits	No deficits	No deficits	MC + CST	5 mA	Oligodendroglioma

*MRC score to assess muscle streght is defined as: 0 of 5 (0/5) = no contraction; 1 of 5 (1/5) = flicker or trace of contraction; 2 of 5 (2/5) = active movement, with gravity eliminated; 3 of 5 (3/5) = active movement against gravity; 4 of 5 (4/5) active movement against gravity and resistance; 5 of 5 (5/5) normal power.

#16	39	M	Left	Left, fronto-opercular	No deficits	No deficits	No deficits	MC + LC + CST + AF	6 mA	Diffuse astrocytoma
Controls										
#1	26	F	Left	Left, deep fronto-temporo-insular	No deficits	Right arm monoparesis (3/5)	Right arm monoparesis (3/5)	–	5 mA	Diffuse astrocytoma
#2	53	M	Left	Left, temporal	No deficits	No deficits	No deficits	–	–	Oligodendroglioma
#3	77	M	Left	Right, frontal	No deficits	No deficits	No deficits	–	7 mA	Oligoastrocytoma
#4	61	M	Left	Left, fronto-temporo-parietal	No deficits	Severe global aphasia, right f-b-c hemiparesis (3/5)	Severe global aphasia, right f-b-c hemiparesis (4/5)	–	4 mA	Diffuse astrocytoma
#5	41	F	Left	Right, fronto-parietal with invasion of corpus callosum	Left b-c hemiparesis (4/5)	Left b-c hemiparesis (3/5)	Left b-c hemiparesis (3/5)	–	5 mA	Diffuse astrocytoma
#6	55	M	Left	Right, fronto-temporo-parietal	Left f-b-c hemiparesis (1/5)	Left f-b-c hemiparesis (1/5)	Left f-b-c hemiparesis (1/5)	–	7 mA	Diffuse astrocytoma
#7	41	M	Left	Right, fronto-temporo-insular	No deficits	No deficits	No deficits	–	9 mA	Diffuse astrocytoma
#8	44	F	Left	Right, temporal	No deficits	No deficits	No deficits	–	7 mA	Diffuse astrocytoma
#9	45	F	Left	Right, frontal parasagittal	No deficits	Left b-c hemiparesis (3/5)	Left b-c hemiparesis (4/5)	–	5 mA	Diffuse astrocytoma
#10	75	F	Left	Left, frontal	Right b-c hemiparesis (4/5)	Right b-c hemiparesis (4/5)	Right b-c hemiparesis (4/5)		7 mA	Diffuse astrocytoma
#11	61	F	Right	Right, parietal	Left f-b-c hemiparesis (3/5)	Left f-b-c hemiparesis (2/5)	Left f-b-c hemiparesis (2/5)	–	6 mA	Diffuse astrocytoma
#12	76	M	Left	Right, temporo-parietal with insular invasion	No deficits	No deficits	No deficits	–	7 mA	Diffuse astrocytoma

(continued)

Table 1 (continued)

Patients	Age (years)	Sex	Dominant hemisphere	Tumor location	Motor and/or language deficits on admission	Motor and language deficits on first postop day	Motor and language deficits at discharge	Type of nTMS-based mapping	MEP intensity and type at which resection was stopped (mA)	Histological diagnosis
#13	55	M	Left	Left, frontal	No deficits	No deficits	No deficits	–	8 mA	Oligodendroglioma
#14	46	F	Left	Left, fronto-parietal	Right b-c hemiparesis (3/5)	Right b-c hemiparesis (2/5)	Right b-c hemiparesis (2/5)	–	5 mA	Oligoastrocytoma
#15	26	M	Left	Left, temporo-insular	Coma, decerebration	Coma, decerebration	No deficits	–	–	Diffuse astrocytoma
#16	78	F	Right	Left, fronto-parietal	No deficits	Right b-c hemiparesis (4/5)	Right b-c hemiparesis (4/5)	–	7 mA	Diffuse astrocytoma

nTMS navigated transcranial magnetic stimulation, *DTI* diffusion tensor imaging, *f-b-c* facio-brachio-crural, *b-c* brachio-crural, *MC* motor cortex, *LC* language cortex, *CST* corticospinal tract, *AF* arcuate fasciculus, *MEP* motor evoked potential response that suggested the absence of functionally relevant language areas in the region surrounding the tumor. The nTMS map was used as the seeding region for the somatotopic DTI tractography of the CST in 13 patients (81 %) (Fig. 1a–d). An nTMS-based reconstruction of the AF was performed in all 3 patients who showed a positive response during speech area mapping, using the nTMS map as the seeding region (Fig. 1e–h). In the remaining 4 cases, standard tractography of the AF was used.

Impact on Surgery

In 12 patients (75 % of cases), the nTMS-based functional mapping added useful information for the preoperative planning. In particular, in 9 patients (56 %) the mapping was able to increase the surgeon's knowledge of the functional anatomy and surgical risks, modifying the previously hypothesized plan based on standard MR. In 3 patients (19 %) the mapping increased the surgeon's awareness of the functional anatomy and surgical risks without modifying the hypothesized plan. Lastly, in the remaining 4 patients (25 %) the information confirmed the expected functional anatomy and surgical risks (Fig. 2a).

Impact on Intraoperative Neurophysiological Monitoring of Motor Pathway

The intraoperative use of the nTMS-based functional mapping guided DCS and DSS, and allowed for an easier identification of the motor cortex and CST. In particular, the availability of real-time verification of the CST location through the neuronavigation increased the surgeon's confidence in anatomo-functional landmarks, significantly reducing the DSS intensity at which resection was terminated in the study group as compared with the controls (5.38 ± 0.96 mA vs 6.35 ± 1.39 mA; $p = 0.04$) (Table 1). Moreover, in all cases, we observed a concordance between the type of recorded MEP and the functionally different CST fiber bundle that was stimulated (arm, face, leg) (Fig. 3).

Impact on Functional Outcome

The control group consisted of 16 patients (8 male; 8 female; mean age 53.75 ± 4.22 years). Twelve patients were affected by diffuse astrocytomas (75 %), 2 by oligodendrogliomas (12.5 %), and the remaining 2 by oligoastrocytomas (12.5 %). Statistical analysis showed the study group and controls were homogenous for demographic and relevant clinical characteristics. On the first postoperative day, 3 patients in the study group (19 %) vs. 7 patients (44 %) in the control group suffered from a worsening of motor and/or speech performance . The comparison of this parameter between the two groups showed no statistically significant difference (Fig. 2b). At discharge, only 1 patient in the study group showed persistence of the worsened language performance (6 %). In the control group, 7 patients still presented a worsened motor and/or language performance (44 %), 8 patients were unchanged (50 %), and 1 presented an improvement of the overall neurological status (6 %). With regard to the functional outcomes (unchanged/improved vs. worsened) the difference between the two groups was statistically significant ($p = 0.03$) (Fig. 2c).

Discussion

In the present study we demonstrated that nTMS-based preoperative mapping provided useful information on the spatial relationship between tumors and the adjacent functional structures, allowing for the preservation of motor and language functions, which may result in improved neurological outcome. Several studies have demonstrated that nTMS is an effective technique for the noninvasive preoperative functional mapping of motor and language areas in patients affected by brain tumors [6, 10], including LGGs [8]. nTMS has shown higher accuracy as compared with functional MRI (fMRI) [3] and it showed similar precision to DCS [7] in the mapping of the motor cortex. In addition, it has been demonstrated that nTMS is able to distinguish between true-eloquent and non-eloquent areas, to extend surgical indications, to increase surgical resection, to improve functional outcome, and to improve progression-free survival [4]. These findings are similar to those reported for intraoperative techniques and can be explained by the fact that nTMS has a solid neurophysiological basis that allows for easier and more objective functional mapping of the motor cortex. The mapping of areas involved in language by nTMS can be more difficult. However, recent evidence reported a good correlation between nTMS and intraoperative language mapping [6]. Picht et al., in 2013, have demonstrated that the added value of nTMS language mapping is its high negative predictive value, which can be used to exclude the proximity of tumors to eloquent language regions [6]. However,

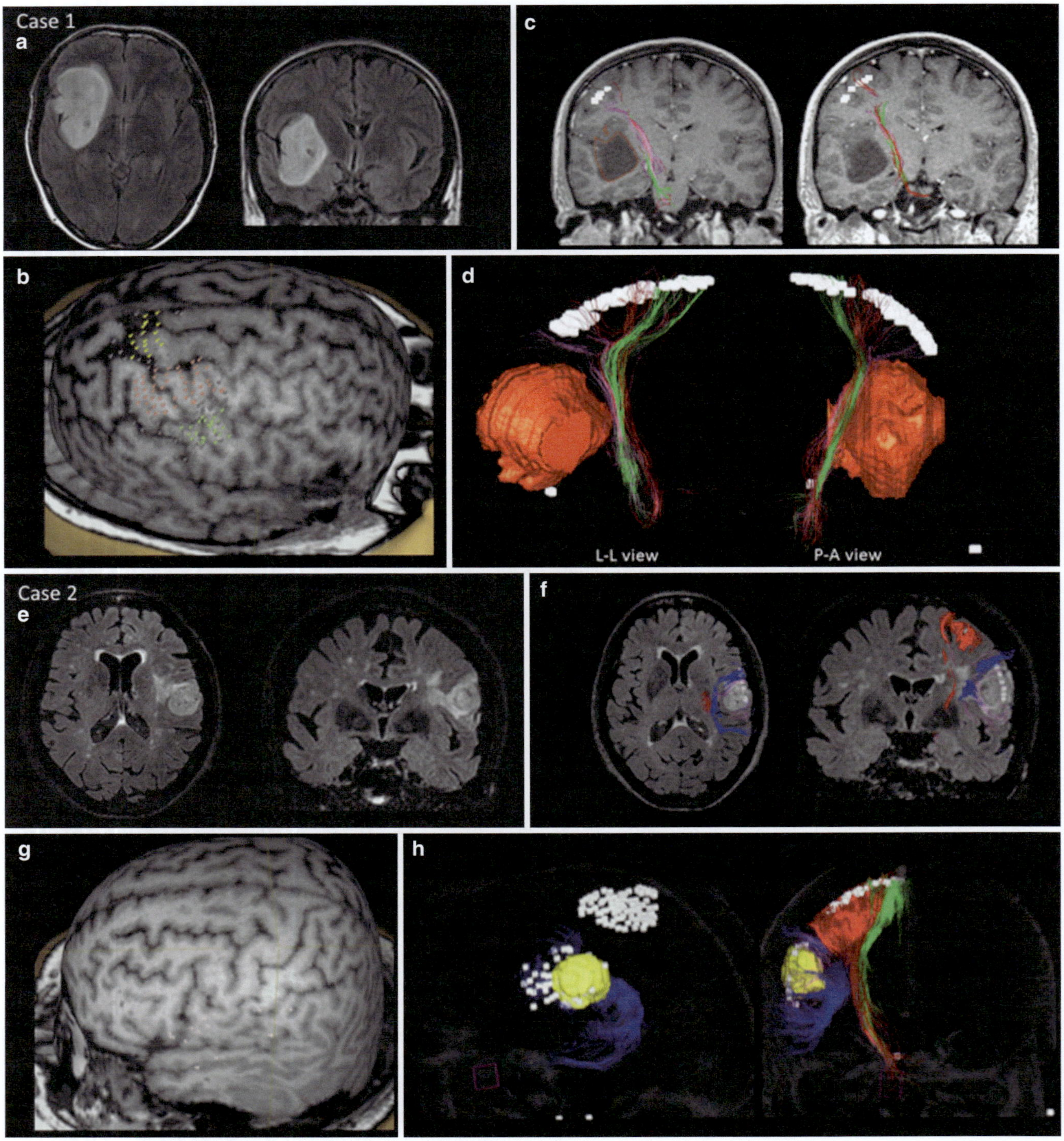

Fig. 1 Case 1. Right fronto-temporo-insular diffuse astrocytoma. (**a**) FLAIR sequences. (**b**) Somatotopic navigated transcranial magnetic stimulation (nTMS) map of the motor cortex (*yellow spots*=leg; *red*=arm; *green*=face). (**c**) Preoperative planning, showing the nTMS map of the motor cortex (*white spots*) and the functionally different corticospinal tract (CST) fibers (*green*=leg; *red*=arm; *pink*=face). (**d**) Somatotopic nTMS-based tractography of the CST, showing the proximity of the tumor (*orange*) to the leg and face fibers. Case 2. Left fronto-opercular oligodendroglioma. (**e**) FLAIR sequences. (**f**) nTMS map of face motor and language cortex (*white spots*) and the functionally different CST fibers and the arcuate fasciculus (AF; *blue*). The cortical representation of the face is overlapped with the error spots of language mapping and both are located above the tumor. (**g**) nTMS language map. (**h**) nTMS-based tractography of the CST and AF. CST face fibers and AF are surrounding the tumor (*yellow*). Arm and leg fibers are located more posteriorly

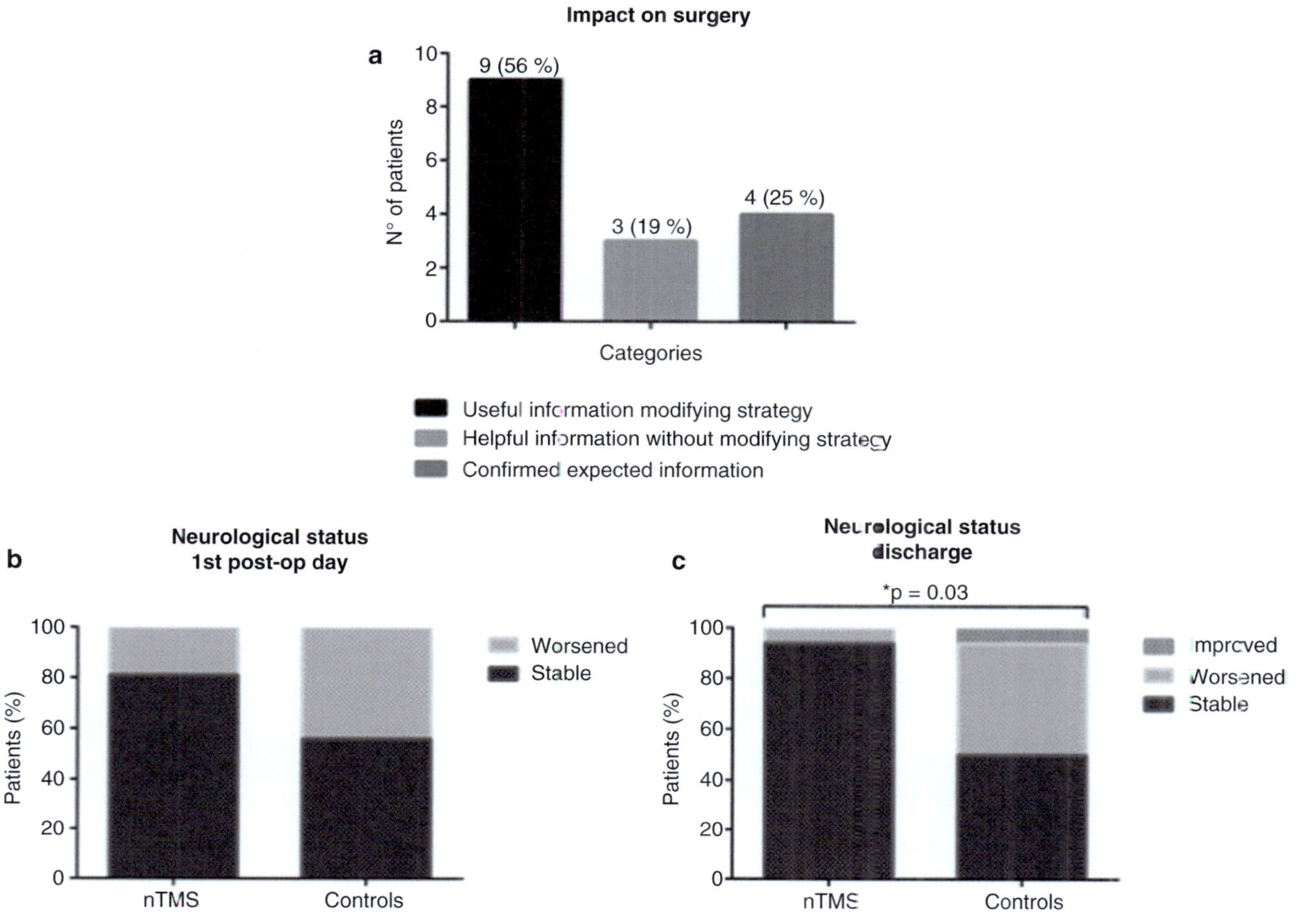

Fig. 2 (**a**) Impact of the nTMS-based functional mapping on (**a**) surgical strategy; (**b**) neurological outcome on first postoperative day, and (**c**) neurological outcome at discharge

although motor and/or language deficits may originate from cortical damage, they also, and perhaps more often, originate from injury to the subcortical structures. Thus, complete preoperative functional planning needs to include a reconstruction of the main subcortical white matter tracts. DTI tractography can be combined with nTMS to provide a reconstruction of the motor network that is more accurate than the standard technique [2, 5]. In our experience the intraoperative use of the DTI-nTMS technique allowed us to perform tailored approaches and oriented intraoperative neurophysiological mapping, and to reduce the time needed for the identification of functional structures (Fig. 3). In particular, the concordance between the nTMS-based functional reconstruction of the motor pathway and DSS findings demonstrated its reliability, thus confirming its value as a new strategy for combined neurophysiology-based image-guided surgery. Collectively, the functional information provided by our preoperative mapping was able to increase the surgeon's awareness of functional anatomy and surgical risks in 75 % of cases, modifying the previously hypothesized strategy in 56 %. Moreover, its use was associated with the stability of motor and language outcomes at discharge (15 of 16 patients) as compared with findings on admission, and with a decrease of postoperative deficits as compared with the control group (44 vs. 6 %; $p = 0.03$).

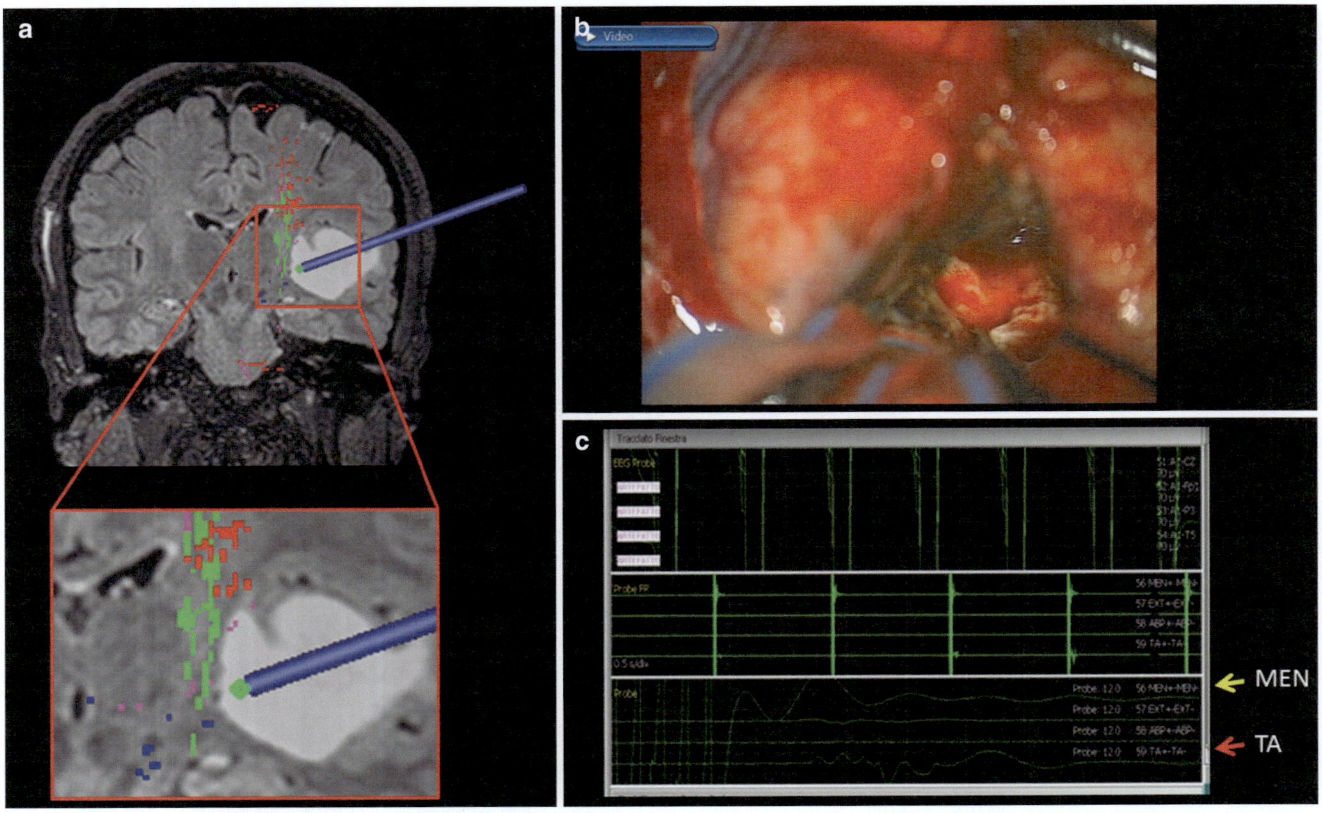

Fig. 3 Right temporo-insular diffuse astrocytoma. Intraoperative use of the nTMS-based mapping. (**a**) Screenshot from neuronavigation during direct subcortical stimulation (DSS). The stimulation probe (*blue stylet*) is in the surgical cave, in proximity to the leg (*green*) and face (*pink*) CST fibers. *Red spots* represent arm fibers. (**b**) Simultaneous microscopic view during DSS and (**c**) neurophysiological findings showing a motor response from leg muscle (TA, tibialis anterior, *red arrow*) and, after few milliseconds, from face muscle (MEN, mentalis, *yellow arrow*)

Conclusions

Functional nTMS-based reconstruction of language and motor pathways provides useful information for the customized planning of surgery for LGGs invading eloquent areas. The combination of nTMS with "functional" DTI tractography increases the surgeon's awareness of the anatomy and surgical risks, allowing them to plan the best surgical strategy in regard to a radical resection; the use of this technique leads to the sparing of functional boundaries, increases confidence in intraoperative mapping, and reduces the incidence of postoperative motor and language deficits.

Conflict of Interest Statement

Disclosure The authors declare that they have no conflicts of interest.

References

1. Berger MS, Ojemann GA (1992) Intraoperative brain mapping techniques in neuro-oncology. Stereotact Funct Neurosurg 58:153–161
2. Conti A, Raffa G, Granata F, Rizzo V, Germano A, Tomasello F (2014) Navigated transcranial magnetic stimulation for "somatotopic" tractography of the corticospinal tract. Neurosurgery 10(Suppl 4):542–554; discussion 554
3. Forster MT, Hattingen E, Senft C, Gasser T, Seifert V, Szelenyi A (2011) Navigated transcranial magnetic stimulation and functional magnetic resonance imaging: advanced adjuncts in preoperative planning for central region tumors. Neurosurgery 68:1317–1324; discussion 1324–1315
4. Frey D, Schilt S, Strack V, Zdunczyk A, Rosler J, Niraula B, Vajkoczy P, Picht T (2014) Navigated transcranial magnetic stimulation improves the treatment outcome in patients with brain tumors in motor eloquent locations. Neuro Oncol 16:1365–1372
5. Krieg SM, Buchmann NH, Gempt J, Shiban E, Meyer B, Ringel F (2012) Diffusion tensor imaging fiber tracking using navigated brain stimulation – a feasibility study. Acta Neurochir (Wien) 154:555–563

6. Picht T, Krieg SM, Sollmann N, Rosler J, Niraula B, Neuvonen T, Savolainen P, Lioumis P, Makela JP, Deletis V, Meyer B, Vajkoczy P, Ringel F (2013) A comparison of language mapping by preoperative navigated transcranial magnetic stimulation and direct cortical stimulation during awake surgery. Neurosurgery 72:808–819
7. Picht T, Schmidt S, Brandt S, Frey D, Hannula H, Neuvonen T, Karhu J, Vajkoczy P, Suess O (2011) Preoperative functional mapping for rolandic brain tumor surgery: comparison of navigated transcranial magnetic stimulation to direct cortical stimulation. Neurosurgery 69:581–588; discussion 588
8. Picht T, Schulz J, Vajkoczy P (2013) The preoperative use of navigated transcranial magnetic stimulation facilitates early resection of suspected low-grade gliomas in the motor cortex. Acta Neurochir (Wien) 155:1813–1821
9. Pouratian N, Asthagiri A, Jagannathan J, Shaffrey ME, Schiff D (2007) Surgery insight: the role of surgery in the management of low-grade gliomas. Nat Clin Pract Neurol 3:628–639
10. Rizzo V, Terranova C, Conti A, Germano A, Alafaci C, Raffa G, Girlanda P, Tomasello F, Quartarone A (2014) Preoperative functional mapping for rolandic brain tumor surgery. Neurosci Lett 583:136–141
11. Sanai N, Berger MS (2010) Intraoperative stimulation techniques for functional pathway preservation and glioma resection. Neurosurg Focus 28:E1
12. Schucht P, Ghareeb F, Duffau H (2013) Surgery for low-grade glioma infiltrating the central cerebral region: location as a predictive factor for neurological deficit, epileptological outcome, and quality of life. J Neurosurg 119:318–323
13. Tomasello F, Conti A (2014) The promise of multimodal image guidance in neurosurgery. World Neurosurg 82:e183–e184

Intraoperative Neurophysiological Monitoring in Spine Surgery: A Significant Tool for Neuronal Protection and Functional Restoration

Antonino Scibilia, Giovanni Raffa, Vincenzo Rizzo, Angelo Quartarone, Massimiliano Visocchi, Antonino Germanò, and Francesco Tomasello

Abstract Although there is recent evidence for the role of intraoperative neurophysiological monitoring (IONM) in spine surgery, there are no uniform opinions on the optimal combination of the different tools. At our institution, multimodal IONM (mIONM) approach in spine surgery involves the evaluation of somatosensory evoked potentials (SEPs) and motor evoked potentials (MEPs) with electrical transcranial stimulation, including the use of a multipulse technique with multiple myomeric registration of responses from limbs, and a single-pulse technique with D-wave registration through epi- and intradural recording, and free running and evoked electromyography (frEMG and eEMG) with bilateral recording from segmental target muscles. We analyzed the impact of the mIONM on the preservation of neuronal structures and on functional restoration in a prospective series of patients who underwent spine surgery. We observed an improvement of neurological status in 50 % of the patients. The D-wave registration was the most useful intraoperative tool, especially when MEP and SEP responses were absent or poorly recordable. Our preliminary data confirm that mIONM plays a fundamental role in the identification and functional preservation of the spinal cord and nerve roots. It is highly sensitive and specific for detecting and avoiding neurological injury during spine surgery and represents a helpful tool for achieving optimal postoperative functional outcome.

Keywords Intraoperative neurophysiological monitoring • Spine surgery • Multimodal IONM • Functional outcome

Introduction

Spinal surgery places the spinal cord (SC) and nerve roots (NRs) at risk of injury [1]. Intraoperative neurophysiological monitoring (IONM) allows for the identification and the real-time verification of the functional integrity of neuronal structures. The early detection of damage, at a reversible stage, allows for the prompt correction of the cause and avoids permanent deficits. Different techniques have been employed for monitoring long SC pathways and NRs. The most reliable are somatosensory evoked potentials (SEPs), motor evoked potentials (MEPs), D-wave registration, and electromyography (EMG). The accuracy of IONM is increased when these tools are employed simultaneously in a multimodal approach [2]. Surprisingly, the literature does not provide clear evidence of this accuracy. Different papers report the use of several combinations of IONM techniques, without suggesting the most accurate and reliable approach. The aim of this study was to describe the multimodal IONM (mIONM) approach in spinal surgery employed at our institution, assessing its accuracy as a diagnostic tool during surgery and its correlation with a postoperative neurological outcome.

Materials and Methods

We prospectively collected data from patients who underwent spine surgery using the mIONM approach at the Neurosurgical Department of the University of Messina

A. Scibilia • A. Germanò • F. Tomasello
Department of Neurosurgery, University of Messina, Messina, Italy

G. Raffa
Neurosurgery and Department of Clinical and Experimental Medicine, University of Messina, Messina Italy

A. Quartarone
Neurology, University of Messina, Messina, Italy

V. Rizzo, MD (✉)
Neurology, University of Messina, Consolare Valeria 1, Messina 98125, Italy
e-mail: enzo.rizzo@gmail.com

M. Visocchi
Institute of Neurosurgery, Medical School, Catholic University of Rome, Rome, Italy

M. Visocchi et al. (eds.), *Trends in Reconstructive Neurosurgery*, Acta Neurochirurgica Supplement 124,
DOI 10.1007/978-3-319-39546-3_38,

from May 2013 to May 2015. The mIONM approach included the use of SEPs, MEPs, D-wave recording, and free-running and evoked EMG (frEMG-eEMG). The combination was designed for each patient according to the type of pathology. Data were recorded and digitally archived using an IONM workstation (NIM Eclipse; Medtronic, Jacksonville, USA). IONM alerts were extracted from the IONM report. Motor, sensory, and urinary dysfunctions were recorded pre- and postoperatively, at discharge, and at 6 months after surgery. Postoperative neurological deficit was defined as a new or significant worsening of motor, sensory, and urinary symptoms. Motor neurological status was assessed according to the Medical Research Council (MRC) grading system. Sensory outcome, bladder continence, and pain were assessed by the use of the scale proposed by Pratheeshe et al. [3]. Anesthesia was induced by total intravenous anesthesia (TIVA). Remifentanil, at 0.10–0.20 mg/kg/min, was used for induction, with 0.25–0.40 mg/kg/min used for maintenance; for Propofol, 3–4 mg/ml, were used for induction and 4-5 mg/ml were used for maintenance [1]. Mean arterial blood pressure for all patients was kept at >75 mmHg. SEPs were elicited from the posterior tibial and median nerves (pulse duration 0.58 ms, frequency 4.1 Hz, and intensity 15-40 mA). Recordings were performed via corkscrew-like electrodes in the scalp at CZ'-FZ (legs) and C3'/C4'-FZ (arms) according to the 10–20 international electroencephalogram (EEG) system. A 50 % drop in amplitude and/or a 10 % prolongation in latency were considered significant [4]. For MEP recording, a multipulse technique with transcranial electrical stimulation and recordings from limb muscles was used. Two short constant voltage trains (600 V) of, respectively, three and five square-wave stimuli of 0.5 ms duration and inter-train interval (ISI) of 20 ms were delivered at a repetition rate of up to 1 Hz. These stimuli were provided through corkscrew-like electrodes, placed at C1 and C2 scalp sites, according to the 10–20 EEG system. MEPs were recorded by needle electrodes inserted into muscles of the upper (biceps and first dorsal interosseous) and lower (tibialis anterior and vastus lateralis) extremities. During tethered cord (TC) surgeries, additional muscle recording needle electrodes (gastrocnemius and external anal sphincter) were placed. A single-pulse technique was used for D-wave recordings; a single transcranial electrical stimulus (pulse duration 0.5 ms, frequency 0.5 Hz, and voltage 600 V) was applied, using the same scheme as for MEPs. The D-wave was recorded by an electrode placed in the epi- or subdural space, caudally to the level of surgery. A significant MEP alert (multipulse technique) was defined as an abrupt disappearance of responses. For D-wave registration the criterion used was a decrement of amplitude >50 %. The surgeon was informed of alerts for any decrease of D-wave amplitude of between 30 % and 50 % of the baseline value. In those cases in which the D-wave recording was not used, surgeons were informed of any abrupt decrease in peak-to-peak MEP amplitude of more than 50 % for more than three successive trials [5]. The frEMG was recorded from upper and lower limbs by subdermal monopolar needle electrodes. The eEMG was used in those patients in whom mapping of functional neural elements was crucial to perform surgical maneuvers, such as during some surgeries for SC and/or spinal NR tumors, as well as TC surgeries. Stimulation was delivered through a monopolar probe connected to a constant current stimulator that conveyed monophasic square-wave pulses (short train of five square-wave stimuli of 0.1 ms duration and after-stimulus interval [ASI] of 1 s delivered at a repetition rate of up to 1 Hz; intensity 0.25–4 mA). The accuracy of the IONM technique was assessed by analysis of its specificity, sensitivity, and positive and negative predictive values (PPVs and NPVs). Statistical significance was defined as $p<0.05$. GraphPad Prism version 6.00 for Windows (GraphPad Software, La Jolla, California, USA, www.graphpad.com) was used for the data analysis. The association between IONM changes and new postoperative neurological deficits was analyzed by using the Fisher exact test.

Results

Patient Demographics

Sixteen patients (median age 41.9 years, range 12-65 years; 8 males and 8 females) underwent spinal surgery using the mIONM approach (Table 1). Surgery was carried out for tumors in 11 cases, for TC syndrome in 4 cases, and for a vascular lesion in 1 case (Table 2). Lesions involved the cervical spine in 12 % (2 of 16), dorsal spine in 32 % (5 of 16), and lumbosacral spine in the remaining 56 % of cases (9 of 16). Three of the 11 tumors (27 %) were intramedullary (IMSCT), while 83 % (8 of 11) were intradural extramedullary (IDEMSCT). A preoperative motor deficit was observed

Table 1 Patients' characteristics

	Value
Age (years)	Mean 41.9, range 12-65
Sex	M/F 8: 8
Pathology	Tumors 69 % (11 of 16 pts)
	Tethered cord surgery 25 % (4 of 16 pts)
	Vascular lesions 6 % (1 of 16 pts)
Level	Cervical 12 % (2 of 16 pts)
	Dorsal 32 % (5 of 16 pts)
	Lumbosacral 56 % (9 of 16 pts)
Total no. of patients	16

in 75 % of the patients (12 of 16). Fifty-six percent of the patients (9 of 16) showed preoperative sensory deficits and 37.5 % (6 of 16) were affected by urinary dysfunction. The IONM modalities performed included SEP, MEP, and frEMG recordings in all patients, while D-wave and eEMG were recorded in 44 % (7 of 16) and 62 % of cases (10 of 16), respectively.

Neurophysiological Alerts and Outcome

No complications were observed with mIONM, as already described in the literature [1]. The surgeon was alerted to a significant change in mIONM in 25 % of the surgeries (4 of 16; 1 patient had an EMG activity burst and 3 had significant MEP and/or D-wave changes). No monitoring changes were observed in 75 % of cases (12 of 16) and none of these patients had new postoperative deficits. SEPs were recorded in 75 % (12/16) of the surgeries (Table 2). No patients had significant changes in SEPs during surgery. SEPs had a sensitivity of 0 %, specificity of 100 %, PPV of 0 %, and an NPV of 92 %, without any statistically significant correlation with new postoperative deficits (Table 3). MEPs were recordable in all cases except in one case of TC surgery, due to a severe preoperative motor impairment (Table 2). In 19 % of the surgeries (3 of 16), a major MEP alert was documented, without returning to baseline throughout the surgery. Following the neurophysiological alert, surgeons modified the surgical strategy; for example, by releasing distraction or changing the tactics used for debulking. Two (12 %) of these MEP changes were true positives, correlating with a new postoperative deficit. One of these two patients was operated for an intramedullary lipoma of the conus medullaris associated with an arachnoid cyst without TC syndrome. In the postoperative period she developed a motor weakness (4/5 on the [MRC] scale) in the right leg that recovered after 1 month. The other patient, affected by a T7–T8 intramedullary metastasis, had preoperative paraparesis (4/5 on the MRC scale) that worsened in the postoperative period (3/5). This deficit was permanent at the 6-month examination. The remaining case of MEP alert occurred during surgery for a T5–T6 meningioma (WHO I). The MEPs disappeared during surgery but the patient did not experience new postoperative deficits (false-positive alert). There was a significant association between MEP alerts and the presence of a new postoperative deficit ($p<0.05$) (Table 3). In our series, we observed a sensitivity of 100 % and a specificity of 92 % for MEP alerts. The PPV was 67 %, and the NPV was 100 %. The D-wave was recorded in 44 % of procedures (7 of 16) as, in the remaining cases, the level of spinal pathology was located below T11. The D-wave was recordable in all 7 of these patients (1 IMSCT, 5 IDEMSCT, 1 vascular lesion). During six of these procedures, the D-wave showed stable values or changes that were considered as not significant. On the contrary, in one case (T7–T8 intramedullary metastasis from esophageal carcinoma) (Fig. 1) there was a significant D-wave amplitude decrease (>50 %) with a worsening of the preoperative paraparesis (MRC 3/5). In another case (T5–T6 meningioma), (Fig. 2) a stable D-wave was the only intraoperative neurophysiological parameter able to predict the absence of new postoperative deficits, despite an intraoperative loss of MEPs. In our series D-wave had sensitivity, specificity, and PPV and NPV of 100 %, although correlation with the presence of a new postoperative deficit was not statistically significant. The frEMG activity was recorded in all patients. In one case, (cauda equina ependymoma) an intraoperative burst activity was recorded and the patient presented new postoperative transient L3 radicular pain. The accuracy analysis showed that frEMG had sensitivity, specificity, and PPV and NPV of 100 % (Table 3). In our case series, eEMG was used in 10 (62 %) patients and was able to map NRs in 100 % of cases. In our series the mIONM approach was able to detect neurological injury in 19 % (3 of 16) of cases, with sensitivity, specificity, and PPV and NPV of 100 %. After surgery, we observed a statistically significant improvement of the MRC scores at 6 months (Student's t-test, $p<0.05$). In detail, the MRC score was improved in 50 % of the patients (8 of 16), worsened in 6 % (1 of 16), and unchanged in 44 % (7 of 16). None of the patients showed a severe postoperative neurological deficit.

Discussion

Recently, new recommendations have been published about the safety, efficacy, and interpretation of mIONM in spine surgery [6, 7]. However, there is still a debate about the combination of techniques to be used for the best functional outcome after spine surgery [8]. In the present study, we report our preliminary results on the use of specific mIONM techniques during spine surgery, demonstrating a really high accuracy of mIONM and a significant correlation between its use and a good postoperative outcome.

The first IONM technique described consisted of SEP recording. However, several reports [1, 9] have documented the inadequacy of SEP recording when assessing the functional integrity of the motor pathway in the SC. MEPs have been considered to be a more reliable technique for the monitoring of motor pathways. Different warning criteria have been proposed for MEPs, and they have changed from the simple presence or absence of responses [4, 7], to changes in the thresholds [10] that elicit muscle MEPs, to a

Table 2 IONM data and outcomes

		Preoperative assessment			IONM						Outcome					
											Motor examination (MRC score)					
	Pathology	Motor examination (MRC score)	Sensory examination	Urinary examination	SEP	MEP	D-Wave	fr-EMG	e-EMG	Intraoperative alert	Discharge	6 Months	Sensory examination	Urinary examination	Final outcome	N.D.D.
1	C0–C2 Meningioma (WHO I)	Tetraparesis (4/5), dysphagia	Negative	Negative	X	X	X	X		No	5	5	Negative	Negative	Improved	No
2	CMJ Cavernous angioma	Right hemiparesis (3/5), dysphagia	Hypoesthesia 4 limbs	Negative	Absent	X	X	X	X	No	4	5	Unchanged	Negative	Improved	No
3	T7–T8 esophageal cancer intramedullary metastasis	Paraparesis (4/5)	Hypoesthesia from D7	Negative	Absent	X	X	X		MEP disappearance, D-wave: amplitude decrease >50 %	3	3	Unchanged	Negative	Worsened	Permanent (Paraparesis)
4	T8–T9 meningioma (WHO I)	Paraparesis (4/5)	Negative	Urinary incontinence	X	X	X	X		No	4	5	Negative	Unchanged	Improved	No
5	T9 meningioma (WHO I)	Paraparesis (4/5)	Negative	Negative	X	X	X	X		No	5	5	Negative	Negative	Improved	No
6	T5–T6 meningioma (WHO I)	Paraparesis (4/5)	Hypoesthesia from D6	Urinary incontinence	X	X	X	X		MEP disappearance (D-wave: stable)	5	5	Unchanged	Unchanged	Improved	No
7	T9 meningioma (WHO I)	Negative (5/5)	Negative	Negative	X	X	X	X		No	5	5	Negative	Negative	Unchanged	No
8	D12–L1 hemangioblastoma	Paraparesis (4/5)	Hypoesthesia from D12	Negative	X	X		X	X	No	5	5	Unchanged	Negative	Improved	No
9	Tethered cord syndrome, L5–S1 lipoma	Left leg weakness (4/5)	Negative	Urinary incontinence	X	X		X	X	No	4	5	Negative	Improved	Improved	No
10	Tethered cord syndrome	Paraparesis (4/5)	Perineal hypoesthesia,	Urinary incontinence	X	X		X	X	No	4	5	Unchanged	Unchanged	Improved	No
11	Tethered cord syndrome, L1–L4 lipoma	Paraparesis (left leg 3, right leg 1)	Negative	Urinary incontinence	X	Poorly recordable		X	X	No	3 left leg 1 right leg	3 left leg 1 right leg	Negative	Unchanged	Unchanged	No

12	Tethered cord syndrome (Chiari 2 malformation)	Paraparesis (4/5)	Perineal hypoesthesia	Urinary incontinence	Absent	X	X	X	No	4	4	Unchanged	Unchanged	Unchanged	No
13	L1 and L4 Schwannoma	Right foot dorsal flexion weakness (4/5)	Negative	Negative	X	X	X	X	No	5	5	Negative	Negative	Improved	No
14	D12 lipoma and arachnoid cyst	Negative (5/5)	Left leg hypoesthesia	Negative	Poorly recordable	X	X	X	MEP disappearance (right GM)	4 (right leg)	5	Unchanged	Negative	Unchanged	Transient (Right leg weakness 4/5 MRC for 1 month)
15	L5 Schwannoma	Negative (5/5)	Negative	Negative	X	X	X	X	No	5	5	Negative	Negative	Unchanged	No
16	L3 Ependymoma WHO II	Negative (5/5)	Negative	Negative	X	X	X	X	frEMG burst irritation	5	5	Radicular	Negative	Unchanged	Transient (Radicular pain for 1 month)

IONM intraoperative neurophysiological monitoring, *MEP* motor evoked potential, *SEP* somatosensory evoked potential, *frEMG* free running electromyography, *eEMG* evoked EMG, *GM* gastrocnemius muscle, *CMJ* cervicomedullary junction, *MRC* Medical Research Council

50 % amplitude variation [5, 11], or 70 % amplitude variation [12], or a stable decrease in response. However, the presence/absence criterion seems to be the best choice, especially when a combined recording of the epidural D-wave is possible [1, 13]. The D-wave recording was developed later [14]. Its amplitude is a direct measure of the number of functioning fast-conducting fibers in the corticospinal tracts, and it is actually considered the gold standard for the assessment of the integrity of the corticospinal tract [7]. It has the advantage of not being influenced by blood pressure, heart rate, temperature, partial pressure of alveolar carbon dioxide, anesthetic drugs [15], or severe neurological impairment, factors that can make MEP and SEP interpretation difficult or even impossible. The D-wave can be recorded also when MEPs disappear during surgery. Different reports have described that when the D-wave decreases by less than 50 % and muscle MEPs are lost, the patient may suffer from a transient deficit, which will finally resolve [7, 13]. When comparing D-wave recordings and MEPs, the latter show a lower sensitivity and an increased number of false-positive cases. In our series, there was an optimal correlation between MEPs and postoperative motor outcome, with a sensitivity of 100 %, specificity of 92 %, PPV of 67 %, and NPV of 100 %. Specificity and PPV increased to 100 % when we considered those cases in which D-wave registration was also used. As regards EMG, there are few studies that address its role in detecting early motor tract injury. Sala et al. [16] reported the possibility of using frEMG criteria to improve the reliability of IONM during SC surgery. In our series frEMG revealed sensitivity, specificity, and PPV and NPV of 100 %, although without any statistically significant correlation with postoperative deficits. Other studies have also found similar results. Sutter et al. [2] described the results of an mIONM approach in 1017 spinal surgeries. The sensitivity of mIONM was 89 % and the specificity 99 %. They concluded that the high reported sensitivity and specificity justified the statement

Table 3 Contingency table comparing the association between mIONM, SEP, MEP, D-wave, free running EMG alerts, and new postoperative neurological deficits

	mIONM alert	No mIONM alert
Number of patients	3	12
New deficit	3	0
No new deficit	0	13
	SEP alert	No SEP alert
Number of patients	0	13
New deficit	0	1
No new deficit	0	12
	MEP alert	No MEP alert
Number of patients	3	12
New deficit	2	0
No new deficit	1	12
	D-wave alert	No D-wave alert
Number of patients	1	6
New deficit	1	0
No new deficit	0	6
	fr-EMG alert	No fr-EMG alert
Number of patients	1	15
New deficit	1	0
No new deficit	0	15

mIONM multimodal IONM, *MEP* motor evoked potential, *SEP* somatosensory evoked potential, *frEMG* free running electromyography, *eEMG* evoked EMG

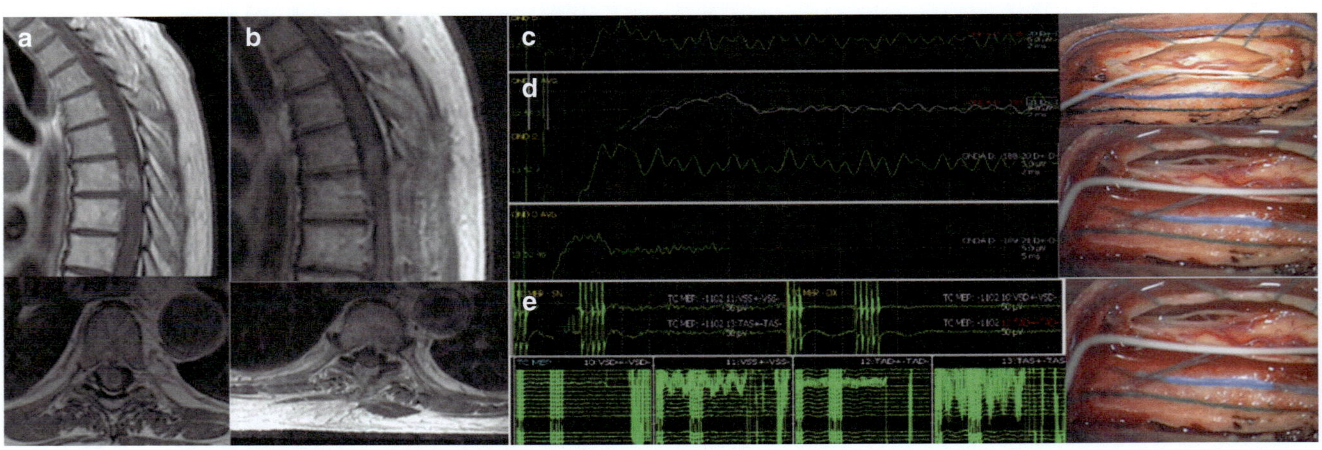

Fig. 1 Intraoperative neurophysiological monitoring (IONM) during surgery for T7–T8 intramedullary metastasis from esophageal carcinoma. During surgery there was a significant decrease in D-wave amplitude, while motor evoked potential (MEP) disappeared. The patient presented with preoperative paraparesis (Medical Research Council [MRC] score 4/5 bilaterally) that worsened in the postoperative period (MRC score 3/5 in the right arm). (**a**). Preoperative magnetic resonance imaging (MRI); sagittal and axial. (**b**). Postoperative MRI; sagittal and axial. (**c**). D-wave registration before tumor removal. (**d**). D-wave registration after tumor removal. (**e**). MEP at the end of surgery

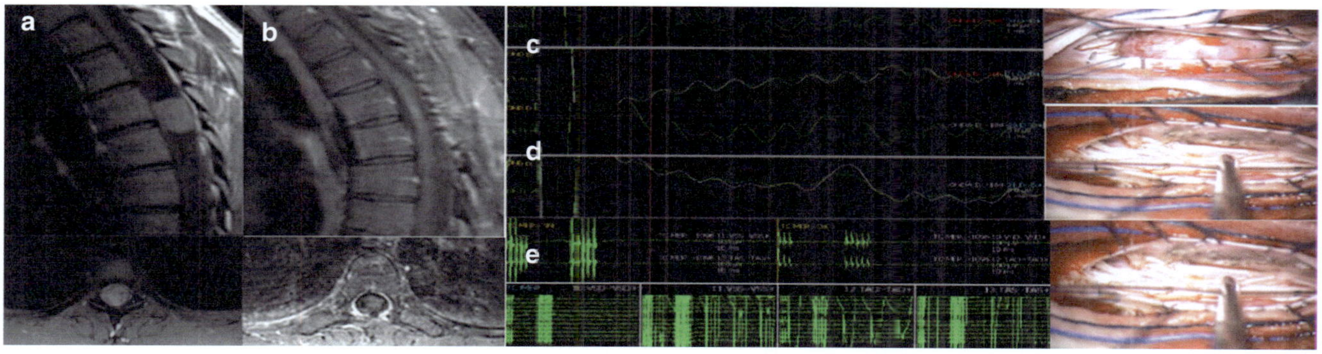

Fig 2 IONM during surgery carried out for T5–T6 meningioma (WHO I). During surgery the MEP disappeared while D-wave registration showed stable values throughout the whole surgery. The patient presented with preoperative paraparesis (MRC score 4/5 bilaterally) that improved during the postoperative period (MRC score 5/5). (**a**). Preoperative MRI; sagittal and axial. (**b**). Postoperative MRI; sagittal and axial. (**c**). D-wave registration before tumor removal. (**d**). D-wave registration after tumor removal. (**e**). MEP at the end of surgery

that mIONM (MEPs, D-wave, EMG, SEPs) performed by experienced neurologists/neurophysiologists should become an integral part of all spinal surgical procedures where potential complications are to be expected.

In the present study, the mIONM approach, by using specific combinations tailored for each case (MEPs, D-wave, SEPs, EMG), was significantly able to detect neurological injury in 19 % (3/16) of cases, with the specificity, sensitivity, and NPV and PPV of the mIONM alert being 100 % ($p<0.01$). These findings are in line with the literature evidence that mIONM can record, and therefore predict, an injury to the SC and NR. Conversely, there is much weaker evidence that mIONM can prevent such injury, improving postoperative outcome. A potential advantage of utilizing mIONM is that, with its use, one can adequately implement immediate actions (e.g., reduce retraction, stop resection, move surgical manipulation to a different area, provide warm irrigation, etc. [4]) to prevent injury and postoperative neurological deficits.

In our series, statistical analysis showed that the use of the mIONM approach was significantly associated with a better functional outcome ($p<0.05$), suggesting a fundamental role of our protocol not only for detecting but also for avoiding postoperative neurological deficits.

Conclusion

The present study confirms the role of mIONM as an essential tool in the operative workup of all spine surgeries. Single monitoring procedures such as MEPs, SEPs, or continuous EMG are definitely not sufficient to account for the complex function of the SC and NRs. mIONM is highly sensitive and specific for detecting neurologic injury during spine surgery and could represent a significant tool to preserve and restore neuronal structures and to achieve an optimal postoperative functional outcome.

Conflict of Interest Statement The authors declare that they have no conflict of interest.

References

1. Costa P, Bruno A, Bonzanino M, Massaro F, Caruso L, Vincenzo I, Ciaramitaro P, Montalenti E (2007) Somatosensory- and motor-evoked potential monitoring during spine and spinal cord surgery. Spinal Cord 45(1):86–91. Epub 2006 May 2
2. Sutter M, Eggspuehler A, Grob D, Jeszenszky D, Benini A, Porchet F, Mueller A, Dvorak J (2007) The diagnostic value of multimodal intraoperative monitoring (MIOM) during spine surgery: a prospective study of 1,017 patients. Eur Spine J 16 Suppl 2:S162–S170. Epub 2007 Jul 31
3. Pratheesh R, Babu KS, Rajshekhar V (2014) Improvement in intraoperative transcranial electrical motor-evoked potentials in tethered cord surgery: an analysis of 45 cases. Acta Neurochir 156:723–731. doi:10.1007/s00701-014-1999-7
4. Sala F, Palandri G, Basso E, Lanteri P, Deletis V, Faccioli F, Bricolo A (2006) Motor evoked potential monitoring improves outcome after surgery for intramedullary spinal cord tumors: a historical control study. Neurosurgery 58(6):1129–1143
5. Clark AJ, Ziewacz JE, Safaee M, Lau D, Lyon R, Chou D, Weinstein PR, Ames CP, Clark JP 3rd, Mummaneni PV (2013) Intraoperative neuromonitoring with MEPs and prediction of postoperative neurological deficits in patients undergoing surgery for cervical and cervicothoracic myelopathy. Neurosurg Focus 35(1):E7
6. Nuwer MR, Emerson RG, Galloway G, Legatt AD, Lopez J, Minahan R, Yamada T, Goodin DS, Armon C, Chaudhry V, Gronseth GS, Harden CL, American Association of Neuromuscular and Electrodiagnostic Medicine (2012) Evidence-based guideline update: intraoperative spinal monitoring with somatosensory and transcranial electrical motor evoked potentials. J Clin Neurophysiol 29(1):101–108
7. Macdonald DB, Skinner S, Shils J, Yingling C, American Society of Neurophysiological Monitoring (2013) Intraoperative motor evoked potential monitoring – a position statement by the American Society of Neurophysiological Monitoring. Clin Neurophysiol 124(12):2291–2316
8. Gavaret M, Jouve JL, Péréon Y, Accadbled F, André-Obadia N, Azabou E, Blondel B, Bollini G, Delécrin J, Farcy JP, Fournet-Fayard J, Garin C, Henry P, Manel V, Mutschler V, Perrin G, Sales

de Gauzy J, The French Society of Spine Surgery (SFCR) (2013) Intraoperative neurophysiologic monitoring in spine surgery. Developments and state of the art in France in 2011. Orthop Traumatol Surg Res 99:S319–S327
9. Nuwer MR, Dawson EG, Carlson LG, Kanim LE, Sherman JE (1995) Somatosensory evoked potential spinal cord monitoring reduces neurologic deficits after scoliosis surgery: results of a large multicenter survey. Electroencephalogr Clin Neurophysiol 96(1):6–11
10. Calancie B, Molano MR (2008) Alarm criteria for motor-evoked potentials: what's wrong with the "presence-or-absence" approach? Spine (Phila Pa 1976) 33(4):406–414
11. Kelleher MO, Tan G, Sarjeant R, Fehlings MG (2008) Predictive value of intraoperative neurophysiological monitoring during cervical spine surgery: a prospective analysis of 1055 consecutive patients. J Neurosurg Spine 8(3):215–221
12. Kobayashi S, Matsuyama Y, Shinomiya K, Kawabata S, Ando M, Kanchiku T, Saito T, Takahashi M, Ito Z, Muramoto A, Fujiwara Y, Kida K, Yamada K, Wada K, Yamamoto N, Satomi K, Tani T (2014) A new alarm point of transcranial electrical stimulation motor evoked potentials for intraoperative spinal cord monitoring: a prospective multicenter study from the Spinal Cord Monitoring Working Group of the Japanese Society for Spine Surgery and Related Research. J Neurosurg Spine 20:102–107
13. Costa P, Peretta P, Faccani G (2013) Relevance of intraoperative D wave in spine and spinal cord surgeries. Eur Spine J 22(4): 840–848
14. Hicks RG, Burke DJ, Stephen JP (1991) Monitoring spinal cord function during scoliosis surgery with Cotrel-Dubousset instrumentation. Med J Aust 154:82–86
15. Traynelis VC, Abode-Iyamah KO, Leick KM, Bender SM, Greenlee JD (2012) Cervical decompression and reconstruction without intraoperative neurophysiological monitoring. J Neurosurg Spine 16:107–113
16. Sala F, Bricolo A, Faccioli F, Lanteri P, Gerosa M (2007) Surgery for intramedullary spinal cord tumors: the role of intraoperative (neurophysiological) monitoring. Eur Spine J 16(Suppl 2):S130–S139. Epub 2007 Jul 26

Combination of Magnetic Resonance Imaging and Electrophysiological Studies in Lumbar Disc Herniation

Wenxiang Zhong[§], Jichao Wang[§], Wenchuan Zhang, Pengfei Liu, Massimiliano Visocchi, and Shi-Ting Li

Abstract *Objective* We aimed to study the clinical value of magnetic resonance imaging (MRI) and electrophysiological studies in the diagnosis of lumbar disc herniation and in the evaluation of the therapeutic effect of discectomy. *Methods* In this study, 265 patients with LDH were treated with discectomy after assessment by the Japanese Orthopedic Association (JOA) score, MRI, and electrophysiological studies. All the patients were followed-up for 6 years. The effects of the operation were assessed by determining the angle between the nerve root canal and disc protrusion (AN value), the stenotic ratio of the spinal canal, the width of the lateral recess, motor conduction velocity (MCV), sensory conduction velocity (SCV), and nerve action potential (NAP) before and after operation. *Results* The AN value, stenotic ratio of the spinal canal, and the width of the lateral recess of protruding intervertebral discs showed significant differences from these values for the patients' unaffected intervertebral discs ($P<0.05$). The MCV, SCV, and NAP of the affected limb showed significant differences from these values for the patients' unaffected limbs ($P<0.05$). In all the patients the values for these indicators showed significant differences before and after operation ($P<0.05$). *Conclusion* MRI and electrophysiological studies can be used in the diagnosis of lumbar disc herniation, and in the evaluation of the effect of surgery.

[§]Author contributed equally with all other contributors.

W. Zhong, MD • W. Zhang, MD (✉) • P. Liu, MD • S.-T. Li
Department of Neurosurgery, Xinhua Hospital,
1665 Kongjiang Road, Shanghai, China
e-mail: zhangwench88@hotmail.com

J. Wang, MD
Xinjiang Uygur Autonomous Region People's Hospital,
Urumchi, China

M. Visocchi
Institute of Neurosurgery, Medical School,
Catholic University of Rome, Rome, Italy

Keywords Lumbar disc herniation • Discectomy • Magnetic resonance imaging • Electrophysiological studies

Introduction

Lumbar disc herniation (LDH) is one of the commonest causes of lower back pain and sciatica, and some patients also present with numbness, paresthesia, and paralysis [13, 17]. Some earlier studies showed that magnetic resonance imaging (MRI) was a reliable way to diagnose LDH. However, the diagnosis of LDH with MRI needs experienced observers and lacks quantitative indices. In the present study, we revealed the clinical application of MRI combined with electrophysiological studies in patients with LDH.

Materials and Methods

Patients

This study was approved by the Xinhua Hospital Ethics Review Board; 265 patients (138 males and 127 females, aged from 18 to 76 years) were surgically treated in the Neurosurgery Department of Xinhua Hospital between January 2010 and December 2012. The main criteria for inclusion were adult age, clearly evidenced LDH, and sufficient surgical indication.

The diagnosis of LDH was made according to: (1) typical symptoms such as lower back pain and sciatica; and (2) neurological examination revealing abnormal reflexes, positive straight leg-raising test, and paresthesia in a lower extremity. Exclusion criteria were: (1) injury, tuberculosis, and tumor of the sacroiliac joint; (2) history of trauma or operation of Lumbar; (3) obvious spinal deformity; and (4) pelvic outlet syndrome.

M. Visocchi et al. (eds.), *Trends in Reconstructive Neurosurgery*, Acta Neurochirurgica Supplement 124,
DOI 10.1007/978-3-319-39546-3_39,

Surgical Procedure

Discectomy was performed under general anesthesia and patients were positioned prone. A midline incision (10 cm) was made and the unilateral paravertebral muscle? was dissected. Then laminotomy was done to expose the herniated disc and the relevant spinal nerve. The herniated disc was removed to decompress the spinal nerve root. The incision was finally closed in layers.

MRI Evaluation

A MAGNETOM Sonata 1.5 T system (Siemens, Erlangen, Germany) was employed for MRI evaluation. The MRI sequence we used was TR 460 ms/TE 42 ms, 3.2-cm field of view, 288 × 192 matrix for T1 weighted image (WI), and TR 2000 ms/TE 120 ms, 3.2-cm field of view, 384 × 256 matrix for T2WI. Section thickness was 4 mm for all the images.

The deviated angle of prominence (AN value) is the angle between the nerve root and disc protrusion. A tangent line to the lateral part of the protrusion was made, and the angle between the tangent line and medial border of the yellow ligament was defined as the AN value. The spinal stenosis ratio was calculated from sagittal images and defined as the sagittal diameter of the herniated disc/total sagittal diameter of the spinal canal × 100 %. The width of the lateral recess was measured on the axial layer of the superior border of the vertebral pedicle.

Electrophysiological Studies

An electromyography (EMG) device (DK 1 2740; Medtronic, Copenhagen, Denmark) was used to accomplish the electrophysiological studies. Parameters were: skin temperature 30 C, room temperature 25 C, and a quiet indoor environment. Stimulation and recording were done using surface electrodes. Bilateral motor nerve conduction velocities (MNCVs) of the tibial nerve and sensory nerve conduction velocities (SNCVs) of the common peroneal nerve and superficial peroneal nerve were recorded.

Results

All the patients were assessed by the Japanese Orthopedic Association (JOA) score [1], MRI, and electrophysiological studies before and 6 years after discectomy. The AN value, spinal stenosis ratio in the sagittal plane, and the width of the lateral recess all showed significant differences between the patients' unaffected and herniated discs ($P < 0.05$) (Table 1). The sensory conduction velocity (SCV), motor conduction velocity (MCV), and nerve action potential (NAP) showed significant differences between patients' intact and affected legs ($P < 0.05$) (Table 2).

In all the patients, the JOA score, AN value, spinal stenosis ratio in the sagittal plane, and the width of the lateral recess had improved significantly 6 years after discectomy ($P < 0.05$) (Table 1). The SCV, MCV, and NAP values of the affected legs had increased 6 years after discectomy ($P < 0.05$) (Table 1). There were no significant differences among L4-5, L5-S1, and L4-S1 patients in either pre- or postoperative JOA scores, AN values, spinal stenosis ratios in the sagittal plane, width of the lateral recess, SCV, MCV, or NAP ($P > 0.05$).

Discussion

Recently, lumbar MRI has become the first choice for the diagnosis of LDH [5, 12, 14]. Kim et al. [12] reported that MRI achieved an accuracy rate of 97% in the diagnosis of LDH. MRI has the advantage of three-dimensional imaging, which can provide images of the spine, intervertebral discs, spinal canal, and nerve roots in the coronal, sagittal, and axial planes without radiation. So MRI has been applied as the first choice in the diagnosis of LDH [2, 9–11, 16].

However, most earlier MRI studies focused on image diversity and lacked quantitative indexes to evaluate patients' situations objectively. The present study was undertaken to assess compression of the spinal nerve root by determining the AN value, the spinal stenosis ratio in the sagittal plane, and the width of the lateral recess before and after discectomy, and electrophysiological studies were also applied to provide guidance for the surgery and to evaluate the surgical effects.

Electrophysiological studies have been widely used as the gold standard in peripheral nerve diseases. The dermatomal somatosensory evoked potential (DSEP) is also used in herniation studies, although sensitivities ranged in different studies [4, 8, 15, 19].

This study showed that the AN value, spinal stenosis ratio in the sagittal plane, and the width of the lateral recess had significant differences between the patients' unaffected and herniated discs before discectomy ($P < 0.05$), indicating that the situation of a spinal nerve root compressed by a herniated disc can be shown by these MRI indexes, among which the AN value is the most sensitive. Even when the absolute size of the herniated disc is tiny, AN values can show significant differences between affected and unaffected discs. So the AN value can be used to diagnose LDH earlier than other

Table 1 Pre- and postoperative JOA scores and MRI measurements

	JOA score		AN value (°)		Spinal stenosis ratio		Width of lateral recess (mm)	
	Preoperative	Postoperative	Preoperative	Postoperative	Preoperative	Postoperative	Preoperative	Postoperative
L4-5	12.74 ± 3.25	28.85 ± 2.12	5.87 ± 1.38	29.45 ± 4.58	32–59 %	5–7 %	2.92 ± 0.87	5.38 ± 1.69
L5-S1	12.42 ± 3.53	27.13 ± 2.36	5.66 ± 1.27	26.83 ± 3.73	38–64 %	6–8 %	2.74 ± 0.79	5.06 ± 1.43
L4-S1	12.45 ± 3.46	28.53 ± 2.05	5.82 ± 1.29	27.43 ± 3.95	39–57 %	5–8 %	2.84 ± 0.74	5.34 ± 1.65

AN angle between the nerve root canal and disc protrusion, *JOA* Japanese Orthopedc Association, *MRI* magnetic resonance imaging

Table 2 Pre- and postoperative electrophysiological studies

	MCV (m/s)		SCV (m/s)		NAP (mV)	
L4-5	Preoperative	Postoperative	Preoperative	Postoperative	Preoperative	Postoperative
Tibial nerve	36.5±3.7	57.4±4.8	34.6±3.7	54.1±4.8	4.4±0.9	7.0±1.8
Common peroneal nerve	38.4±4.4	56.7±4.3	36.2±3.4	56.3±4.2	5.2±1.5	6.7±1.3
Superficial peroneal nerve	39.3±3.5	60.3±5.1	34.8±3.6	57.4±4.0	5.5±1.3	8.9±3.1
L5-S1	Preoperative	Postoperative	Preoperative	Postoperative	Preoperative	Postoperative
Tibial nerve	40.1±4.5	63.2±5.8	36.2±4.2	62.7±5.4	5.1±1.3	7.8±2.5
Common peroneal nerve	41.3±4.9	60.4±5.3	38.7±4.4	60.8±5.2	5.9±1.8	7.5±2.2
Superficial peroneal nerve	42.1±4.2	64.3±5.8	38.3±4.1	60.9±4.8	6.1±1.7	9.7±3.5
L4-S1	Preoperative	Postoperative	Preoperative	Postoperative	Preoperative	Postoperative
Tibial nerve	39.3±4.0	62.1±5.2	36.5±3.7	61.2±5.4	4.7±1.2	7.4±2.1
Common peroneal nerve	40.6±4.7	58.3±4.7	37.8±4.0	58.6±4.7	5.5±1.6	7.3±1.9
Superficial peroneal nerve	40.8±3.6	62.6±5.6	37.3±3.6	59.6±4.2	5.6±1.4	9.6±3.2

MCV motor conduction velocity, *SCV* sensory conduction velocity, *NAP* nerve action potential

MRI indexes. The spinal stenosis ratio in the sagittal plane shows not only a herniated disc but also a stenosed spinal canal. The lateral recess is the most stenosed part of the spinal canal, so a compressed spinal nerve root can be shown through the width of the lateral recess.

SCV, MCV, and NAP showed significant differences between patients' intact and affected legs before discectomy ($P<0.05$); this is because distal nerves in the lower limb, such as the common peroneal nerve, tibial nerve, and superficial peroneal nerve, will be impacted when the spinal nerve root is compressed. So electrophysiological studies, as well as the patients' clinical symptoms, can be used to indicate the affected side; these studies are valuable for designing the laminar window and they increase the neurosurgeon's confidence. The preoperative and postoperative function of the lower limbs can be assessed by electrophysiological studies and compared with images provided by MRI.

In our study, there were no significant differences in AN value, spinal stenosis ratio in the sagittal plane, width of the lateral recess, SCV, MCV, and NAP among L4-5, L5-S1, and L4-5+L5-S1 patients before discectomy ($P>0.05$). This finding indicates that MRI and electrophysiological studies are repeatable and reliable, and can be applied to different kinds of LDH.

The JOA score is a composite index that incorporates patients' clinical manifestations, such as the main complaint, physical signs, and activities of daily living; accordingly, the JOA score can be used to evaluate the patient's condition. In our study, the JOA scores were significantly different before and after discectomy ($P<0.05$). Discectomy is a mature and effective way to treat LDH [6, 7, 18, 20–22], and in our study we found that discectomy was reliable enough to be employed as an intervention.

The AN value, spinal stenosis ratio in the sagittal plane, and the width of the lateral recess improved significantly after discectomy ($P<0.05$). Spinal stenosis and lateral recess stenosis are the most important causes of root compression, as shown by the spinal stenosis ratio in the sagittal plane and the width of the lateral recess [3]. Thus, these MRI indexes can not only provide assessment of the compressed spinal nerve root, but they can also be used to evaluate the surgical effect of discectomy morphologically. The SCV, MCV, and NAP in our study also improved significantly after discectomy ($P<0.05$); this was because the sensory and motor function of the distal nerves in the lower limb could recover after decompression of the spinal nerve root. And this finding showed that electrophysiological studies could be added to follow-up to observe the recovery of lower limb function.

We conclude that a combination of MRI and electrophysiological studies is valuable for the diagnosis of lumbar disc herniation, and for the surgical design and evaluation of the clinical effect of the surgery.

Conflict of Interest Statement All authors certify that they have no affiliations with or involvement in any organization or entity with any financial interest (such as honoraria; educational grants; participation in speakers' bureaus; memberships, employment, consultancies, stock ownership, or other equity interest; and expert testimony or patent-licensing arrangements), or non-financial interest (such as personal or professional relationships, affiliations, knowledge or beliefs) in the subject matter or materials discussed in this manuscript.

References

1. Azimi P, Mohammadi HR, Montazeri A (2012) An outcome measure of functionality and pain in patients with lumbar disc herniation: a validation study of the Japanese Orthopedic Association (JOA) score. J Orthop Sci 17(4):341–345
2. Babar S, Saifuddin A (2002) MRI of the post-discectomy lumbar spine. Clin Radiol 57(11):969–981
3. Bartynski WS, Lin L (2003) Lumbar root compression in the lateral recess: MR imaging, conventional myelography, and CT myelography comparison with surgical confirmation. Am J Neuroradiol 24:348–360
4. Beyaz EA, Akyüz G, Us O (2009) The role of somatosensory evoked potentials in the diagnosis of umbosacral radiculopathies. Electromyogr Clin Neurophysiol 49(4):131–142
5. Bundshuh CV, Stein L, Slusser JH, Schinco FP, Ladaga LE, Dillon JD (1990) Distinguishing between sear and recurrent disc herniated disc in postoperative patients: value of contrast enhanced CT and MR imaging. J Am J Neuroradio 11:949–958
6. Buttermann GR (2004) Treatment of lumbar disc herniation: epidural steroid injection compared with discectomy – a prospective, randomized study. J Bone Joint Surg Am 86A(4):670–679
7. Dumitru D, Dreyfuss P (1996) Dermatomal/segmental somatosensory evoked potential evaluation of L5/S1 unilateral/unilevel radiculopathies. Muscle Nerve 19(4):442–449
8. Desai A, Bekelis K, Ball PA, Lurie J, Mirza SK, Tosteson TD, Zhao W, Weinstein JN (2012) SPORT: do outcomes vary across centers for surgery for lumbar disc herniation? Neurosurgery 71(4):833–842
9. Florist R, Spallone A, Aref TY, Rizzo A, Apruzzese A, Mulas M, Castriota Scanderbeg A, Simonetti G (1997) Early postoperative MRI findings following surgery for herniated lumbar disc Part II: a Gadolinium-enhanced study. Acta Neurochir 139(12):1101–1107
10. Jackson RP, Cain JE Jr, Jacobs RR, Cooper BR, McManus GE (1989) The neuroradiographic diagnosis of lumbar herniated nucleus pulposus: II. A comparison of computed tomography (CT), myelography, CT-myelography, and magnetic resonance imaging. Spine (Phila Pa 1976) 14(12):1362–1367
11. Janssen ME, Bertrand SL, Joe C, Levine MI (1994) Lumbar herniated disk disease: comparison of MRI, myelography, and postmyelographic CT scan with surgical findings. Orthopedics 17(2):121–127
12. Kim KY, Kim YT, Lee CS, Kang JS, Kim YJ (1993) Magnetic resonance imaging in the evaluation of the lumbar herniated intervertebral disc. Int Orthop 17(4):241–244
13. Kumar R, Kalra SK, Vaid VK, Mahapatra AK (2011) Bilateral acute foot drop following lumbar disc herniation——a case report. J Indian Med Assoc 109(11):834–835
14. Mullin WJ, Heithoff KB, Gilbert TJ, Renfrew DL (2000) Magnetic resonance evaluation of recurrent disc herniation is gadolinium necessary. Spine 25(12):1493–1499
15. Pape E, Eldevik OP, Vandvik B (2002) Diagnostic validity of somatosensory evoked potentials in subgroups of patients with sciatica. Eur Spine J 11(1):38–46
16. Pfirrmann CW, Dora C, Schmid MR, Zanetti M, Hodler J, Boos N (2004) MR image-based grading of lumbar nerve root compromise due to disk herniation: reliability study with surgical correlation. Radiology 230(2) 583–588
17. Radcliff K, Hilibrand A, Lurie JD, Tosteson TD, Delasotta L, Rihn J, Zhao W, Vaccaro A, Albert TJ, Weinstein JN (2012) The impact of epidural steroid injections on the outcomes of patients treated for lumbar disc herniation: a subgroup analysis of the SPORT trial. J Bone Joint Surg Am 94(15):1353–1358
18. Thomas KC, Fisher CG, Boyd M, Bishop P, Wing P, Dvorak MF (2007) Outcome evaluation of surgical and nonsurgical management of lumbar disc protrusion causing radiculopathy. Spine (Phila Pa 1976) 32(13):1414–1422
19. Walk D, Fisher MA, Dotmdoulakis SH, Hemmati M (1992) Somatosensory evoked potentials in the evaluation of lumbosacral radiculopathy. Neurol 42:1197–1202
20. Weinstein JN, Lurie JD, Tosteson TD, Skinner JS, Hanscom B, Tosteson AN, Herkowitz H, Fischgrund J, Cammisa FP, Albert T, Deyo RA (2006) Surgical vs nonoperative treatment for lumbar disk herniation: the Spine Patient Outcomes Research Trial (SPORT) observational cohort. JAMA 296(20):2451–2459
21. Weinstein JN, Lurie JD, Tosteson TD, Tosteson AN, Blood EA, Abdu WA, Herkowitz H, Hilibrand A, Albert T, Fischgrund J (2008) Surgical versus nonoperative treatment for lumbar disc herniation: four-year results for the Spine Patient Outcomes Research Trial (SPORT). Spine (Phila Pa 1976) 33(25):2789–2800
22. Weinstein JN, Tosteson TD, Lurie JD, Tosteson AN, Hanscom B, Skinner JS, Abdu WA, Hilibrand AS, Boden SD, Deyo RA (2006) Surgical vs nonoperative treatment for lumbar disk herniation: the Spine Patient Outcomes Research Trial (SPORT): a randomized trial. JAMA 296(20):2441–2450

Use of High-Resolution Ultrasonography in Anterior Subcutaneous Transposition of the Ulnar Nerve for Cubital Tunnel Syndrome

Min Yang[$], Jichao Wang[$], Xiaosheng Yang, Wenxiang Zhong, Qiufeng Ma, Shiting Li, and Wenchuan Zhang

Abstract

Objectives

Cubital tunnel syndrome (CTS) is the most common form of ulnar nerve entrapment. In this study, ultrasonography (US) was used not only for diagnosis but also for operation. US findings could be used to establish the diagnosis of CTS and could demonstrate the pathological anatomy in the cubital tunnel region to guide anterior subcutaneous transposition of the ulnar nerve.

Methods

Sixty-two patients with clinical and electrophysiological evidence of ulnar nerve entrapment were included. All patients received ultrasonographic examination and anterior subcutaneous transposition of the ulnar nerve. The maximal diameter of the ulnar nerve (MDU) was measured in longitudinal views and the range of the hypoechoic area around the nerve was observed. The cross-sectional area (CSA) was also measured on transverse scans. The actual MDU was measured during operation.

Results

The actual MDU was 6.4 ± 0.4 mm, measured during operation. The preoperative MDU was 3.1 ± 0.2 mm. The MDU values recorded in CTS patients were greater than those in normal subjects. The range of the hypoechoic area observed on longitudinal US scans was 2.9–5.6 mm (mean, 4.1 ± 0.4 mm).

Conclusions

High-resolution US can be used not only in the diagnosis of CTS, also for providing effective preoperative evaluation for the anterior subcutaneous transposition of the ulnar nerve in CTS.

Keywords Anterior subcutaneous transposition • High-resolution ultrasonography • Cubital tunnel syndrome • Ulnar nerve

Introduction

Cubital tunnel syndrome is the most common form of ulnar nerve entrapment and the second most common entrapment neuropathy of the upper limb [1, 2]. The clinical manifestations may include pain along the medial aspect of the elbow and tenderness along the groove posterior to the medial epicondyle, dysesthesia, paresthesia, numbness, muscle atrophy, and/or motor dysfunction in the distribution of the ulnar nerve [3].

The diagnosis of cubital tunnel syndrome is made from the clinical signs and symptoms, and is generally confirmed by electrodiagnostic testing. Recently, ultrasonography has been described as a tool to confirm the diagnosis of ulnar neuropathy at the elbow [4–8]. It has been shown that ultrasonography, in addition to electrodiagnostic studies, improves the reliability of this diagnosis [6].

Many different operative techniques have been described for the treatment of cubital tunnel syndrome. Anterior subcutaneous transposition is one of several different surgical techniques used in treating this syndrome. It was first described by Curtis in 1998 as releasing the ulnar nerve from all potential sites of entrapment and moving it anterior to the motion axis of the elbow, thereby relieving tension [9]. Subcutaneous ulnar nerve transposition has yielded predictably good results in a majority of patients in several studies [10–12].

[$]Author contributed equally with all other contributors.

M. Yang • X. Yang • W. Zhong • Q. Ma • S. Li (✉)
W. Zhang, MD, PhD (✉)
Department of Neurosurgery, Xinhua Hospital of Shanghai Jiaotong University, No 1665, Kongjiang Road, Shanghai 200092, China
e-mail: zhangwench88@hotmail.com

J. Wang
Department of Neurosurgery, The People's Hospital of Xinjiang Uygur Autonomous Region, Urumqi, China

M. Visocchi et al. (eds.), *Trends in Reconstructive Neurosurgery*, Acta Neurochirurgica Supplement 124,
DOI 10.1007/978-3-319-39546-3_40,

In the present study, ultrasonography was used not only in diagnosis but also during operation. The purpose of this study was to determine whether ultrasonographic findings could be used to establish the diagnosis of cubital tunnel syndrome and whether these findings could demonstrate the pathological anatomy in the cubital tunnel region, so as to guide the anterior subcutaneous transposition of the ulnar nerve.

Patients and Methods

Patients

From 2005 to 2010, 62 patients (25 male and 38 female patients) with a mean age of 56 years (range, 40–82 years) with clinical and electrophysiological evidence of ulnar nerve entrapment were included in this study. The mean duration of symptoms before surgery was 17 months (range, 4–48 months). The right side was involved in 39 patients and the left in 23. The dominant side was affected in 2 patients.

The patients had persistent pain and progressive motor and sensory deficits, despite conservative treatment. All the patients had persistent pain and progressive numbness radiating along the ulnar aspect of the forearm to the little and ring fingers. Tinel's sign at the elbow was also present in all the patients. Some patients presented with weakness or atrophy of the flexor carpi ulnaris and flexor digiti minimi profundus. Significant cervical spine and shoulder diseases were excluded by clinical examination and magnetic resonance imaging (MRI), as necessary. Patients with serious acute forearm injury and brachial plexus injury were also excluded from the study. All the patients were carefully examined clinically by the authors, and were diagnosed with cubital tunnel syndrome by electrodiagnostic testing and ultrasonography.

Ultrasonography

To avoid inter-observer bias, ultrasonographic examination of the ulnar nerve in all patients was performed by the same radiologist (who is well trained in ultrasonographic studies of the peripheral nerves in the arm), who was blinded to the clinical and electrodiagnostic data.

Ultrasonography was performed with a Philips iU 22 scanner (Eindhoven, The Netherlands) and a 7–12 MHz linear array transducer. Ultrasonographic examinations of the ulnar nerve were carried out in the region of the cubital tunnel at the level of the medial epicondyle. Measurements were repeated at 2 cm proximal to the medial epicondyle of the humerus to 2 cm distal to the epicondyle, where the nerve entered the flexor carpi ulnaris muscle, both with the elbow in full extension and in full flexion at 90°. The diameter of the ulnar nerve was determined within the echogenic rim surrounding the nerve as described before [6–8]. The maximal diameter of the ulnar nerve (MDU) was measured in longitudinal views (Fig. 1) and the range of the hypoechoic area around the nerve was also observed (Fig. 1). The cross-sectional area (CSA) was also measured on transverse scans by drawing a best-fitting ellipsoid over the ulnar nerve (Fig. 2).

Surgical Technique: Anterior Subcutaneous Transposition

All of the procedures were performed by the corresponding author, Professor Doctor Wenchuan Zhang. The technique originally described by Eaton et al. [13] was utilized in all patients. All the anterior subcutaneous transposition procedures were performed under brachial plexus block and a tourniquet was placed high on the brachium. The medial epicondyle is marked as a reference point on the elbow, which is flexed to 45°. At a point 1 to 1.5 cm anterior to the medial epicondyle, a second mark is made to indicate where the fascial sling will be sutured to the deep dermis, because once the incision is made the skin retracts, distorting the reference points. A curvilinear 10-cm incision is made midway between the epicondyle and the olecranon, along the sulcus for the ulnar nerve. The skin, with superficial and deep fascia, is cut. In the deep layer of subcutaneous tissue, one or more branches of the medial brachial and antebrachial cutaneous nerves are variably located, and they must be preserved during the blunt dissection. The ulnar nerve is readily palpable proximal to the cubital tunnel, where it lies posterior to the medial intermuscular septum. After the cubital tunnel was dissected, five common entrapment sites, including the arcade of Struthers, the medial intermuscular septum, Osborne's ligament, and the fascial origins of the flexor digitorum superficialis and flexor carpi ulnaris (FCU), were decompressed [4], and then the ulnar nerve was identified in the arm.

Wherever possible, concomitant vessels and branches of the ulnar nerve were preserved [5]. As the nerve is dissected distally, care must be taken to retain the accompanying longitudinal venae comitantes, thereby preserving critical longitudinal blood supply to the nerve, especially since certain small segmental vessels must be sacrificed to allow for anterior transposition of the nerve. The isolated ulnar nerve was transposed underneath the subcutaneous tissue. Finally, a fascial relaxing suture was used to make a subcutaneous bed for the nerve, and the skin was closed successively with 3-0 nonabsorbable sutures.

With the skin stretched back toward its original position, the position of the nerve and its fasciodermal sling is evaluated

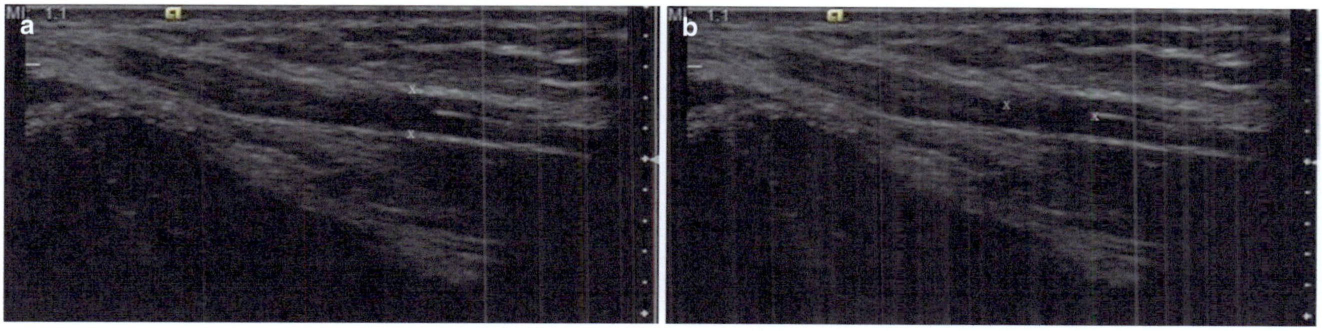

Fig. 1 Preoperative longitudinal ultrasonographic scan of the ulnar nerve. (**a**) The maximal diameter of the ulnar nerve (MDU) was measured under the mark shown (between + signs). (**b**) The range of the hypoechoic area around the nerve was also observed (between + signs)

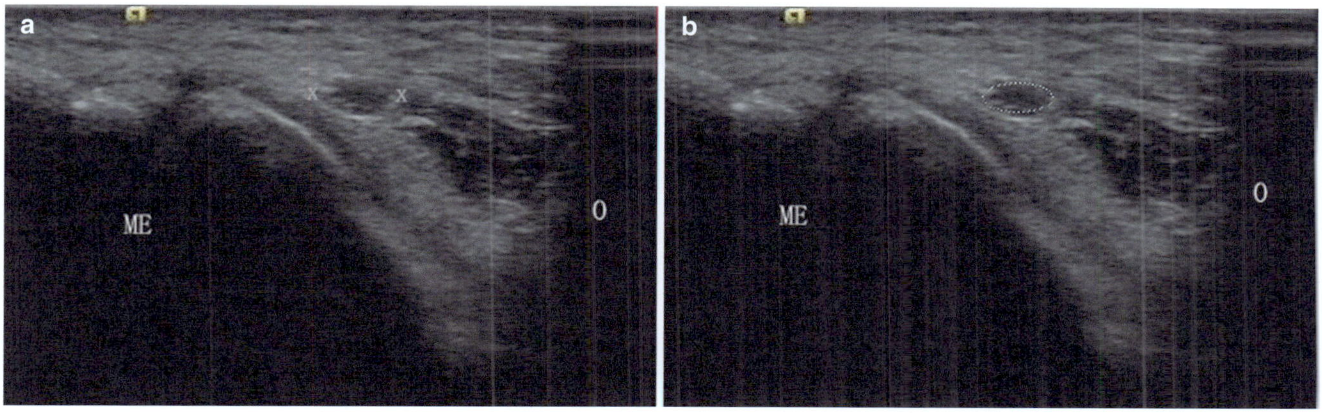

Fig. 2 Preoperative transverse ultrasonographic scan of the ulnar nerve. (**a**) The maximal diameter of the ulnar nerve (MDU) was measured under the mark shown (between + signs). (**b**) The cross-sectional area (CSA) was also measured on the transverse scans by drawing a best-fitting ellipsoid over the ulnar nerve

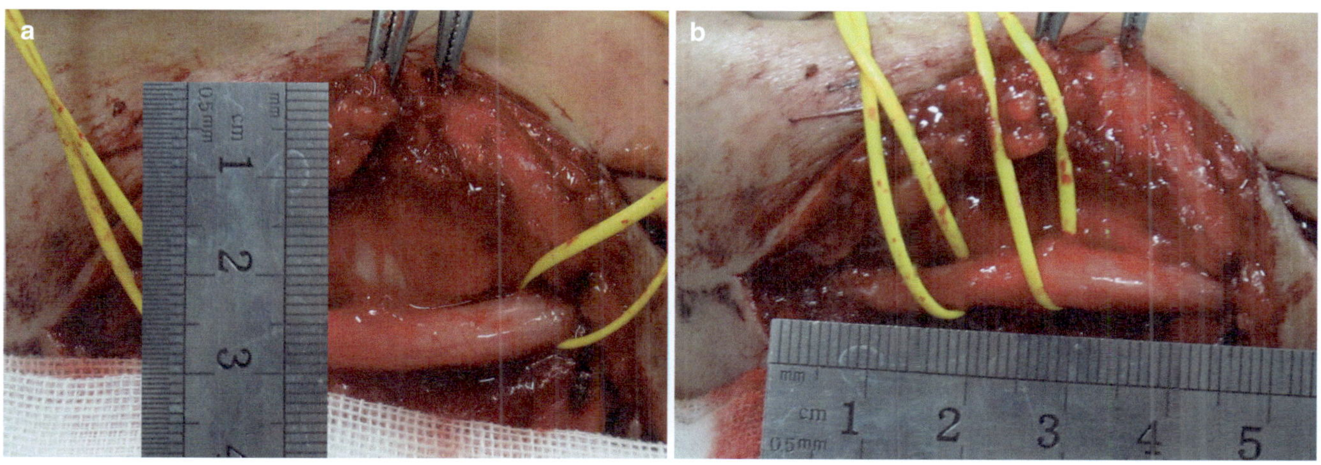

Fig. 3 Intraoperative view of the ulnar nerve after decompression. The actual MDU was measured during operation

through the full arc of elbow motion. Adjustments in the dermal attachment are rarely necessary. The subcutaneous tissue and the skin are closed with absorbable suture. A soft mobile dressing is placed about the elbow, and a sling is provided for comfort.

The actual MDU was measured by rule during the operation (Fig. 3).

Statistical Analysis

Statistical analysis was done using SPSS 11.5 (Lead Technologies, Shanghai, China). For testing differences a paired Student's t-test was used. Significance was reached when $p < 0.05$. Data are presented as means ± standard deviation. If appropriate, these are followed by a range or a p value.

Results

All patients underwent an anterior subcutaneous transposition of the ulnar nerve at the elbow. The procedures were uneventful. The injured ulnar nerve in cubital tunnel syndrome revealed swelling and thickening, and the hypoechoic area was recorded by high-resolution ultrasonography. The actual diameter of the injured ulnar nerve measured during operation was 6.4±0.4 mm (range: 4.9–7.6 mm). On longitudinal ultrasonographic scans the preoperative MDU was 3.1±0.2 mm (range: 1.9–4.0 mm). The mean difference from the actual intraoperatively measured diameter was 3.3±0.2 mm ($p<0.001$). Both the actual diameter and the preoperative MDU recorded by high-resolution ultrasonography were greater in the patients with cubital tunnel syndrome than these values in normal subjects. The range of the hypoechoic area observed by ultrasonography on longitudinal ultrasonographic scans was 2.9–5.6 mm (mean 4.1±0.4 mm).

Discussion

Many different operative techniques have been described for the treatment of cubital tunnel syndrome and none has been shown to consistently produce results superior to those of the others. Subcutaneous transposition of the ulnar nerve was first described by Curtis in 1998 as releasing the ulnar nerve from all potential sites of entrapment and moving it anterior to the motion axis of the elbow, thereby relieving tension [9]. Gelberman et al. found that traction on the ulnar nerve during flexion of the elbow was a major cause of increased intraneural pressure [14]. They concluded that operative procedures that decompress the ulnar nerve without transposing it do not effectively relieve the symptoms caused by neural traction. Subcutaneous transposition of the ulnar nerve, compared with other operative procedures, has been shown to lead to a higher degree of patient satisfaction and relief of symptoms, with no notable complication. In our study, 62 patients with cubital tunnel syndrome recovered from the symptoms with anterior subcutaneous transposition and 55 patients showed objective improvement. Complications were not found in any patients. Subcutaneous transposition avoids detaching the flexor pronator muscle from its origin, thus placing this procedure at a distinct surgical advantage over submuscular transposition for the throwing athlete. It has been found that the subcutaneous transposition of the ulnar nerve was associated with shorter incision, shorter operative time, less postoperative pain, fewer postoperative complications, and better outcome compared with the submuscular transposition [15].

Anterior subcutaneous transposition of the ulnar nerve requires a long longitudinal incision to be made posterior to the medial epicondyle, extending several centimeters proximal and distal to the condyle. In the present study, after the cubital tunnel was dissected, five common entrapment sites, including the arcade of Struthers, the medial intermuscular septum, Osborne's ligament, and the fascial origins of the flexor digitorum superficialis and FCU, were decompressed [16]. Thus, in this surgical procedure, it is often necessary to decompress the nerve for a distance of up to 10 cm or more [17]. Proponents of a simple decompression argue that transposition of the ulnar nerve requires extensive dissection and puts the vascularity of the nerve at risk, whereas a simple decompression preserves the vital blood supply to the nerve. Additional disadvantages of anterior subcutaneous transposition include longer operative times and decomposition of the anatomic location of the nerve and, therefore, potentially longer postoperative recovery times [18]. Apart from devascularization of the ulnar nerve, risks specific to subcutaneous transposition also include the potential creation of other areas of impingement by altering the course of the ulnar nerve, possible subluxation of the ulnar nerve behind the medial epicondyle causing a recurrence of symptoms, and possible damage caused by the surgical incision to the posterior branch of the medial antebrachial cutaneous nerve, leading to further distressing or painful paresthesias [9].

Previous studies have shown that high-resolution ultrasonography can easily visualize the ulnar nerve at the elbow [6–8, 19]. In addition, as we found in this study, ultrasonography allowed for the visualization and measurement of the inlet and outlet of the cubital tunnel, in both elbow extension and flexion. The ability to generate dynamic images of the arm in different positions is one advantage of ultrasonography over other imaging modalities when visualizing the ulnar nerve and cubital tunnel [19].

However, we found that the maximal diameter of the ulnar nerve on the longitudinal ultrasonographic scan was smaller than the actual diameter measured intraoperatively. A statistically significant difference existed between the maximal diameter on the ultrasonographic scan and the actual diameter. This phenomenon was also reported by others [20]. It is suggested that it occurred because preoperative measurements did not include the thickness of the epineurium and perineural fibrous tissue, or that it was due to an error caused by differences in the ultrasonographic velocity in the tissue [4]. But this phenomenon did not interfere with our preoperative evaluation. Ultrasonography was found to improve the reliability of the diagnosis of ulnar neuropathy at the elbow [21–23]. In patients with cubital tunnel syndrome we found that the MDU on the longitudinal ultrasonographic scan was greater than that in normal persons. In the present study, the hypoechoic area corresponding with the fascicles of the ulnar nerve was recorded by ultrasonography, and the range of the hypoechoic area was correlated with the severity of the ulnar nerve injury. We note that, in patients with cubital tunnel syndrome, ultrasonography could help to determine the type of operation and the position and the range of decompression of the injured ulnar nerve.

Risks specific to subcutaneous transposition include devascularization of the ulnar nerve, as noted above, and injury to a branch of the ulnar nerve.We found that, to avoid devascularization of the ulnar nerve as far as possible and to avoid extreme decompression of the nerve, the range of the hypoechoic area of the injured ulnar nerve shown by high-resolution ultrasonography was helpful for performing anterior subcutaneous transposition. High-resolution ultrasonography can also help to confirm the most serious entrapment position and range of decompression of the injured ulnar nerve during operation. It can help to protect the branch of the decompressed injured ulnar nerve and reduce vascular injury, promoting the recovery of the injured nerve. Therefore, we conclude that high-resolution ultrasonography can provide effective preoperative evaluation for the anterior subcutaneous transposition of the ulnar nerve for cubital tunnel syndrome.

Conflict of Interest Statement The authors declare that there are no conflicts of interest.

References

1. Bednar MS, Blair SJ, Light TR (1994) Complications of the treatment of cubital tunnel syndrome. Hand Clin 10:83–92
2. Apfelberg DB, Larson SJ (1973) Dynamic anatomy of the ulnar nerve at the elbow. Plast Reconstr Surg 51:79–81
3. Mitsionis GI, Manoudis GN, Paschos NK, Korompilias AV, Beris AE (2010) Comparative study of surgical treatment of ulnar nerve compression at the elbow. J Shoulder Elbow Surg 19:513–519
4. Bartels RH, Meulstee J, Verhagen WI, Luttikhuis TT (2008) Ultrasound imaging of the ulnar nerve: correlation of preoperative and intraoperative dimensions. Clin Neurol Neurosurg 110:687–690
5. Beekman R, Van Der Plas JP, Uitdehaag BM, Schellens RL, Visser LH (2004) Clinical, electrodiagnostic, and sonographic studies in ulnar neuropathy at the elbow. Muscle Nerve 30:202–208
6. Beekman R, Schoemaker MC, Van Der Plas JP et al (2004) Diagnostic value of high-resolution sonography in ulnar neuropathy at the elbow. Neurology 62:767–773
7. Park GY, Kim JM, Lee SM (2004) The ultrasonographic and electrodiagnostic findings of ulnar neuropathy at the elbow. Arch Phys Med Rehabil 85:1000–1005
8. Wiesler ER, Chloros GD, Cartwright MS, Shin HW, Walker FO (2006) Ultrasound in the diagnosis of ulnar neuropathy at the cubital tunnel. J Hand Surg Am 31:1088–1093
9. Robertson C, Saratsiotis J (2005) A review of compressive ulnar neuropathy at the elbow. J Manipulative Physiol Ther 28:345
10. Black BT, Barron OA, Townsend PF, Glickel SZ, Eaton RG (2000) Stabilized subcutaneous ulnar nerve transposition with immediate range of motion. Long-term follow-up. J Bone Joint Surg Am 82-A:1544–1551
11. Osterman AL, Davis CA (1996) Subcutaneous transposition of the ulnar nerve for treatment of cubital tunnel syndrome. Hand Clin 12:421–433
12. Rettig AC, Ebben JR (1993) Anterior subcutaneous transfer of the ulnar nerve in the athlete. Am J Sports Med 21:836–839; discussion 9–40
13. Eaton RG, Crowe JF, Parkes JC 3rd (1980) Anterior transposition of the ulnar nerve using a non-compressing fasciodermal sling. J Bone Joint Surg Am 62:820–825
14. Gelberman RH, Yamaguchi K, Hollstien SB et al (1998) Changes in interstitial pressure and cross-sectional area of the cubital tunnel and of the ulnar nerve with flexion of the elbow. An experimental study in human cadavera. J Bone Joint Surg Am 80:492–501
15. Jaddue DA, Saloo SA, Sayed-Noor AS (2009) Subcutaneous vs submuscular ulnar nerve transposition in moderate cubital tunnel syndrome. Open Orthop J 3:78–82
16. Hoffmann R, Siemionow M (2006) The endoscopic management of cubital tunnel syndrome. J Hand Surg Br 31:23–29
17. Zhong W, Zhang W, Zheng X, Li S, Shi J (2011) Comparative study of different surgical transposition methods for ulnar nerve entrapment at the elbow. J Int Med Res 39:1766–1772
18. Caputo AE, Watson HK (2000) Subcutaneous anterior transposition of the ulnar nerve for failed decompression of cubital tunnel syndrome. J Hand Surg Am 25:544–551
19. Yoon JS, Kim BJ, Kim SJ et al (2007) Ultrasonographic measurements in cubital tunnel syndrome. Muscle Nerve 36:853–855
20. Okamoto M, Abe M, Shirai H, Ueda N (2000) Diagnostic ultrasonography of the ulnar nerve in cubital tunnel syndrome. J Hand Surg Br 25:499–502
21. Beekman R, Visser LH, Verhagen WI (2011) Ultrasonography in ulnar neuropathy at the elbow: a critical review. Muscle Nerve 43:627–635
22. Kutlay M, Colak A, Simsek H et al (2009) Use of ultrasonography in ulnar nerve entrapment surgery – a prospective study. Neurosurg Rev 32:225–232; discussion 32
23. Volpe A, Rossato G, Bottanelli M et al (2009) Ultrasound evaluation of ulnar neuropathy at the elbow: correlation with electrophysiological studies. Rheumatology (Oxford) 48:1098–1101

Neuropsychological Assessment in the Differential Diagnosis of Idiopathic Normal Pressure Hydrocephalus. An Important Tool for the Maintenance and Restoration of Neuronal and Neuropsychological Functions

Carmela Sindorio, Rosaria Viola Abbritti, Giovanni Raffa, Stefano M. Priola, Antonino Germanò, Massimiliano Visocchi, and Maria C. Quattropani

Abstract Idiopathic normal pressure hydrocephalus (iNPH) is a progressive clinical syndrome that includes gait disturbances, urinary incontinence, and cognitive impairment. iNPH shows similarities to other neurodegenerative disorders, primarily Alzheimer's Disease (AD). Definition of the neuropsychological profile of iNPH and the qualitative analysis of systematic mistakes made in cognitive tests could represent a valid method for systematizing possible specific markers of iNPH dementia and differentiating it from other dementias. To evaluate the role and the efficacy of a neuropsychological protocol, designed at our institution, based on psychometric analysis and qualitative assessment, in the differential diagnosis of iNPH from AD dementia, we prospectively enrolled 12 patients with suspected iNPH, 11 patients with AD, and 10 healthy controls (HC) who underwent neuropsychological assessment. The assessment was done with the Mini Mental State Examination (MMSE), Mental Deterioration Battery (MDB), Frontal Assessment Battery (FAB), and the Deux Barrage Test. Evaluation in the iNPH group was performed before extended lumbar drainage (ELD), 48 h after ELD, and 1 week and 3 months after the insertion of a ventriculoperitoneal shunt (VPS). Statistical analysis demonstrated the cognitive profile of iNPH, which was mainly characterized by executive function and immediate verbal memory impairment compared with AD. Additionally, the neuropsychological markers were different between the two groups. The qualitative analysis of systematic mistakes made on the tests demonstrated differences in cognitive performances between the iNPH, AD, and HC cohorts. Neuropsychological assessment and qualitative evaluation could represent a useful tool for achieving effective management and restoration of functions in patients with iNPH.

Keywords Idiopathic normal pressure hydrocephalus • Neuropsychological assessment • Qualitative analysis • Restoration

Introduction

Idiopathic normal pressure hydrocephalus (iNPH) accounts for 2 %–10 % of all forms of dementia and 40 % of adult hydrocephalus [9]. The correct identification of iNPH, frequently hidden in the setting of coexisting diseases, and considering that 1 % of the population aged ≥65 years old shows ventriculomegaly without symptoms [10], is critical for maintaining neuronal and neuropsychological integrity and restoration [12]. However, the criteria used to select patients for treatment remain unclear [1]. The large amount of data that has emerged from recent series suggest that the cognitive profile of iNPH is a complex result of the impairment of several areas, which leads to specific alterations in executive functions, working memory, speed processing information, attention, learning and memory, and visuospatial functions, similar

C. Sindorio, PsyD • G. Raffa, MD
Neurosurgery, Department of Clinical and Experimental Medicine, University of Messina, Messina, Italy

R.V. Abbritti, MD (✉)
Neurosurgery, University of Messina, Messina, Italy
e-mail: rv.abbritti@hotmail.it

S.M. Priola, MD • A. Germanò, MD
Neurosurgery, University of Messina, Messina, Italy

M. Visocchi, MD
Insitute of Neurosurgery, Medical School, Catholic University of Rome, Rome, Italy

M.C. Quattropani, PsyD
Department of Human and Social Sciences, University of Messina, Messina, Italy

M. Visocchi et al. (eds.), *Trends in Reconstructive Neurosurgery*, Acta Neurochirurgica Supplement 124,
DOI 10.1007/978-3-319-39546-3_41, © Springer International Publishing Switzerland 2017

to the cognitive profile in Alzheimer's Disease (AD) [8, 9]. The modern concept of iNPH cognitive disorders is of a dysexecutive syndrome with episodic and immediate memory dysfunction [3, 5]. There is a lack of specific diagnostic criteria needed to systematically define neuropsychological markers that would be useful for achieving early diagnosis and outlining the specific neuropsychological profile of iNPH. The aim of this study was to evaluate the role and the efficacy of a neuropsychological protocol, designed at our institution, based on psychometric analysis and qualitative assessment, in the differential diagnosis of iNPH from AD dementia.

Materials and Methods

The Institutional Ethics Board of the University of Messina approved the study, and informed consent was obtained from each patient and/or their relatives. We prospectively enrolled 12 patients with clinically and neuroradiologically suspected iNPH. Inclusion criteria were: age ≥65 years, clinical triad (gait disturbances, dementia, and urinary incontinence), ventriculomegaly on magnetic resonance imaging (MRI), and other neuroradiological characteristics. Patients underwent a neuropsychological evaluation, with qualitative analysis to assess any systematic mistakes. Results were compared with those for 11 AD patients and 10 healthy controls (HC). The neuropsychological assessment was performed by C.S. and M.Q. (clinical neuropsychologists) on admission, 48 h after extended lumbar drainage (ELD) positioning, and postoperatively (1 week, and 1 and 3 months after VPS), when applicable. Patients who responded positively to preoperative tests for iNPH diagnosis were submitted to ventriculoperitoneal shunting with a programmable valve (Codman Hakim Medos; Codman & Shurtleff, Inc., 325 Paramount Drive, Raynham, MA 02767 0350, USA).

Neuropsychological Assessment

Quantitative Analysis The neuropsychological protocol adopted was chosen for its wide use in the neuropsychological community to assess dementia disorders, as it included several batteries for the assessment of general cognitive status, short- and long-term memory, episodic memory, immediate visual memory, constructive praxia, reasoning, and executive functions. The Mini Mental State Examination (MMSE), consisting of 30 items, allowed the exploration of temporospatial orientation, memory, attention, calculation, language (comprehension, repetition, denomination, reading, and writing), and constructive praxia. The highest score of 30 was modified in relation to age and education. The Mental Deterioration Battery (MDB) was divided into verbal and nonverbal tasks, including neuropsychological tests to detect the deterioration of different cognitive areas: memory, intellectual function, language, executive functions, and constructive praxia. The MDB included seven subtests for immediate and delayed recall, the Rey Auditory Verbal Learning Test (RAVLT), for evaluating semantic and phonological fluency, phrase construction, and immediate visual memory; and Raven's Colored Progressive Matrices test (PM 47), which involves copy drawings, and copy drawings with landmarks. The Deux Barrage Test was used to evaluate divided attention. The Frontal Assessment Battery (FAB) enabled the assessment of executive functions.

Qualitative Analysis In order to distinguish between cognitive impairments in iNPH and AD, we employed the following markers for AD diagnosis: in RAVLT, the absence of the primacy effect derived from a verbal learning task, the presence of the recency effect, the absolute decay of memory trace, and the tendency to produce false alarms during delayed recognition of the same word list; in Raven's Colored Progressive Matrices test, the tendency to choose globalistic or odd responses and positional preference mistakes; in the copy drawings test, the occurrence of the closing-in phenomenon; and in the Deux Barrage Test, inaccuracy in task execution. When the abovementioned markers were mostly presented, we were able to confirm the AD diagnosis, and to exclude those clinically suspected of having iNPH.

Statistical Analysis

Statistical analysis was performed with GraphPad Prism Software (GraphPad Software, Inc., La Jolla, CA 92037, USA). For the descriptive analysis of neuropsychological scores we used nonparametric analysis of variance (ANOVA); we used the Fisher test to compare the frequencies of systematic mistakes in iNPH and AD patients and the paired Student's *t*-test to evaluate the effect of ventriculoperitoneal shunting on cognitive functions in iNPH patients. Additionally, we used the SPSS (Statistical Package for the Social Science; SPSS Inc., Chicago, IL,

USA) [23] to evaluate the covariance of cognitive functions in iNPH patients following surgical treatment. $p<0.05$ was considered statistically significant.

Results

Patient Characteristics We prospectively enrolled 12 patients (8 male; 4 female) with suspected iNPH; mean age 70±6 years, mean educational level 9±4 years. In order to compare psychometric results and systematic mistakes, we also enrolled 11 patients (7 male; 4 female) with AD; mean age 76±5 years, mean educational level 8±5 years; and 10 HC volunteers (4 male; 6 female), mean age 72±8 years, mean educational level 11±5 years. Seven iNPH patients who responded positively to preliminary tests, underwent ventriculoperitoneal (VP) shunting within 2 weeks after assessment.

Quantitative Results Table 1 summarizes the mean±standard deviation values and statistics of the neuropsychological scores for the iNPH, AD, and HC groups on admission. As compared with the HC group, iNPH and AD patients showed significant impairments of different cognitive functions, including MMSE, short- and long-term memory (RAVLT), reasoning (PM 47), and semantic and verbal fluency, language, and constructive praxia. AD patients showed impairments of episodic memory, immediate visual memory, and executive functions. The ANOVA showed statistically significant differences between the iNPH and AD groups in MMSE, long-term memory, episodic memory, immediate visual memory, language, constructive praxia, and executive functions. Table 2 shows the frequencies of different mistakes in the iNPH and AD patients. The iNPH patients presented significant differences, compared with the AD group, in primacy effect, tendency to produce false alarms during delayed recognition of words, globalistic responses, odd responses, inaccuracy on the Deux Barrage Test, and the occurrence of the closing-in phenomenon. Table 3 shows the effect of VP shunting on cognitive performances. In detail, we observed a significant improvement in short- and long-term memory, immediate visual memory, and reasoning in these patients.

Correlational Analysis When performing the correlational analysis of the neuropsychological scores, we did not find significant differences between the cognitive profiles of iNPH and AD patients. In the AD group the correlation coefficient showed a statistically significant between general

Table 1 Summary of the neuropsychological scores in iNPH, AD, and HC groups on admission

	iNPH	AD	HC	Cutoff
Neuropsychological assessment				
Mini Mental State Examination	18.8 (±6.7) ** ###	16.5(±4) ****	27(±2)	0–30
Mental assessment battery				
RAVLT immediate	25.8 (±5) **	22.4(±5) ****	39 (±9.3)	>23.53
RAVLT delayed	4.87(±2.4) *** #	3(±2) ****	10(±2)	>4 69
RAVLT recognition	9(±2)	6(±3)	13.4(±1.5)	0–15
Episodic memory	8(±3) #	4.6(±2.2) **	11(±4.4)	0–28
Immediate visual memory	15.7(±2.7) ##	12.6(±4.6) ***	19(±3)	0–22
Raven's Colored Progressive Matrices	18(±5.4) *	16(±6.4) ***	27(±5.3)	>18.96
Semantic verbal fluency	11. 3(±3) *	9.3(±3.5) ***	18(±5)	>7.25
Phonological verbal fluency	18.3(±12) **	17.2(±10) **	35.3(±9.4)	>17.35
Phrase construction	13.7(±8) ** ##	5(±5) ****	23(±3)	0–25
Copy drawing	5.5(±3) * #	4(±2.7) ***	9.3(±1.2)	>7.17
Copy drawing with landmarks	54.5(±7.2) **	35.3(±26) ***	66.5(±4)	>61.85
Frontal Assessment Battery	9(±4) #	9(±5.6) *	14(±3.4)	>12.03

Data are presented as means±SD

iNPH idiopathic normal pressure hydrocephalus, *AD* Alzheimer's Disease, *HC* healthy control, *RAVLT* Rey Auditory Verbal Learning Test

* = $p<0.05$, ** = $p<0.01$, *** = $p<0.005$, **** = $p<0.0001$ when comparing iNPH and AD vs HC, # = $p<0.05$, ## = $p<0.01$, ### = $p<0.005$ when comparing iNPH vs AD

Table 2 Frequencies of different mistakes in iNPH and AD patients

	iNPH	AD	Fisher test
Primacy effect	10	2	$p=0.03$
Recency effect	10	10	–
Tendency to produce intrusions during free recall of words	10	24	–
Tendency to produce false alarms during delayed recognition of words	21	95	$p=0.001$
Globalistic responses	0	22	$p=0.0003$
Odd responses	4	42	$p=0.01$
Positional preference responses	134	53	–
Inaccuracy on the Deux Barrage test	1	8	$p=0.0001$
Occurrence of closing-in phenomenon	1	53	$p=0.001$

Table 3 Neuropsychological performance in 12 iNPH patients on admission, and after surgical treatment in 7 patients

iNPH	Pre ELD	Post ELD	Post VPS	Post 1 month	Post 3 months	Cutoff
Neuropsychological assessment						
Mini Mental State Examination	18.8(±6.7)	23.5 (±3.3)	24(±3)	24(±3)	25.4(±4)	0–30
Mental assessment battery						
RAVLT immediate	25.8 (±5)	30.3 (±5)	39(±13) *	42.3(±16.5) #	40(±19)	>28.53
RAVLT delayed	4.8(±2.4)	5.5(±2.4)	8(±4.5)	8(±3.5) ##	6(±4.2)	>4.69
RAVLT recognition	9(±2)	12.3(±3)	13.6(±1.5)	12.2(±3)	12.4(±3)	0–15
Episodic memory	8(±3)	9(±0.8)	11(±5.4)	12(±6)	18.2(±12.4)	0–28
Immediate visual memory	15.7(±2.7)	16(±3)	19(±2) *	19.4(±1.5)	20(±1.3)	0–22
Raven's Colored Progressive Matrices	18(±5.4)	24(±3)	24.4(±7)	26(±4) #	27.5(±3)	>18.96
Semantic verbal fluency	11(±3.3)	10.5(±3)	23(±20)	13(±4.7)	15.4(±6)	>7.25
Phonological verbal fluency	18.3(±12)	18.6(±11.5)	17.5(±13)	20.4(±16)	19(±12)	>17.35
Phrase construction	13.7(±8)	11.2(±2.5)	18.6(±7.5)	18(±6)	20.3(±9)	0–25
Copy drawing	5.5(±3)	6.2(±4)	8(±4.3)	9(±2)	6(±2)	>7.17
Copy drawing with landmarks	54.5(±7.2)	55(±19.4)	63(±9)	63(±8)	61(±5)	>61.85
Frontal Assessment Battery	9(±4)	10.5(±2.6)	11(±5)	9(±3.5)	13(±2)	>12.03

Data are presented as mean ± SD
Pre ELD: pre external lumbar drainage, *Post ELD*: post external lumbar drainage, *Post VPS*: post ventriculoperitoneal shunt
* = $p<0.05$ when comparing pre- and 1 week postoperative performances, # = $p<0.05$, ## = $p<0.01$, when comparing pre- and 1-month postoperative performance

cognitive dysfunction, memory, praxia, and executive function impairment (*rho 0.814;* $p<0.01$). As compared with AD patients, the iNPH group showed a significant association between executive variables and memory abilities (*rho 0.798;* $p<0.05$). In the iNPH group, we observed a significant cognitive improvement after ELD in immediate verbal memory and semantic phonological verbal fluency (*rho 0.829;* $p<0.05$), and in divided attention and praxia (*rho 0.926*; $p<0.01$). The improvement of immediate verbal memory, as assessed 1 week postoperatively, was significantly related to delayed verbal memory (*rho 0.900;* $p<0.05$).

Discussion

In the present study we assessed the role of a neuropsychological protocol, combined with the qualitative analysis of systematic neuropsychological mistakes, in the differential diagnosis of iNPH from AD, as compared with results in the HC group. For this purpose we employed specific markers to exclude the AD syndrome, and we evaluated the frequencies of these markers in the iNPH patients. Moreover, the effect of surgery on postoperative cognitive neuropsychological restoration was evaluated. We have demonstrated that the psychometric tests cannot be considered as a sufficient tool for differentiating AD from iNPH patients. Conversely, the combination of neuropsychological markers and psychometric tests was able to achieve an effective differential diagnosis between iNPH and AD.

iNPH represents a complex syndrome for which several authors have attempted to systematize criteria, in order to obtain an effective differential diagnosis from other neurodegenerative disorders or comorbidities [7, 11, 16–19]. iNPH, and its neuropsychological profile, are, to date, still not clarified [6, 8, 14, 20], and a detailed characterization of the cognitive dysfunction in iNPH, especially in view of the specific neuropsychological patterns and differentiation of iNPH from AD, is crucial both for a correct diagnosis [4, 15, 21] and for obtaining neuropsychological restoration following treatment [2, 4, 22]. The neurocognitive profile of patients with suspected iNPH was mainly characterized by the impairment of executive functions and short-term memory [13], whereas in AD patients, the neurocognitive profile was mainly characterized by alterations of general cognitive status, short- and long-term memory, praxia, and executive functions. In detail, we found significant differences between the iNPH and AD groups in MMSE, long-term memory, episodic memory, immediate visual memory, language, constructive praxia, and executive functions. The qualitative analysis of systematic mistakes, made during the assessment, demonstrated statistically significant differences between our groups. iNPH differed from AD patients in the following markers: primacy effect, tendency to produce false alarms during delayed recognition of words, globalistic responses, odd responses, inaccuracy on the Deux Barrage, and the occurrence of the closing-in phenomenon. As compared with results in AD patients, scores in iNPH patients showed a significant association with executive variables and memory abilities. Changes in neuropsychological performances were demonstrated after ELD, 1 week after the operation, and at 1 and 3 months postoperatively.

The results of the present study are not unexpected and are in line with those already published in the literature [6, 8]. We recognize that the limited series in the present study does not allow us to draw definitive conclusions. However, the neuropsychological assessment based on psychometric scores and qualitative analysis of neuropsychological patterns may represent a useful tool for making a correct differential diagnosis between iNPH and AD, and for achieving the restoration of neuronal and neuropsychological functions after treatment. These results may encourage an extension in the use of such a protocol to define the cognitive profile of iNPH.

Conflict of Interest Statement The authors declare that they have no conflict of interest.

References

1. Bergsneider M, Black MP, Klinge P, Marmarou A, Relkin N (2005) Surgical management of idiopathic normal-pressure hydrocephalus. Neurosurgery 57:29–39
2. Campos Gomes Pinto F, Saad F, Fernades de Oliveira M, Pereira RM, Letsaske de Miranda F, Benevenuto Tornai J, Romao Lopes MI, Carvalhal Ribas ES, Valinetti E, Teixeira MJ (2013) Role of endoscopic third ventriculostomy and ventriculoperitoneal shunt in idiopathic normal pressure hydrocephalus: preliminary results of a randomized clinical trial. Neurosurgery 72:845–854
3. Devito EE, Pickard JD, Salmond CH, Iddon JL, Loveday C, Sahakian BJ (2005) The neuropsychology of normal pressure hydrocephalus (NPH). Br J Neurosurg 19(3):217–224
4. Gainotti G, Marra C, Villa G, Parlato V, Chiarotti F (1998) Sensitivity and specificity of some neuropsychological markers of Alzheimer dementia. Alzheimer Dis Assoc Disord 12(3):152–162
5. Golz L, Ruppert FH, Meir U, Lemcke J (2014) Outcome of modern shunt therapy in patients with idiopathic normal pressure hydrocephalus 6 years postoperatively. J Neurosurg 121(4):771–775
6. Hellstrom P, Edsbagge M, Archer T, Tisel M, Tullberg M, Wikkelso C (2007) The neuropsychology of patients with clinically diagnosed idiopathic normal pressure hydrocephalus. Neurosurgery 61:1219–1228
7. Hellstrom P, Klinge P, Tans J, Wikkelso C (2012) A new scale for assessment of severity and outcome in iNPH. Acta Neurol Scand 126:229–237
8. Hellstrom P, Klinge P, Tans J, Wikkelso C (2012) The neuropsychology of iNPH: findings and evaluation of tests in the European multicentre study. Clin Neurol Neurosurg 114:130–134
9. Iddon JL, Pickard JD, Cross JJ, Griffiths PD, Czosnyka M, Sahakian BJ (1999) Specific patterns of cognitive impairment in patients with idiopathic normal pressure hydrocephalus and Alzheimer's disease: a pilot study. J Neurol Neurosurg Psychiatry 67:723–732
10. Iseki C, Kawanami T, Nagasawa H, Wada M, Koyama S, Kikuchi K, Arawaka S, Kurita K, Daimon M, Mori E, Kato T (2009) Asymptomatic ventriculomegaly with features of idiopathic normal pressure hydrocephalus on MRI (AVIM) in the elderly: a

prospective study in a Japanese population. J Neurol Sci 277(1–2):54–57
11. Ishikawa M, Hashimoto M, Kuwana N, Mori E, Miyake H, Wachi A, Takeuchi T, Kazui H, Koyama H (2008) Guidelines for management of idiopathic normal pressure hydrocephalus: guidelines from the Guidelines Committee of Idiopathic Normal Pressure Hydrocephalus, the Japanese Society of Normal Pressure Hydrocephalus. Neurol Med Chir (Tokyo) 48(Suppl):S1–S23
12. Johanson CE (2011) Chapter 33. Production and flow of cerebrospinal fluid. In: Winn HR. Youmans: neurological surgery. Elsevier, Philadelphia
13. Kanno S, Saito M, Hayashi A, Uchiyama M, Hiraoka K, Nishio Y, Hisanaga K, Mori E (2012) Counting-backward test for executive function in idiopathic normal pressure hydrocephalus. Acta Neurol Scand 126:279–286
14. Kaplan E (1988) A process approach to neuropsychological assessment. In: Boll T, Bryant BK (eds) Clinical neuropsychology and brain function: research, measurement, and practice. American Psychological association, Washington, DC, pp 129–167
15. Kiefer M, Unterberg A (2012) The differential diagnosis and treatment of normal-pressure hydrocephalus. Dtsch Aztebt Int 109:15–26
16. Klinge P, Marmarou A, Bergsneider M, Relkin N, Black MP (2005) Outcome of shunting in idiopathic normal-pressure hydrocephalus and the value of outcome assessment in shunted patients. Neurosurgery 57:S40–S52
17. Lamar M, Libon DJ, Ashley AV, Lah JJ, Levey AI, Goldstein FC (2010) The impact of vascular comorbidities on qualitative error analysis of executive impairment in Alzheimer's disease. J Int Neuropsychol Soc 16(1):77–83
18. Marmarou A, Bergsneider M, Klinge P, Relkin N, Black MP (1998) The value of supplemental prognostic tests for the preoperative assessment of idiopathic normal-pressure hydrocephalus. Neurosurgery 57:S2-17–S2-28
19. Mori E, Ishikawa M, Kato T, Kazui H, Miyake H, Miyajima M, Nakajima M, Hashimoto M, Kurijama N, Tokuda T, Ishii K, Kaijima M, Hirata Y, Saito M, Arai H (2012) Guidelines for management of idiopathic normal pressure hydrocephalus: second edition. Neurol Med Chir (Tokyo) 52:775–809
20. Relkin N, Marmarou A, Klinge P, Bergsneider M, Black MP (2005) Diagnosing idiopathic normal-pressure hydrocephalus. Neurosurgery 57:S2-4–S2-16
21. Saito M, Nishio Y, Kanno S, Uchiyama M, Haiashi A, Takagi M, Kikuchi H, Yamasaki H, Shimonura T, Iizuka O, Mori E (2011) Cognitive profile of idiopathic normal pressure hydrocephalus. Dement Geriatr Cogn Dis Extra 1:202–211
22. Solana E, Sauquillo J, Junquè C, Quintana M, Poca MA (2012) Cognitive disturbances and neuropsychological changes after surgical treatment in cohort of 185 patients with idiopathic normal pressure hydrocephalus. Arch Clin Neuropsychol 27:304–317
23. SPSS Inc. Released (2007) SPSS for Windows, Version 16.0. SPSS Inc, Chicago

Does Navigation Improve Pedicle Screw Placement Accuracy? Comparison Between Navigated and Non-navigated Percutaneous and Open Fixations

Gualtiero Innocenzi, Simona Bistazzoni, Manuela D'Ercole, Giovanni Cardarelli, and Francesco Ricciardi

Abstract

Background

The aim of our study was to assess how a preoperative computed tomography (CT)-based navigation system affected the correctness and safety of transpedicular screw insertion, compared with standard techniques.

Method

Between January 2012 and February 2014, 203 patients underwent thoracic and lumbar fixation, with open and percutaneous techniques; 218 screws were implanted through an open navigated technique (1.0 Spine & Trauma 3d ver. 2.0 BrainLab, Feldkirchen Germany) in 43 patients; 220 screws were inserted with an open free-hand technique in 45 patients; 230 screws were implanted in 56 patients using percutaneous CT-based navigation; and 236 screws were inserted in 59 patients using a percutaneous fluoroscopy-guided technique. To our knowledge, this is the first work comparing these four different techniques. The position of each screw was evaluated on CT scan reconstruction and classified according to a four-point grading scale (grade 0: no breach, grade 1: breach < 2 mm, grade 2: breach between 2 and 4 mm; grade 3: breach >4 mm). Statistical analysis was assessed by two-way analysis of variance (ANOVA) *t* test, while the Fisher least significant difference (LSD) method was employed to determine statistical significance.

Results

Statistical analysis showed a significant difference in accuracy between the open CT-based navigation and the percutaneous CT-based navigation techniques (P= 0.0263) and between the open CT-based navigation and the percutaneous fluoroscopy-guided techniques (P=0.0258): a particular difference was observed in anterior misplacement between open CT-based navigation and the percutaneous fluoroscopy-guided technique (P= 0.0153).

Conclusions

Our results confirm the advantages of the navigation technique, which ensures greater accuracy, in open as well as percutaneous procedures.

Keywords Percutaneous pedicular screw fixation • Free-hand pedicular screw fixation • Fluoroscopy-guided pedicle screw fixation • Spinal CT-based navigation system • Pedicle screw accuracy

Introduction

Pedicle screw fixation techniques are progressively improving in terms of less invasivity and greater safety and accuracy. Advances in percutaneous techniques allow a less traumatic approach resulting in improved short- as well as long-term results. Safety in transpedicular screw fixation especially concerns violation of the cortical bone of the pedicles and vertebral bodies that can potentially lead to the damage of neurovascular structures. Misplacement of the screws can also threaten the grip of the implant [11].

Many attempts have been made in order to improve insertion accuracy. Among these, computed tomography (CT)-based navigation seems to be the most reliable, due to visualization of the precise anatomy, as well as reduced radiation exposure.

The aim of this study was to compare the efficacy and reliability of an open free-hand technique, an open navigated technique (BrainLab® System), a percutaneous CT-based navigation technique, and a percutaneous fluoroscopy-guided technique.. To our knowledge this is the first work comparing these four different techniques applied in the same period by the same surgical team.

G. Innocenzi, MD (✉) • S. Bistazzoni, MD • M. D'Ercole, MD • G. Cardarelli, MD • F. Ricciardi, MD
IRCCS Neuromed, Via Atinense 18, Pozzilli, IS 86077, Italy
e-mail: nch@neuromed.it

DOI 10.1007/978-3-319-39546-3_42,

Materials and Methods

Between January 2012 and February 2014, 203 patients (115 females, 88 males) underwent thoracic and lumbar fusion with percutaneous (Viper 2 System, DePuy Synthes Spine, Raynham, MA) and open pedicle screw (Expedium 5.5 System, DePuy Synthes Spine, Raynham, MA) instrumentation carried out by the same neurosurgical team.

Indications for instrumentation were for degenerative pathologies (166 patients; spondylolisthesis with or without stenosis, post-laminectomy syndrome, twice-recurrent disc herniation), unstable fractures (25 patients), tumors (10 patients), aneurysmal cyst (1 patient), and inflammatory lesion (1 patient).

Two hundred and eighteen pedicle screws were implanted in 43 patients (21 females, 22 males) of average age 56.5 years (range 20–79 years) using open CT-based navigation (1.0 Spine & Trauma 3D ver. 2.0 BrainLab®) (ON group); 220 screws were implanted in 45 patients (21 females, 24 males) of average age 53.8 years (range 19–75 years) using an open free-hand technique (O group); 230 screws were implanted in 56 patients (35 female, 21 males) of average age 60.2 years (range 25–81 years) using percutaneous CT-based navigation (1.0 Spine & Trauma 3D ver. 2.0 BrainLab®) (PN group); and 236 screws were implanted in 59 patients (38 females, 21 males) of average age 59 years (range 16–78 years) using a percutaneous fluoroscopy-guided technique (P group).

Distribution of screws in all groups is shown in Table 1.

All patients were operated in the prone position on a carbon-top radiolucent table.

For patients in in the ON and PN groups a preoperative CT scan (CT scan 24-multislice GE Healthcare, Little Chalfont, UK) was performed in the prone position and transferred to the computer navigation platform which reconstructed data to provide real-time intraoperative three-dimensional images of the vertebra.

In the O and P groups, screws were implanted according to the Roy-Camille technique [28]. The procedure in the P group was assessed using the C-arm in a step-by-step fashion, while in the O group, we performed just a final check at the end of the procedure with the C-arm.

Postoperative CT scans with sagittal and coronal reconstruction were performed in each patient. The position of each screw was reviewed by a neurosurgeon and a radiologist uninvolved in the procedure and classified according to a four-point grading scale: grade 0 (screws fully contained in the pedicle); grade 1 (perforating screws, up to 2 mm misplacement); grade 2 (perforating screws, between 2 and 4 mm misplacement); and grade 3 (perforating screws, greater than 4 mm misplacement) [25] (Fig. 1).

Statistical relationships between various groups were assessed by two-way analysis of variance (ANOVA) *t*-test, while the Fisher least significant difference (LSD) method was employed for determining statistical significance. Significance was defined as $P<O=0.05$.

Results

Thoracic pedicle screws were not included in the analysis, because the low number did not allow any statistical analysis.

Screw distribution, according to the Neo classification [24], is shown in Table 2.

Table 1 Distribution of screws

Level	Group ON	Group O	Group PN	Group P
D4				2
D5		2		
D6		2	2	2
D8			4	
D10	2		2	2
D11			2	4
D12	2	2	6	12
L1		2	2	12
L2	10	8	10	20
L3	28	18	26	16
L4	64	64	84	72
L5	80	78	82	78
S1	32	44	10	16
Total	218	220	230	236

ON open computed tomography (CT)-based navigation, *O* open free-hand, *PN* percutaneous CT-based navigation, *P* percutaneous fluoroscopy-guided

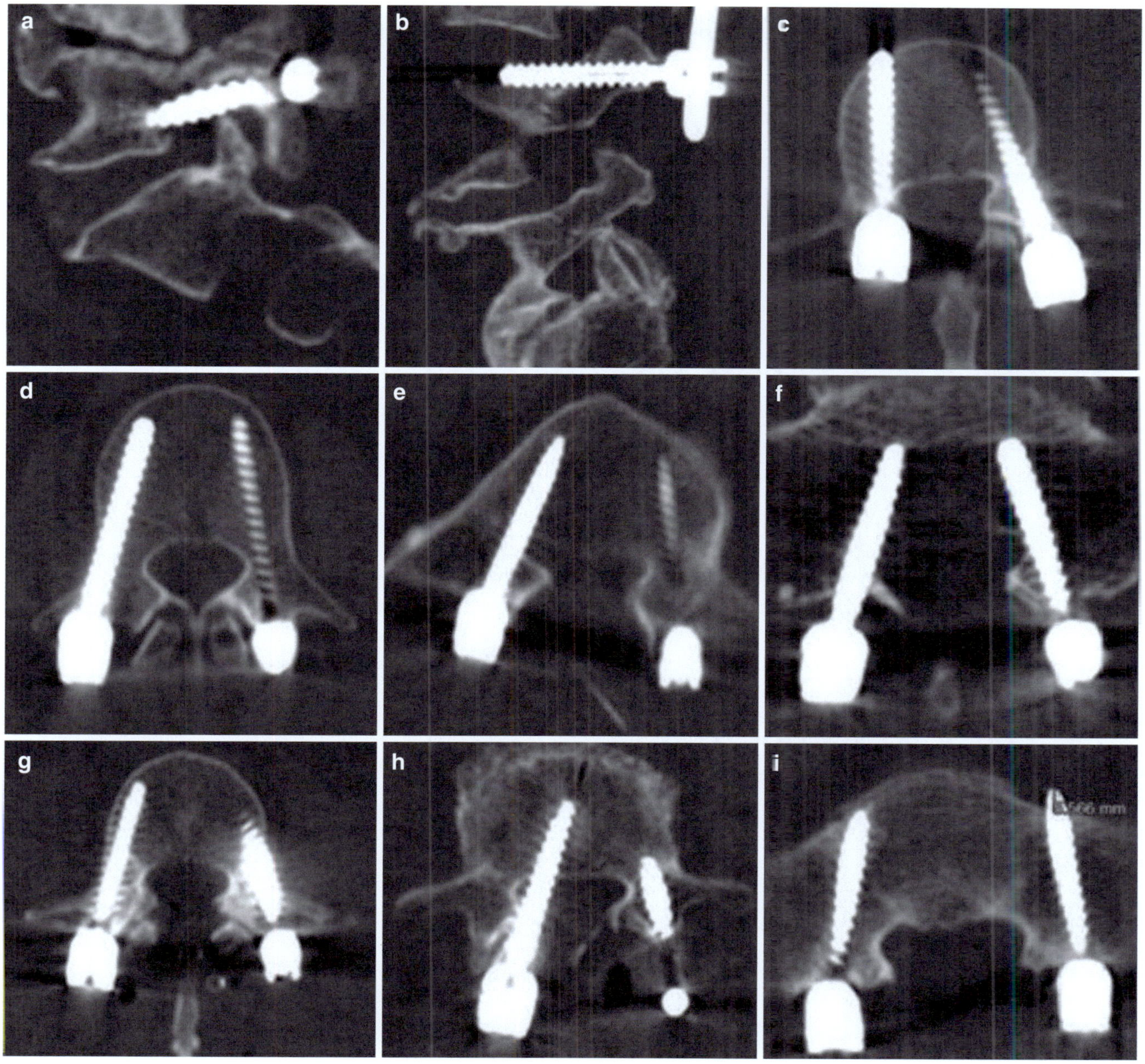

Fig. 1 Examples of grade 1 screw misplacement (**a** caudal, **b** cranial, **c** anterior, **d** lateral, **e** medial); grade 2 misplacement (**f** medial, **g** lateral); and grade 3 misplacement (**h** medial, anterior)

Statistical analysis of collected data showed a significant difference in accuracy between the ON and PN groups (*P*: 0.0263) and between the ON and *P* groups (*P*: 0.0258): a particular difference was observed in anterior misplacement between the ON and *P* groups (*P*: 0.0153) (Table 3).

We also considered each single vertebra, but we found no significant differences (Table 4).

Complications are shown in Table 5.

Discussion

Surgical landmarks and fluoroscopy have been used routinely for pedicle screw insertion, but a number of studies reveal inaccuracies in placement using these conventional techniques (inaccuracies range from 14 to 55 %, with as many as 7 % of these misplaced screws resulting in neurological injuries) [1, 21–23, 25, 38].

Table 2 Direction of breaches in lumbar vertebrae

Screw placement	Group ON	Group O	Group PN	Group P
Lateral axial/coronal				10 (4.7 %)
Grade 1			6 (2.8 %)	3 (1.4 %)
Grade 2	4 (1.9 %)	2 (0.9 %)	1 (0.5 %)	1 (0.5 %)
Grade 3	1 (0.5 %)	2 (0.9 %)	2 (0.9 %)	
Medial axial/coronal	1 (0.5 %)		8 (3.7 %)	7 (3.3 %)
Grade 1			5 (2.3 %)	3 (1.4 %)
Grade 2			2 (0.9 %)	8 (3.7 %)
Grade 3				
Anterior axial	4 (1.9 %)	11 (5.1 %)	8 (3.7 %)	10 (4.7 %)
Grade 1		1 (0.5 %)		4 (1.9 %)
Grade 2		2 (0.9 %)		6 (2.8 %)
Grade 3				
Caudal sagittal/coronal	5 (2.3 %)			2 (0.9 %)
Grade 1				1 (0.5 %)
Grade 2				
Grade 3				
Cranial sagittal/coronal	1 (0.5 %)			1 (0.5 %)
Grade 1				
Grade 2				
Grade 3				

ON open CT-based navigation, *O* open free-hand, *PN* percutaneous CT-based navigation, *P* percutaneous fluoroscopy-guided

Table 3 Statistical analysis of differences in lumbosacral screw misplacement according to placement method

	Total misplaced screws (p value)	Anterior misplacement (p value)	Lateral misplacement (p value)	Medial displacement (p value)	Caudal misplacement (p value)	Cranial misplacement (p value)
Group ON vs Group O	0.580	0.214	0.742	0.423	0.423	0.423
Group ON vs Group PN	**0.0263**	0.423	0.184	0.0848	0.423	0.423
Group ON vs Group P	**0.0258**	**0.0153**	0.188	0.0599	0.635	1
Group O vs Group PN	0.204	0.0742	0.370	0.102	—	—
Group O vs Group P	0.0641	0.321	0.289	0.0591	0.225	0.423
Group PN vs Group P	0.0742	0.0742	0.370	0.729	0.225	0.423

ON open CT-based navigation, *O* open free-hand, *PN* percutaneous CT-based navigation, *P* percutaneous fluoroscopy-guided, statistical significative P values in bold type

Table 4 Statistical analysis for each vertebra

	L2	L3	L4	L5	S1
Group ON vs Group O	0.182	0.229	1	0.769	0.291
Group ON vs Group PN	1	0.664	0.0917	0.839	0.340
Group ON vs Group P	0.127	0.508	0.252	0.948	0.426
Group O vs Group PN	0.182	0.0917	0.138	0.391	0.323
Group O vs Group P	0.124	0.860	0.201	1	0.373
Group PN vs Group P	0.206	0.495	0.435	0.866	0.0577

ON open CT-based navigation, *O* open free-hand, *PN* percutaneous CT-based navigation, *P* percutaneous fluoroscopy-guided

Table 5 Complications

	Group ON	Group O	Group PN	Group P
Transient neurological deficit	1	1		
Superficial wound infection (debridement and antibiotics)		2		
Superficial wound infection (antibiotics)	2			
Deep wound infection	1			
Prolongation of stabilization	1	2		1
Replacement because of pain			1 (after 1 week) L4 medial displacement grade 3	1 (intraoperative) D12 medial displacement grade 3 1 (after 1 week) L5 medial displacement grade 3 1 (after 2 months) L5 medial displacement grade 3 2 (after 2 days) L5 medial displacement grade 3 and L4 medial displacement grade 3

ON open CT-based navigation, *O* open free-hand, *PN* percutaneous CT-based navigation, *P* percutaneous fluoroscopy-guided

Spinal navigation was introduced in 1995 in order to improve the accuracy of pedicle screw insertion and thereby minimize the risk of neurovascular injuries [25]. In a number of published studies, the use of image guidance has been reported to consistently reduce pedicle breaches to less than 5 % [14, 15, 25, 26, 31, 37].

Our accuracy rate for lumbar and sacral vertebrae with the open free-hand technique was 91.1 %, which is comparable to the literature data (range from 69 to 94 %) [10]; we found no case of grade 3 medial misplacement; only 2 cases (0.9 %) showed an anterior breach of more than 4 mm, and this was not accompanied by neurological or vascular injury.

The open CT-based navigation technique showed an accuracy of 92.5 %, which is comparable to the literature data (range from 72.03 to 95.68 %) [32, 35]. We found no case of grade 3 misplacement. In only one case we observed a transient neurological deficit.

Our accuracy rate for lumbar and sacral vertebrae with the percutaneous fluoroscopy-guided technique was 73.8 %. Accurate minimally invasive pedicle screw placement is complicated, however, by the obscuring of normal anatomical landmarks [40]. Errors in placement are therefore a primary concern, with one study reporting almost 10 % of patients needing revision surgery [27]. We reported 16 cases (7.5 %) of grade 3 misplacement, of which 8 cases (3.7 %) showed medial breach, 6 cases (2.8 %) anterior breach, 1 case (0.5 %) lateral breach, and 1 case (0.5 %) inferior breach. Only four medial screws were replaced for irritative pain in the period from the first postoperative day to 2 months after surgery. We found no permanent neurological deficit or construction failure related to screw misplacement.

The percutaneous CT-based navigation technique showed an accuracy of 85 %. Jako et al. [40] reported an accuracy rate of 64.9 % for screws placed with electromagnetic field guidance vs. 40 % for screws placed with fluoroscopy. We reported four cases (1.9 %) of grade 3 misplacement, of which two cases (0.9 %) showed medial breach and two cases (0.9 %) lateral breach. Only one medial screw was relocated for irritative pain, after 1 week. We found no permanent neurological deficit or construction failure related to screw misplacement.

The potential for neurological risk is due to the intrinsic anatomy inherent to screw placement and anatomical variability among patients. Gertzbein et al. [11] postulated a 4-mm safe zone for medial misplaced screws in the lower back region, this being without neurological complications. Even in patients with medial and lateral grade 3 misplacement no neurological deficits were observed this may be explained in the light of the degenerative spine, which has a different threshold for nerve root irritation.

In our series, the accuracy of placement over all lumbar and sacral segments was slightly better in the open CT-based navigation group than in the conventional open free-hand group, but this difference was not statistically significant (*P*: 0.580).

However, we noted an advantage with statistical significance for the open CT-based navigation technique compared with the percutaneous CT-based navigation (*P*: 0.0263) and percutaneous fluoroscopy-guided techniques (*P*: 0.0258)

One important finding of our study is that screws positioned with the open CT-based navigation technique compared with the percutaneous fluoroscopy-guided technique (*P*: 0.0153) tended to perforate the cortex anteriorly, especially for L5 (2/4 L5screws in group ON, 12/20 L5 screws in group *P*), probably due to difficult evaluation of fluoroscopic bidimensional images in the lateral view and due to the shape of L5, which is similar to S1 with a reduction of the lateral diameter of the body.

Gelalis et al. [10] found that screws placed with CT navigation guidance seemed to perforate the lateral cortex more often, differently from screws placed with a free-hand technique, which tended to perforate the cortex medially. Our data did not confirm such a statistically significant difference, but we noted a more medial trajectory in the percutaneous groups than in the open groups.

In comparison with the percutaneous technique, we found that the open technique showed five cases of wound infection, due to the significant exposure of the posterior bony elements of the spine and significant amounts of blood loss [16, 40].

These data confirm the superiority of the navigation technique, which ensures greater accuracy, above all for the percutaneous procedure, probably because of the poor quality of the fluoroscopy images (especially for obese patients); our data also reinforce the superiority of percutaneous vs. open techniques with regard to less damage to the surrounding muscles, less blood loss, less postoperative back pain, and shorter recovery time.

Although image-guided surgical techniques have resulted in lower perforation rates (ranging from 9.3 to 14.3 %), these technologies have their limitations [2–9].

Inaccuracies could also be associated with lack of correspondence between preoperative CT, acquired in the standard supine position, and the intraoperative prone position, especially in cases of severe instability and isthmic lysis. In order to reduce these inaccuracies, we started to perform preoperative CT scanning in the prone position, mimicking the position in the operation room. Although intraoperative CT scanning could be useful in screw placement, this device is not still available in every operation room, due to its high cost. Moreover, it is mandatory to consider the radiation exposure for operative staff, which is significantly higher with intraoperative CT scanning than with standard and neuronavigated techniques.

We also noted an inaccuracy in a patient with intraspinous devices from a previously implanted fixation system, due to anatomical distortion.

Intraoperative CT scanning has recently been introduced to bridge the gap between preoperative and intraoperative position-dependent changes [5, 12, 17–21, 23]. Moreover, this CT scanning offers the possibility of monitoring and visualizing pedicle screws immediately after their placement [26, 29, 30, 33–36, 39, 41]. Bydon et al. [5] maintain that the intraoperative CT scanner is much more sensitive for detecting unfavorably placed screws than conventional intraoperative fluoroscopy or radiography, and dramatically lowers the threshold for screw revision.

An advantage of a navigation system is the decreased radiation exposure for the patient, surgeon, and all the operating room staff, especially when compared with the percutaneous fluoroscopy-guided technique, where we usually perform at least five to six fluoroscopic scans for each pedicular screw. The average dose in a single scan is 0.10 mGy for patients weighing up to 75 kg and 0.21 mGy for those over 75 kg. The radiation exposure of the patient is obviously reduced with a navigation system, and also the radiation exposure of surgeons is greatly reduced, considering the addition of all such exposures during all the surgeries performed in their careers; and of course the reduced number of fluoroscopic scans is associated with a reduction in surgery time.

Another advantage of the Brainlab® system compared with fluoroscopy is that it allows simultaneous and multiplanar visualization of the spinal anatomy, which helps in virtually tracking surgical instruments in relation to the displayed anatomy in real time [13].

Conclusion

Our results confirm the superiority of the navigation technique, which ensures greater accuracy, above all for the percutaneous procedure. A significant reduction in radiation exposure was also noted in our percutaneous navigation group. The free-hand technique is safe and accurate when it is in the hands of an experienced surgeon. In our opinion the navigation system is a valuable tool for spine surgeons, especially for complex cases.

Conflict of Interest Statement We declare that we have no conflict of interest.

References

1. Amiot LP, Lang K, Putzier M, Zippel H, Labelle H (2000) Comparative results between conventional and computer-assisted pedicle screw installation in the thoracic, lumbar, and sacral spine. Spine (Phila Pa 1976) 25:606–614
2. Assaker R, Reyns N, Vinchon M, Demondion X, Louis E (2001) Transpedicular screw placement: image-guided versus lateral-view fluoroscopy: in vitro simulation. Spine (Phila Pa 1976) 26:2160–2164
3. Austin MS, Vaccaro AR, Brislin B, Nachwalter R, Hilibrand AS, Albert TJ (2002) Image-guided spine surgery: a cadaver study comparing conventional open laminoforaminotomy and two image-guided techniques for pedicle screw placement in posterolateral fusion and nonfusion models. Spine (Phila Pa 1976) 27:2503–2508
4. Benzel EC, Rupp FW, McCormack BM, Baldwin NG, Anson JA, Adams MS (1995) A comparison of fluoscopy and computed tomography-derived volumetric multiple exposure transmission holography for the guidance of lumbar pedicle screw insertion. Neurosurgery 37:711–716
5. Bydon M, Xu R, Amin AG, Macki M, Kaloostian P, Sciubba DM, Wolinsky JP, Bydon A, Gokaslan ZL, Witham TF (2014) Safety and efficacy of pedicle screw placement using intraoperative computed tomography: consecutive series of 1148 pedicle screws. J Neurosurg Spine 21:320–328

6. Choi WW, Green BA, Levi AD (2000) Computer-assisted fluoroscopic targeting system for pedicle screw insertion. Neurosurgery 47:872–878
7. Ebmeier K, Giest K, Kalff R (2003) Intraoperative computerized tomography for improved accuracy of spinal navigation in pedicle screw placement of the thoracic spine. Acta Neurochir Suppl 85:105–113
8. Farber GL, Place HM, Mazur RA, Jones DE, Damiano TR (1995) Accuracy of pedicle screw placement in lumbar fusions by plain radiographs and computed tomography. Spine (Phila Pa 1976) 20:1494–1499
9. Fu TS, Wong CB, Tsai TT, Liang YC, Chen LH, Chen WJ (2008) Pedicle screw insertion: computed tomography versus fluoroscopic image guidance. Int Orthop 32:517–521
10. Gelalis ID, Paschos NK, Pakos EE, Politis AN, Arnaoutoglou CM, Karageorgos AC, Ploumis A, Xenakis TA (2012) Accuracy of pedicle screw placement: a systematic review of prospective in vivo studies comparing free hand, fluoroscopy guidance and navigation techniques. Eur Spine J 21:247–255
11. Gertzbein SD, Robbins SE (1990) Accuracy of pedicular screw placement in vivo. Spine (Phila Pa 1976) 15:11–14
12. Holly LT, Foley KT (2007) Image guidance in spine surgery. Orthop Clin North Am 38:451–461
13. Holly LT (2006) Image-guided spinal surgery. Int J Med Robot 2:7–15
14. Hott JS, Deshmukh VR, Klopfenstein JD, Sonntag VKH, Dickman CA, Spetzler RF, Robert F, Papadopoulos SM (2004) Intraoperative Iso-C C-arm navigation in craniospinal surgery: the first 60 cases. Neurosurgery 54:1131–1137
15. Ito Y, Sugimoto Y, Tomioka M, Hasegawa Y, Nakago K, Yagata Y (2008) Clinical accuracy of 3D fluoroscopy-assisted cervical pedicle screw insertion. Clinical article. J Neurosurg Spine 9:450–453
16. Johnson RG, Murphy M, Miller M (1989) Fusions and transfusions. An analysis of blood loss and autologous replacement during lumbar fusions. Spine 14:358–362
17. Kamimura M, Ebara S, Itoh H, Tateiwa Y, Kinoshita T, Takaoka K (2000) Cervical pedicle screw insertion: assessment of safety and accuracy with computer-assisted image guidance. J Spinal Disord 13:218–224
18. Kim YJ, Lenke LG, Cheh G, Riew KD (2005) Evaluation of pedicle screw placement in the deformed spine using intraoperative plain radiographs: a comparison with computerized romography. Spine (Phila Pa 1976) 30:2084–2088
19. Kosmopoulos V, Schizas C (2007) Pedicle screw placement accuracy: a meta-analysis. Spine (Phila Pa 1976) 32:E111–E120
20. Kotani Y, Abumi K, Ito M, Takahata M, Sudo H, Ohshima S, Minami A (2007) Accuracy analysis of pedicle screw placement in posterior scoliosis surgery: comparison between conventional fluoroscopic and computer-assisted technique. Spine (Phila Pa 1976) 32:1543–1550
21. Laine T, Lund T, Ylikoski M, Lohikoski J, Schlenzka D (2000) Accuracy of pedicle screw insertion with and without computer assistance: a randomised controlled clinical study in 100 consecutive patients. Eur Spine J 9:235–240
22. Laine T, Makitalo K, Schlenzka D, Tallroth K, Poussa M, Alho A (1997) Accuracy of pedicle screw insertion: a prospective CT study in 30 low back patients. Eur Spine J 6:402–405
23. Laine T, Schlenzka D, Makitalo K, Tallroth K, Nolte LP, Visarius H (1997) Improved accuracy of pedicle screw insertion with computer-assisted surgery. A prospective clinical trial of 30 patients. Spine (Phila Pa 1976) 22:1254–1258
24. Neo M, Sakamoto T, Fujibayashi S, Nakamura T (2005) The clinical risk of vertebral artery injury from cervical pedicle screws inserted in degenerative vertebrae. Spine (Phila Pa 1976) 30:2800–2805
25. Quinones-Hinojosa A, Robert Kolen E, Jun P, Rosenberg WS, Weinstein PR (2006) Accuracy over space and time of computer-assisted fluoroscopic navigation in the lumbar spine in vivo. J Spinal Disord Tech 19:109–113
26. Richter M, Cakir B, Schmidt R (2005) Cervical pedicle screws: conventional versus computer-assisted placement of cannulated screws. Spine (Phila Pa 1976) 30:2280–2287
27. Ringel F, Stoffel M, Stuer C, Meyer B (2006) Minimally invasive transmuscular pedicle screw fixation of the thoracic and lumbar spine. Neurosurgery 59:ONS361–ONS367
28. Roy-Camille R, Saillant G, Berteaux D, Salgado V (1976) Osteosynthesis of thoraco-lumbar spine fractures with metal plates screwed through the vertebral pedicles. Reconstr Surg Traumatol 15:2–16
29. Sakai Y, Matsuyama Y, Nakamura H, Katayama Y, Imagama S, Ito Z, Ishiguro N (2008) Segmental pedicle screwing for idiopathic scoliosis using computer-assisted surgery. J Spinal Disord Tech 21:181–186
30. Sakamoto T, Neo M, Nakamura T (2004) Transpedicular screw placement evaluated by axial computed tomography of the cervical pedicle. Spine (Phila Pa 1976) 29:2510–2515
31. Shin BJ, James AR, Njoku IU, Hartl R (2012) Pedicle screw navigation: a systematic review and meta-analysis of perforation risk for computer-navigated versus freehand insertion. A review. J Neurosurg Spine 17:113–122
32. Silbermann J, Riese F, Allam Y, Reichert T, Koeppert H, Gutberlet M (2011) Computer tomography assessment of pedicle screw placement in lumbar and sacral spine: comparison between freehand and O-arm based navigation techniques. Eur Spine J 20:875–881
33. Slomczykowski M, Roberto M, Schneeberger P, Ozdoba C, Vock P (1999) Radiation dose for pedicle screw insertion. Fluoscopic method versus computer-assisted surgery. Spine (Phila Pa 1976) 24:975–983
34. Tian NF, Huang QS, Zhou P, Zhou Y, Wu RK, Lou Y, Xu HZ (2011) Pedicle screw insertion accuracy with different assisted methods: a systematic review and meta-analysis of comparative studies. Eur Spine J 20:846–859
35. Tian NF, Xu HZ (2009) Image guided pedicle screw insertion accuracy: a meta-analysis. Int Orthop 33:895–903
36. Tjardes T, Shafizadeh S, Rixen D, Paffrath T, Bouillon B, Steinhausen ES, Bäthis H (2010) Image-guided spine surgery: state of the art and future directions. Eur Spine J 19:25–45
37. Tormenti MJ, Kostov DB, Gardner PA, Kanter AS, Spiro RM, Okonkwo DO (2010) Intraoperative computed tomography image-guided navigation for posterior thoracolumbar spinal instrumentation in spinal deformity surgery. Neurosurg Focus 28(3):E11
38. Vaccaro AR, Rizzolo SJ, Balderston RA, Allardyce TJ, Garfin SR, Dolinskas C, An HS (1995) Placement of pedicle screws in the thoracic spine Part II: an anatomical and radiographic assessment. J Bone Joint Surg Am 77:1200–1206
39. Verma R, Krishan S, Haendlmayer K, Mohsen A (2010) Functional outcome of computer-assisted spinal pedicle screw placement: a systematic review and meta-analysis of 23 studies including 5992 pedicle screws. Eur Spine J 19:370–375
40. Von Jako R, Finn MA, Yonemura KS, Araghi A, Khoo LT, Carrino JA, Perez-Cruet M (2011) Minimally invasive percutaneous transpedicular screw fixation: increased accuracy and reduced radiation exposure by means of a novel electromagnetic navigation system. Acta Neurochir 153:589–596
41. Xu R, Ebraheim A, Shepherd ME, Yeasting RA (1999) Thoracic pedicle screw placement guided by computed tomographic measurements. J Spinal Disord 12:222–226

The Significance of Abnormal Muscle Response Monitoring During Microvascular Decompression for Hemifacial Spasm

Ming Xing Liu, Jun Zhong, Lai Xia, Ning-Ning Dou, Hui Sun, Bin Li, Massimiliano Visocchi, and Shi-Ting Li

Abstract

Background

Despite the wide adoption of the abnormal muscle response (AMR) to electrical stimulation of the facial nerve during microvascular decompression (MVD) surgery, the value of AMR in the prognosis of the postoperative outcome is still controversial. In order to better use this intraoperative electrophysiology, it is necessary to further address the relationship between AMR and postoperative results.

Methods

Three hundred and thirty-two patients with hemifacial spasm (HFS) in whom MVD surgery was performed and in whom AMR was available were finally enrolled in this study. The intraoperative AMR changes were classified as amplitude ≥ 50 %, <50 %, and disappearance. These changes were retrospectively analyzed in association with intraoperative findings and postoperative outcomes. The follow-up period ranged from 11 to 62 months, with an average of 34.1 months.

Results

Among the 332 patients with a typical AMR wave recorded at the beginning of the operation, the AMR disappeared in 305, and amplitude was <50 % in 11 and ≥50 % in 16. Of those with AMR disappearance plus those with amplitude < 50 %, 98.4 % achieved relief on the first postoperative day and at the latest follow-up, while of those with amplitude ≥50 %, 18.8 % and 25 %, respectively, achieved relief on the first postoperative day and at the latest follow-up ($P<0.01$). Accordingly, a more than 50 % decrease of AMR amplitude may predict a good prognosis. The accuracy, sensitivity, and specificity of AMR monitoring were 97.5 %, 99 %, and 72.2 %, respectively.

Conclusions

AMR could be a good tool for successful MVD in patients with HFS when a rational analysis is conducted in association with the intraoperative findings. Persistence of AMR may imply that the real offending vessel was missed. If the entire facial nerve root is cleared of any vessel, a remaining AMR amplitude of less than 50 % might be acceptable. Otherwise, neurocombing is suggested before finishing the operation.

Keywords Abnormal muscle response • Hemifacial spasm • Microvascular decompression • Intraoperative monitoring • Amplitude

M.X. Liu, MD • J. Zhong, MD, PhD (✉) • L. Xia, MD, MS
N.-N. Dou, MD, MS • H. Sun, MD, PhD • B. Li, MD, PhD
S.-T. Li, MD, PhD
Department of Neurosurgery, XinHua Hospital, Shanghai JiaoTong University School of Medicine,
1665 Kongjiang Road, Shanghai 200092, China
e-mail: ZhongMDPhD@sjtu.edu.cn

M. Visocchi, MD
Institute of Neurosurgery, Medical School, Catholic University of Rome, Rome, Italy

Introduction

Hemifacial spasm (HFS) is a syndrome of unilateral facial nerve hyperactive dysfunction. The etiology of the disorder has been generally agreed to be the result of vascular compression of the facial nerve [6, 12, 24]. Nowadays, microvascular decompression (MVD) is the standard operative procedure for the treatment of HFS [5, 17, 20]. However, the results of MVD are not always satisfactory, because it is difficult to identify the offending vessels intraoperatively in some atypical HFS patients [1, 19]. Recently, intraoperative monitoring of an abnormal muscle response (AMR) to electrical stimulation of the facial nerve has been used in order to improve the postoperative outcome [16, 22].

The AMR was first described by Møller and Jannetta in 1986, who demonstrated a positive correlation between intraoperative resolution of AMR and clinical outcome in patients undergoing MVD for hemifacial spasm [7, 8]. However, debate exists regarding the reliability of intraoperative AMR monitoring as an indicator of postoperative

M. Visocchi et al. (eds.), *Trends in Reconstructive Neurosurgery*, Acta Neurochirurgica Supplement 124,
DOI 10.1007/978-3-319-39546-3_43, © Springer International Publishing Switzerland 2017

outcome [4, 14, 16]. Sometimes, the AMR disappeared after decompression, but the spasms existed postoperatively [2, 3]. In some cases, the AMR persisted with alteration of its amplitude, but the symptoms improved after the operation [9, 13]. Therefore, the relationship between AMR and post-operative outcome needs to be further addressed.

Patients and Methods

Patients

Patients' data were obtained from the MVD operations performed by Dr. Zhong in the period between 2010 and 2014. After exclusion of those with unavailable AMR or those lost to follow-up, 332 patients were finally enrolled in this study. They were 114 males and 218 females, with a mean age of 52.2 ± 10.2 years. The right side was affected in 168, the left in 159, and both sides in 5. The time since onset was 65.3 ± 57.2 months.

Intraoperative AMR Monitoring

With the methods proposed by Møller and Jannetta [16, 7], AMR recordings were achieved from the mentalis muscle by electrical stimulation of the temporal branch of the facial nerve and from the orbicularis oculi muscles by stimulation of the marginal mandibular branch, using an evoked potential system (Medtronic Keypoint 4; Dantec, Skovlunde, Denmark). After induction of anesthesia using a short-duration muscle relaxant, bipolar subdermal needle electrodes were inserted 0.5–1 cm apart subcutaneously on the affected side. Electrical stimulation, consisting of 0.2 ms square-wave pulses, was adjusted to supramaximal strength, and the frequency was 0.5 Hz. Electrical stimulation and electromyographic recordings were filtered through a 5 Hz to 3 kHz bandpass (gain: 500 mV/division; analysis time: 50 ms). Usually, a stable AMR was recorded at a stimulation intensity level of 15 mA. To avoid nerve fatigue, the AMR was evoked at 5-min intervals before dura opening. Once the dura was opened, the AMR was recorded continuously until the end of the operation.

Surgical Strategy

A retro-mastoid microsurgical approach was performed while the patient was in the lateral decubitus position [17, 18, 21]. After the edge of the sigmoid sinus was identified, the dura mater was opened. The dissection was started from the caudal cranial nerves. With the arachnoid being opened, gentle retraction of the cerebellum was used to expose the entire intracranial facial nerve. Basically, the root exit zone (REZ) of the facial root was checked first. If the AMR disappeared with the separation of the compressing artery, the process was finished. If the AMR was persistent, an entire root dissection and a thorough decompression of the nerve would be carried out until the AMR amplitude dropped by more than 50 %; otherwise a neurocombing procedure would be added (Fig. 1).

Outcome Evaluation

The postoperative results were evaluated on the first day after the surgery and on the latest follow-up day. The follow-up period ranged from 11 to 62 months, with an average of 34.1 months. The postoperative result was regarded as "relief" when the spasms disappeared completely or the symptoms (frequency and degree of the attack) improved by more than 75 %; the postoperative result was regarded as "no relief" when the spasms decreased by less than 25 % or remained unchanged.

Statistical Analysis

Data processing was performed using commercially available software (SAS, version V8, North Carolina America). The Chi-square test and grouped *t*-test were used to assess the correlation between the intraoperative AMR and the clinical outcome. The level for statistical significance was a probability value of less than 0.05.

Results

AMR Findings

Among the 332 patients with a typical AMR wave recorded at the beginning of the operation, 6 had a negative AMR after durotomy and cerebrospinal fluid (CSF) drainage. With exposure of the REZ and transposition of the compressing artery, the AMR vanished in 176 patients. For the other 150 patients with remaining AMR, the entire VII cranial nerve root was dissected. After all the contacting vessels had been moved away, the AMR disappeared in 118 patients. Among the remaining 32

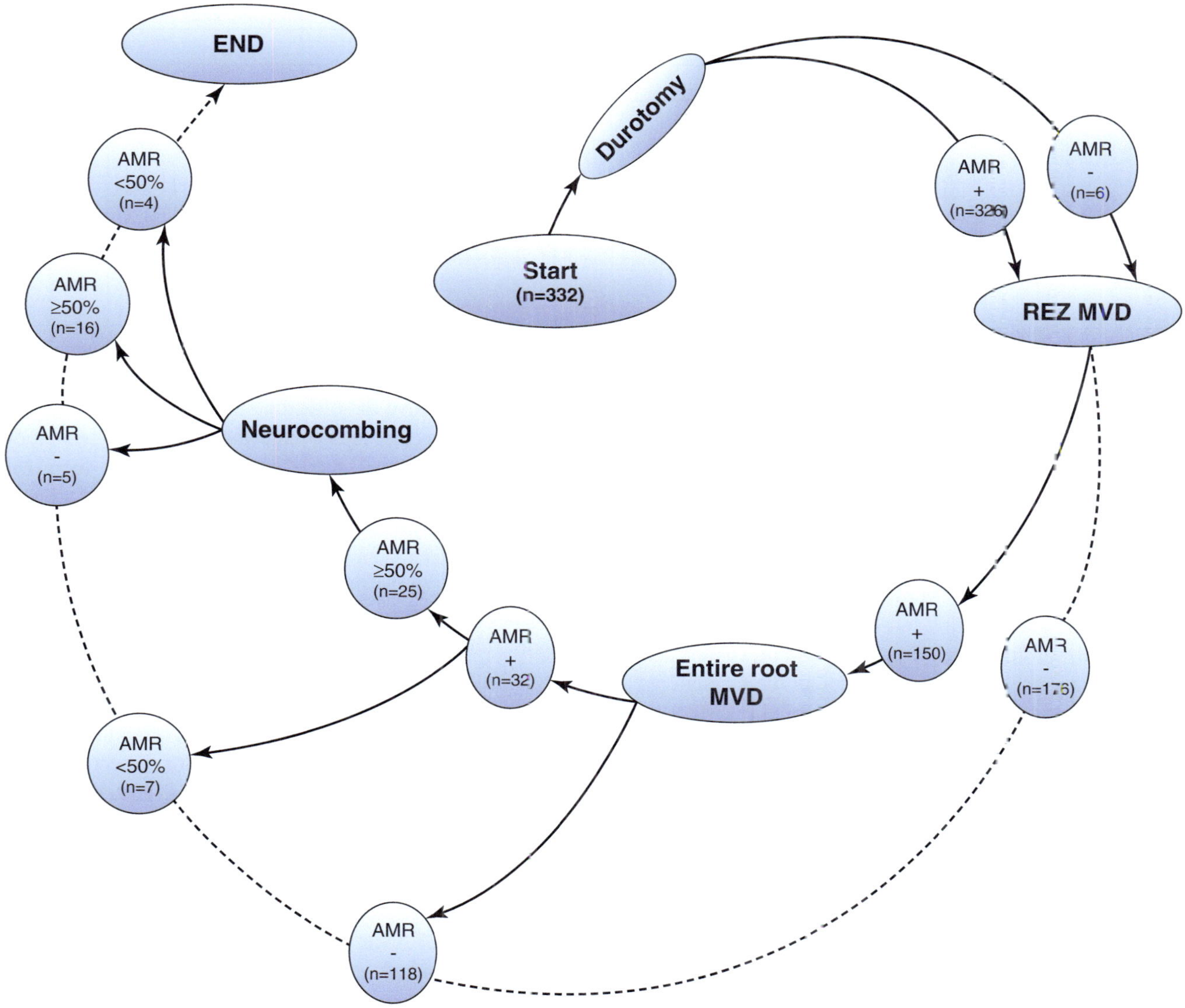

Fig. 1 The AMR findings during surgery

patients with positive AMR, the amplitude changed to <50 % of the baseline in 7, while it changed to ≥50 % in 25. For the latter group, neurocombing was performed. Finally, there were still 16 patients with AMR amplitude of ≥50 % of the baseline at the end of the operation (Fig. 1).

Correlation Between AMR and Postoperative Outcomes

In this series, AMR disappeared completely in 305 patients (91.9 %) and its amplitude decreased to less than 50 % of the baseline in 11 (3.3 %), while its amplitude remained at more than 50 % of the baseline in 16 (4.8 %). Of those with AMR < 50 %, 98.4 % achieved relief on the first postoperative day and the final follow-up day, while of those with AMR ≥ 50 %, relief was achieved in 18.8 % and 25 %, respectively, at these times ($P<0.01$). Accordingly, a more than 50 % decrease of AMR amplitude may predict a good prognosis. The sensitivity, specificity, and accuracy of AMR monitoring during MVD were 99 %, 72.2 % and 97.5 %, respectively. The false-negative and false-positive rates were 0.96 % and 27.8 %, respectively (Table 1).

Complications

There was no mortality and no severe complication occurred postoperatively, with the exception of transient conductive-type hearing impairment in three patients

Table 1 Patient distribution in abnormal muscle response (AMR) ≥ or <50 % groups according to postoperative outcomes

	Postoperative		Follow-up		
Group	Relief	Non-relief	Relief	Non-relief	Total
AMR < 50 %	311	5	311	5	316
AMR ≥ 50 %	3	13	4	12	16
Total	314	18	315	17	332
P value	<0.05		<0.05		

AMR < 50 %, patients with AMR disappearance plus those with AMR amplitude decreased by over 50 %; *AMR ≥ 50 %*, patients with AMR amplitude decreased by no more than 50 %

(0.9 %), immediate facial weakness in three (0.9 %), and delayed facial palsy in six (1.8 %).

Discussion

Several investigators have attempted to determine whether AMR monitoring during MVD leads to positive outcomes. Some authors believed that intraoperative monitoring of the AMR was useful for identifying the offending vessels and for confirming a successful decompression of the facial nerve [3, 13]. It has been reported that the chance of cure in those with AMR disappearance was 4.2 times that in patients with AMR persistence [11]. An analysis of 1301 cases with a 3- to 6-month follow-up showed that complete disappearance of AMR was associated with a higher postoperative cure rate. The sensitivity, specificity, positive predictive value, and negative predictive value were calculated as 80 %, 39 %, 89 %, and 24 %, respectively [14]. Some authors have stated that AMR may be an unreliable predictor of long-term outcome, since they found that those in the non-AMR-disappeared group also had relief from HFS after MVD. For example, in a 90-patient series, AMR disappeared during surgery in 80, of whom 5 had persistent HFS, and 1 developed a recurrence of HFS. Of the 10 patients with persistent AMR despite effective MVD, 8 achieved complete resolution [4].

Nevertheless, we believe that AMR was a good tool to navigate the MVD process when it was used properly. Actually, this AMR is very sensitive, and could change as soon as the offending artery was separated from the facial nerve. That is the reason why the AMR disappeared in some cases even before the offending artery was visualized. Change of AMR may happen during the process of durotomy with CSF drainage or retraction of the cerebellum [10]. Evidently the AMR is not reliable until you find a compressing artery. As a matter of fact, the AMR is not used to search for the culprit but to confirm your finding. Sometimes the disappearance of AMR may be delayed for a couple of minutes, which could lead to a false-negative result. Therefore, in order to achieve a better result with fewer complications, our surgical strategy of MVD is to dissect the caudal REZ of the facial nerve root first, because, in this area, the chance of finding the offending vessel is more than 90 % according to the literature [22]. Accordingly, we suggest, if you find an apparent compressing artery with a dent in the nerve in REZ, you just move the offending artery away and put some Teflon between the artery and the nerve for separation. Afterwards, you check the electrophysiology; if the AMR wave is abolished you can get ready to finish the operation. In this way, you may obtain resolution of HFS in most cases with minimal risk (our data exhibited fewer complications compared with data in the literature [19, 22]). However, the monitoring should not be ceased at this stage. If the AMR reappears, a double check is necessary. In that case, one should not hesitate to reopen the incision, because the offending artery may have contacted with the nerve again while the retracted cerebellum was released. On the other hand, if an offending artery is not discovered in the REZ, then an entire dissection of the intracranial nerve root is strongly suggested. You should separate all the contacting vessels away from the nerve. After that, if the AMR remains, a combing of the facial nerve could be added. In consideration of a persistent AMR after a thorough decompression, we recommended a new criterion: a more than 50 % decrease of AMR amplitude (Fig. 1). Our study showed this criterion has high accuracy.

To support our new criterion, we need to mention the mechanism of HFS. Currently, the pathophysiology of HFS can be classified in accordance with two main hypotheses [12, 15, 23]: (1) it may be induced by ectopic excitation from the compressed nerve fibers; or (2) it might result from the hyperexcitability of facial motor neurons. However, debate has been continuing as to whether the AMR originates from a peripheral or a central site. We believe that ectopic impulses emerging from the compressed nervous fibers could be the main cause of HFS, as the AMR disappeared immediately in most of our cases. Nevertheless, it is conceivable that cross transmission at the compression site may combine with subsequent kindling at the nucleus level, which could explain the remaining AMR with decreased amplitude, if the facial nucleus was more involved in the pathogenesis of HFS than the cross transmission site.

Conclusion

AMR could be a good tool for a successful MVD in patients with HFS. However, a rational analysis of this electrophysiology in relation to the intraoperative findings is necessary. In principle, the disappearance of AMR when an apparent compressing artery in the caudal REZ of the facial nerve is moved away may be an indicator of a satisfactory postoperative outcome. Persistence of AMR may imply that the real offending vessel was missed. If the entire facial nerve root has been cleared of any vessel, a remaining AMR amplitude of less than 50 % might be acceptable. Otherwise, neurocombing is suggested before finishing the operation.

Acknowledgments This study was supported by the National Natural Science Foundation of China (#81471317) and the Science & Technology Committee of Shanghai Municipality (124119a0800).

Conflict of Interest All authors certify that they have NO affiliations with or involvement in any organization or entity with any financial interest (such as honoraria; educational; participation in speakers' bureaus; membership, employment, consultancies, stock ownership, or other equity interest; and expert testimony or patent- licensing arrangements), or non-financial interest (such as personal or professional relationships, affiliations, knowledge or beliefs) in the subject matter or materials discussed in this manuscript.

References

1. Abbruzzese G, Berardelli A, Defazio G (2011) Hemifacial spasm. Handb Clin Neurol 100:675–680
2. Fukuda M, Oishi M, Takao T, Hiraishi T, Sato Y, Fujii Y (2012) Monitoring of abnormal muscle response and facial motor evoked potential during microvascular decompression for hemifacial spasm. Surg Neurol Int 3:118
3. Hirono S, Yamakami I, Sato M, Kado K, Fukuda K, Nakamura T, Higuchi Y, Saeki N (2014) Continuous intraoperative monitoring of abnormal muscle response in microvascular decompression for hemifacial spasm; a real-time navigator for complete relief. Neurosurg Rev 37:311–319, discussion 319–320
4. Li J, Zhang Y, Zhu H, Li Y (2012) Prognostic value of intra-operative abnormal muscle response monitoring during microvascular decompression for long-term outcome of hemifacial spasm. J Clin Neurosci 19:44–48
5. Li S, Hong W, Tang Y, Ying T, Zhang W, Li X, Zhu J, Zhong J, Hua X, Xu S, Wan L, Wang X, Yang M, Li Y, Zheng X (2010) Re-operation for persistent hemifacial spasm after microvascular decompression with the aid of intraoperative monitoring of abnormal muscle response. Acta Neurochir (Wien) 152:2113–2118
6. Li Y, Zheng X, Hua X, Ying T, Zhong J, Zhang W, Li S (2013) Surgical treatment of hemifacial spasm with zone-4 offending vessel. Acta Neurochir (Wien) 155:849–853
7. Møller AR, Jannetta PJ (1985) Hemifacial spasm: results of electrophysiologic recording during microvascular decompression operations. Neurology 35:969–974
8. Moller AR, Jannetta PJ (1987) Monitoring facial EMG responses during microvascular decompression operations for hemifacial spasm. J Neurosurg 66:681–685
9. Mooij JJ, Mustafa MK, van Weerden TW (2001) Hemifacial spasm: intraoperative electromyographic monitoring as a guide for microvascular decompression. Neurosurgery 49:1365–1370, discussion 1370–1361
10. Niijima KH, Yonekawa Y, Kaku Y (1990) Disappearance of hemifacial spasm following spinal drainage: a case report. No Shinkei Geka 18:577–580
11. Sekula RF Jr, Bhatia S, Frederickson AM, Jannetta PJ, Quigley MR, Small GA, Breisinger R (2009) Utility of intraoperative electromyography in microvascular decompression for hemifacial spasm: a meta-analysis. Neurosurg Focus 27, E10
12. Sindou MP, Polo G, Fischer C, Vial C (2006) Neurovascular conflict and hemifacial spasm. Suppl Clin Neurophysiol 58:274–281
13. Tobishima H, Hatayama T, Ohkuma H (2014) Relation between the persistence of an abnormal muscle response and the long-term clinical course after microvascular decompression for hemifacial spasm. Neurol Med Chir (Tokyo) 54:474–482
14. von Eckardstein K, Harper C, Castner M, Link M (2014) The significance of intraoperative electromyographic "lateral spread" in predicting outcome of microvascular decompression for hemifacial spasm. J Neurol Surg B Skull Base 75:198–203
15. Yamakami I, Oka N, Higuchi Y (2007) Hyperactivity of the facial nucleus produced by chronic electrical stimulation in rats. J Clin Neurosci 14:459–463
16. Ying TT, Li ST, Zhong J, Li XY, Wang XH, Zhu J (2011) The value of abnormal muscle response monitoring during microvascular decompression surgery for hemifacial spasm. Int J Surg 9:347–351
17. Zhong J (2012) An ideal microvascular decompression technique should be simple and safe. Neurosurg Rev 35:137–140, author reply 140
18. Zhong J, Li ST, Zhu J, Guan HX (2011) Is entire nerve root decompression necessary for hemifacial spasm? Int J Surg 9:254–257
19. Zhong J, Li ST, Zhu J, Guan HX, Zhou QM, Jiao W, Ying TT, Yang XS, Zhan WC, Hua XM (2012) A clinical analysis on microvascular decompression surgery in a series of 3000 cases. Clin Neurol Neurosurg 114:846–851
20. Zhong J, Zhu J, Li ST, Guan HX (2011) Microvascular decompressions in patients with coexistent hemifacial spasm and trigeminal neuralgia. Neurosurgery 68:916–920, discussion 920
21. Zhong J, Zhu J, Li ST, Li XY, Wang XH, Yang M, Wan L, Guan HX (2010) An analysis of failed microvascular decompression in patients with hemifacial spasm: focused on the early reoperative findings. Acta Neurochir (Wien) 152:2119–2123
22. Zhong J, Zhu J, Sun H, Dou NN, Wang YN, Ying TT, Xia L, Liu MX, Tao BB, Li ST (2014) Microvascular decompression surgery: surgical principles and technical nuances based on 4000 cases. Neurol Res 36:882–893
23. Zhou QM, Zhong J, Jiao W, Zhu J, Yang XS, Ying TT, Zheng XS, Dou NN, Wang YN, Li ST (2012) The role of autonomic nervous system in the pathophysiology of hemifacial spasm. Neurol Res 34:643–648
24. Zhu J, Li ST, Zhong J, Ying TT, Guan HX, Yang XS, Zhou QM, Jiao W (2012) Microvascular decompression for hemifacial spasm. J Craniofac Surg 23:1385–1387

One-Pot Aqueous Synthesization of Near-Infrared Quantum Dots for Bioimaging and Photodynamic Therapy of Gliomas

Ming-Xing Liu, Jun Zhong, Ning-Ning Dou, Massimiliano Visocchi, and Guo Gao

Abstract

Background

As the early detection and total destruction of gliomas are essential for longer survival, we attempted to synthesize a quantum dot (QD) that is capable of recognizing glioma cells for imaging and photodynamic therapy.

Methods

Using a one-pot aqueous approach, near infrared-emitting CdTe was produced. After detection of its physicochemical characteriistics, it was conjugated with RGD. The emission images were observed with confocal microscopy. To test its toxicity, CdTe-RGD at various concentrations was separately added to a human glioma cell line (U251) and a mouse embryo fibroblast cell line (3T3) (control) for incubation in dark conditions. To test its photodynamic effect, the U251 and 3T3 cells were then irradiated for 5-60 min, using a 632.8-nm laser.

Results

This QD ($\Phi = 3.75$ nm, photoluminescence (PL) peak wavelength = 700 nm, photoluminescence quantum yield (PLQY) = 20 %), was a spherical crystal with excellent monodispersity. Under a confocal microscope, U251 cells were visualized, but not the 3T3 cells. In dark conditions, the survival rates of both U251 and 3T3 cells were above 85 %. After laser irradiation, the survival rate of U251 cells decreased to 37 ± 1.6 % as the irradiation time and the CdTe-RGD concentration were increased.

Conclusions

With good physicochemical characteriistics and low toxicity, this QD-RGD has broad prospects for use in the biomedical imaging and photodynamic therapy of gliomas.

Keywords Near-infrared • Glioma • Bioimaging • Photodynamic therapy • Quantum dots

M.-X. Liu, MD • J. Zhong, MD, PhD (✉) • N.-N. Dou, MD
Department of Neurosurgery, XinHua Hospital,
Shanghai JiaoTong University School of Medicine,
1665 Kongjiang road, Shanghai 200092, China
e-mail: ZhongMDPhD@sjtu.edu.cn

M. Visocchi, MD
Insitute of Neurosurgery, Medical School,
Catholic University of Rome, Rome, Italy

G. Gao, MD
Department of Bio-Nano Science and Engineering,
Key Laboratory for Thin Film and Microfabrication Technology
of Ministry of Education, Institute of Micro/Nano Science
and Technology, Shanghai Jiao Tong University,
Shanghai 200240, China

Introduction

Glioma is one of the most common primary brain tumors, accounting for about 30–40 % of all intracranial tumors [3]. Current imaging modalities are unable to take full advantage of the unique nature of protein expression in the tumor cell for early diagnosis. Due to the invasive nature of gliomas, their treatment is still a challenge today [3]. The traditional remedies include surgical excision, chemotherapy, and radiotherapy [20, 22], with surgical excision being the primary choice. However, it is difficult to remove a glioma completely because of the tumor's invasiveness [8, 18]. Therefore, the development of a targeting multimodel agent specific for glioma may revolutionize the diagnosis and treatment of the disease.

Because of their unique chemical and optical properties, quantum dots (QDs) may be a potential nano-material for the bioimaging and killing of glioma cells. QDs have been shown to remain brightly stable and emissive even after long periods of excitation, whereas organic dyes are photobleached quickly. The flexibility of optical tuning obtained by changing the size and composition of QDs allows them to emit in the near infrared (NIR) region, which is optimal for imaging through tissue. QDs can also

M. Visocchi et al. (eds.), *Trends in Reconstructive Neurosurgery*, Acta Neurochirurgica Supplement 124,
DOI 10.1007/978-3-319-39546-3_44, © Springer International Publishing Switzerland 2017

be used for photodynamic therapy (PDT), which is a noninvasive treatment for tumors. In a photodynamic reaction, the QD is irradiated by light to an excited state. When the excited QD returns to its fundamental singlet state, energy transfers to vicinal compounds to kill cells by the generation of reactive oxygen species (ROS) [7, 19]. Recent studies have demonstrated that a QD could kill pancreatic carcinoma cells and HeLa cells effectively, thus showing the prospective use of PDT for gliomas [16, 23]. The use of targeted QDs has developed as a strategy for both bioimaging and targeted therapy. QDs have been conjugated to different biological molecules, such as DNA, RNA, protein, peptides, and sugars, for targeting [12, 15]. When it comes to glioma, various targeted biological molecules have been used [6, 12]. The RGD peptide can combine the integrin $\alpha_v\beta_3$, which is overexpressed in glioma cells. Here, cyclic RGD can be chosen as a targeting ligand. The use of NIR QDs for bioimaging and PDT has been widely researched, but the use of QD-RGD for gliomas has barely been studied [6, 11]. In the present study, we prepared NIR CdTe conjugated with RGD for bioimaging and PDT.

Materials and Methods

Materials

None of the chemicals in this study were previously purified. The QDs were synthesized from $CdCl_2$ (99.0 %), Na_2TeO_3 (97 %), 3-mercaptopropionic acid (MPA; 99 %), folic acid (FA), $NaBH_4$ (96 %), 1-ethyl-3-(3-dimethylaminopropyl) carbodiimide, and trisodium citrate (99 %), which were all purchased from Sinopharm Co. (Shanghai, China). The RGD was conjugated to the QD using carbodiimide hydrochloride (EDC) and N-hydroxysulfosuccinimide (NHS), which were purchased from Aldrich (Sigma); RGD (95 %) was purchased from Shanghai Top-peptide Bio Co., Ltd (Shanghai, China).

Preparation of CdTe/CdS Nanoprobe

CdTe/CdS nanoprobes were synthesized by the one-pot aqueous approach, as described in the literature [2, 5]. Briefly, 0.1099 g (0.6 mmol) of $CdCl_2$ was dissolved in 100 ml of deionized water, and 74 μl of MPA and 0.02 g of trisodium citrate were added to this solution in succession. After 10-min stirring, the pH of the mixed solution was adjusted to 11.0. Then, 4.4 mg (0.02 mmol) of Na_2TeO_3 and 20 mg of $NaBH_4$ were added to the mixture under stirring. Finally, the mixture was refluxed at 90°C, under open-air conditions. Different emission spectra were obtained by varying the time of reflux. QD solution/ethanol at a volume ratio of 1: 3 was centrifuged at 8,000 rpm. RGD was conjugated with MPA-coated CdTe using EDC-NHS chemistry. CdTe-RGD conjugates were obtained by adding c (RGDfK) solution to the activated solution of CdTe QD. The mixture was reacted under stirring at room temperature overnight. To remove the excess of c (RGDfK) and coupling reagents after the reaction with the QD, the reaction solution was dialyzed (3 kDa cutoff) for 10 min at 5,000 rpm.

Characterization of QD

The morphology and size of the QD were characterized by high-resolution transmission electron microscopy (HRTEM) on a JEM-2100F (JEOL, Tokyo, Japan). The luminescence emission spectra were measured using a Hitachi FL-4600 spectrofluorometer (Hitachi, Tokyo, Japan). UV–visible (vis) spectra were recorded with a Varian Cary 50 spectrophotometer (Varian Inc., Palo Alto, CA, USA). The photoluminescence quantum yield (PLQY) of the QD was measured through comparison using Rhodamine G6 as the standard organic dye.

Cell Culture

A human glioma cell line (U251) was used for recognition by QD-RGD. A mouse embryo fibroblast (3T3 cells) cell line was used for the control. Cells were cultured at 37°C (5% CO_2) in Dulbecco's modified Eagle's medium (DMEM), supplemented with 100 mg/ml penicillin G, 10% fetal bovine serum (FBS), 1.8 mg/ml $NaHCO_3$, and 100 mg/ml streptomycin sulfate.

In Vitro Imaging

Confocal microscopic imaging was carried out using a Leica TCS SP5 (Solms Germany). The cells were plated onto 14-mm glass coverslips in a 12-well tissue culture plate and were allowed to adhere for 24 h. Afterward, QD-RGD (about 10 μg/ml) was added to the U251 and 3T3 cells and they were incubated for 30 min. As a contrast, QD was also added to both cell lines. QD emission images were obtained using a 590-nm long pass filter.

Dark Cytotoxicity

The cytotoxic effect of QD in U251 and 3T3 cells was measured by MTT assay. Briefly, U251 and 3T3 cells were plated in 96-well microtitration plates at an initial density of 10,000 cells per well and incubated for 24 h at 37°C (5% CO_2) Then the cells were treated with QDs at various concentrations (range from 3 to 24 µg/ml) without light. After 24-h incubation at 37°C, cell survival was measured by MTT assay as described previously.

Photodynamic Activity

U251 and 3T3 cells were plated in 96-well microtitration plates; 48 h after plating, the cells were exposed to photoactive QDs at various concentrations (range from 3 to 24 µg/ml). After incubation at 37°C for 24 h in the dark, the cells were washed three times with cold phosphate-buffered saline (PBS) and fresh RPMI medium was added before cell irradiation. Irradiation was carried out at 632.8 nm with energy density of 30 mW/cm^2 for various times (5–60 min). Cell viability was measured 24 h after photosensitization by MTT assay. The cell viabilities of both dark cytotoxicity and PDT were measured three times.

Statistical Analysis

We conducted a factor analysis study based on a factorial design to prove the correlation between PDT effectiveness and QD concentrations and irradiation times. Each experiment was repeated three times and the results were presented as means ± SD. The statistical significance was supported by an analysis of variance (ANOVA) using SAS 8.0 Software, North Carolina America, and significance was considered at $P<0.05$.

Results

Physicochemical Characterization of QD

The photoluminescence (PL) peak wavelength of the QD was gradually shifted from 500 nm to 800 nm with the prolongation of reflux time. The variations of absorption and PL spectra showed the enlargement of QD size with continuous refluxing. After 3 h of reflux, QDs with excellent NIR-emitting PL properties were acquired. The PLQY of QD increased first and then decreased with the prolongation of reflux time. The maximum PLQY was 37 % at the PL peak wavelength of 610 nm. It was noted that there was still a high PLQY of 20 % in the NIR spectra range. The diameter of the QD with PL peak wavelength of 700 nm was 3.75 nm. Under TEM, the spherical QD appeared with a crystalline structure and possessed excellent monodispersity (Fig. 1). The average size of the QDs was small (4 nm) enough for bioimaging applications.

Targeting for Human Glioma Cells

In the QD group, non-RGD conjugated QD was used on both U251 and 3T3 cells it was found that no significant cell shape was recognized. In the QD-RGD group, QD-RGD was used on both U251 and 3T3 cells, but only the U251 cell structures were visible (Fig. 2).

Cytotoxicity

The survival rates of U251 and 3T3 cells incubated with QD-RGD without light were above 85 %, which showed the low dark cytotoxicity of QD-RGD. MTT assays were performed in both U251 and 3T3 cells, to assess QD cytotoxicity, with 632.8 nm, 30 mW/cm^2 irradiation for various times (5–60 min) and various concentrations (3–24 µg/ml) We conducted a factor analysis study based on a factorial design. In the light cytotoxicity group, the survival rates of U251 cells decreased to 37 ± 2 % with the increasing of irradiation time and concentration (Fig. 3); however, the survival rates of 3T3 cells did not decrease significantly. ANOVA results showed the effectiveness of PDT was associated with cell type, irradiation time and concentration ($p<0.01$).

Discussion

To date, various QD probes have been developed for NIR-fluorescence imaging [6, 9], while obtaining high-quality NIR-emitting QDs has remained a challenging research objective due to stability and biocompatibility. Consequently, it is essential to design an aqueous method to obtain NIR-emitting QDs for high-sensitivity targeted bioimaging [1]. Deng et al. had developed an aqueous method for the synthesis of stable and bright NIR-emitting water-soluble CdTe/CdS nanocrystals [5]. However, the approach was complicated and the nanocrystals were unstable in open air. Chen and colleagues reported an improved and easier aqueous

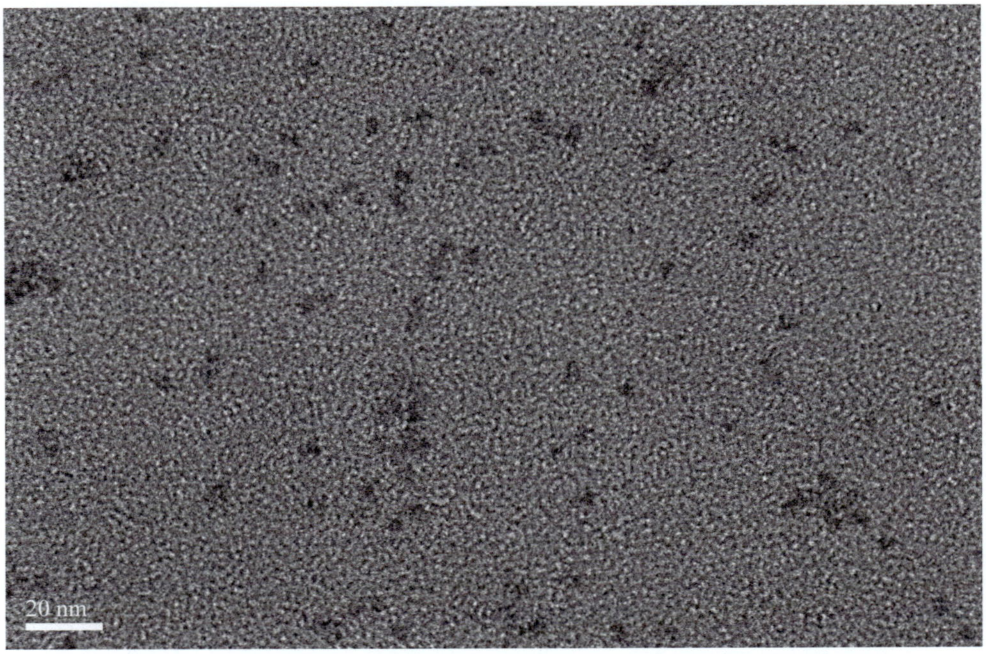

Fig. 1 TEM images of CdTe/CdS quantum dots (QDs), with photoluminescence at 700-nm wavelength. The spherical QDs appeared with crystalline structures and possessed excellent monodispersity. *TEM* transmission electron microscopy)

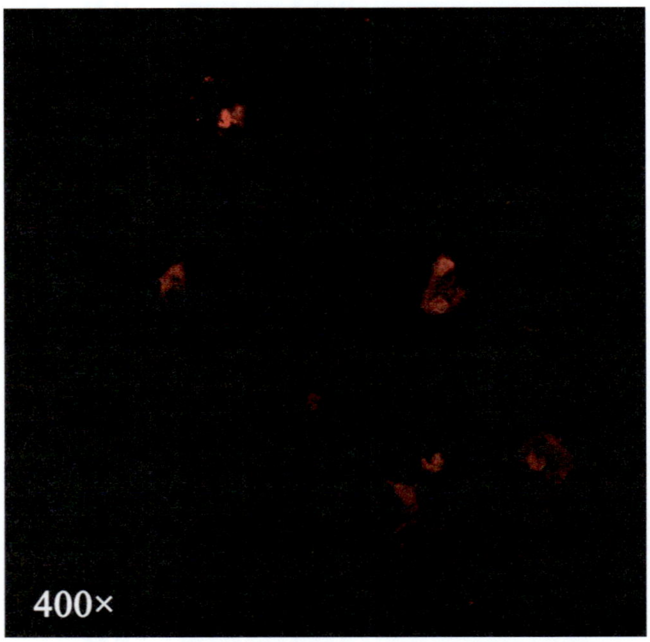

Fig. 2 QD-RGD targeting of human glioma cells (in the CdTe-RGD group, the U251 cell structures were visible). *U251* human glioma cell line

synthesis of highly fluorescent and ultrasmall size CdTe/CdS [2]. The reaction they reported can be carried out with a single procedure in the open air because Na_2TeO_3 can be readily reduced by $NaBH_4$ to generate Te^{2-}. After 6-h reflux, the research produced NIR-emitting QDs with a PL peak wavelength of 700 nm. In the present study, we used the one-pot aqueous approach outlined above to produce a high PL quantum yield in a QD with excellent physicochemical characteristics. We adjusted the molar ratio of Cd^{2+}/TeO_3^{2-} to 30: 1 and pH to 11; as a result, we produced high-quality NIR-emitting QDs more quickly.

The RGD peptide can specifically recognize integrin, which is restrictively expressed on the cell surfaces of malignant glioma cells [10]. However, few studies have reported NIR-emitting QDs conjugated with RGD for both bioimaging and PDT in U251 cells. As can be seen distinctly in the present study, the QD-RGD could label U251 cells (Fig. 2), but not the fibroblast 3T3 negative control cells. Besides, neither the U251 nor the 3T3 cells could be labeled by QDs without the conjugation of RGD. This labeling process was achieved by the interaction between U251 and the RGD attached to the QD surface. Accordingly, the QD-RGD can recognize U251 cells specifically and sensitively.

Ion leakage from the core of QDs has been reported, especially the leakage of Cd^{2+}, which has become a significant problem because the Cd^{2+} may kill cells [21]. However, one study reported that stable coating with a shell and capping may effectively prevent ion leakage and, additionally, protect the core from air oxidation [17]. Due to its stable structure, the QD we synthesized was verified to be safe for biomedical use. QDs offer great promise in PDT applications [13, 14]. And NIR QDs can be used to penetrate tissue to depths of several centimeters, thereby allowing access to deep-seated tumors, which makes them ideal agents for PDT applications. Researchers have found statistically significant ROS

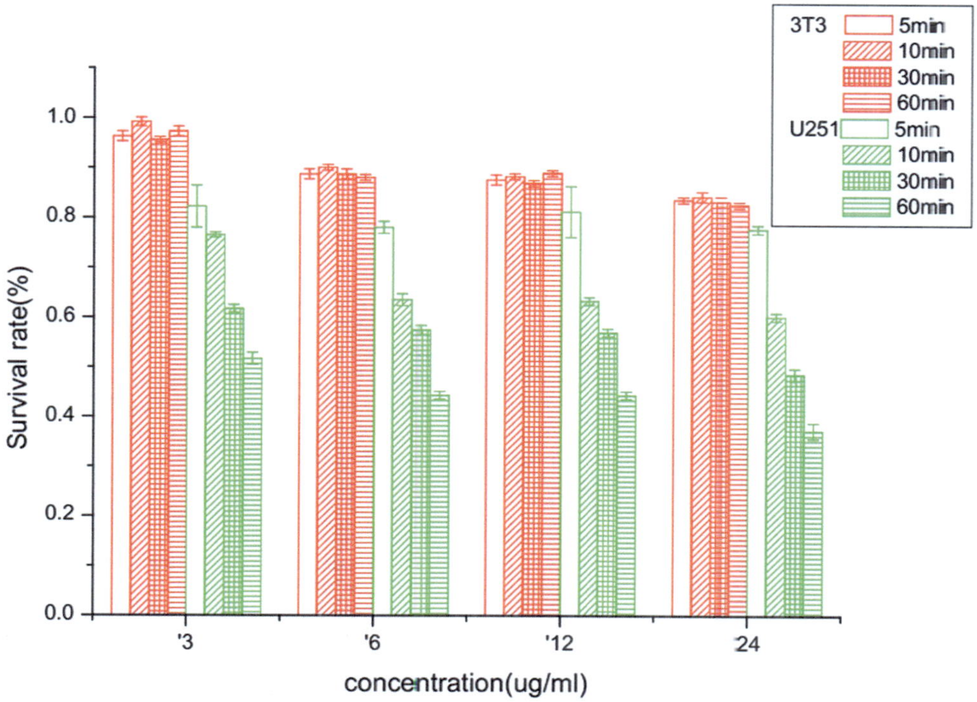

Fig. 3 The dark and photodynamic cytotoxicity of QD-RGD; with increasing irradiation time and concentration, the survival rate of U251 cells decreased. The survival rate of U251 cells was related to the irradiation time and QD-RGD concentration ($p<0.01$))

production from QDs. It appears that QDs with CdSe and CdTe are very efficient at ROS generation [4]. Here, irradiation was carried out at 632.8 nm with energy density of 30 mW/cm^2 for various times (30 s to 10 min) and at various concentrations (3–24 µg/ml). With the increasing of irradiation time and concentration, the survival rate of U251 cells decreased (Fig. 3). The survival rate of U251 cells was significantly related to the irradiation time and concentration ($p<0.01$), which proved the PDT of QD-RGD can kill U251 cells effectively.

In PDT, the photosensitizing agent transfers its triplet state energy to nearby oxygen molecules to form reactive singlet oxygen ($1O^2$) species, which cause cytotoxic reactions in the cells. The increasing popularity of PDT is largely due to its selectivity: photosensitizer, light, and oxygen are simultaneously necessary.

Conclusions

Using the one-pot aqueous approach, CdTe/CdS-RGD QDs were synthesized. This type of QD has properties of NIR emission, as well as low toxicity and good physicochemical characteristics. This QD could be used for biomedical imaging and photosensitizing.

Acknowledgments This study was supported by the National Natural Science Foundation of China (#81471317) and the Science & Technology Committee of Shanghai Municipality (124119a0800).

Conflict of Interest All authors certify that they have NO affiliations with or involvement in any organization or entity with any financial interest (such as honoraria; educational; participation in speakers' bureaus; membership, employment, consultancies, stock ownership, or other equity interest; and expert testimony or patent- licensing arrangements), or non-financial interest (such as personal or professional relationships, affiliations, knowledge or beliefs) in the subject matter or materials discussed in this manuscript.

References

1. Cassette E, Helle M, Bezdetnaya L, Marchal F, Dubertret B, Pons T (2013) Design of new quantum dot materials for deep tissue infrared imaging. Adv Drug Deliv Rev 65:719–731
2. Chen LN, Wang J, Li WT, Han HY (2012) Aqueous one-pot synthesis of bright and ultrasmall CdTe/CdS near-infrared-emitting quantum dots and their application for tumor targeting in vivo. Chem Commun (Camb) 48:4971–4973
3. Chintagumpala M, Gajjar A (2015) Brain tumors. Pediatr Clin North Am 62:167–178
4. Cooper DR, Dimitrijevic NM, Nadeau JL (2010) Photosensitization of CdSe/ZnS QDs and reliability of assays for reactive oxygen species production. Nanoscale 2:114–121
5. Deng Z, Schulz O, Lin S, Ding B, Liu X, Wei X, Ros R, Yan H, Liu Y (2010) Aqueous synthesis of zinc blende CdTe/CdS magic-core/thick-shell tetrahedral-shaped nanocrystals with emission tunable to near-infrared. J Am Chem Soc 132:5592–5593
6. Fatehi D, Baral TN, Abulrob A (2014) In vivo imaging of brain cancer using epidermal growth factor single domain antibody bio-conjugated to near-infrared quantum dots. J Nanosci Nanotechnol 14:5355–5362

7. Ge J, Lan M, Zhou B, Liu W, Guo L, Wang H, Jia Q, Niu G, Huang X, Zhou H, Meng X, Wang P, Lee CS, Zhang W, Han X (2014) A graphene quantum dot photodynamic therapy agent with high singlet oxygen generation. Nat Commun 5:4596
8. Gerard CS, Straus D, Byrne RW (2014) Surgical management of low-grade gliomas. Semin Oncol 41:458–467
9. Kuzyniak W, Adegoke O, Sekhosana K, D'Souza S, Tshangana SC, Hoffmann B, Ermilov EA, Nyokong T, Hopfner M (2014) Synthesis and characterization of quantum dots designed for biomedical use. Int J Pharm 466:382–389
10. Liu X, Cui W, Li B, Hong Z (2012) Targeted therapy for glioma using cyclic RGD-entrapped polyionic complex nanomicelles. Int J Nanomedicine 7:2853–2862
11. Lu Y, Zhong Y, Wang J, Su Y, Peng F, Zhou Y, Jiang X, He Y (2013) Aqueous synthesized near-infrared-emitting quantum dots for RGD-based in vivo active tumour targeting. Nanotechnology 24:135101
12. Master A, Malamas A, Solanki R, Clausen DM, Eiseman JL, Sen Gupta A (2013) A cell-targeted photodynamic nanomedicine strategy for head and neck cancers. Mol Pharm 10:1988–1997
13. Morosini V, Bastogne T, Frochot C, Schneider R, Francois A, Guillemin F, Barberi-Heyob M (2011) Quantum dot-folic acid conjugates as potential photosensitizers in photodynamic therapy of cancer. Photochem Photobiol Sci 10:842–851
14. Rakovich A, Savateeva D, Rakovich T, Donegan JF, Rakovich YP, Kelly V, Lesnyak V, Eychmuller A (2010) CdTe Quantum Dot/Dye Hybrid System as Photosensitizer for Photodynamic Therapy. Nanoscale Res Lett 5:753–760
15. Ruan J, Song H, Qian Q, Li C, Wang K, Bao C, Cui D (2012) HER2 monoclonal antibody conjugated RNase-A-associated CdTe quantum dots for targeted imaging and therapy of gastric cancer. Biomaterials 33:7093–7102
16. Steponkiene S, Valanciunaite J, Skripka A, Rotomskis R (2014) Cellular uptake and photosensitizing properties of quantum dot-chlorin e6 complex: in vitro study. J Biomed Nanotechnol 10:679–686
17. Su Y, He Y, Lu H, Sai L, Li Q, Li W, Wang L, Shen P, Huang Q, Fan C (2009) The cytotoxicity of cadmium based, aqueous phase – synthesized, quantum dots and its modulation by surface coating. Biomaterials 30:19–25
18. Tate MC (2015) Surgery for gliomas. Cancer Treat Res 163:31–47
19. Tekdas DA, Durmus M, Yanik H, Ahsen V (2012) Photodynamic therapy potential of thiol-stabilized CdTe quantum dot-group 3A phthalocyanine conjugates (QD-Pc). Spectrochim Acta A Mol Biomol Spectrosc 93:313–320
20. Venur VA, Peereboom DM, Ahluwalia MS (2015) Current medical treatment of glioblastoma. Cancer Treat Res 163:103–115
21. Wang Q, Fang T, Liu P, Min X, Li X (2011) Study of the bioeffects of CdTe quantum dots on Escherichia coli cells. J Colloid Interface Sci 363:476–480
22. Wick W, Wiestler B, Platten M (2015) Treatment of anaplastic glioma. Cancer Treat Res 163:89–101
23. Zhou M, Ni QW, Yang SY, Qu CY, Zhao PC, Zhang JC, Xu LM (2013) Effects of integrin-targeted photodynamic therapy on pancreatic carcinoma cell. World J Gastroenterol 19:6559–6567

Awake Craniotomy for Tumor Resection: Further Optimizing Therapy of Brain Tumors

H. Maximilian Mehdorn, Felix Schwartz, and Juliane Becker

Abstract In recent years more and more data have emerged linking the most radical resection to prolonged survival in patients harboring brain tumors. Since total tumor resection could increase postoperative morbidity, many methods have been suggested to reduce the risk of postoperative neurological deficits: awake craniotomy with the possibility of continuous patient-surgeon communication is one of the possibilities of finding out how radical a tumor resection can possibly be without causing permanent harm to the patient.

In 1994 we started to perform awake craniotomy for glioma resection. In 2005 the use of intraoperative high-field magnetic resonance imaging (MRI) was included in the standard tumor therapy protocol. Here we review our experience in performing awake surgery for gliomas, gained in 219 patients.

Patient selection by the operating surgeon and a neuropsychologist is of primary importance: the patient should feel as if they are part of the surgical team fighting against the tumor. The patient will undergo extensive neuropsychological testing, functional MRI, and fiber tractography in order to define the relationship between the tumor and the functionally relevant brain areas. Attention needs to be given at which particular time during surgery the intraoperative MRI is performed. Results from part of our series (without and with ioMRI scan) are presented.

Keywords Awake craniotomy • Glioma surgery • Neuropsychology • Intraoperative MRI

H.M. Mehdorn, MD, PhD (✉) • F. Schwartz, MD • J. Becker, PhD
Department of Neurosurgery, University Hospitals Kiel,
Sternwartenweg 1a, Kiel D – 24105, Germany
e-mail: mehdorn@nch.uni-kiel.de

Introduction

In recent years more and more data, publications, and reviews have emerged that link the most radical tumor resection to prolonged survival in patients harboring intracranial brain tumors of various histological grades [1, 18, 24, 27]. While well recognizing the importance of tumor biology as shown e.g. by MGMT status [14]. The frequent use of fluorescence-guided resection of malignant brain tumors [7, 8, 13, 19, 31] is a good example showing the dilemma of radical tumor resection versus an increased risk of damage to the eloquent areas of the brain, with subsequent new or increased neurological deficits. One of the major objections to radical tumor resection using fluorescence guidance has been the risk of increased neurological deficits. Intraoperative neuronavigation has been suggested to be helpful for radical tumor resection [18]. However, this benefit was disputed [3, 34] and the indication for image-guided surgery was thought to be biased by others [21]. In our experience, image guidance is at least helpful for access just over the tumor site when planning a small awake craniotomy (AC), in order to reduce the skin incision and surgical trauma, and not so much for resection radicality control, as, for example, in skull base meningiomas [23]. Preoperative evaluation of patients using functional magnetic resonance imaging (fMRI) techniques and the calculation of fiber tracts in a three-dimensional way [6]—mostly the pyramidal tract and arcuate fibers—from diffusion/tensor weighted images (DWI/DTI) has been suggested [11], but it is well known that brain shift occurs immediately after the opening of the dura mater and the subarachnoid space, and this factor is even more relevant with continued glioma resection [4]. Intraoperative MRI (ioMRI) [15, 17, 24, 26, 28], computed tomography (CT) [2, 12], and ultrasound [20, 29] have become tools to compensate for brain shift and improve resection control. Initially, when

M. Visocchi et al. (eds.), *Trends in Reconstructive Neurosurgery*, Acta Neurochirurgica Supplement 124,
DOI 10.1007/978-3-319-39546-3_45, © Springer International Publishing Switzerland 2017

neuronavigation was introduced in our department in 1994, the modern techniques of preoperative fMRI imaging of eloquent areas were not yet available to us, as they are nowadays [22], so with the idea of performing as radical a resection as possible, we have relied, since 1994, on the technique of AC and we continued to employ this technique when ioMRI scanning became available in our department. We have previously published our clinical results of AC in glioma surgery [30] and the utility of ioMR imaging during high-grade glioma surgery [24]; here we would like to highlight only the most important data from our experience in using both techniques over the past 20 years, in 219 patients.

Materials and Methods

Between February 1994 and November 2005, using AC, we operated on 107 patients with intracranial gliomas of various grades (in part of the series described by Pinsker [30], 22 glioblastoma patients and 30 glioma grades 2–3 were included); there were three metastases and one cavernoma located within the eloquent areas of the brain,i.e., the motor region and the speech areas, respectively. Within this time period and while being under our care, 28 of the entire group of glioma patients subsequently developed symptomatic or asymptomatic recurrent gliomas, or had shown, on early postoperative MRI scans, some tumor remnants (e.g., those in the Amino levulinic acid (ALA)-study in the white-light arm) and were operated again to remove the tumor remnants, early redo surgery being one of our options to offer to the patient before the advent of ioMRI [33]. Twelve of these patients had undergone the first surgery under general anesthesia and AC was suggested for the second operation because of the direction of tumor growth, while 13 patients had already undergone an AC as the first operation and agreed to an AC for the second operation.

Based on this experience and being confident about the technique, we continued to use the AC technique in adequately selected patients when we included, in September 2005, the ioMRI imaging technique in our neurosurgical setting; we were well aware of the additional neuropsychological burden being placed on the patient, and we were aware that this would require even more attention before and during surgery [9, 10]. Between December 2005 and March 2015, 108 additional patients, aged 15–77 years, were suggested as candidates and accepted the offer of undergoing glioma surgery while being awake, with the use of ioMRI control [25]. Of the 8 patients who developed a recurrent tumor during the postoperative follow-up time under our care and for whom repeat surgery was suggested, 6 patients agreed to undergo AC using ioMRI for tumor section control. Of the 2 remaining patients, 1 patient had undergone the first surgery under general anesthesia and AC was suggested for the recurrence, to which he agreed. One patient rejected the offer of AC and did not have surgery for the recurrent tumor.

Results

As we have shown previously [30], comparison of the results of surgeries performed between 1998 and 2002 for patients with gliomas located in eloquent areas showed, in an exemplary fashion, that for tumors located in the motor area, AC with intraoperative electrophysiological and neuropsychological guidance resulted in an increased radical surgical resection while reducing the risk of neurological deficit. This became even more important when ALA-fluorescence and, later, ioMRI scanning were included to further increase the amount of tumor tissue resected. Nevertheless, we had, in one young patient with a temporo-mesial glioma, intraoperatively progressive hemiplegia, which developed during hemostasis on a lenticulostriate artery after complete tumor removal. Also, early in our series, an intraoperative seizure that occurred before the removal of a precentral glioma led to severe hemiparesis, due to poor anesthesiological reaction to the initial phase of the seizure. This led us to become even further aware of possible complications and their prevention, e.g., by leaving a margin of brain tissue for hemostasis near eloquent structures, observing the suggested distance from the tumor to the pyramidal tract [16, 32, 35], and using ice-cold sponge application to the brain surface when a focal seizure started.

In the second part of the series, after the introduction of ioMRI scanning, we had to be more careful in the preparation of patients, because they had to lie flat for 15–35 min in the MRI machine tunnel, draped with sterile drapes, with ventilation being possible only through a funnel-type construction of these drapes. So it became a rule for the surgeon to stay in the MRI-operating room (OR) when the patient was scanned in order to be able to intervene if any problems occurred. The MRI sequences had to be interrupted only twice due to patient discomfort.

Comparing the percentages of repeat surgeries between our two groups—from 1995 to 2005 and from 2005 to 2015—there was a slight preponderance in redo surgeries for tumor remnants in the first period (12 of 107 = 11.2 %), compared with 8 of 108 = 7.4 % in the second period (1 of which was an early redo surgery). The intervals between the surgeries in both groups depended mainly on the histological grade. While there was no striking difference in histologies between our two series (see text and Table 1), patients with low-grade gliomas in the second series did not experience recurrences as frequently as those in the first series: 1 vs 5 patients har-

Table 1 Histology of gliomas operated using awake craniotomy and intraoperative magnetic resonance imaging (MRI) control (N = 106)

Astrocytoma II	10
Astrocytoma III	22
Glioblastoma multiforme (GBM)	46
Recurrent GBM	6
Gliosarcoma	2
Oligoastrocytoma II	3
Oligoastrocytoma III	5
Oligodendroglioma I	4
Oligodendroglioma III	6
Others	2

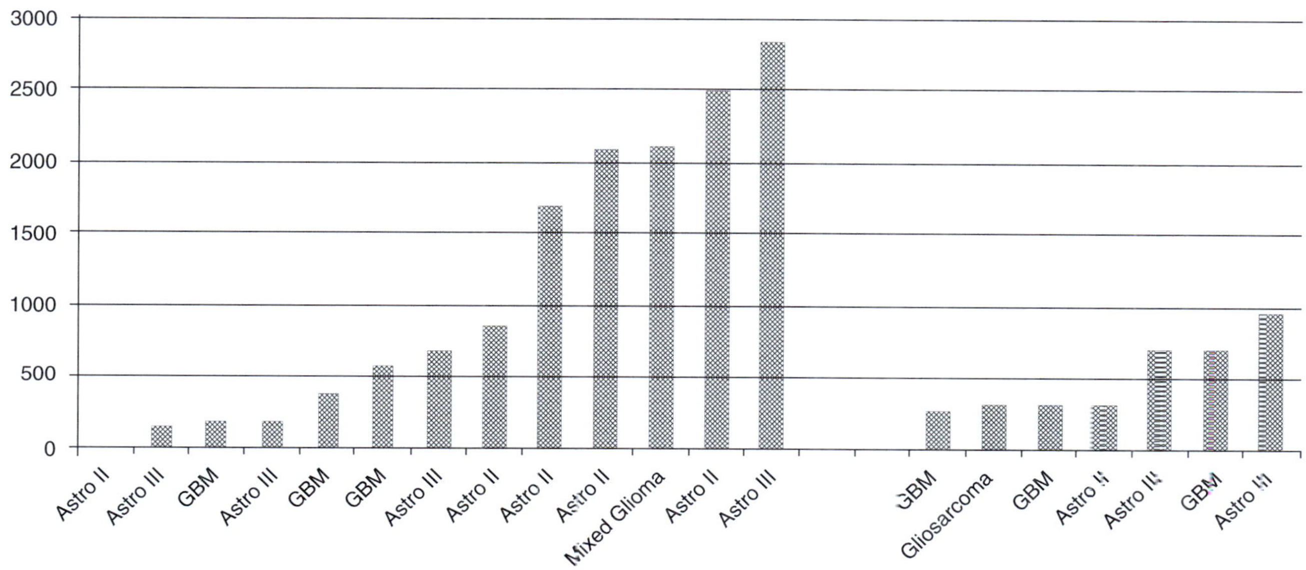

Fig. 1 Intervals between first and second awake craniotomy surgeries, in days, without (*left*) and with (*right*) intraoperative magnetic resonance imaging (ioMRI). There was less recurrence, and recurrences were almost all only in high-grade gliomas, when ioMRI was used. *Note*: each bar represents a single patient

boring an astrocytoma II (Fig. 1). This is congruent with the results of a recently published multicenter study on ioMRI in low-grade glioma surgery [5].

Discussion

AC has become routine in the armamentarium of modern neurosurgical clinics. Also, the combination of AC and ioMRI scanning has become routine in specialized centers. Long-term follow-up of patients will show whether this additional burden on the patients and the surgeon and their team, as well as the increased surgical time (also related to increased cost), is warranted by longer survival and higher quality of life, as would be expected, a question that can only be solved by careful data pooling and multicenter studies. It has been suggested by some surgeons to perform awake surgery for nearly all brain tumor patients in order to save money on anesthesia costs, and then discharge the patient the day after surgery.

The high number of patients in our series who agreed to undergo repeat AC for recurrent glioma shows the "comfort" that they had experienced during the first operation and that they were able to balance the benefit of surgery in the awake state against the psychological discomfort, despite the psychological burden [10]. It needs to be stressed that discomfort during a second AC is worse than that during the first AC, since, in most instances, the dura is firmly attached to the bone flap, which causes pain at operation even if one adds what would be considered a sufficient amount of local anesthetic around the scar. Also, the patient is usually more attentive, since they are more experienced with what could happen and they try to compare the first and second surgeries. Another possibility of reducing the discomfort during a second surgery would certainly be to use more propofol during the initial steps of the surgery. This, however, would prevent the patient from being completely awake when they need to be tested. Furthermore, it has been our experience that surgery, even if smoothly performed, needs to be finished rapidly, because the patient's cooperative capacity is not endless [9].

Conclusion

Awake craniotomy with ioMRI resection control offers the best possible neurosurgical treatment for gliomas in eloquent areas. Patients harboring these tumors should be referred to experienced centers where modern adjunctive therapy can be applied, the neurosurgical part being the first and most important step in a patient's experience undergoing several treatment steps.

Conflict of Interest Statement We declare that we have no conflict of interest.

References

1. Almeida JP, Chaichana KL, Rincon-Torroella J, Quinones-Hinojosa A (2015) The value of extent of resection of glioblastomas: clinical evidence and current approach. Curr Neurol Neurosci Rep 15(2):517. doi:10.1007/s11910-014-0517-x
2. Barbagallo GM, Palmucci S, Visocchi M, Paratore S, Attinà G, Sortino G, Albanese V, Certo F (2015) Portable intraoperative computed tomography scan in image-guided surgery for brain high-grade gliomas: analysis of technical feasibility and impact on extent of tumor resection. Neurosurgery. [Epub ahead of print].
3. Barone DG, Lawrie TA, Hart MG (2014) Image guided surgery for the resection of brain tumours. Cochrane Database Syst Rev (1):CD009685. doi:10.1002/14651858.CD009685.pub2.
4. Berman JI, Berger MS, Chung SW, Nagarajan SS, Henry RG (2007) Accuracy of diffusion tensor magnetic resonance imaging tractography assessed using intraoperative subcortical stimulation mapping and magnetic source imaging. J Neurosurg 107(3):488–494
5. Coburger J, Merkel A, Scherer M, Schwartz F, Gessler F, Roder C, Pala A, König R, Bullinger L, Nagel G, Jungk C, Bisdas S, Nabavi A, Ganslandt O, Seifert V, Tatagiba M, Senft C, Mehdorn M, Unterberg AW, Rössler K, Wirtz CR (2015) Low-Grade glioma surgery in intraoperative magnetic resonance imaging I: results of a multicenter retrospective assessment of the German study group for intraoperative magnetic resonance imaging. Neurosurgery 2016; 78: 775–786.
6. Coenen VA, Krings T, Mayfrank L, Polin RS, Reinges MH, Thron A, Gilsbach JM (2001) Three-dimensional visualization of the pyramidal tract in a neuronavigation system during brain tumor surgery: first experiences and technical note. Neurosurgery 49(1):86–92; discussion 92–3
7. Eljamel S, Petersen M, Valentine R, Buist R, Goodman C, Moseley H, Eljamel S (2013) Comparison of intraoperative fluorescence and MRI image guided neuronavigation in malignant brain tumours, a prospective controlled study. Photodiagnosis Photodyn Ther 10(4):356–361. doi:10.1016/j.pdpdt.2013.03.006, Epub 2013 Apr 17
8. Feigl GC, Ritz R, Moraes M, Klein J, Ramina K, Gharabaghi A, Krischek B, Danz S, Bornemann A, Liebsch M, Tatagiba MS (2010) Resection of malignant brain tumors in eloquent cortical areas: a new multimodal approach combining 5-aminolevulinic acid and intraoperative monitoring. J Neurosurg 113(2):352–357. doi:10.3171/2009.10.JNS09447
9. Goebel S, Nabavi A, Schubert S, Mehdorn HM (2010) Patient perception of combined awake brain tumor surgery and intraoperative 1.5-T magnetic resonance imaging: the Kiel experience. Neurosurgery 67(3):594–600. doi:10.1227/01.NEU.0000374870.46963.BB; discussion 600
10. Goebel S, Stark AM, Kaup L, von Harscher M, Mehdorn HM (2011) Distress in patients with newly diagnosed brain tumours. Psychooncology 20(6):623–630. doi:10.1002/pon.1958, Epub 2011 Mar 30
11. Gulati S, Berntsen EM, Solheim O, Kvistad KA, Håberg A, Selbekk T, Torp SH, Unsgaard G (2009) Surgical resection of high-grade gliomas in eloquent regions guided by blood oxygenation level dependent functional magnetic resonance imaging, diffusion tensor tractography, and intraoperative navigated 3D ultrasound. Minim Invasive Neurosurg 52(1):17–24. doi:10.1055/s-0028-1104566, Epub 2009 Feb 26
12. Gumprecht H, Lumenta CB (2003) Intraoperative imaging using a mobile computed tomography scanner. Minim Invasive Neurosurg 46(6):317–322
13. Hadjipanayis CG, Widhalm G, Stummer W (2015) What is the surgical benefit of utilizing 5-aminolevulinic acid for fluorescence-guided surgery of malignant gliomas? Neurosurgery 77(5):663–673. doi:10.1227/NEU.0000000000000929
14. Iaccarino C, Orlandi E, Ruggeri F, Nicoli D, Torricelli F, Maggi M, Cerasti D, Pisanello A, Pedrazzi G, Froio E, Crafa P, D'Abbiero N, Michiara M, Ghadirpour R, Servadei F (2015) Prognostic value of MGMT promoter status in non-resectable glioblastoma after adjuvant therapy. Clin Neurol Neurosurg 132:1–8. doi:10.1016/j.clineuro.2015.01.029. Epub 2015 Feb 7.
15. Knauth M, Wirtz CR, Tronnier VM, Aras N, Kunze S, Sartor K (1999) Intraoperative MR imaging increases the extent of tumor resection in patients with high-grade gliomas. Am J Neuroradiol 20(9):1642–1646
16. Krishnan R, Raabe A, Hattingen E, Szelényi A, Yahya H, Hermann E, Zimmermann M, Seifert V (2004) Functional magnetic resonance imaging-integrated neuronavigation: correlation between lesion-to-motor cortex distance and outcome. Neurosurgery 55(4):904–914; discusssion 914–5.
17. Kuhnt D, Bauer MH, Nimsky C (2012) Brain shift compensation and neurosurgical image fusion using intraoperative MRI: current status and future challenges. Crit Rev Biomed Eng 40(3):175–185
18. Kurimoto M, Hayashi N, Kamiyama H, Nagai S, Shibata T, Asahi T, Matsumura N, Hirashima Y, Endo S (2004) Impact of neuronavigation and image-guided extensive resection for adult patients with supratentorial malignant astrocytomas: a single-institution retrospective study. Minim Invasive Neurosurg 47(5):278–283
19. Li Y, Rey-Dios R, Roberts DW, Valdés PA, Cohen-Gadol AA (2014) Intraoperative fluorescence-guided resection of high-grade gliomas: a comparison of the present techniques and evolution of future strategies. World Neurosurg 82(1–2):175–185. doi:10.1016/j.wneu.2013.06.014. Epub 2013 Jul 9.
20. Lindner D, Trantakis C, Renner C, Arnold S, Schmitgen A, Schneider J, Meixensberger J (2006) Application of intraoperative 3D ultrasound during navigated tumor resection. Minim Invasive Neurosurg 49(4):197–202
21. Litofsky NS, Bauer AM, Kasper RS, Sullivan CM, Dabbous OH, Glioma Outcomes Project Investigators. (2006) Image-guided

resection of high-grade glioma: patient selection factors and outcome. Neurosurg Focus 15;20(4):E16

22. Mahvash M, Maslehaty H, Jansen O, Mehdorn HM, Petridis AK (2014) Functional magnetic resonance imaging of motor and language for preoperative planning of neurosurgical procedures adjacent to functional areas. Clin Neurol Neurosurg 123:72–77. doi:10.1016/j.clineuro.2014.05.011, Epub 2014 May 29
23. Mehdorn HM, Schrader B, Nabavi A, Hempelmann R (2000) Neuronavigation in the region of the skull base. Laryngorhinootologie 79(7):404–411. German.
24. Mehdorn HM, Schwartz F, Dawirs S, Hedderich J, Dörner L, Nabavi A (2011) High-field iMRI in glioblastoma surgery: improvement of resection radicality and survival for the patient? Acta Neurochir Suppl 109:103–106. doi:10.1007/978-3-211-99651-5_16
25. Nabavi A, Goebel S, Doerner L, Warneke N, Ulmer S, Mehdorn M (2009) Awake craniotomy and intraoperative magnetic resonance imaging: patient selection, preparation, and technique. Top Magn Reson Imaging 19(4):191–196. doi:10 1097/RMR.0b013e3181963b46
26. Nimsky C, Ganslandt O, Cerny S, Hastreiter P, Greiner G, Fahlbusch R (2000) Quantification of, visualization of, and compensation for brain shift using intraoperative magnetic resonance imaging. Neurosurgery 47(5):1070–1079; discussion 1079–80
27. Olubiyi OI, Ozdemir A, Incekara F, Tie Y, Dolati P, Hsu L, Santagata S, Chen Z, Rigolo L, Golby AJ (2015) Intraoperative magnetic resonance imaging in intracranial glioma resection: a single-center, retrospective blinded volumetric study. World Neurosurg 84(2):528–536. doi:10.1016/j.wneu.2015.04.044, Epub 2015 May 1
28. Parney IF, Goerss SJ, McGee K, Huston J 3rd, Perkins WJ, Meyer FB (2010) Awake craniotomy, electrophysiologic mapping, and tumor resection with high-field intraoperative MRI. World Neurosurg 73(5):547–551. doi:10.1016/j.wneu.2010.02.003
29. Petridis AK, Anokhin M, Vavruska J, Mahvash M, Scholz M (2015) The value of intraoperative sonography in low grade glioma surgery. Clin Neurol Neurosurg 131:64–68. doi:10 1016/j.clineuro.2015.02.004, Epub 2015 Feb 11
30. Pinsker MO, Nabavi A, Mehdorn HM (2007) Neuronavigation and resection of lesions located in eloquent brain areas under local anesthesia and neuropsychological-neurophysiological monitoring. Minim Invasive Neurosurg 50(5):281–284
31. Roder C, Bisdas S Ebner FH, Honegger J, Naegele T, Ernemann U, Tatagiba M (2014) Maximizing the extent of resection and survival benefit of patients in glioblastoma surgery: high-field iMRI versus conventional and 5-ALA-assisted surgery. Eur J Surg Oncol 40(3):297–304. doi: 10.1016/j.ejso.2013.11 022, Epub 2013 Dec 19
32. Sanai N, Berger MS (2010) Intraoperative stimulation techniques for functional pathway preservation and glioma resection. Neurosurg Focus 28(2), E1. doi:10.3171/2009.12.FOCUS09266
33. Schucht P, Murek M, Jilch A, Seidel K, Hewer E, Wiest R, Raabe A, Beck J (2013) Early re-do surgery for glioblastoma is a feasible and safe strategy to achieve complete resection of enhancing tumor. PLoS One 13;8(11):e79846. doi:10.1371/journal.pone.0079846. eCollection 2013.
34. Willems PW, Taphoorn MJ, Burger H, Berkelbach van der Sprenkel JW, Tulleken CA (2006) Effectiveness of neuronavigation in resecting solitary intracerebral contrast-enhancing tumors: a randomized controlled trial. J Neurosurg 104(3):360–368
35. Zhu FP, Wu JS, Song YY, Yao CJ, Zhuang DX, Xu G, Tang WJ, Qin ZY, Mao Y, Zhou LF (2012) Clinical application of motor pathway mapping using diffusion tensor imaging tractography and intraoperative direct subcortical stimulation in cerebral glioma surgery: a prospective cohort study. Neurosurgery 71(6):1170–1183. doi:10.1227/NEU.0b013e318271bc61; discussion 1183–4

Radioguided Occult Lesion Localization in Deep Schwannomas of the Peripheral Nerves: Results of a Preliminary Case Series

Giorgio Stevanato, Giovanni Monteleone, Donatella Sgubin, Michele Sicolo, Giulia Pastorello, Luca Vazzana, and Franco Guida

Abstract

Background

The detection of small deep schwannomas of the peripheral nerves has been increasing since the the use of precise neuroimaging techniques has become more widespread; however, although nonpalpable lesions can be well defined by images, it is often difficult to identify them during the surgical procedure. The authors report seven cases of nonpalpable small deep schwannomas surgically treated after their identification using the radioguided occult lesion localization (ROLL) technique.

Methods

Seven men, whose ages ranged from 34 to 70 years (mean 52 years), presented with symptomatic nonpalpable peripheral nerve lesions; two cases involved the sciatic nerve, two the femoral nerve, two the radial nerve, and one the tibial nerve. Before the operation, all the patients were studied by ultrasonography and magnetic resonance imaging (MRI); 1 h before the surgery 3–5 MBq of ^{99m}Tc labeled with human albumin macroaggregates was injected into the lesion. A gamma detection probe permitted the preoperative and intraoperative detection of the nonpalpable schwannomas.

Conclusions

The ROLL technique provides good support for identifying small lesions of the peripheral nerves both preoperatively and intraoperatively. This technique permits the use of minimally invasive approaches performed with local anesthesia, with good cosmetic results and acceptance by the patients.

Keywords Occult lesion • Peripheral nerve schwannoma • ROLL

Introduction

Schwannomas, also called neurilemmomas or peripheral neurinomas, are benign tumors arising from the perineurium of peripheral nerves [4]. The term *neurinoma* was coined in 1910 by Verocay to indicate a tumor that originated from Schwann cells and that was histologically different from a neurofibroma [9]. Schwannomas constitute 5 % of all benign soft-tissue tumors; men and women are equally affected, most frequently in their third decade of life [4, 6]. The most frequent localization is in the lower extremity, followed by upper extremity, dorsal roots, and retroperitoneum.

To determine the diagnosis of peripheral nerve neurinomas, magnetic resonance imaging (MRI) can clearly define the lesion's characteristics. The mass is usually isointense relative to skeletal muscles on T1-weighted images, and eccentric relative to the involved nerve that is displaced to the periphery; a capsule can be visualized; thus, the margins are well defined. On T2-weighted images the lesion is hyperintense. Schwannomas usually show enhancement after gadolinium injection [1, 3]. Clinically, these lesions can cause pain and motor or sensory deficits in the territory of the involved nerve [9]. On examination of the patient, a peripheral neurinoma can be visible and palpated as a round, well-defined mass that can be mobilized from side to side but not along the axis of the nerve [5]. Tinel's sign is very often evocable over the mass [9] Microsurgical excision with an interfascicular dissection technique is the most effective treatment, and complete recovery can be obtained; biopsies are no longer indicated [9].

G. Stevanato (✉) • D. Sgubin • G. Pastorello
L. Vazzana • F. Guida
Neurosurgery Unit, Dell'Angelo Hospital,
Via Paccagnella 11, Mestre-Venezia 30172, Italy
e-mail: giorgio.stevanato@ulss12.ve.it

G. Monteleone
Department of Biomedicine and Preventive Medicine, University "Tor Vergata", Rome, Italy

M. Sicolo
Nuclear Medicine Unit, Dell'Angelo Hospital,
Mestre-Venezia, Italy

M. Visocchi et al. (eds.), *Trends in Reconstructive Neurosurgery*, Acta Neurochirurgica Supplement 124,
DOI 10.1007/978-3-319-39546-3_46, © Springer International Publishing Switzerland 2017

The surgical strategy can be optimized by the use of radioguided occult lesion localization (ROLL). This technique was developed in 1997 by Paganelli et al. [8] to identify sentinel nodes in breast cancer and small nonpalpable lesions. Starting from these considerations, it was possible to extend the method to detect small nonpalpable lesions of the peripheral nerves.

Methods

Seven patients with deep nonpalpable suspicious peripheral nerve schwannomas previously studied by ultrasound and MRI underwent surgical excision after the preoperative localization of the lesion by the ROLL technique (Table 1). The radiotracer used was ^{99m}Tc labeled with human albumin macroaggregates (^{99m}Tc MAA), with a diameter of 10–150 μm (TechneScan LyoMAA; Mallinckrodt Medical, Petten, The Netherlands). One hour before the surgery 3–5 MBq of ^{99m}Tc MAA in 0.2 ml of saline solution was injected into the lesion under ultrasound guidance (Fig. 1). The site was marked with a dermographic pen. Front and lateral images of the lesion were acquired with a gamma detection system Neoprobe (Dublin, Ohio, U.S.A.) and the patients were then brought to the operating room.

The patients underwent preoperative detection of the lesion, using a sterile gamma probe to identify the gamma rays emanating from the radioactive tracer injected into the neurinoma (Fig. 2).

The probe was connected to an instrument that analyzes the signals and transduces them into a digital scale together with acoustic signal. The intensity of the signal is directly proportional to the radioactivity detected and represent the relative distance of the probe to the lesion.

After the identification of the "hot spot", the neurosurgeon performed a small tailored incision. During surgery the lesion's positions were checked to guide their isolation. Using a microsurgical technique, the neurosurgeon dissected unaffected nerve fascicles from the tumor capsule and the lesions were then resected (Fig. 3).

Results

The diagnosis of peripheral Schwannoma was confirmed in all seven patients.

Complete removal of the lesion was performed in all cases and all patients presented with resolution of the pain. In one patient a preoperative motor deficit had been described (muscle strength 4/5 on the Medical Research Council [MRC] scale) and improvement was reported at 18-months follow up; of five patients who presented with preoperative sensory deficits, complete recovery had occurred in four patients at 10–18 months. All the treated patients were pain-free after the surgery (Table 1).

In six of the seven patients the surgical procedures were performed with local anesthesia.

Table 1 Characteristics, surgical treatment, and outcome of seven patients with small nonpalpable schwannomas of the peripheral nerves

n°	Sex/age (years)	Site (nerve)	Size (cm)	Preoperative status	Anesthesia	Histology	Postoperative status	Follow-up (months)
1	M/56	Sciatic	4×3	Pain	General	Schwannoma	Pain-free	Full recovery (12)
2	M/43	Tibial	3.5×3	Pain, motor and sensory deficits	Local	Schwannoma	Pain-free, improved motor and sensory deficits	Full recovery (18)
3	M/34	Femoral	2.3×1.2	Pain	Local	Schwannoma	Improved	Full recovery (18)
4	M/63	Radial	0.7×1.6	Pain, sensory deficit	Local	Schwannoma	Improved pain and sensory deficit	Full recovery (10)
5	M/70	Radial	1.4×0.9	Pain, sensory deficit	Local	Schwannoma	Pain-free improved sensory deficit	Slight sensory deficit (18)
6	M/56	Sciatic	1.4×2.6	Pain, sensory deficit	Local	Schwannoma	Pain-free improved sensory deficit	Full recovery (12)
7	M/42	Femoral	1.8×1.3	Slight pain, sensory deficit	Local	Schwannoma	Pain-free stable sensory deficit	Full recovery (18)

M male

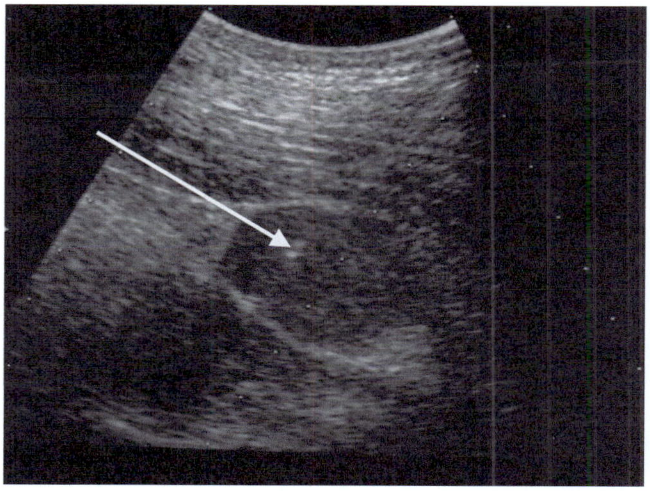

Fig. 1 Before surgery the radiotracer was injected into the lesion (*arrow*) under ultrasound guidance

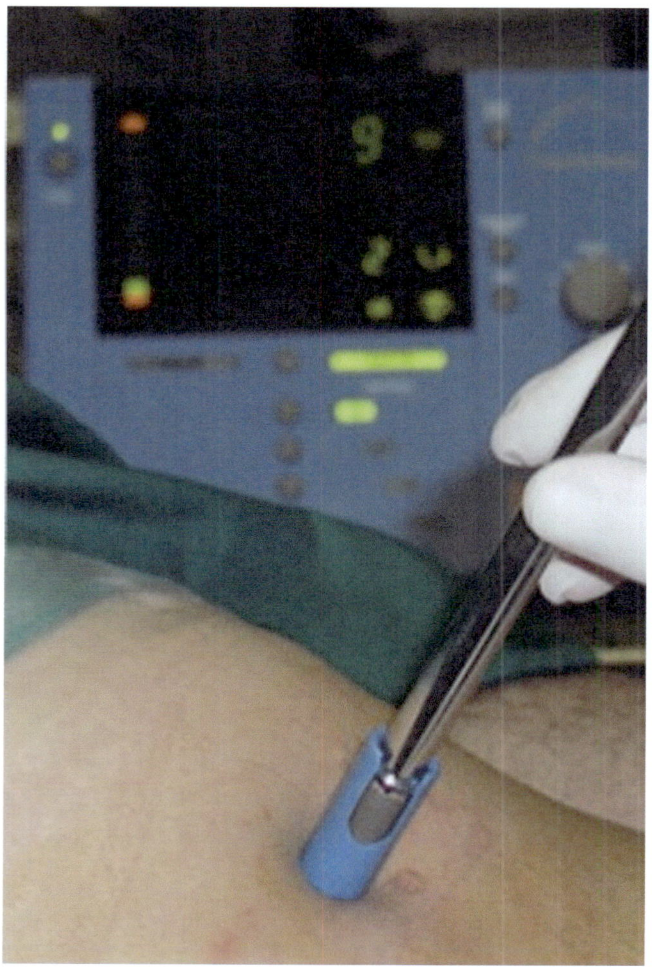

Fig. 2 Detection of the radiotracer, performed with a sterile probe. Precise localization of the lesion (case 1 in Table 1) was possible with a minimally invasive approach

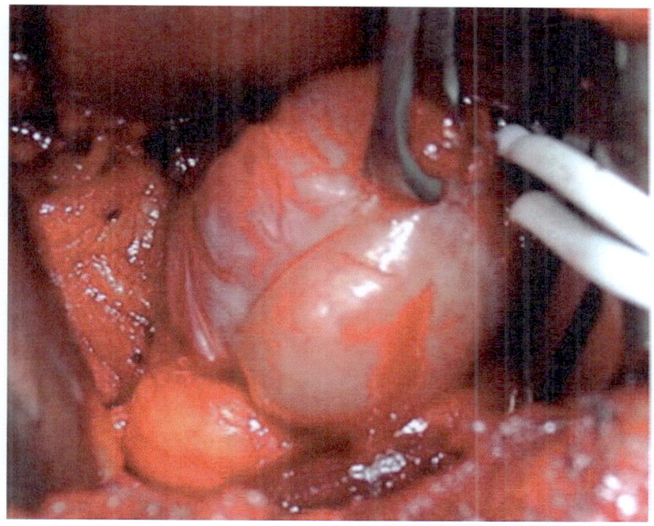

Fig. 3 During surgery the lesion position was checked to guide its isolation

Discussion

Small and deep lesions of the peripheral nerve can be identified in the early stages using new neuroimaging techniques. However, despite the excellent visualization of these lesions so provided, it is often difficult to find them during surgery, and invasive interventions can be necessary for their removal.

In this series, we used the ROLL technique – a well known method to localize sentinel nodes and small lesions in breast cancer – to detect small and deep peripheral nerve schwannomas preoperatively and during surgery.

The use of the ROLL technique to detect occult breast tumors was described by Paganelli et al. in 1997 [8], but this technique has not been reported previously for the detection of peripheral nerve schwannomas.

In a previous study, it was shown that there was no movement of the radiotracer from the lesion; thus, the radiotracer's captation can be considered as a guide for the surgeon to identify occult lesions during operation [2]. Starting from these considerations, in our institution the ROLL technique was used for the first time for preoperatively and intraoperatively identifying deep and/or small peripheral nerve lesions, in seven patients.

One hour before the surgery, a radiotracer (^{99m}Tc MAA) was injected into the lesion under ultrasound guidance; image acquisition with a gamma camera was required and then the surgical excision was made. Using a sterile probe that detected the ^{99m}Tc MAA radiation, the lesion was localized, and consequently it was possible to plan a small incision for a minimally invasive surgical approach (Fig. 2). The surgical procedures were performed with local anesthesia in six of the seven patients. Thus, we conclude that the gamma detection system permits the easy identification of lesions in the superficial layers and also those in deep layers during surgery.

We have to report that the preoperative ^{99m}Tc MAA intralesional injection done under ultrasound guidance can be painful (four cases). Due to the short half-life of ^{99m}Tc, the procedure is safe for the patient, for the surgeon, and for the operating room personnel. The complete removal of the lesion injected with the radiotracer also removes the radioactivity from the patient's body.

These results of a preliminary case series suggest that the ROLL technique is safe and could be used for the removal of multiple deep peripheral nerve schwannomas during the same surgical procedure. We note that the surgeon can manage the sterile probe by himself/herself in order to localize the lesion during open surgery.

This ROLL technique requires interdisciplinary cooperation, but it provides good support for preoperatively and intraoperatively identifying small lesions of the peripheral nerves. It permits a minimally invasive approach, performed with local anesthesia, with a shorter operative time, and minor costs, together with good cosmetic results and better patient acceptance.

Conflict of Interest Statement The authors declare that they have no conflicts of interest.

References

1. Beaman FD, Kransdorf MF, Menke DM (2004) Schwannoma: radiologic-pathologic correlation. RadioGraphics 24:1481–1497
2. De Cicco C, Cremonesi M, Luini A, Bartolomei M, Grana C, Prisco G, Galimberti V, Calza P, Viale G, Veronesi U, Paganelli G (1998) Lymphoscintigraphy and radioguided biopsy of the sentinel axillary node in breast cancer. J Nucl Med 39:2080–2084
3. Hems TE, Burge PD, Wilson DJ (1999) The role of magnetic resonance imaging in the management of peripheral nerve tumours. J Hand Surg [Br] 22:57–60
4. Lee JA, Boles CA (2004) Peripheral Schwannoma lacking enhancement on MRI. Am J Roentgenol 182:534–535
5. Lewis D, Hart D (1930) Tumors of peripheral nerves. Ann Surg 92:961–983
6. Murphey MD, Smith WS, Smith SE, Kransdorf MJ, Temple HT (1999) From the archives of the AFIP Imaging of musculoskeletal neurogenic tumors: radiologic-pathologic correlation. Radiodraphics 19:1253–1280
7. Oberle J, Kahamba J, Richter HP (1997) Peripheral nerve Schwannomas – an analysis of 16 patients. Acta Neurochir 139:949–953
8. Paganelli G, De Cicco C, Luini A (1997) Radioguided surgery in non-palpable breast lesions. Eur J Nucl Med 24 (Suppl):893
9. Paganelli G, Luini A, Veronesi U (2002) Radioguided occult lesion localization (ROLL) in breast cancer: maximizing efficacy, minimizing mutilation. Ann Oncol 13:1839–1840

EMG-Guided Percutaneous Placement of Cement-Augmented Pedicle Screws for Osteoporotic Thoracolumbar Burst Fractures

Domenico Gerardo Iacopino, Francesco Certo, Francesca Graziano, Luigi Basile, Carlo Gulì, Massimiliano Visocchi, Alfredo Conti, and Rosario Maugeri

Abstract

Background

Percutaneous techniques have increasingly gained popularity in recent years. The application of technological innovation, including neuromonitoring techniques, has the potential to increase the safety and efficacy of these procedures.

Methods

Thirty patients suffering from osteoporotic dorsolumbar burst fracture were prospectively enrolled in this study. The patients underwent percutaneous fenestrated pedicle screw fixation augmented with polymethylmethacrylate (PMMA) injection. A novel surgeon-dedicated neuromonitoring device was used in order to increase the safety and the accuracy of the screw insertion. A second group of 30 patients who did not undergo neuromonitoring during percutaneous pedicle screw placement, matched for demographic characteristics, constituted the control group.

Findings

A total of 296 screws were inserted. All treated patients had a good outcome, documented by an improvement in visual analogue scale (VAS) scores. Excellent trajectories were achieved in all patients. Cobb's angle and anterior vertebral height were satisfactorily restored in all study group patients. Three misplaced screws in three patients and a case of PMMA leakage without neurological deficits were observed in the control group, whereas no complication was recorded in the study group ($p=0.03$).

Conclusions

Neuromonitoring in cement-augmented percutaneous pedicle screw placement appears to improve surgeon confidence during surgery, reducing the risk of screw misplacement or cement leakage.

Keywords Neuromonitoring • Burst fracture • Cement augmentation • Percutaneous pedicle screw fixation • Minimally invasive spinal surgery

Introduction

Percutaneous pedicle screw placement has significantly changed relevant aspects of spinal surgery [16, 17]. The advantages of this technique are widely recognized, and include shorter surgical times, reduced blood loss, reduced injury to the paravertebral muscular structures, and quicker postoperative recovery for the patients [13, 22, 23]. Despite the success of this minimally invasive technique, several authors have reported on the limits of these methods [21]. The main difficulty for the surgeon dealing with percutaneous procedures is the lack of anatomical landmarks and the consequent risk of screw misplacement. An increased use of intraoperative fluoroscopic control measures has been often performed for overcoming this difficulty, but with a greater than before exposure to radiations for the patient and for the surgeons. However, recently, the implementation of percutaneous techniques with new technologies, such as neuronavigation and three-dimensional (3D) fluoroscopy, has greatly improved the feasibility of these procedures, reducing the percentage of screw misplacements [5, 15, 20]. Intraoperative neurophysiological monitoring, applied to the percutaneous placement of the screws, is becoming a further technological support to improve the accuracy of screw positioning. The possibility of monitoring electromyography (EMG) responses

D.G. Iacopino, MD, PhD (✉) • F. Certo, MD • F. Graziano, MD
L. Basile, MD • C. Gulì, MD • R. Maugeri, MD, PhD
Department of Experimental Biomedicine and Clinical Neurosciences, School of Medicine, Neurosurgical Clinic, University of Palermo, Italy,
Via del Vespro 129, Palermo 90100, Italy
e-mail: gerardo.iacopino@gmail.com

M. Visocchi, MD, PhD
Institute of Neurosurgery, Medical School, Catholic University of Rome, Rome, Italy

A. Conti, MD, PhD
Department of Neurosurgery, University of Messina, Italy, Messina, Italy

M. Visocchi et al. (eds.), *Trends in Reconstructive Neurosurgery*, Acta Neurochirurgica Supplement 124,
DOI 10.1007/978-3-319-39546-3_47, © Springer International Publishing Switzerland 2017

during the insertion of tools and screws into vertebral pedicles and bodies provides updated, real-time information about changes in neuroelectrical activity brought about by the procedure [3, 8, 9]. Recently, new software and neuromonitoring devices have been developed to improve the applicability of EMG during spine surgery. Thanks to these innovations, an immediate and intuitive interpretation of the EMG trace is available to the surgeon, even without the presence of an experienced neurophysiologist in the operating theater. In the context of percutaneous techniques, another recent innovation is the use of cannulated screws, through which the cement can be introduced into the vertebral body [18]. This device appears to be particularly useful in patients suffering from osteoporosis, in order to increase the tightness of the screws and to prevent pullout [2]. However, despite improvements in the design of the instruments and surgical techniques and the advent of image-guided techniques, a cortical pedicle perforation may occur during the percutaneous placement of the screws, increasing the risks of impingement on adjacent neural or extravertebral structures, and, more likely, increasing the risk of polymethylmethacrylate (PMMA) leakage during the following augmentation. The aim of this study was to analyze the results obtained in patients who underwent cement-augmented pedicle screw placement under neurophysiological monitoring performed with a novel surgeon-controlled device (NIM-Eclipse® Nerve Monitoring System; Medtronic®, Medtronic, Memphis, TN, USA).

Material and Methods

Study Design

We prospectively enrolled 30 patients (12 male, 18 female; median age 68.9 years), chronically suffering from osteoporosis, admitted to the neurosurgical clinic of the University of Palermo for the surgical treatment of traumatic thoracolumbar (Th12-L3) burst fracture causing low back pain. Inclusion criteria encompassed: absence of neurological deficits or radiculopathy; bone fragment occupying the spinal canal to less than 30 % of the anterior-posterior diameter; and no indication for surgical decompression. Patients underwent thoracolumbar percutaneous fixation under electrophysiological neuromonitoring. Demographic characteristics are summarized in Table 1. The visual analogue scale (VAS) score was recorded before and 30 days after the surgery. Preoperative radiological imaging consisted of X-rays, computed tomography (CT) with multiplanar reconstruction, and magnetic resonance imaging (MRI) of the lumbar spine. CT scans were also performed postoperatively. Fractures were classified according to the Magerl classification [13]. Kyphosis angle and anterior vertebral height were evaluated before and after (30 days) the operation on CT scan (Table 2). Data for a control group of 30 patients with osteoporotic fractures who underwent surgical fixation without the aid of neuromonitoring, matched for basic demographic characteristics (Table 1), was obtained from the institutional registry and retrospectively analyzed.

Surgical Procedure

The surgical steps of the neuromonitoring-assisted and cement-augmented pedicle screw fixation are detailed below.

All patients underwent a single-level percutaneous cannulated PMMA cement-augmented pedicle screw fixation, using the Longitude® System (Medtronic®) (Fig. 1). After the induction of general anesthesia and before patient positioning on the surgical table, needles for neuromonitoring were placed according to the electrode placement instructions that were provided with the NIM-Eclipse System. Based on the vertebral levels to be treated, the electrical activity of the rectus abdominis, vastus lateralis, quadriceps femoris, medial gastrocnemius, and extensor hallucis longus (both sides) was monitored. Two transcranial needles were also inserted subcutaneously in the scalp, for motor evoked potential (MEP) recording. Each electrode was connected to the workstation, and baseline electromyography (EMG) was recorded. We calculated that the entire electrode positioning and connecting procedures lasted about 10 min. The patient was placed in the prone position, maintaining the physiological alignment of the spinal curves, on a radiolucent table. In this step it is crucial both to try to achieve an initial reduction of the kyphosis and to ensure that the subcutaneous needles are not displaced by the positioning maneuvers. After patient positioning, EMG recording was restarted. The comparison with the first EMG recording allowed us to detect all variations in electrical activity arising from the position on the table. Firstly, the vertebral pedicles where the percutaneous screws had to be implanted were detected with anteroposterior (AP) fluoroscopic scanning. This allowed us to mark, on the skin surface, the entry point of the percutaneous device, considering that it is preferable to use an inlet positioned laterally to the outer margins of the pedicles, to promote a convergent trajectory of the screws. A 12-mm linear skin incision was performed and the dorsolumbar fascia was sectioned. A Jamshidi needle, adequately modified, was then connected to the stimulating system and introduced through each skin incision for pedicle hole preparation (Fig. 1a). We used the needle as a pedicle probe, tracing the best and safest direction for the screw. Usually, when possible, we preferred a more lateral pedicular entry point to allow a convergent trajectory, in order to place the tips of the screws in close proximity to the median line. This seemed to improve the resistance of the implanted systems and to prevent

Table 1 Summary of demographic and clinical features

	Group A	Group B	P
Age, years; mean(range)	68.9 (55–77)	62.8 (61–69)	n.s.
Sex	12 M, 18 F	14 M, 16 F	n.s.
Vertebral level	6 pts Th 12 16 pts L1 4 pts L2 4 pts L3	10 pts Th 12 10 pts L1 8 pts L2 2 pts L3	n.s.
Fracture type	8 pts A 3.2 22 pts A 3.3	12 pts A 3.2 18 pts A 3.3	n.s.

Group A study group, *Group B* control group, *n.s.* not significant, *M* male, *F* female, *pts* patients

Table 2 Preoperative, postoperative, and follow-up clinical and radiological evaluations

Variable	Group A value – mean (range)	Group B value – mean (range)
Preoperative VAS	7.87 (7–9)	7.33 (6–9)
Postoperative VAS	4.27 (3–6)	4.13 (2–6)
6-Month VAS	1.73 (1–3)	2.00 (1–3)
12-Month VAS	1.40 (1–2)	1.60 (1–3)
Preoperative Cobb angle kyphosis	20.66° (14–26)	19.20° (13–28)
Postoperative Cobb angle kyphosis	8.93° (6–14)	13.13° (4–12)
6-Month kyphosis angle	7.53° (6–11)	11.07° (5–11)
12-Month kyphosis angle	7.53° (6–11)	11.07° (5–11)
Preoperative anterior vertebral height	17.9 mm (15–23)	18.00 mm (14–22)
Postoperative anterior vertebral height	24.00 mm (20–27)	24.67 mm (22–28)
6-Month anterior vertebral height	23.87 mm (19–27)	24.53 mm (22–28)
12-Month anterior vertebral height	23.93 mm (19–27)	24.46 mm (22–28)

VAS visual analogue scale

pullout. During this phase, monitoring is useful to detect any changes in EMG activity. If EMG abnormalities are recorded, the trajectory must be modified. Subsequently a Kirschner wire (K-wire) was introduced through the needle. The K-wire reached the vertebral body and was maintained in place in order to guide the introduction of progressive tubular retractors used to facilitate the introduction of the screw. Finally, under fluoroscopic guidance, the cannulated screw was placed and the K-wire removed. Once the screw was positioned, a special ball-tip probe was inserted through the cannula, for a final test. The probe is provided with two buttons to access the functions of the workstation. The device also allows the performance of so-called screw integrity test stimulation, which combines automatic electrical stimulation and EMG response analysis to speed-test pedicle screw placement. The surgeon can control this test directly from the operative field, contacting and stimulating the screw with the probe (Fig. 1b). The test begins by stimulating at the start intensity (usually 3 mA). For each increase in stimulus intensity, the EMG response is measured and compared with default response criteria, and the response either passes or fails based on the pass/fail criteria. A stimulation threshold greater than 15 mA indicates adequate screw positioning. A good response on this test predicts the absence of electric current leakage from the bone structures and this means that the pedicle is intact. A summary of the responses of all screws tested was recorded on the workstation, then printed and stored in the clinical folder. When all the transpedicular screws had been implanted, PMMA cement was injected through the cannulated screws (Fig. 1c). We introduced about 1.5 ml of PMMA cement for each screw under fluoroscopic guidance. During cement introduction, the EMG trace was carefully monitored to promptly detect any variation arising from this procedure. The last surgical step consisted of positioning and fixing the rods. The Longitude® System provides a free-hand inserter and reduction screw extenders that are designed to allow a stabilizing rod to be passed through a small incision over the higher treated level (Fig. 1d). After the last fluoroscopic control, the rods were locked and fixed to the screws after an adequate compression, in order to guarantee reduction of the fracture and the correct alignment of the spine (Fig. 2).

Results

No significant differences in age, sex, fracture site, and cause of injury were observed between the two groups (Table 1). No surgery-related complications were observed either intraoperatively or postoperatively. All operations in Group A (study group) were performed using the aforementioned technique. The average operative time ranged from 90 to 190 min. All

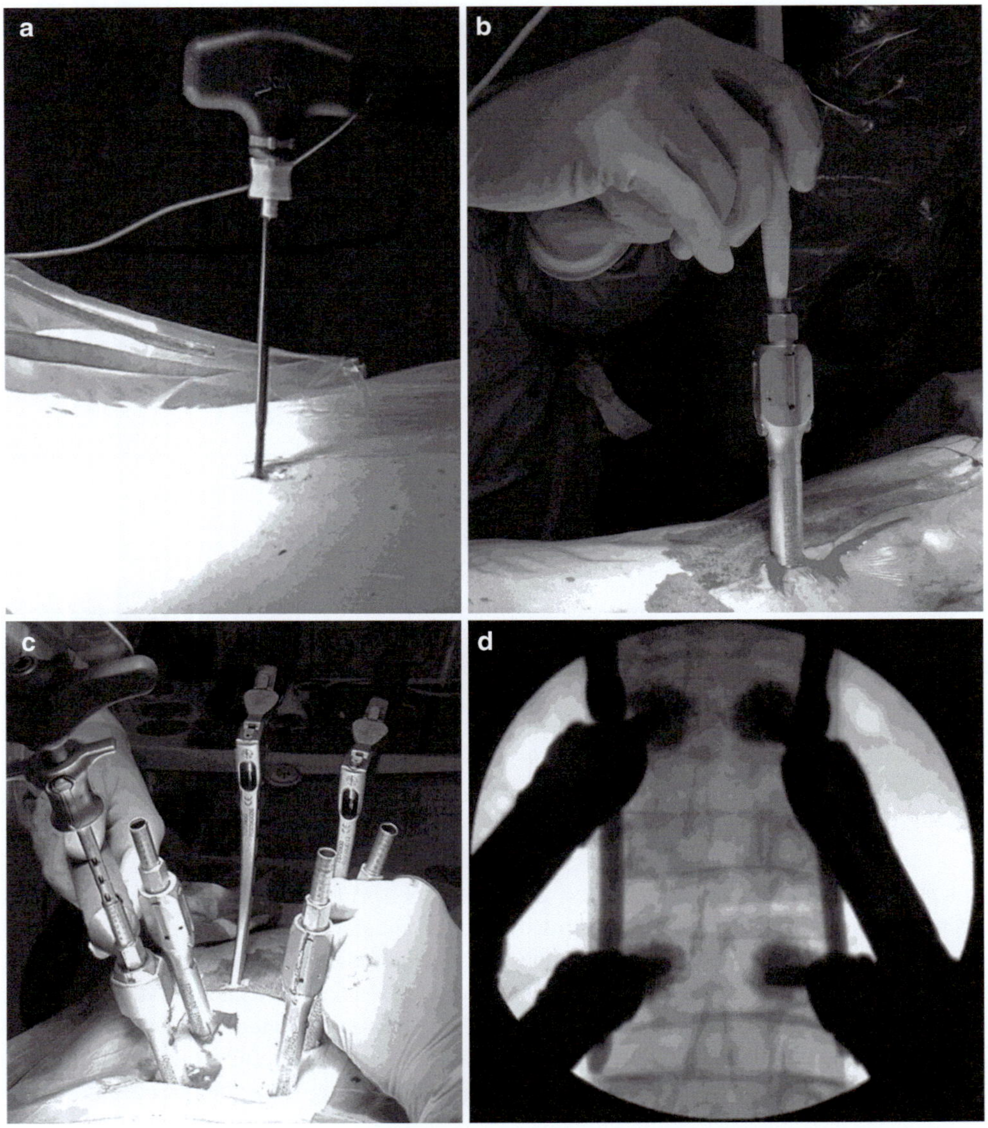

Fig. 1 Surgical procedure: the combined use of surgeon-detected neuromonitoring and polymethylmethacrylate (PMMA) augmentation of the screws is shown. Firstly a modified Jamshidi needle connected with the workstation is inserted (**a**); after the positioning of each screw, the integrity of the pedicle is tested by introducing the probe into the cannulated screw (**b**). Finally, the PMMA augmentation is performed under fluoroscopic guidance (**c**) and the rods are inserted percutaneously with a free-hand inserter (**d**)

patients were able to walk on the first day after surgery and were discharged 3 days after the surgery. Improvements in the VAS score were significant for all patients. Preoperatively, VAS scores were 7.87 and 7.33 in Group A and Group B, respectively, and these scores were significantly decreased in both groups during the follow-up period ($p<0.05$) (Table 2). In the group of patients who underwent neuromonitoring, a total of 296 screws were inserted; here we observed stimulation thresholds ranging between 28 and 30 mA at the last control. In one case we observed a threshold of 11 mA, but after slightly changing of the orientation of the screw we obtained a threshold of 29 mA. Based on the method proposed by Zdichavsky et al. [24, 25], an objective evaluation of screw positioning was performed by an independent radiologist blinded to the clinical outcome. All screws used in the study group were well positioned (1a or 1b according to Zdichavsky et al.), whereas three screws in three different patients in the control group were misplaced (2b and 3b according to Zdichavsky et al.). In one case the misplaced screw was repositioned because the patient complained of sciatic pain that was related to neural conflict (Table 3). CT scans also revealed satisfactory augmentation of osteoporotic vertebral bodies, without leakage, in all patients included in the study group (Fig. 3). In contrast, one case of cement leakage was observed in the control group. The rate of complications including screw misplacement and PMMA leakage was significantly higher ($p=0.03$) in the group of patients who did not receive intraoperative neuromonitoring (controls). Reduction of the

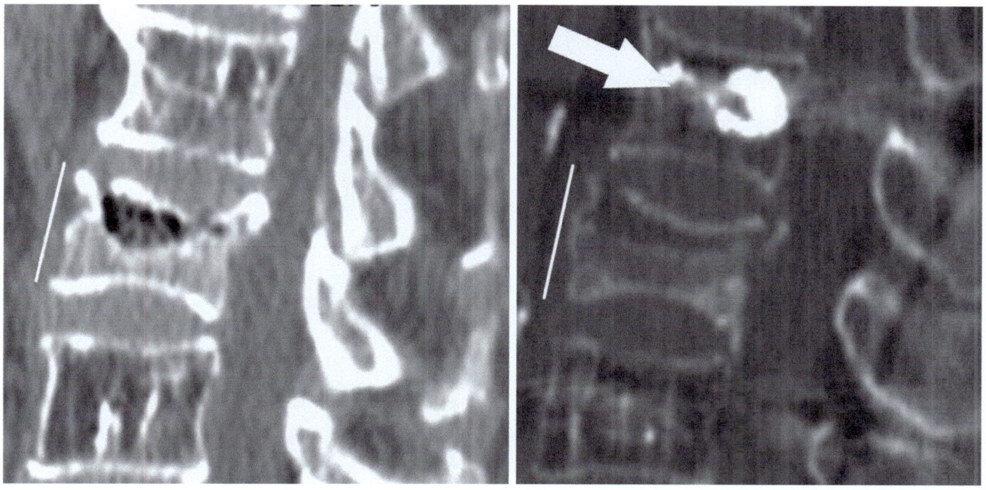

Fig. 2 Case 2, L2 fracture: preoperative (*left*) and postoperative (*right*) sagittal reconstructed computed tomography (CT) scans with comparison between preoperative and postoperative anterior vertebral height. PMMA augmentation is visible on L1 (*arrow*)

Table 3 Screw positioning: radiological evaluation according to Zdichavsky

	Group A	Group B
1a	130	120
2a	166	151
3a	–	–
1b	–	–
2b	–	2
3b	–	1
Total number of screws inserted	296	274

VAS visual analogue scale

kyphosis angle and restoration of the normal vertebral body height were measured and documented as reported in Table 2. Clinical and radiological follow-up was performed at 6 and 12 months, respectively. All patients underwent clinical evaluation and AP and lateral X-ray scans. The same parameters as those analyzed immediately after surgery were evaluated at follow-up, as reported in Table 2. We did not observe any adjacent segment disease or displacement of the implanted devices and cement. Satisfactory vertebral alignment was documented in all patients in the study group.

Discussion

Our results suggest that neuromonitoring guidance may increase safety and surgeon confidence in the positioning of PMMA cement-augmented screws in osteoporotic patients affected by thoracolumbar burst fracture. The use of neuromonitoring reduced surgery-related complications, including PMMA leakage and screw misplacement. All patients in the study group had immediate correction of kyphosis and pain reduction. In the past decade, many authors have debated the benefits of percutaneous techniques over traditional open fixation. Several studies have pointed out the advantages of percutaneous techniques, including less muscle damage, shorter hospital stay, reduced postoperative pain, faster recovery of ability to work, and overall better functional recovery [14, 16, 17, 19]. The value of percutaneous techniques has also been demonstrated by biomechanical studies demonstrating the high mechanical performance of percutaneous devices as compared with “open” systems [11]. The most obvious limitation of percutaneous methods concerns the absence of anatomical landmarks visible to the surgeon. In fact, these procedures require a specific learning curve, even when performed by experienced spine surgeons. In different series the actual percentage of screw misplacement during percutaneous procedures ranges from 11% to 15% [19]. In 2011, Kim et al. retrospectively examined the accuracy of the insertion of 488 percutaneous screws in 110 patients, and reported a 12.5 % (61 screws) rate of cortical encroachment, and frank penetration in 54 screws (11.1 %) [10]. Two patients (0.4 %) with medial penetration underwent revision surgery. Injection of a PMMA bone cement through cannulated screws following the

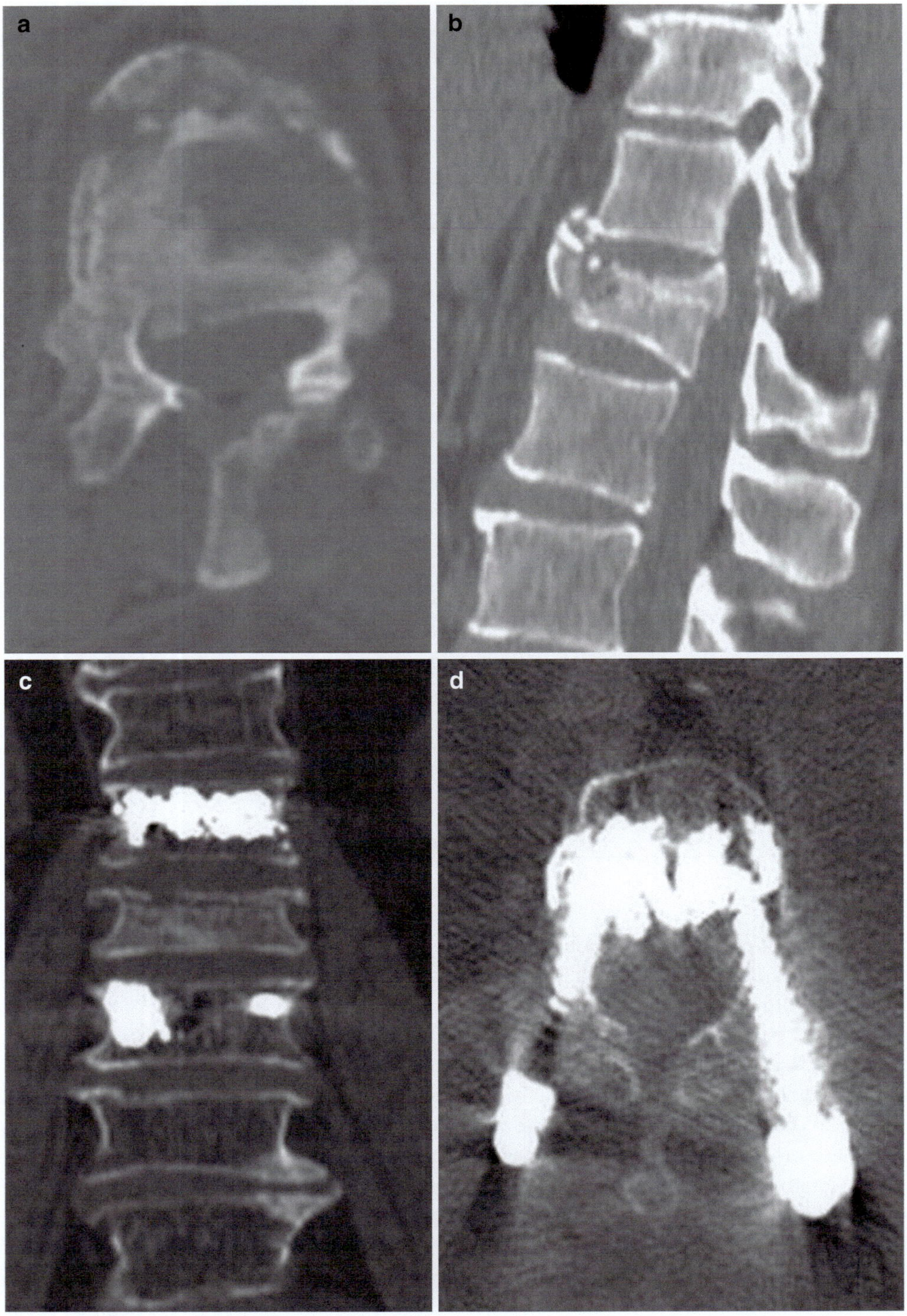

Fig. 3 Case 1, Th12 fracture: preoperative axial (**a**) and sagittal (**b**) reconstructed CT scans show burst fracture of Th12 with posterior wall partially pushed within the canal and initial kyphotic deformity. Postoperative images depict PMMA augmentation of adjacent vertebrae (**c**, **d**)

positioning inside the pedicle and the vertebral body was reported in a conventional open approach, performed in order to reduce the complications related to the bone-implant interface (pullout of screw, implant fracture) [1, 2, 4, 7]. Recently Lubansu et al. described the use of PMMA cement-augmented screws during percutaneous fixation in osteoporotic patients [12]. Intraoperative neuromonitoring may provide useful support to obtain the best positioning of the screw and to avoid inadvertent PMMA leakage. Neuromonitoring offers constantly updated data on nerve function and also guarantees the integrity of the pedicle. In 1995, Glassman et al. first demonstrated the accuracy of intraoperative EMG monitoring of pedicle screw placement, verified by postoperative CT scans [6]. A total of

512 screws were implanted in 90 patients, using intraoperative EMG monitoring. These authors concluded that a stimulation threshold higher than 15 mA provided 98 % confidence that the screw was within the pedicle. A stimulation threshold between 10 and 15 mA suggests re-exploration of the pedicle, even if in most cases this threshold is associated with adequate screw position. A stimulation threshold lower than 10 mA is generally associated with significant cortical perforation. The screw integrity test provided by the NIM-Eclipse System appears to be a simple surgeon-controlled stimulation test that is based on the same parameters as those suggested by Glassman and colleagues. The surgeon can control the start of the test with the probe and the response can be easily interpreted, even without a neurophysiologist in the operating room.

Conclusion

The simultaneous utilization of neuromonitoring and PMMA-augmentation of percutaneous screws has not been reported previously. Although our results are limited by the small number of patients included, they suggest that neurophysiological monitoring represents a significant adjunct to avoid screw misplacement and displacement of the cement.

Conflict of Interest Statement The Authors report no declarations of interest. No funds were received in support of this work.

References

1. Becker S, Chavanne A, Spitaler R, Kropik K, Aigner N, Ogon M et al (2008) Assessment of different screw augmentation techniques and screw designs in osteoporotic spines. Eur Spine J 17:1462–1469
2. Chang MC, Liu CL, Chen TH (2008) Polymethylmethacrylate augmentation of pedicle screw for osteoporotic spinal surgery: a novel technique. Spine (Phila Pa 1976) 33:E317–24
3. Duffy MF, Phillips JH, Knapp DR, Herrera-Soto JA (2010) Usefulness of electromyography compared to computed tomography scans in pedicle screw placement. Spine (Phila Pa 1976) 35:E43–8
4. Fransen P (2007) Increasing pedicle screw anchoring in the osteoporotic spine by cement injection through the implant. Technical note and report of three cases. J Neurosurg Spine 7:366–369
5. Fraser J, Gebhard H, Irie D, Parikh K, Härtl R (2010) Iso-C/3-dimensional neuronavigation versus conventional fluoroscopy for minimally invasive pedicle screw placement in lumbar fusion. Minim Invasive Neurosurg 53:184–190
6. Glassman SD, Dimar JR, Puno RM, Johnson JR, Shields CB, Linden RD (1995) A prospective analysis of intraoperative electromyographic monitoring of pedicle screw placement with computed tomographic scan confirmation. Spine (Phila Pa 1976) 20:1375–1379
7. Hu MH, Wu HT, Chang MC, Yu WK, Wang ST, Liu CL (2011) Polymethylmethacrylate augmentation of the pedicle screw: the cement distribution in the vertebral body. Eur Spine J 20:1281–1288
8. Isley MR, Zhang XF, Balzer JR, Leppanen RE (2012) Current trends in pedicle screw stimulation techniques: lumbosacral, thoracic, and cervical levels. Neurodiagn J 52:100–175
9. Joglekar SB, Mehbod AA (2012) Surgeon's view of pedicle screw implantation for the monitoring neurophysiologist. J Clin Neurophysiol 29:482–488
10. Kim MC, Chung HT Cho JL, Kim DJ, Chung NS (2011) Factors affecting the accurate placement of percutaneous pedicle screws during minimally invasive transforaminal lumbar interbody fusion. Eur Spine J 20:1635–1643
11. Lehmann W, Ushmaev A, Ruecker A, Nuechtern J, Grossterlinden L, Begemann PG et al (2008) Comparison of open versus percutaneous pedicle screw insertion in a sheep model. Eur Spine J 17:857–863
12. Lubansu A, Rynkowski M, Abeloos L, Appelboom G, Dewitte O (2012) Minimally invasive spinal arthrodesis in osteoporotic population using a cannulated and fenestrated augmented screw: technical description and clinical experience. Minim Invasive Surg. doi:10.1155/2012/507826
13. Magerl F, Aebi M, Gertzbein SD et al (1994) A comprehensive classification of thoracic and lumbar injuries. Eur Spine J 3:184–201
14. Mobbs RJ, Sivabalan P, Li J (2011) Technique, challenges and indications for percutaneous pedicle screw fixation. J Clin Neurosci 18:741–749
15. Nakashima H, Sato K, Ando T, Inoh H, Nakamura H (2009) Comparison of the percutaneous screw placement precision of isocentric C-arm 3-dimensional fluoroscopy-navigated pedicle screw implantation and conventional fluoroscopy method with minimally invasive surgery. J Spinal Disord Tech 22:468–472
16. Oppenheimer JH, DeCastro I, McDonnell DE (2009) Minimally invasive spine technology and minimally invasive spine surgery: a historical review. Neurosurg Focus 27, E9
17. Payer M (2011) "Minimally invasive" lumbar spine surgery: a critical review. Acta Neurochir (Wien) 153:1455–1459
18. Samartzis D, Shen FH, Perez-Cruet MJ, Anderson DG (2007) Minimally invasive spine surgery: a historical perspective. Orthop Clin North Am 38:305–326
19. Schizas C, Michel J, Kosmopoulos V, Theumann N (2007) Computer tomography assessment of pedicle screw insertion in percutaneous posterior transpedicular stabilization. Eur Spine J 16:613–617
20. Shin BJ, James AR, Njoku IU, Härtl R (2012) Pedicle screw navigation: a systematic review and meta-analysis of perforation risk for computer-navigated versus freehand insertion. J Neurosurg Spine 17:113–122
21. Smith ZA, Sugimoto K, Lawton CD, Fessler RG (2014) Incidence of lumbar spine pedicle breach after percutaneous screw fixation: a radiographic evaluation of 601 screws in 151 patients. J Spinal Disord Tech. 27(7):358–63. doi:10.1097/BSD.0b013e31826226cb. PubMed PMID:22688617.
22. Villavicencio AT, Burneikiene S, Bulsara KR, Thramann JJ (2006) Perioperative complications in transforaminal lumbar interbody fusion versus anterior-posterior reconstruction for lumbar disc degeneration and instability. J Spinal Disord Tech 19:92–97
23. Wu RH, Fraser JF, Härtl R (2010) Minimal access versus open transforaminal lumbar interbody fusion: meta-analysis of fusion rates. Spine (Phila Pa 1976) 35:2273–2281
24. Zdichavsky M, Blauth M, Knop C, Graessner M, Herrmann H, Krettek C, Bastian L (2004) Accuracy of pedicle screw placement in thoracic spine fractures. Part I: inter- and intra-observer reliability of the scoring system. Eur J Trauma 30:234–240
25. Zdichavsky M, Blauth M, Knop C, Lotz J, Krettek C, Bastian L (2004) Accuracy of pedicle screw placement in thoracic spine fractures. Part II: a retrospective analysis of 278 pedicle screws using computed tomographic scans. Eur J Trauma 30:241–247

Transpedicular Approach to Thoracic Disc Herniaton Guided by 3D Navigation System

Gualtiero Innocenzi, Manuela D'Ercole, Giovanni Cardarelli, Simona Bistazzoni, Francesco Ricciardi, Francesco Marzetti, and Francesco Sasso

Abstract

Background

The choice of surgical approach for thoracic disc herniation should consider the location on the axial plane and the consistency of the herniated disc. Calcified midline disc herniations are difficult to remove with a transpedicular approach because of limitations due to blind spots; so they are usually treated via a transthoracic approach, although this entails a high risk of thoracopulmonary injuries.

Methods

In this work we present two cases of calcified midline thoracic disc herniations treated with a transpedicular approach, improved by using a three-dimensional (3D) neuronavigation system to verify the extent of removal on the blind side.

Results

Postoperative computed tomography (CT) scans demonstrated that this original technical innovation, in the two present cases, allowed us to reach the side opposite the disc herniation and to assess the extent of resection at the end of the procedure.

Conclusions

The employment of a neuronavigation system in the transpedicular approach allowed safe and effective removal of calcified midline thoracic disc herniations. We did not observe any postoperative neurological worsening, onset of spinal instability, or other adverse events.

Keywords Thoracic disc herniation • Thoracic transpedicular approach • Spinal neuronavigation

G. Innocenzi • M. D'Ercole (✉) • G. Cardarelli • S. Bistazzoni
F. Ricciardi • F. Marzetti
Department of Neurosurgery, IRCCS Neuromed,
Via Atinense 18, Pozzilli, IS 86077, Italy
e-mail: manueladercolemd@gmail.com

F. Sasso
Department of Orthopaedics and Traumatology,
University of Naples "Federico II", Naples, Italy

Introduction

Surgical treatment of calcified thoracic disc herniations is challenging because of the probability of spinal cord damage. Many approaches have been developed to obtain satisfactory decompression and to overcome the significant rate of neurological damage associated with a strictly posterior laminectomy. These approaches are categorized as anterior (transthoracic [17], transsternal, and thoracoscopic [8, 12]), lateral (lateral extracavitary and costotransversectomy [10, 20]), and posterolateral (transpedicular [5] and transfacet pedicle-sparing [22]).

The approach should be selected individually in each case, considering the location on the axial plane and the consistency of the herniated disc. In our experience, the transpedicular approach, first described by Patterson and Arbit in 1978 [16], is a convincing technique with a lower rate of morbidity than that of the anterior and lateral approaches, although these approaches could be more effective in cases of calcified midline disc herniations. Here we describe an original technical innovation that allowed us to check the resection of a calcified median disc done by a posterolateral thoracic approach, guided by a three-dimensional (3D) navigation system.

Materials and Methods

Patient Population

The posterolateral thoracic approach was used in two patients. The herniated discs were located, respectively, at the T8–T9 and T7–T8 levels.

Case 1 was a 53-year-old woman with dorsal pain not responsive to medical care, initial gait disturbance, and leg paresthesia. Magnetic resonance imaging (MRI) showed a

M. Visocchi et al. (eds.), *Trends in Reconstructive Neurosurgery*, Acta Neurochirurgica Supplement 124,
DOI 10.1007/978-3-319-39546-3_48, © Springer International Publishing Switzerland 2017

left paramedian T8–T9 disc herniation compressing the spinal cord, with a T2 hyperintense signal at that level (Fig. 1a and b). A computed tomography (CT) scan showed a calcified disc herniation (Fig. 1c and d). Motor evoked potentials (MEPs) and somatosensory evoked potentials (SSEPs) before surgery were slightly altered.

Case 2 was a 62-year old woman with severe paraparesis mainly in the right leg with impairment of standing and walking, sensory loss at a low thoracic level, and urinary retention. MRI showed a median disc herniation with compression of the spinal cord and T2 hyperintensity at the T7–T8 level. A CT scan showed calcific consistency of the disc. Preoperative MEPs and SSEPs showed only slight activity in the left leg.

Preoperative Preparation

Before surgery, a CT scan is performed to acquire the images that will be matched with the intraoperative field to create a 3D image base for navigation. The procedure is performed under total intravenous anesthesia. Inhalational anesthetic agents are not used in conjunction with electrophysiological monitoring because they interfere with electrical conduction.

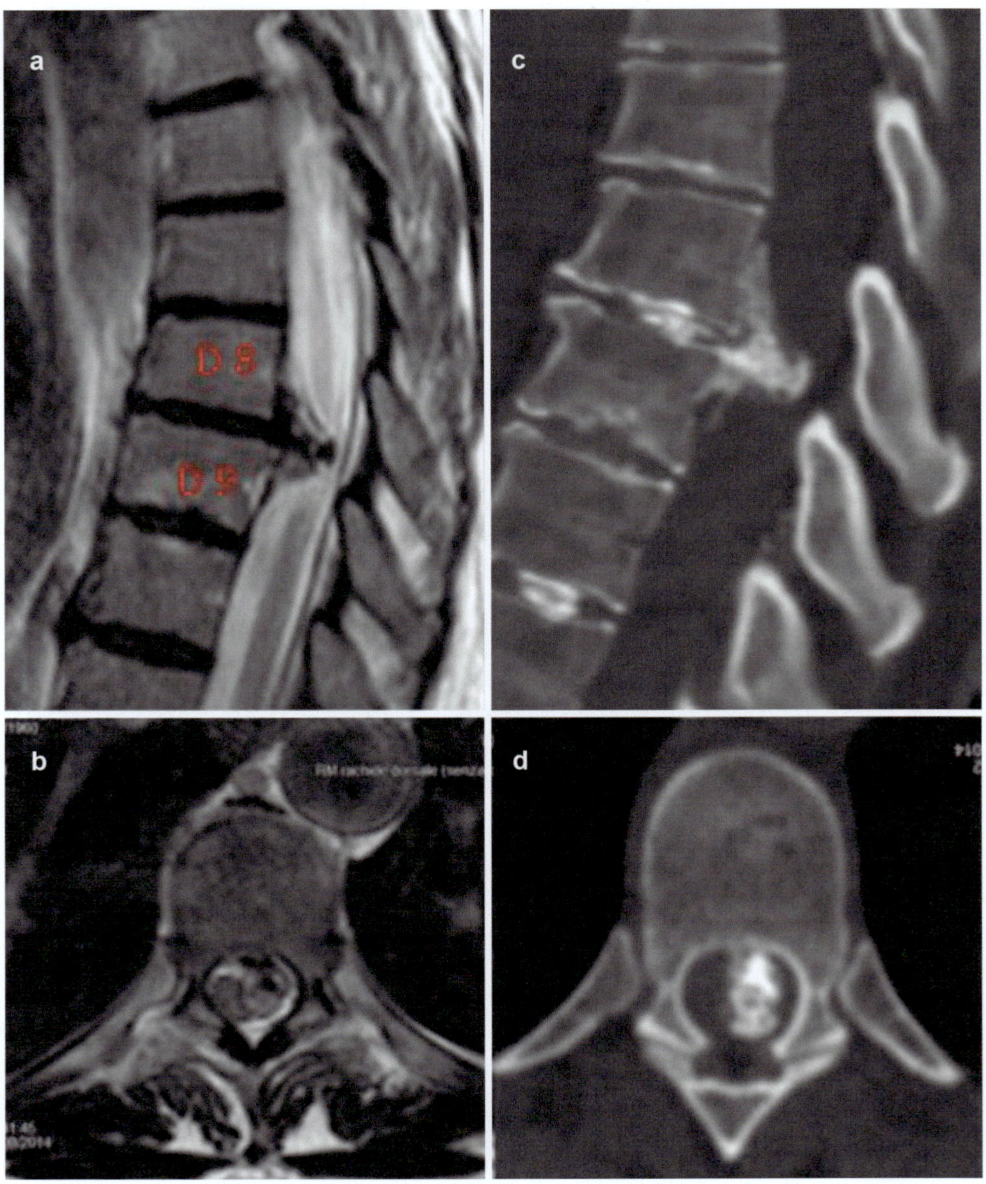

Fig. 1 Case 1: preoperative magnetic resonance imaging (MRI) in sagittal (**a**) and axial (**b**) planes. Computed tomography (CT) scans in sagittal (**c**) and axial (**d**) planes show calcified midline disc herniation

We used a modified park-bench position, with the side of the surgical approach facing upwards. Localization of the correct level was obtained by anteroposterior and lateral fluoroscopy after the positioning of the patient on the operation table.

Electrophysiological Monitoring

Electrodes were applied to the patient's limbs and scalp for continuous intraoperative spinal cord monitoring of SSEPs and transcranial MEPs.

Surgical Procedure

A midline linear skin incision was made, extended approximately 4 cm over the spinous processes at the level adjacent to the disc herniation. The paraspinous muscles were reflected to expose the medial portion of the transverse processes and facet joints at that level. The reference system for navigation (BrainLab Munich, Germany) was inserted on the spinous process below the level of surgery. Then we proceeded to match the selected vertebra with the preoperative images. In case 1, we removed—alternating drilling and the use of a Kerrison punch—the hemilamina of D8 and D9, the medial facet of D8, and the lateral facet of D9, and we did a partial pediculectomy of D9. The depth of the pedicle resection was established by the transition from the cancellous bone of the pedicle to the posterior cortical bone of the vertebral body. The rib was kept untouched. In case 2 we performed the same procedure at the upper level (D7–D8). Under microscopic view, we started removing osteophytes from the endplates and then we progressively reduced and debulked the compression, using an ultrasonic bone curette with a 30-degree angled tip (Misonix Farmingdale, NY). The disc space was incised and a large cavity was created using curettes, the Kerrison punch, and the ultrasonic bone curette, working in a lateral-to-medial direction. During the resection, we sometimes checked the extent of the removal by placing the navigation system pointer, in an oblique way, on the contralateral side of the surgical field, which was covered by the spinal cord. The residual thin layer of calcification on the contralateral side, not adherent to the dura, was dislodged with Penfield dissectors. The shell strictly attached to the dural plane was left. At the end of the procedure we checked the extent of removal by moving the pointer as described above, and by further assessment with endoscopy (Fig. 2 and 3).

Results

Case 1 During the surgical procedure we observed the depression of MEPs and SSEPs after moving the shell of the calcified disc herniation that was attached to the dura, so we decided to fracture the shell and leave it in situ. At the end of the procedure an improvement in the monitoring parameters was observed. In the early postoperative period this patient did not present with any neurological worsening. The postoperative CT scan showed the removal of the herniated disc and the dislocation of the shell attached to the spinal cord (Fig. 3).

Case 2 During surgery no alterations of the monitoring parameters were observed. After 9 months of rehabilitation therapy the patient was able to maintain a standing position and to walk with support and she had achieved urinary control. Postoperative MRI and CT scans showed the removal of the herniated disc.

Discussion

Incidental thoracic disc herniations are common, but symptomatic ones are rare [20]. Surgery is usually not recommended for asymptomatic patients, because the great majority of thoracic disc herniations remain stable or decrease in size, without the onset of neurological symptoms [6, 24]. However, no predictors currently exist to suggest which patients will develop symptoms [4]. In a review of ten surgical series, Bilsky [4] found that indications for surgery were myelopathy in 70 %, intractable radiculopathy in 24 %, and back pain in 6 % of cases.

The choice of the surgical approach for thoracic disc herniation depends on the location on the axial plane and on the consistency of the herniated disc.

There is general agreement in the literature about the preference for anterior transthoracic approaches to treat calcified midline thoracic disc herniations [2, 3, 7, 9, 13, 14, 19–21]. Transthoracic approaches have a high risk of bronchopulmonary complications [25], which has not been completely overcome by the introduction of thoracoscopy [1, 23]. However, posterolateral approaches do not provide adequate exposure of the ventral surface of the spinal cord without manipulation, with a blind angle between the midline dura and the calcified disc herniation, representing a critical phase of the surgical procedure [13, 18, 19]. This problem has been partially overcome by the use of a transforaminal approach, which provides tangential exposure of the ventral dura, albeit through a narrow passageway. In our experience, these drawbacks could be bypassed applying neuronavigation and endoscopic systems in

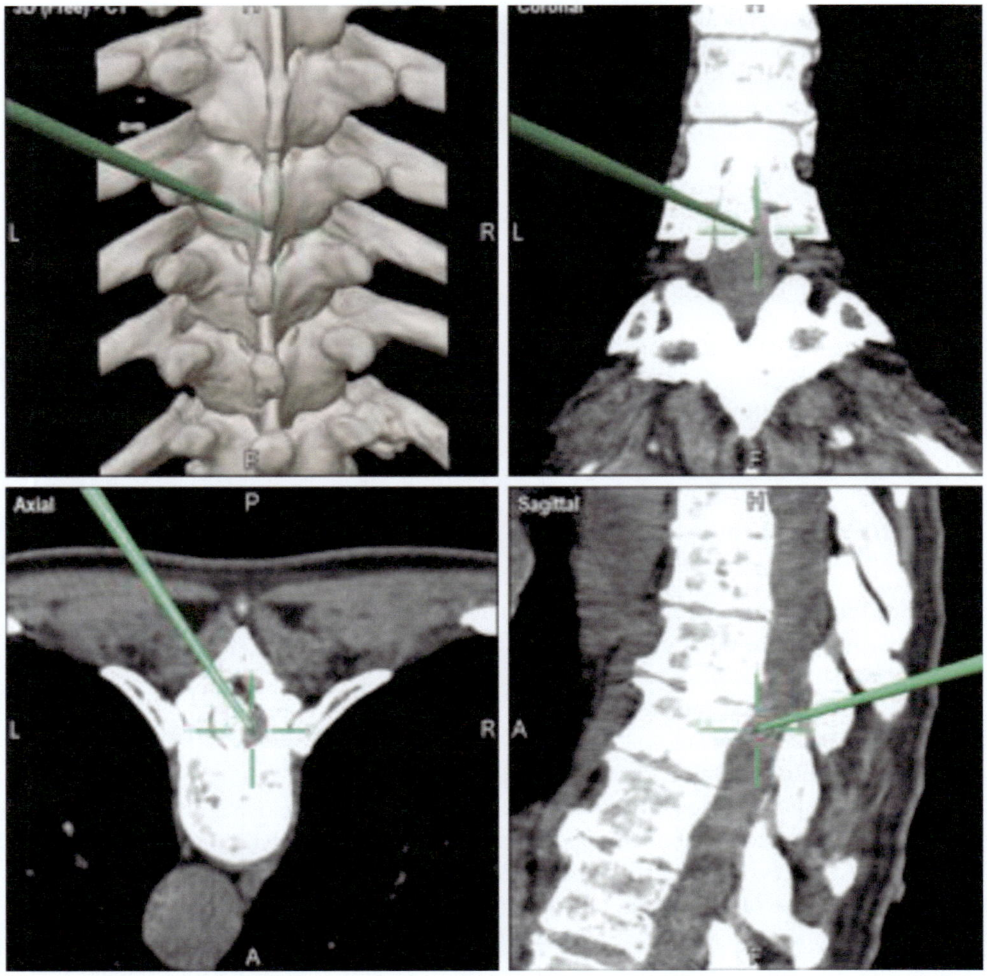

Fig. 2 Case 1: removal of calcified thoracic disc herniation guided by navigation system

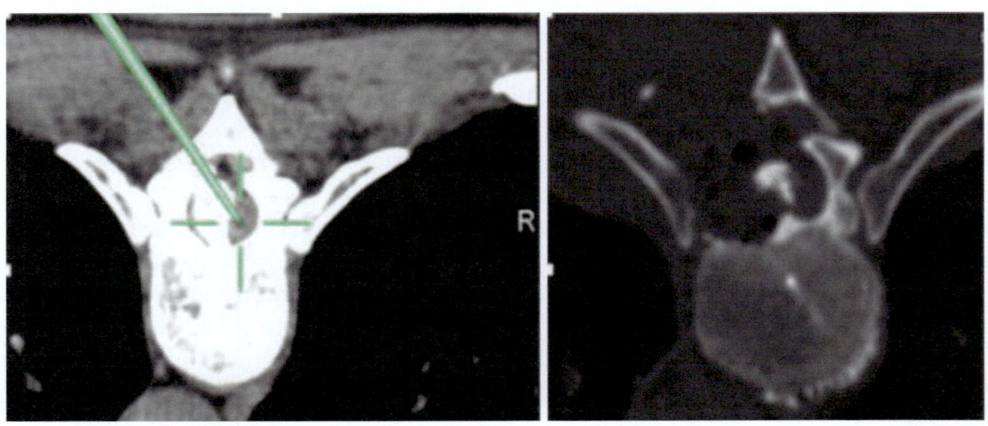

Fig. 3 Case 1: comparison between the position of the pointer on the blind side during surgery and postoperative CT scan

standard posterior approaches. These simple and easily available tools allow the quick verification of the intraoperative position, in order to obtain complete removal of the midline calcified disc herniation, even beyond the standard surgical corridor. The application of angled endoscopy in surgery for thoracic disc herniation has already been described, and it is currently a standard procedure in our department [15]. Neuronavigation systems, however, have been reported only in the thoracoscopic approach [11]; to our knowledge, the present study reports the first employment of such a system in the posterior approach to the thoracic spine to be described in the literature. However, even when these systems are used, the complete removal of calcified disc herniations can be difficult because of the presence of tight adhesions between the calcification and the dura. In some cases a transdural extent of the herniated disc is also possible. Such conditions increase the risk of cerebrospinal fluid (CSF) fistula and neural injuries. In such conditions a safe and feasible surgical target could be considered to be the emptying of the disc herniation, leaving an external shell, still obtaining satisfactory decompression [15].

Conclusions

The preliminary clinical experience reported here suggests that the employment of tools such as a neuronavigation system in the transpedicular approach to thoracic disc herniations allows us to compensate for the lack of visualization and to overcome the limitations of this surgical corridor in midline calcified disc herniations. With neurophysiological monitoring, the dural sac and spinal cord were never retracted and the modality provided added safety . We did not observe any postoperative neurological worsening, onset of spinal instability, or other adverse events. On the basis of this experience, we consider that we will continue to employ this technique.

Conflict of Interest Statement The authors declare that they have no conflicts of interest.

References

1. Anand N, Regan JJ (2002) Video-assisted thoracoscopic surgery for thoracic disc disease: classification and outcome study of 100 consecutive cases with a 2-year minimum follow-up period. Spine 27:871–879
2. Arnold PM, Johnson PL, Anderson KK (2011) Surgical management of multiple thoracic disc herniations via a transfacet approach: a report of 15 cases. J Neurosurg Spine 15:76–81
3. Arts MP, Bartels RHMA (2014) Anterior or posterior approach of thoracic disc herniation? A comparative cohort of mini-transthoracic versus transpedicular discectomies. Spine J 14(8):1654–1662
4. Bilsky MH (2000) Transpedicular approach for thoracic disc herniations. Neurosurg Focus 9(4):e3
5. Bilsky MH, Patterson RH (1999) Transpedicular approaches In: Benzel EC, Stillerman CB (eds) The thoracic spine. Quality Medical, St. Louis, pp 311–322
6. Blumenkopf B (1988) Thoracic intervertebral disc herniations: diagnostic value of magnetic resonance imaging. Neurosurgery 23:36–40
7. Bransford R, Zhang F, Bellabarba C et al (2010) Early experience treating thoracic disc herniations using a modified transfacet pedicle-sparing decompression and fusion. J Neurosurg Spine 12:221–231
8. Connelly CS, Horowitz MB, Moossy JJ, Julian T et al (1994) Thoracic discectomy using video assisted thoracoscopy. Spine 19:1082–1086
9. Delfini R, Di Lorenzo N, Ciappetta P et al (1996) Surgical treatment of thoracic disc herniation: a reappraisal of Larson's lateral extracavitary approach. Surg Neurol 45:517–522 (discussion 522–523)
10. Fujimura Y, Nakamura M, Matsumoto M (1997) Anterior decompression and fusion via the extrapleural approach for thoracic disc herniation causing myelopathy. Keio J Med 46:173–176
11. Johnson JP, Drazin D, King WA, Kim TT (2014) Image guided navigation and video assisted thoracoscopic spine surgery: the second generation. Neurosurg Focus 36(3):E8
12. Manges PA (1998) Video-assisted thoracoscopic discectomy and fusion. AORN J 67 940–945, 998
13. Nie H-F, Liu K-X (2013) Endoscopic transforaminal thoracic foraminotomy and discectomy for the treatment of thoracic disc herniation. Minim Invasive Surg 2013:264105
14. Oda I, Abumi K, LU D et al (1996) Biomechanical role of the posterior elements, costovertebral joints, and rib cage in the stability of the thoracic spine. Spine 21:1423–1429
15. Paolini S, Tola S, Missori P, Esposito V, Cantore GP (2015) Endoscope-assisted resection of calcified thoracic dsc herniations. Eur Spine J 25(1):200–206
16. Patterson RH Jr, Arbit E (1978) A surgical approach through the pedicle to protruded thoracic discs. J Neurosurg 48(5):768–772
17. Perot PL Jr, Munro DD (1969) Transthoracic removal of midline thoracic disc prostrusions causing spinal cord compression. J Neurosurg 31:452–458
18. Regev GJ, Salame K, Behrbalk E et al (2012) Minimally invasive transforaminal, thoracic microscopic discectomy: technical report and preliminary results and complications. Spine J 12(7):570–576
19. Smith JS, Eichholz KM, Shafizadeh S et al (2013) Minimally invasive thoracic microendoscopic diskectomy: surgical technique and case series. World Neurosurg 80:421–427
20. Stillerman CB, Chen TC, Couldwell WT et al (1998) Experience in the surgical management of 82 symptomatic herniated thoracic discs and review of the literature. J Neurosurg 88:623–633
21. Stillerman CB, Chen TC, Day JD et al (1995) The transfacet pedicle-sparing approach for thoracic disc removal: cadaveric morphometric analysis and preliminary clinical experience. J Neurosurg 83:971–976
22. Stillerman CB, Chen TC, Couldwell WT et al (1999) Transfacet pedicle-sparing approach. In: Benzel EC, Stillerman CB (eds) The thoracic spine. Quality Medica, St. Louis, pp 338–345, J Neurosurg 48:768–772, 1978
23. Wait SD, Fox DJ, Kenny KJ, Dickman CA (2012) Thoracoscopic resection of symptomatic herniated thoracic discs: clinical results in 121 patients. Spine 37:35–40
24. Williams MP, Cherryman GR, Husband JE (1989) Significance of thoracic disc herniation demonstrated by MR imaging. J Comput Assist Tomograph 13:211–214
25. Yoshihara H, Yoneoka D (2014) Comparison of in-hospital morbidity and mortality rates between anterior and nonanterior approach procedures for thoracic disc herniation. Spine 39:E728–E733

Retraction Note to: Anterior Cervical Discectomy and Fusion with a Compressive C-JAWS Staple

Lei Xia, Ming-Xing Liu, Jun Zhong, Ning-Ning Dou, and Massimiliano Visocchi

Retraction Note to:
Chapter 23 in: M. Visocchi et al. (eds.), *Trends in Reconstructive Neurosurgery*, Acta Neurochirurgica Supplement 124, DOI 10.1007/978-3-319-39546-3_23

This chapter has been retracted at the request of the author in agreement with the editor. The article has been already accepted by a journal, and it can be searched in pubmed: Xia L, Liu MX, Zhong J. et al. Anterior cervical discectomy & fusion with a compressive staple of C-JAWS [J]. British Journal of Neurosurgery. 2016, Jun 22:1–5.

The updated original online version for this chapter can be found at DOI 10.1007/978-3-319-39546-3_23

Erratum to: Clinical and Neuropsychological Outcome After Microsurgical and Endovascular Treatment of Ruptured and Unruptured Anterior Communicating Artery Aneurysms: A Single-Enter Experience

Andrea Pietrantonio, Sokol Trungu, and Antonino Raco

M. Visocchi et al. (eds.), Chapter 27 in: *Trends in Reconstructive Neurosurgery*, Acta Neurochirurgica Supplement 124,
DOI 10.1007/978-3-319-39546-3_27

The order of the author's names was wrong in chapter 27. The correct names are:

1) Andrea Pietrantonio
2) Sokol Trungu
3) Antonino Raco

The updated original online version for this chapter can be found at
DOI 10.1007/978-3-319-39546-3_27

Author Index

M. Visocchi et al. (eds.), *Trends in Reconstructive Neurosurgery*, Acta Neurochirurgica Supplement 124,
DOI 10.1007/978-3-319-39546-3, © Springer International Publishing Switzerland 2017

Subject Index

M. Visocchi et al. (eds.), *Trends in Reconstructive Neurosurgery*, Acta Neurochirurgica Supplement 124,
DOI 10.1007/978-3-319-39546-3, © Springer International Publishing Switzerland 2017

W

MIX
Papier aus verantwortungsvollen Quellen
Paper from responsible sources
FSC® C105338

If you have any concerns about our products,
you can contact us on
ProductSafety@springernature.com

In case Publisher is established outside the EU,
the EU authorized representative is:
Springer Nature Customer Service Center GmbH
Europaplatz 3, 69115 Heidelberg, Germany

Printed by Libri Plureos GmbH
in Hamburg, Germany